Arild Stubhaug

Niels Henrik Abel and his Times

Springer

Berlin
Heidelberg
New York
Barcelona
Hong Kong
London
Milan
Paris
Singapore
Tokyo

Arild Stubhaug

NIELS HENRIK ABEL
and his Times

Called Too Soon by Flames Afar

Translated from the Norwegian
by Richard H. Daly

With 51 Figures, 13 in Colour

 Springer

ARILD STUBHAUG
Grimsrudveien 88
3442 Hyggen, Norway

Translator:
RICHARD H. DALY
Sternsrudlia 1
1294 Oslo, Norway

Original publication (in Norwegian):
Et foranskutt lyn: Niels Henrik Abel og hans tid, second edition 1996
© 1996 H. Aschehoug & Co. (W. Nygaard), Oslo
ISBN 82-03-16697-0

Publisher's note: Springer-Verlag thanks Elisabet Middelthon of MUNIN, Oslo, for the excellent collaboration and financial support for the translation, which made the publication of this book possible. Thanks go also to Cathrine Janicke, Åsta Brenna and Harald Engelstad of H. Aschehoug & Co., Oslo.

The *background illustration on the endpapers* shows a part of the wallpaper pattern from Abel's death room at Froland (modified after a photo by Dannevig, Arendal). Moreover, the aforementioned photo is completely reproduced and printed as a color figure (Fig. 51) on p. 568 with a more detailed description.

ISBN 3-540-66834-9 Springer-Verlag Berlin Heidelberg New Nork

Cataloging-in-Publication Data applied for.
Die Deutsche Bibliothek – CIP-Einheitsaufnahme
Stubhaug, Arild: Niels Henrik Abel and his times: called too soon by flames afar / Arild Stubhaug. Transl. by R. Daly. – Berlin; Heidelberg; New York; Barcelona; Hongkong; London; Mailand; Paris; Singapur; Tokio: Springer, 2000
 ISBN 3-540-66834-9

Springer-Verlag is a company in the BertelsmannSpringer publishing group.
© Springer-Verlag Berlin Heidelberg 2000
Printed in Germany

The use of general descriptive names, registered names, trademarks, etc. in this publication does not imply, even in the absence of a specific statement, that such names are exempt from the relevant protective laws and regulations and therefore free for general use.

Cover Design: Erich Kirchner, Heidelberg, Germany
Production: PRO EDIT GmbH, D-69126 Heidelberg, Germany
Printed on acid-free paper   SPIN: 10699500   46/3143Re – 5 4 3 2 1 0

Contents

List of Illustrations

PART I

Introduction

Fig. 1. Niels Henrik Abel, sculpture by Gustav Lærum, 1902. Originally cast in plaster of Paris for the 1902 competition held on the occasion of the centennial of Abel's birth; now located in the main house of Froland Ironworks, Arendal. Photo: Dannevig, Arendal.

1.

A Short Life

Niels Henrik Abel died on April 6, 1829, not yet 27 years of age. Two weeks later in the obituary column of *Den Norske Rigstidende*, Sivert Smith, owner of Froland Ironworks announced that "the highly esteemed Niels Henrik Abel, widely famous for his mathematical excellence, and Docent (Professor) at the Royal Frederik University, following 12 weeks of serious illness, has died at my house..." In a column beside the death notice, and in the section entitled "Paid Contributions", was a seven stanza poem entitled "Niels Henrik Abel."

Abel had spent his last months at Sivert Smith's house in the company of his fiancée and friends from the ironworks; it was one of the most prosperous homes of the time. They had all intended to celebrate Christmas together. Abel had been looking forward to the Christmas holidays and at first was full of joy at the gathering of friends. But later, as the all too serious illness overcame his energies and he understood that his life was coming to an end, his deep feelings toward friends, mathematics and his conjugal future became inextricably intertwined.

He had certainly known for a couple of years that he was ill, but he had been too absorbed by his mathematical project to pay sufficient attention to his health. To the contrary, he stepped up his work tempo; this scramble between health and illness had had the effect of quickening to the extreme the tempo of his mathematic insights.

The young candidate, as he was called at Sivert Smith's Froland, was attended by the district's best doctor, one of the founders of Norway's constitution, Alexander Christian Møller from Arendal. But despite this attention, the sickbed became a deathbed. Before spring arrived that year, Niels Henrik Abel succumbed to "galloping consumption".

Already before he had even become a university student, Niels Henrik Abel was in the process of becoming one of the most promising men of science, but no one in Norway had sufficient mathematical knowledge to understand the significance of his stringent proofs. At the Royal Frederik University in

Christiania (Oslo)[1] there were several professors and acquaintances who in
the course of the previous decade, had watched Niels Henrik Abel mature
from an average cathedral school graduate to a mathematician of a level they
were neither interested in, nor had mastery over. But they were aware of his
renown, and knew that he had become a respected scholar who published his
work in one of Europe's leading mathematical journals. What had happened
to the boy? How had it transpired, friends and acquaintances asked, that the
young man, Niels Henrik Abel, in the course of just a few years had transformed himself from someone ordinary and unremarkable, into someone
extraordinary and incomprehensible? One wonders how they might have
explained this transformation.

It was clear to many that someone unique had departed on that April day
in 1829; however, there was no public offering of personal recollections. The
poem beside the announcement of death in *Den Norske Rigstidende* was
written by Conrad Nicolai Schwach, the country's most respected lyric poet
of the 1820s. He had an eye for epoch-making events, and was bright enough
to give expression to the refinement of taste of his day, but he had no
understanding of mathematics, and only knew Niels Henrik Abel by general
reputation. Consequently, the twenty-eight lines of verse are nothing more
than a bundle of phrases taken from the arsenal of well-wrought images and
poetic turns-of-phrase current at the time.

At the beginning of May, the weekly, *Den Norske Huusven,* printed a
twelve-stanza poem by a young theological student, Hans Christian Hammer,
probably one of those who had come to Abel for private lessons during the
autumn of 1827, and who had been overwhelmed at the news of young Abel's
death. But these verses reveal nothing other than a boundless admiration for
the genius Abel, who through "the silvery blue ether had fathomed the depths
of the pale nymphs' exquisite dance."

But the public obituary and words of remembrance that Norway was to
give Abel did not appear until more than six months after his death, and were
written by university lecturer[2] Bernt Michael Holmboe. As a new young
teacher at the Christiania Cathedral School in 1818, B. M. Holmboe had
discovered Abel's mathematical genius and, as part of his pedagogical duty,
had quickly initiated him into the classic works of mathematics. A lifelong
companionship developed between Abel and Holmboe. However, the fact that
Holmboe received the permanent position of *lektor* at the University that

---

[1] Christiania was officially renamed Oslo in 1925.

[2] The *lektor* was a junior post in Norway's new university, but one which perhaps had greater
status than the current position of *lecturer* in the Anglo-American academic world.

could have fallen to Abel, did not in the least seem to dampen Abel's devotion to his gifted teacher. But in November, 1829, when Holmboe released his Abel obituary for publication, the public was presented with a somewhat pallid and perfunctory picture of the uncommon friend and human being that Niels Henrik Abel had been.

Many had awaited Abel's obituary. But the country's star of science and scholarship, and he who for many reasons was perhaps the most appropriate to honour Abel's memory, Professor Christopher Hansteen, was away on a lengthy expedition to Siberia. Hansteen had received generous support to pursue his theory that there were four magnetic poles to be found, and after long years of painstaking and meticulous measurements carried out with the help of many of his students, including Abel, Hansteen reasoned that the northern hemisphere's second magnetic pole must lie somewhere in Siberia. Hansteen had gotten no further than St. Petersburg before he sent a message home saying that he needed more money in order to carry out his field research. And since the expedition was being carried out on behalf of the nation, the Norwegian cabinet granted the request, and King Karl Johan was unable to do other than give his reluctant approval. Hansteen was Professor of Applied Mathematics and Astronomy, and in addition to his studies of the magnetic properties of the globe, he was responsible for the national almanac. He measured the heights of the mountain peaks, and calculated precise geographical positions for the different places that were to be mapped in the new nation of Norway.

In Professor Hansteen's absence, the task of writing an obituary for Niels Henrik Abel fell to Lektor Holmboe. A space was reserved in *Magazinet for Naturvidenskaberne* [Magazine of the Natural Sciences], the country's very first journal of the natural sciences, which was now in its sixth year. Holmboe's obituary was finally printed at the end of November, but it also appeared for sale as a special offprint brought out by the book publisher, Christopher Grøndahl for 16 shillings under the title, *A Short Account of the Life and Work of Niels Henrik Abel.* But such was the public curiosity about Niels Henrik Abel that on November 21st, even before the magazine and the offprint were out, the newspaper *Patrouillen* printed an extract.

He was born on August 5, 1802, when his father, Søren Georg Abel was the pastor at Finnøy, near Stavanger on the southwest coast of Norway. His mother, Anne Marie Simonsen, was the daughter of a shipping merchant from Risør in southern Norway, who grew up surrounded by the luxury and abundance which the patrician merchants in that part of the country were able to display during the closing decades of the eighteenth century. Niels Henrik was the second son of this young ecclesiastical couple. When Niels Henrik was born, the vicar was thirty and his wife was twenty-one years of age.

Before Niels Henrik had reached his second birthday his father had been appointed to the parish of Gjerstad, not far from where Niels Henrik's mother grew up. Here, Søren Georg took over the ministry from his father, Hans Mathias Abel. Niels Henrik grew up in Gjerstad with his brothers and sisters: one older and three younger brothers, and a sister of whom he was particularly fond. He was taught by his father, and later by a young man, Lars Thorsen, who in time became the permanent village teacher. In the fall of 1815, Niels Henrik was enrolled at Christiania's Cathedral School, as (in the terminology of the day) a disciple. During the first years at school he attracted no special attention, until the summer of 1818, when Holmboe had begun to challenge the disciples' own capabilities by setting them small algebraic and geometric assignments; thereupon, Abel's mathematical talents came to the fore. It soon became necessary to set him special assignments, and with an ardour and self-assurance approaching genius, he moved beyond elementary mathematics and in a short time, appropriated everything that Holmboe knew; thereafter he continued to work on his own.

His father, who, in addition to his active and manifold ecclesiastical activities had become a Member of Parliament in the extraordinary parliament of 1814, and again in the session of 1818, died in 1820. In both his vocations, the parish vicar, Søren Georg Abel, had distinguished himself in different ways, perhaps mostly in ways that incited negative responses.

Niels Henrik Abel became a university student with middling results in all subjects apart from mathematics. Although he lacked support from home, Abel soon received lodgings at the student residence, "Regentsen", where some of the University's teachers subsidized his keep. Abel was now ready to become, and was discussed as, a rare mathematical talent, and a person who could achieve for the new university, and for the new country of Norway, a welcome renown in foreign lands. In the summer of 1823 he received money from the Professor of Mathematics, Søren Rasmussen, for a two-month stay in Copenhagen in order to meet leading mathematicians there. It proved to be an inspiring visit, and in addition to scholars, scientists, family and acquaintances, he also met a young girl, Christine Kemp, to whom, a year later, he became engaged.

To the University Senate in Christiania it was clear that Niels Henrik Abel must go abroad to gain further education, to learn more, and become what they hoped and believed: "A remarkable Man of Science, a Triumph for the Fatherland, and a Citizen, who with uncommon Ability in his Discipline one day shall abundantly recompense this Help which must now be granted to him." The importance of Abel obtaining the means to travel to the world's foremost mathematical milieu, namely Paris, was abundantly clear to the Senate. But those who controlled the purse strings felt that Abel still needed a couple more years at the University in Christiania to improve his ability with languages: "the Learned Language, and other subjects important to his main field of Mathematics." In the spring of 1824 he received a stipend of 200 *speciedaler*[3] per annum and which was to extend for a period of two years. This sum was granted by the state treasury and should be applied to Abel's period of further study and preparation in his own homeland.

During these years he read all the considerable mathematical literature available in the library of the newly founded university in Christiania, and he himself began to write mathematical treatises. The first to be published was in 1823 in the debut volume of *Magazinet for Naturvidenskaberne*. There was no one at the university who could help him with his science, nor was there on offer a course of studies in higher mathematics. The mathematics offered was of a relatively elementary level, required of all students who would sit what was called the Secondary Degree Examination. Professor Rasmussen regularly taught a rather full course of studies. Here one entered the world of functions, as well learning about curved lines, which were the classical conic sections, arithmetic series which were used in the calculation of the number of cannonballs in a pile, and the new and exciting differential and integral calculus. Yet all these were materials that Rasmussen had already taught to specially interested students, including B. M. Holmboe. And Holmboe had passed his knowledge on to Abel, such that this fresh-faced student probably already knew more mathematics than anyone else in the country.

As a disciple at the Christiania Cathedral School he had worked with one of the period's most popular mathematical problems; namely, how to find the roots of a common quintic equation with the help of five classical methods of calculation (addition, subtraction, multiplication, division, as well as the application of square roots). Now, while he was a university student, Abel solved the problem; that is to say, he showed that it was impossible to find any such solution within the bounds of the classical framework. This was his first great achievement and a large advance in the theory of equations.

In July, 1825, before his stipend ran out, Abel petitioned the king for a travel grant of 600 silver *speciedaler* to be used over a period of two years. The year before, he had published, at his own expense, the evidence for his conclusion that ordinary algebraic solutions to equations to the fifth power were impos-

---

[3] The *speciedaler*, named from the German *thaler,* was a silver coin generally worth 2 *riksdaler* across Scandinavia. It was used in Denmark/Norway until 1813. Thereafter Norway developed its own currency and with the establishment of the mint in 1816, the speciedaler was worth 5 ort (marks) or 120 shillings. When the crown (*krone*) became the standard currency of Norway in 1872/73, the speciedaler was valued at 4 kroner. In the early 1800s, Denmark and Sweden also used the speciedaler and riksdaler, or *riksbankdaler*.

sible. In order to reach an international readership, he had written in French. And to keep the costs as low as possible, he had confined his evidence to six small pages, which he now hoped would act as his passport to Europe's most learned circles.

With the best recommendations from the University, his application for a travel stipend was quickly granted, and in September of the same year, 1825, Abel managed, together with four of the country's other foremost studious youths, to travel to the big outside world. Three of his travel companions were mineralogists. One, Baltazar Mathias Keilhau, later became a professor in Christiania; the second, Nicolaj Benjamin Møller, became mining superintendent at Kongsberg's silverworks; and the third, Nils Otto Tank, became an adventure-seeker and missionary for the Hutterite Brethren, first in Surinam, South America, and later in Wisconsin. The fourth travel companion, Christian Peter Bianco Boeck, became Professor of Veterinary Medicine and founder of the Veterinary College in Christiania.

Abel's planned itinerary was, as stated in his application, to travel to Paris, and on the way there, to visit Europe's most brilliant man of science, Carl Friedrich Gauss at Göttingen. For the others in the entourage, the most interesting places of study and centres of learning lay in southern Germany and northern Italy. For the sake of friendship and to see a little more of the world, as well as to avoid being alone, Abel accompanied the others for some distance. And when he finally reached Paris, the mathematical metropole, almost a year after leaving Christiania, and without having visited Gauss, the stay in the mathematical metropole became a genuine disappointment. His greatest stroke of luck had occurred in Berlin at the beginning of his travels in the autumn of 1825. There Abel met the versatile engineer and natural scientist, August Leopold Crelle, and through him Abel was incorporated into the city's research milieu and the elite of scientific discourse. Upon meeting Abel, Crelle took the plunge in relation to something that had long been in his plans, namely, to start Germany's first mathematical journal: *Journal für die reine und angewandte Mathematik*. The first issue came out in the spring of 1826, and continues to this day. It was in this journal that Abel was to publish most of his works as soon as they were completed. And it was first and foremost Abel's contributions which rapidly gave this publication, commonly known as *Crelle's Journal*, renown across Europe. Abel became known and appreciated in Berlin, and many wanted to obtain a professorial position for Abel in that city. The last message about the university appointment was written by a radiant Crelle on April 8, 1829, without knowing that Abel had died two days earlier.

Then, as mentioned, Abel travelled further south, to Dresden, Prague, Vienna, Trieste and Venice in the spring and summer of 1826 before he set course northward again through Switzerland, reaching Paris in August. There he began to compose the work which he himself considered to be his most important, and which he had been saving precisely for the famous academy of science, the French Institute in Paris. On October 30, 1826, he delivered his treatise to the leading mathematicians, Adrien Marie Legendre and Augustin Louis Cauchy. But this work, which later came to be known as the Paris Treatise, and in mathematical literature is regarded as one of the greatest treatises ever written, sat, unread. This manuscript was retrieved around the time of Abel's death, and the French Institute awarded Abel a grand prize *post mortem* for this work.

Even though the Paris delay was a disappointment, he met a number of the young men of science of that city, he gained access to a vast library, and met the Norwegian artist and portrait painter, Johan Gørbitz. Gørbitz also drew Abel in Paris and this was the only portrait made of Niels Henrik Abel during his lifetime.

The European tour, and the stops in foreign countries, 20 months in all, became *the* big external event of Abel's short life. He left Paris, on his homeward journey, during the week between Christmas and New Year's, and arrived at his friends in Berlin in a destitute and perishing condition in the first days of January, 1827. When he returned home to Norway in May, he found that the position he could have had at the University had been taken by his friend and teacher, B. M. Holmboe. He was aware of this, having heard about it on the way. More difficult to deal with was the fact that he was now unable to get a renewal of the stipend of 200 *speciedaler* that he had had before he went abroad. The academic collegium petitioned the government, saying that there ought to be public financial support so that Abel could continue his studies and his work. But the government declined; there were no funds available for such an endeavour. The collegium then took an unusual step to renew this denied request: they felt they had to come up with 200 *speciedaler*. The higher authorities refused once again, but proposed that the University, as an advance against future salary in one or other position, should pay this from its own budget.

While Niels Henrik had been abroad, one of his younger brothers was in Christiania and left behind a number of debts that Niels Henrik felt obliged to honour. As well, his deceased father's liabilities now fell on Niels Henrik's shoulders. In response to the colossal wave of patriotism and belief in the power of the fatherland that had swept the country, and to the fundraising drive initiated by the "Society for Norway's Wellbeing" in 1811 for the establishment of a university in Norway, Pastor Søren Georg Abel of Gjerstad had been an ardent supporting player. In addition to 100 *riksdaler* cash and 10 *riksdaler* annually, he had committed himself to an annual contribution of a half-cask of barley from the farm and dowager house of Lunde at Gjerstad. In the period after 1814, with chaos and inflation prevailing in the monetary system, a big jump in the number of bankruptcies, and the shift of currencies from *riksdaler* to *speciedaler*, the grain contributions came to be the most difficult to honour. The weight of the half cask of barley which Pastor Abel had pledged to provide, a little less than 70 litres, became a heavy burden. The consequent obligations of his descendants to the university were, by 1827, of the order of 26 *speciedaler*. Niels Henrik's mother lived in wretched conditions on the Lunde farm at Gjerstad and could contribute nothing.

In order to raise some income, Niels Henrik gave private lessons in elementary mathematics, astronomy and mechanics, in conformity with the syllabus for the Secondary Degree Examination, but there was little money in that. In September, 1827, the University granted from its own funds, the 200 *speciedaler* of support money, as had been recommended by the government. Despite this, he had to take out a loan of 200 *speciedaler* from "The Bank of Norway" (Norges Bank) in October, money that Abel never managed to pay back.

While his external worries were mounting, Abel worked intensively on his mathematics and sent a stream of his findings to Berlin. It was however, fortunate that his beloved sister Elisabeth had managed to get a position in the house of the august professor and cabinet minister, Niels Treschow at Tøyen in Christiania, and indeed, it was an honour that he was chosen for membership in the Norwegian Royal Society of the Sciences in Trondheim. As well, at New Year's, 1828, he received a temporary substitute position to replace Professor Hansteen, who was about to set off on his Siberian expedition. This brought about a considerable financial improvement. But perhaps this was too late? Abel was no longer physically strong, and in the course of the fall of 1828, he fell ill, and in a few weeks became bedridden. As Christmas approached he was determined to rise from his bed and visit his fiancée, who for the past half-year had been governess at the Froland Ironworks near Arendal in southern Norway. He set out on the long, exhausting winter journey, and arrived in tolerably good form, celebrated a very pleasant Christmas before being confined to bed shortly thereafter. He began spitting blood, and the end came when he had been bedridden for twelve weeks.

Lektor Holmboe apologized for the obituary not appearing sooner, explaining that he had been abroad, the printing had been delayed, and that he had waited before writing to be sure that he could avail himself of the opinions of foreign mathematicians about Abel's importance and his achievements. Holmboe wanted his obituary to show Abel's incomparable gifts and contributions to mathematics and the admiration that this inspired in the great wide world.

First and foremost Holmboe sought support from Crelle in Berlin and Legendre in Paris, who maintained that Abel in his short life had raised an everlasting monument to himself. Legendre used the words of the Roman poet, Horace, *monumentum ære perennius,* about "a monument as durable as bronze" to describe Abel's work.

Holmboe had been both teacher, friend and confidant to Abel. During Abel's grand tour it was Holmboe who most frequently received letters. One would have thought that Holmboe was the person most capable of explaining who Niels Henrik really was. But he did not want, did not manage, or perhaps thought that it was unseemly, to say anything about how he saw and experienced Niels Henrik Abel in his full human dimensions. One possible explanation is that at that time one found a set of norms and traits of character that any respectable and educated man practiced as a matter of course, and where nothing of an emotional nature was expressed; indeed, the fulfillment of these norms was taken for granted. In light of this attitude, Holmboe's rather bloodless obituary would not be surprising had it not been for the reactions of other Abel friends. One of those who tried to broaden the picture was Christian Boeck, one of Abel's companions on the grand tour. Boeck had already by 1829 been made lektor in veterinary medicine and editor of *Magazinet for Naturvitenskaberne,* which continued to be the country's only scientific journal.

As a supplement to Holmboe's article, Boeck gave his impressions of Abel's life and the vicissitudes of fortune. Boeck maintained from the outset that Abel had not had a strong constitution, and that his weak chest appeared early on to be the bane of his existence. Nevertheless, Boeck added, while they were travelling together Abel was in good cheer and never really ill.

Those who did not know him well thought Abel was high-spirited; indeed, those who met him only sporadically tended to regard him as of a rather rash and frivolous character. But Boeck was of the opinion that in reality Abel's disposition was serious, that he had profound emotions, and that quite often he spoke in a particularly sorrowful manner. He frequently sank into extremely dark moods, even when the ambience was at its most pleasant as, for example, in Berlin when he had met Crelle and been accepted into the circle around him. Boeck said that Abel tried to banish this melancholy of the mind with an assumed gaiety and by presenting a carefree mien.

Abel confided fully in very few. Among his closest friends, according to Boeck, there were few who really knew him or assessed him correctly. Concerning Abel's relations with the fair sex, Boeck said that he loved his fiancée with a tenderness greater than any that his acquaintances had noticed hitherto.

Abel often spoke of his worries about the future. And Boeck's explanation of Abel's recurring melancholy was that the brooding about life's uncertainties completely ground him down. Only in brief flashes did he have any hope of achieving a situation free of worry and insecurity, but normally there was no word or concept that could cheer him up. Only at his desk did he manage to forget about the future. After formulating one mathematical proposition or another he could forget the whole external world for a second, and experience the joy of perfection.

Many years later, when Boeck had become a professor, and was a leading scientist, he explained that almost every night in Berlin Abel lit the candle, got out of bed, and sat down to write or make computations. One time he had solved a problem in his head, but had forgotten the sequence of proof upon which he had built his conclusions. No matter how much he brooded over this, he was unable to recover what he had forgotten. But one night when they had gone to bed and were asleep, Abel suddenly woke them with a whoop of joy! The solution, and the whole sequence had, in a flash, become evident to him.

The aging Professor Boeck no doubt wanted to portray the friend of his youth as a new Archimedes. Archimedes, in the third century before Christ, had leaped out of his bath in Syracuse, and run naked through the streets, shouting "Eureka! Eureka! I have found it! I have found it!" The old Greek had on that occasion discovered the law of buoyancy in liquids. History does not tell us which mathematical correlation Abel made that night in Berlin. Perhaps it was an insight into what the requirements were for a solution to algebraic equations? Or maybe he saw the conditions for the convergence of infinite series?

Among other anecdotes about Abel, one has him, in his student days, springing from his seat during a lecture in Greek by Professor Georg Sverdrup, and to the great astonishment of the whole auditorium, dashing for the door shouting, "I have it! I have it!"

In his supplement to Holmboe's obituary, Boeck stresses that Abel was very dispirited in his last years. Since not so much as a single letter arrived from Crelle to cheer him up, concerning probable and impending employment at the university in Berlin, this is as one would expect. It was humiliating not to be able to work in his own scholarly and scientific field in Norway. He thought it was immensely sad to have to exile himself, to wander off among foreigners and maybe never see his homeland again. Abel was always moved when he discussed this, and vacillated between following up or rejecting possibilities from abroad. According to Boeck, Abel did not expect the university in Christiania would do anything more for him. He had not sat his degree examinations, and could never really hold any title other than student! He knew that he would have no permanent job or income once Professor Hansteen returned from Siberia. He knew that once again it would be difficult to support himself, and any marriage plans with Christine Kemp were unthinkable.

With great panache, Boeck also asserted, in a manner that would be clear to his contemporaries, that the reason Abel had been so low and so melancholy in his last period was precisely because he had not received employment in his homeland. While he was the toast of scientists abroad for his ingenious work and discoveries, and while a foreign state was willing to take care of his daily wellbeing, at home in Norway he received no encouragement and his achievements were awarded no acknowledgement. Boeck maintained that despite his protests to the contrary, Abel found the professorship in Berlin to be an alluring prospect. And only through his own work could Abel maintain and increase the esteem and fame that he had already achieved. But in the course of strenuous efforts in this work, Abel got too little repose, and too little diversion. The intense and excessive studies taxed his nervous system, and this uninterrupted sedentary activity had harmful effects on his health; the final blow was his little winter trip to Froland. That was all it took to make him go under.

And this is the manner in which Boeck brought his description of his friend and travel companion to a conclusion: "His Wish that the soil of the Fatherland should embrace his Dust was fulfilled. Now his Worries have ended; he did not seek Bread in foreign lands, but nonetheless Norway lost its Son."

That Abel's departure was a big loss to the young nation soon became strikingly evident. And the contention that Norwegian officialdom should be blamed for this provoked strong reactions. Lektor B. M. Holmboe felt himself treated with disapproval because it was precisely he who had the position that Abel *could* have had. The University Senate felt obliged to reply, in both of the newspapers *Morgenbladet* and *Patrouillen*. The Senate referred to documentation and discussions that would show that throughout, the University had worked *for* Abel and against the Ministry. By means of both stipends and private contributions the University and its teaching staff had wholeheartedly been on Abel's side. The reason that B. M. Holmboe had been chosen professor over Abel in 1826 was said to be that Abel had at that precise moment begun his grand tour of Europe and that they feared Abel, with his effervescent manner and his advanced knowledge would have difficulties satisfactorily teaching those who would study theology, law, or medicine, and only required a basic minimum of mathematical knowledge. But it was said that during the entire period, the Senate had hoped a position in higher mathematics could be established, and as recently as the previous year, when it appeared that

Abel might receive an appointment in Berlin, had regretted that they did not have any prospects for permanent employment of "this talented young Man." The budgets would simply not cover this.

But criticism of Abel's treatment by public officials did not abate despite this rejoinder from the university. For example, *Patrouillen* pointed out that right at the time that the Senate was protesting that it did not possess sufficient budgetary funds for Abel and his science, it nonetheless hastened to grant a considerable sum from the public budget the moment that Professor Hansteen had requested the same during his Siberian expedition. To wit, Hansteen received 9,500 *speciedaler* from the public purse.

The discourse on whether or not more could have been done for Abel became an ongoing issue in different Norwegian publications. The result was that public memory of Abel remained fresh, and his life became exemplary of what the nation of Norway ought never allow to happen again. This reached its apogee in 1902, when the hundredth anniversary of Abel's birth became an occasion of monumental commemoration. King Oscar issued invitations to a celebration at the royal palace; the citizenry celebrated at the concert hall, Logen (Gamle Losjen); and the students organized the largest torchlight procession ever seen in the streets of Christiania. Speeches were made by the nation's most prominent men: Fridtjof Nansen chaired the arrangement committee and made a call to the youth; the leading poet, Bjørnsterne Bjørnson had written an eleven-stanza cantata in praise of Abel. This was performed by chorus and orchestra at Logen with music composed by Christian Sinding. An echo of Boeck's words from 1829 are found as well in the closing verse: "A westland boy was he/well in his twentieth year./Now he is owned by the world;/but the boy was ours." Mathematicians from around the world gathered at the University in Christiania, and a thick *festschrift* was published subsequently, with articles on Abel's life and work, with his letters and with documents pertaining to his relations with government and university. One of the editors and contributors, the mathematician and children's writer, Elling Holst, also wrote a poem on Abel that was recited by the country's leading actress, Johanne Dybwad, as the epilogue to a celebratory performance of Ibsen's *Peer Gynt* at the National Theatre in Oslo.

Celebrations were also held in the communities of Froland and Gjerstad, and various forms of commemoration were undertaken. A public competition was announced, to create a sculpture of Abel that was to stand to the west side of the entrance to the university's main building. The jury received 19 entries and awarded the prize to the sculptor, Ingebrigt Vik, but after much hemming and hawing, in 1908, it was Gustav Vigeland's inspired Abel monument was installed in the park of the royal palace, later known as the "Abel Garden".

The original arguments, evaluations and opinions persisted about what the nation of Norway should and could have done differently in the Abel affair. What Abel could have done for the esteem and reputation of the nation, had he lived on, received much greater emphasis than did his mathematical work. And as a person, Abel continued to be viewed as an awkward and unfortunate young creature who did not know how to comport himself in society. It was only when seated at his desk that he came into his own. There he was the genius able to transcend ordinary knowledge in a manner that few others had done.

It was also clear in Abel's day that what he had done was unique and highly significant, but his mathematical contributions did not lie in the areas which at that time gave scientific lustre and prestige - namely, the field of mathematical astronomy. In the young nation of Norway, priority of task was given to mapping the country, making accurate calculations of the length and breadth of places, determining the height of mountains and the width of fjords. In this sphere, magnetic cartography and knowledge of northern lattitudes was much more than a national affair. Professor Hansteen's field of interest inspired worldwide interest and his renown as a man of science placed him in high demand. Further afield, in Europe, money and prestige were given to those who calculated and computed the pathways and locations of the heavenly bodies newly revealed by developments in optical technology. By means of his pathbreaking research, the Englishman, Isaac Newton, had contributed to his countrymen becoming the best navigators, which in turn gave Great Britain domination over the seas, as well as giving direction to that particular field of mathematics. Mathematical insights had given rise to practical applications. Napoleon had also assembled the best mathematicians around him in the task of devising weapons and defence works. Abel, however, did not work with anything that could be considered to have practical significance or contemporary application. What application could there be to knowing the convergence criteria for infinite series? And why should someone know about the properties of algebraic equations? No one had given Abel the task of conquering a new field of mathematics. Nevertheless, the insights that he developed, and the methods he devised would later form the basis for a whole spectrum of further research, research which in its way is basic to the technology upon which modern society relies.

Just as his mathematical work appeared obscure to his companions in the Christiania and Norway of his day, so too he seems to have remained obscure and indistinct as a person.

But what could be expected of someone who had not reached the age of 27? What were the premises upon which personal development occurred in a life that was determined largely by such outward factors as family, teachers and other authorities? Being the son of a Norwegian senior public servant, he was, at the outset, driven by a sense of responsibility and expectation. Responsibility for the family fell on his shoulders when his father died, an obligation that weighed heavily both economically and morally. He passionately tried to live up to what was expected of him but did not know if he could manage; nor did he know whether or not he was acting the way he should. It seems that with regard to life, he lacked an overview, he did not know the premises that made for a liveable life. He was overcome by periodic bouts of anxiety and worry, and he failed to find ways of straightening out the problems of everyday human life. He found neither faith nor reason capable of providing a comprehensive understanding of life. Mathematics, however, was the complete reverse: there he found that things previously considered true, could demonstrably be wrong; that which had been taken for granted, could be based on a completely erroneous set of assumptions. This desire to do the right thing, in his non-mathematical life, but to have no idea of what the right thing was, seems to have made him dejected and depressed.

Abel's great escape from all this was the theatre. He attended the theatre as often as he could, and there watched life unfold without his participation. Perhaps in watching the orderly actions on stage he was able to forget for a moment life's lack of logical guidelines. He had no feeling for the music they played before the performances. It was rather the crystal clear activities of human beings that fascinated him.

He had had a strong, domineering father, who had plans and expectations on behalf of his children. In 1807 parish priest Søren Georg Abel published a catechism in which step by step he explained how one could achieve a state of grace in both this and the next life. But no matter how much Niels Henrik wanted to understand his father and his father's unassailable faith, this seemed to remain an insoluble enigma for him. This may also have been the case in relation to his mother who, due to a self-indulgent, growing alcoholism, had more or less removed herself from everyday life. One consequence is perhaps the fact that Niels Henrik seems to have limited to what was socially necessary, his contact with the other sex. Abel became engaged to marry at an unexpectedly early age: his comrades were astonished when he returned to the student residence after the Christmas holidays in 1824 as a man engaged to be married. But perhaps this was precisely his way of avoiding all the talk about women and eroticism that were so much part of student life. There is little to indicate a passionate love underlying the engagement, beyond his friend Boeck's words about great tenderness. The erotic seems never to have been a strong force in his life. His fiancée was a girl none of his friends knew, a girl he had met at a social gathering at his mother's sister's house in Copenhagen. What brought them together was a dance that neither had mastered. The waltz was a novelty and, thus embarassed, the two young people had backed away and stood looking at one another. In the years that followed it was economic factors which - maybe somewhat conveniently - prevented the wedding. No one married before they could set up housekeeping. Nevertheless he felt responsible for his fiancée and he felt guilty that she had to wait so long to be wedded. Was this also one of the diffuse, but all-absorbing demands he felt life had constantly made upon him, or was it something else: an understandable responsibility based on desire and spontaneity?

When Abel finally realized that he was dying, he bequeathed, in a manner, his fiancée to one of his friends. Was this with a feeling of deliverance or of sorrow? "She is not pretty," he is reported to have said, "has red hair and freckles, but is a splendid Human Being." He urged Keilhau, his travel comrade who had just obtained a lectureship at the University to take care of her. Indeed, in January, 1830, they were engaged, and married that autumn. They lived together for the rest of their lives, childless. In the autobiography that Professor Keilhau published in German in 1857, the year before he died, he was constantly thankful to have had Christine at his side. She followed him and helped him on many arduous journeys, and "above all contributed what I myself completely lacked, namely, her wisdom about life."

Had she indeed possessed such a wisdom about life, perhaps Niels Henrik's fate would have been different, if, despite his poor prospects, he had drawn his fiancée more closely into his life. If he had come to understand himself, would he, perhaps like his friend, Keilhau, have confessed: I completely lacked wisdom about life...

The funeral took place on April 13, 1829, at Froland Church. Smith, the owner of the establishment, wanted to give Abel a place in his own family's graveyard, and invited people from the surrounding villages and from Arendal and Grimstad to the service. However, winter had returned before the day of the funeral. The previous day saw the occurrence of a violent snowstorm, the like of which no one could recall so late in the year. Very few managed to arrive on time. Those familiar with the area did not even try to navigate the road from Arendal to Froland. Others who tried were forced to turn back and spent many hours covering only a few kilometres.

Six men from the Froland Works, dressed in black, bore the coffin out of the house. It was adorned with a floral wreath and the name of the deceased. The coffin was placed on a sled and carried the five kilometres to the church, where it was carried in through an archway of spruce boughs raised in Abel's honour. The priest, Christoffer Natvig, made a short speech. Afterwards, the coffin was taken into the churchyard beside the river and, through deep wet April snow, Niels Henrik Abel was lowered into the earth.

Both in Germany and France obituaries on Niels Henrik Abel were published long before people at home in Norway were able to read about this unique human being who had been carried off at such an incomprehensively early age.

Crelle's obituary, which was dated Berlin, June 20, 1829, made the biggest impression, and was published in *Journalen*, the journal that in all ways had meant so much to Abel. Crelle had now become a leading man in the work of the Ministry of Culture with the task of establishing a polytechnical college. He had met Alexander von Humboldt and cooperated closely with this well-known natural science researcher, who at this time had come to Berlin following his great expeditions on the American continent and his many years in Paris. Crelle was beginning to see the fulfillment of his dream: to initiate an epoch in his homeland in which mathematics, as part of the history of the natural sciences, would have its place alongside the epoch-making developments of earlier periods in history. Crelle thought of Frederick the Great of Prussia and Leibniz who had made Berlin the focal point for scientific work and research. Like von Humboldt, who after going through all the individual findings and simplicities now wanted to find a unifying structure, a world harmony, a unifying principle, Crelle too was inspired by the same philosophy, as he expressed in his obituary of Abel: "All Abel's work is shaped by an exceptional brilliance and discernment. He overcomes all obstacles and pushes on to the bottom of things with an apparently irresistible power. He tackles problems with an overwhelming energy, grasps them so thoroughly, and in this way transcends their actual position such that difficulties seem to disappear in front of the victorious power of his genius. And this from a mere youth..."

To show that he was not exaggerating his admiration, Crelle referred to a series of Abel's works that were published in *Journalen*, and he stressed: "Abel's unique talent has become universally recognized recently; in truth, had he been Newton's contemporary, what Newton said about Cotes could also have been said about Abel: 'If he had lived longer, we would have had even more to learn from him.' (Roger Cotes was Professor of Astronomy and Physics at Cambridge and wrote and published the introduction to Newton's major work, *Philosophiae naturalis principia mathematica*.)

In his obituary notice, Crelle described the part of Abel's life that he knew best, and naturally enough, stressed that while Abel had great difficulties getting a position in his homeland, it was the Prussian government that had been the first to offer Abel a permanent position. Crelle also mentioned that a number of the French mathematicians had personally petitioned King Karl

Johan to find a permanent position for Abel, an enquiry that few knew about in Norway, probably not even Abel himself.

But for Crelle, it was not only Abel's great mathematical talent that earned respect for the young man and caused great sorrow over his loss: "Abel distinguished himself equally for the purity and nobility of his character, because his exceptional modesty made him as popular as did his uncommon genius. Envy of the fortunes of others was completely foreign to him. He was far above that voracious craving for money, titles and fame that so often leads to science being misused and employed as a means of achieving fame and distinction. Abel held the sublime truths he pursued in such high esteem that he could not sell them for so low a price. The recompense he found for his exertions lay in the results themselves. A new discovery delighted him, whether made by himself or by others. He was a total stranger to the methods by which he could have secured illustrious positions. He did nothing for his own gain, and everything for the science that he loved. Everything undertaken for the improvement of his welfare came solely from his friends." And Crelle concluded: "Perhaps such selflessness is out of place in this world: Abel sacrificed his life for science without thinking of his own wellbeing. But who can say that such a sacrifice is of less value than that which is made for other lofty goals and is without hesitation awarded the greatest honours? Let us honour the memory of this man who distinguished himself equally with his uncommon talents and the integrity of his character. He was one of those singular beings that nature rarely brings forth in the course of a century!"

This obituary written by Crelle was translated and printed in Christiania's *Morgenbladet* in November. In Paris it was published in the Baron de Ferrusac's *Bulletin universel des sciences et de l'industrie*, most frequently called simply, *Ferrusac's Bulletin.* In the scientific circles that revolved around this journal, Abel, while in Paris, had met Jacques Frédéric Saigey who was at the time responsible for the mathematics and physics section of *Ferrusac's Bulletin*, a magazine which would give the broadest possible orientation to everything astir in the world of the natural sciences. Through Saigey, Abel had gained entry to the Baron's well-equipped private library, a meeting place of the learnèd of Paris. Later, Saigey together with his friend, François-Vincent Raspail began their own journal, *Annales des sciences d'observations*, and in the issue of May, 1829, Saigey brought out his own Abel obituary. The piece contains many imprecisions about Abel's life, but above all, it was a reaction to the unworthy manner in which the young man of science had been treated in Paris by the Academie Française and the French Institute. And on the long term, as late as 1870, Raspail made a fiery speech in the Chamber of Deputies, using Abel's life as an example of how the elders of the world of science greedily accumulated great wealth for their own specific fields while keeping the youth locked out. But the fate that befell Abel, Saigey wrote in May, 1829, befalls all young men who have nothing to recommend them except their own work: "Senior men of science either say that one shall reap what one has sown, or they say that one should work to bring the knowledge of the natural sciences to applicability and practical use." Saigey concluded his obituary on Abel with the call: "Young men of science! Do not listen to any voice other than the one inside that tells you which endeavours best suit your tastes and talents. Follow your natural impulses wherever they lead you...Read and reflect upon the writings of brilliant men, but never become blind disciples or egotistical admirers. Fidelity to fact, and freedom for all points of view, this must be the motto!"

It was said that Abel was of average height, and slight of build. His head and features were well formed, but he had an ashen complexion. His eyes were blue; some described them as warm and sparkling. Many remarked that he was quick to express almost childlike happiness, but could also slip rapidly into despondency. Among friends he was cheerful and wont to sing the odd song from his home region, take the odd drink and smoke a pipe of tobacco, but always in moderation. Nobody doubted in the least that he was far removed from all petty jealousies. And for his part, Abel never indicated what this conduct cost him. Saigey wrote that he felt Abel's appearance had been haggard and dark, and gave the impression of worry and fatigue, and that Abel's mild character fluctuated between modesty and embarassment.

It was said in his home village of Gjerstad that Abel had been a strong swimmer. In swimming competitions with the bigger boys, Niels Henrik was the most frequent winner because he swam like an eel while the others expended all their strength ploughing their way through the water. He was also said to have been very good on skis. He shot down hillsides crusted with hard spring snow, hills that others found difficult enough under ordinary snow conditions. Later, whenever he went home during the holidays he often wandered around the farms and passed for an ordinary, straightforward fellow who was well regarded everywhere. Among other things, people said that he was a skilful weather forecaster. One bright sunny day he had come to the haymakers at Lunde where his mother lived, and said they would have to forego the mid-day break if they wanted to get the hay in while the weather was still dry. And quite rightly so, for a little while later the rain was pouring down.

Abel did not mind going out of his way for a good joke, and could be very quick on the uptake. In a story from Gjerstad it is said that once while he was calculating the distance to the sun, some of the local youths had in an unguarded moment seized the opportunity to change the height of the chair on which he had his instruments. When Abel was ready to resume, his earlier calculations did not tally, and with some powerful swear words he jokingly declared that the sun must have moved a couple of inches!

Actually, it was not really the distance to the sun that he was measuring. It is probable he was standing there with the swinging apparatus designed by Professor Hansteen and measuring the intensity of magnetic fields. And in all likelihood it had something to do with the moon. At any rate, while Abel was waiting for his European travels to begin, he received the task of calculating the moon's gravitational pull, using the pendulum with which Professor Hansteen's measuring apparatus was equipped. The task was no doubt assigned by Professor Hansteen himself, who had come to consider that possibly the pendulum's swinging was affected not only by the earth's magnetism and force of gravity, but also perhaps by the gravitational pull of the moon. Abel measured and calculated that the moon really must have such a measurable effect. It was probably these calculations he was in the process of checking at Gjerstad. Whether it was the sun or the moon, a chair or something else, magnetic fields do not move around, and Abel must have been joining in the local boys' joke and making a rather humorous story out of it.

Hansteen later encouraged Abel to write a paper on these mathematical calculations, and Abel wrote "Regarding the Moon's Influence on the Motion of a Pendulum." Hansteen published this in the spring of 1824 in *Magazin for Naturvidenskaberne*, of which he was editor, and he felt the question was so important that he soon sent Abel's article abroad to see if it could be published in German or French in the journal *Astronomische Nachrichten*, that had just been established by the renowned astronomer, Heinrich Christian Schumacher in Altona (Hamburg). Hansteen hoped that this paper of Abel's would serve as his letter of introduction to Europe's learnèd circles, which Abel would need during his approaching European tour. Schumacher, who had been professor in Copenhagen, and could read what Abel had written, answered right away that he would not publish the paper: Student Abel had calculated the moon's attractional pull on the pendulum to be 60 times greater than it actually was! This had occurred because Abel had forgotten that the moon also attracts the centre of the earth. After all, it was not a question of the moon's absolute attraction on the pendulum, but of the difference between this attraction and the attraction to the centre of the earth, and in practice it was completely without significance." Schumacher concluded: According to Abel's formulas the sun must give a deviation from the true plumb of several arc minutes. So, for the sake of his own reputation, let us speak no more about this."

Hansteen's attempt to get Abel a passport into the science milieu of Europe consequently became a fiasco. And Abel was probably sorry that he had allowed himself to venture out into fields that he had neither mastered, nor was much interested in, and he was afraid that Schumacher had mentioned this blunder to his good friend, the great Gauss at Göttingen. In any case, Abel felt that he had to correct the mistake with a new article in *Magazinet*, which he did. But later, when in Paris, he was recruited as a sort of journalist for *Ferrusac's Bulletin*, he thought once more about his lunar calculations. Abel was to write an assessment of the articles in *Crelle's Journal* for the French journal, in order to bring these works to the attention of the French. Despite the fact that this was perhaps not a task of which Abel was particularly fond he felt it was his duty to do so for Crelle's sake, and besides, it was a point of connection to the learnèd circles of Paris. Consequently Abel made short summaries of his own articles and those of a number of others from *Crelle's Journal* in Berlin, but he took only one lone example from the natural science journal *Magazinet* in Christiania, namely his own corrected calculations on the effect of lunar attraction on the pendulum!

When Abel glided into Froland on a sleigh in December, 1828, it was full winter and cold. He was pale but did not feel ill. He was wearing a black, long-tailed frock coat with particularly broad sleeves. Abel was a treasured guest and was received with open arms. It was also said that Abel had a special affection for the house cat, and on that December day, and to amuse himself and the others, he took the cat into his frock coat, and when it emerged from the wide left sleeve, he took it in his right hand and allowed it to creep back in the opposite direction so that it reappeared on his breast.

Factory owner Sivert Smith had 11 children, ranging in age from 6 to 22; in addition, several of Smith's 16 siblings lived around the estate, where the wheels of production were in full swing producing iron bars and nails, and where several large saws were in operation. There was a confusion of life everywhere. It was said that Abel was like a son to the house, that he went up on the roof to roll snowballs with the boys and they frolicked together in the snow. As for the girls, they felt it was rather grand to have in their midst this strange university candidate who was so brilliant with mathematical calculations. When he had visited the previous summer, they had given him gifts, including six pairs of stockings and they loved to test his arithmetic skills. One of them is said to have asked: When one and a half herring cost one and a half shillings, what do eleven cost? Abel was said to have replied with a smile, that he could not answer such a difficult question, and the girls, having stumped him, were in a state of rapture.

In another anecdote it is the philosopher, Niels Treschow who poses the question: When one and a half herring cost one and a half shillings, what is the price of twelve herring? And when Abel was said to have answered eighteen, the old Treschow rubbed his hands in glee.

Otherwise, time and again the respectable and influential Niels Treschow was an important figure standing in the wings during Niels Henrik Abel's life. Ever since the *fin de siècle* Treschow had been the main architect of teaching and school policy in Norway. And without Treschow's persistent support for specialized subject teachers in secondary schooling, *contra* generalized class-room teachers, Abel would probably not have been exposed to a teacher of mathematics as gifted as Holmboe was at the Christiania Cathedral School in 1818. As cabinet minister, Treschow refused to bow to the parliamentary majority and return to the old system of classroom teachers, something that in practice would have meant that theologists and Latin teachers would once again teach all subjects. In the 1780s, when Treschow was principal of the Helsingør Cathedral School, he had Niels Henrik's father, Søren Georg Abel, as a pupil and guided him such that he passed with distinction the examina-tion for university entrance. Later he became principal of the Christiania Cathedral School, then Professor of Philosophy in Copenhagen, and he was the most obvious choice in the appointment of professors at the new Norwe-gian university in 1813. A year later he was in the King's Council and became head of the Department of Church Affairs. Some thought that this change occurred dangerously quickly, that it was disgraceful to substitute politics for philosophy. In the extraordinary parliamentary session of the autumn, 1814, Søren Georg Abel and Treschow had been at political loggerheads. But in 1818 they stood together against the parliamentary majority in determined defense of the new developments in pedagogy, such as the system of specialized teachers in the higher schools. In 1823 Treschow became vice-chancellor, and in practice, chancellor and leader, of the University. And in this position in 1825 he spoke warmly in support of Abel's request for the travel grant, the 600 *daler* over two years, an award about which, for the most part, there was strong agreement. A year later he had taken Niels Henrik's sister, Elisabeth, into his house, an arrangement that gladdened both old Treschow and his family, as well as Elisabeth Abel.

When Abel returned from abroad in 1827, he visited his sister at the Treschows' home as often as he could. It was probably during this period that Treschow, by way of jest, posed Abel the question of the twelve herring. Otherwise, mathematical calculation was not one of Treschow's strong points, although he had an intense relationship to simple numbers. As a master of the Lodge of Freemasonry in Christiania he tended to expound his insights on the numbers 3, 7 and 9. It was said that in such surroundings he had a remarkable propensity toward the mystical, and to be sure, in his lectures he got his listeners to conceive in numbers all the world's relationships. Apart from this, it was Treschow who had reopened the Freemasons' Lodge when he returned to Christiania as professor. This was no doubt a consequence of his own desire as well as in response to the wish of the Swedish king, Karl XIII, who, during the autumn of 1814, become King of Norway. Voices of criticism maintained that the king's only concern on earth lay precisely in the devising and writing of new rituals of Freemasonry and installing them in the Great Nordic Lodge. Also during this period of the first union, Treschow worked zealously to bring about a union between the crown prince, Karl Johan's son, and one of the daughters of the Danish royal house.

At that period, while Abel would visit his sister at Treschow's house in Christiania's Tøyen district, he was in the depths of economic crisis, but he never asked Treschow for help. Perhaps he felt that Treschow had done enough by securing a position for his sister Elisabeth. Treschow, who often used the term, "Inner Sense," and by this meant an awareness in people of psychic "exceptions", which mechanistic and physical causes were incapable of explaining, nevertheless did not have an inkling about young Abel's inner life. Treschow felt that thought functioned according to a design or protocol that, so to speak, demanded fulfillment. He was well schooled in all of Western philosophy and was intensely preoccupied to find unity in the manifold, a connection and continuity in everything that happens. This, Treschow argued, was a divine sense that manifested itself in the world, but in relation to Abel he seems to have shown neither comprehension nor sensibility.

At Froland, Abel was also known to be a star-gazer. Of course this description was commonly applied to learnèd men of science, but in Abel's case it probably also was applied to him because he would gaze at the evening sky in order to determine what the next day would be like. Besides, there was the known fact that Abel worked with Professor Hansteen, who was in charge of the almanac. One of the workers in the factory complained that the almanac did not tell the truth, especially concerning the weather. And Abel was said to have replied that he could not expect much more for a mere six shillings.

In the baroquely landscaped surroundings of the Froland Works, Candidate and Stargazer Abel, on fine summer evenings, would certainly have been courting his fiancée, Governess Christine, along the byways of the garden and in the summerhouse. It was reported that he had ardently portrayed their future, and said to her: "When we get to Berlin, you are not going to be called Madame, or my wife, but rather *Herr Professor mit seiner Gemahlin.*"

Another anecdote has it that early on in the Christmas holidays, 1828, Abel had danced until, warm and perspiring he had gone out to the hallway to cool off when the coughing began. The next day he had a fever and was confined to bed with an infection of the lungs. He was somewhat better at the beginning of January, in any case by January 6th he was able to work. Since he had not heard from the Paris Institute about the work he had given them, he was afraid that his submission had been lost. Above all else in the world, he wanted to save the main ideas in this work. The Paris paper filled 65 large pages. Now, while he lay in bed at Froland, he wrote two pages demonstrating the first major point in this treatise. This was the last thing he ever wrote.

In 1903, the Swedish mathematician and editor, Gösta Mittag-Leffler, wrote about this great exertion made by Abel at Froland: "the 6th of January, 1829, was a date more memorable in the history of culture than the dates of kings and emperors, or the commemorative dates of specific nations, for this was when Abel wrote down the greatest conceptualization of his life for Crelle's journal." And when in 1882 Mittag-Leffler, with support from the Scandinavian countries, founded *Acta mathematica*, an international mathematical journal, the magazine was introduced with a portrait of Abel.

Reports that Abel was lying ill at Froland Works awoke a certain anxiety in Christiania. And these concerns were not only about the lecturing responsibilities he had undertaken in Professor Hansteen's absence. Many were aware that he had been bedridden for some weeks in the autumn, and those who had seen him reported that he seemed worn out, exhausted. After several such reports had arrived, people began to understand that at the very least the illness would be protracted.

While Hansteen's setting out on his Siberian expedition had improved Abel's acute economic situation, Hansteen's journey had also meant that his spouse, Mrs. Hansteen, had in September, 1828, gathered her children and gone to her nearest relatives in Copenhagen and Sorø. For a long time, and to an increasing degree, Mrs. Hansteen had played one of the most important roles in Abel's everyday life. She was one of the very few with whom he spoke openly. He called her his "second mother," and to be with her, as he expressed it, was to be "among the angels." Mrs. Hansteen held Abel in high regard and she helped him in the practical side of life as far as she was able, given her responsibilities to husband and children.

Before Abel had set out for his European tour he had left money from his stipend with Mrs. Hansteen so that she might give out moderate sums to his brother Peder, who otherwise would have wasted the whole sum on drink and carryings on. Abel also set aside money from the stipend so that his sister Elisabeth could get away from their mother in Gjerstad and move to Christiania. And in Christiania it was Mrs. Hansteen who received Elisabeth, and Elisabeth stayed at Mrs. Hansteen's for six months before she received a position at the Treschows', who were cousins of Professor Hansteen.

While on his European tour Abel wrote several letters to Mrs. Hansteen in which he described his depression and his periods of gloom, about his unease at being alone, and about the joys of experiencing new cities and unknown landscapes. He told about mad fits of work in the midst of everything, and how he was looking forward to getting home and working consistently on all the new ideas that so recently had come to him. Mrs. Hansteen sent encouraging words back, often enclosed within her husband's letters to Abel. "Letter from dear Abel," she wrote, for example, in her journal on March 17, 1827.

Mrs. Hansteen seems to have had a kindred feeling for Abel's mental anxiety. She confided in her journal that she could become "despondent" and "melancholy" after having talked with Abel, and "through the Association of Ideas," enter her own "Labyrinths."

Mrs. Hansteen was born Cathrine Andrea Borch, in Sorø, Denmark, where her Norwegian father, Caspar Abraham Borch from Trondheim, was professor at the Academy. She had married Christopher Hansteen in 1814, who was then lecturer in mathematics at the Frederiksborg Latin School. She was a member of a large family, with five sisters and three brothers. Her father died in 1805 but her mother presided over a very hospitable house in Sorø, and when Abel was in Copenhagen in 1823, one of Mrs. Hansteen's sisters, Henriette Frederichsen, invited him out to Sorø. There Abel also met the youngest of the Borch sisters, Charité, only a year his senior, and Niels Henrik seems to have been attracted to her. In letters to Mrs. Hansteen he often asks her to greet her sisters on his behalf, and "the sweet-tempered Charité." And when the exhausted and travel-worn Abel passed through Copenhagen on the way home from his grand tour in May, 1827, he had again visited Mrs. Hansteen's sisters in that city. And it appears that the portrait that Gørbitz made of him in Paris was left with them, and not with his fiancée, who also met him in Copenhagen.

Following Abel's homecoming at the end of May, 1827, he often visited Mrs. Hansteen. During the month of June she noted in her journal that Abel had visited four times, and similar numbers of visits appear in the following months. Sometimes it was merely a short chat on the steps in front of the Hansteens' house; other times it was over tea at social gatherings, and in between there were occasions when he would pull up a stool and sit at her feet in the parlour. Following a visit on June 18th she became particularly distressed by the thought of Abel's position and the insecurity of his future, and concluded her entry: "Talked late into the evening with Hansteen which consoled me considerably so that I was able to sleep with a lighter Heart." Perhaps what consoled her was the professor's reassurance that Abel would get a teaching assignment and a salary when the professor himself travelled to Siberia.

Mrs. Hansteen suffered from migraines and rheumatism, and moreover it upset her that despite her good circumstances she was not happy. She had given birth to her fourth daughter in December, 1824, who was named Aasta. Later it was this very Aasta Hansteen who became Norway's foremost spokesperson for the rights of women. But Aasta's mother, Andrea Hansteen, felt herself to be a failure; she believed that there was something wrong with her, that she created scandal in society by speaking in a too open and impetuous manner. She ardently wanted to understand the scientific and scholarly achievements of both her husband and Abel, and it came to public attention that she and several other leading ladies of the capital would appear at lectures on physics and the natural sciences.

For Abel, Mrs. Hansteen was probably the mother figure he had always longed for. From his sickbed at Froland during that late winter, he spoke, half delirious, about Mrs. Hansteen, fussed that he had been ill in bed during the autumn, and had not managed to say goodbye before she had gone to Copenhagen. Would things perhaps have turned out differently had she been in Christiania?

When the district's doctor, Møller, had examined Abel in February and had written to the university authorities about his condition, it was no doubt known in the house where his sister Elisabeth was in service, that Abel was suffering from "a strong Infection of the Lungs and considerable Spitting of Blood." Mrs. Treschow conveyed this news to Mrs. Hansteen in Copenhagen in a letter dated February 27, 1829: "poor Niels Henrik is lying dangerous ill at the Froland Works, and was ill when he went there. Perhaps if you had been here in the City he would have been content to stay put in peace and quiet, but he wanted to hide how sick he was, although early on I had detected the Distress that the Pain in his bones occasioned."

At Froland, Abel lay in a bed in a room on the shady side of the house. He wanted so ardently to be healthy again, but he could not ignore the blood; he coughed more and more, became sleepless and was afraid to be alone at night. When he got up he could see the sparkling moonlight on the snow outside. And when at the end he recognized that death was approaching, he complained loudly and in strong language about contemporary science which had not researched and struggled to master his illness. It is said he shouted, "I am struggling for my life!" In between, in calmer moments, he continued to work, at a low level, fading away, "like a sick eagle yearning for the sun."

Before, when he was still in good health, he never liked to be alone, except when working. He became melancholy without company. Apart from going out to the theatre or a revue, which he adored, he loved to play cards, and he often sat at the card table long into the night. He played well, even when the stakes were low. He calculated his chances of winning in a manner that could incur the displeasure of others. But perhaps he found the challenge of achieving the best results in this intersection of randomness and rules, perhaps it was not only winning at cards that he was after, but rather a method for mastering randomness. When playing, he kept his cards very close to his chest, and he usually knew what was in the hands of the other players. It is certain that at the playing table he deployed his cards with brilliance.

But how should one come to grips with life's tasks? How to make one's daily bread from mathematics? This most urgently had to do with getting a permanent position, ensuring one's circumstances, one's working time. But how to get this? The axe had fallen upon almost the whole of the young nation, and Abel had no arguments; he did not understand how to play his cards correctly in this real and decisive positional game. Was it this lack of instinctive self-assertion which friends called, among other things, magnanimity or bashfulness, that caused him to hesitate whenever in daily life there were chances to "take a trick"?

Abel was an engaging and modest young man who in his work room had insights into the calibre of the whole world, but outside this room he did not know how to act.

Abel lived in a time when it was possible to view science in its state of innocence, before the forces of nature had been harnessed, before the interaction between magnetism and electricity made it possible to handle matter to achieve predetermined goals, before electromagnetism and light seriously reshaped the world and its people. But behind many of these radical new developments lay the mathematical analysis that could make the most of these new observations. Abel died before it became evident that the results of scientific insight would come to change living conditions, before it was understood that science could give people a secure existence, a power over nature. That these advances in the course of the 1800s would lead to a *faith* in the ability of science to master life, to achieve a new golden age, was still barely imaginable. That new technological developments, as for example the spinning machine, would involve the formation of new social classes in England, seems not to have been discussed in the circles familiar to Abel. The most revolutionary thing that Abel experienced with regard to the practical results of science, was to see communications radically changed in the summer months of 1827 with the introduction of steam ships on the route between Christiania and Copenhagen, and from Christiania, southwards down the coast to Kristiansand.

And what was the place of mathematics at this time?

In the 1600s the great Galileo Galilei had said that we human beings cannot understand the universe before we learn its language, and that language was mathematics. For Galilei, mathematics meant first and foremost an understanding of geometry. Moreover, the counting system, such as the position system we know today, was new in Europe in Galilei's time, and began to become the basis for a new way of viewing the world. Only that which could be weighed, measured and quantified was understood in the strongest sense. Rationalism with mechanistic explanations began to replace old explanations which considered that the purpose and intention of things was their deepest and most essential secret.

Much had happened since that time; above all, the development of infinitesmal calculations that Newton and Leibniz relied upon at the beginning of the eighteenth century. Differential and integral calculus allowed for a degree of conceptualization and computation using infinitely minute variables that opened huge vistas and possibilities that no one before had even dreamed about. The introduction of infinitesmal variables otherwise had led to a lively discussion, and mathematics in general faced resistance from the more philosophical circles.

It was in revolutionary France that mathematics was seriously treated as an indispensable tool, and Napoleon set great store in having the best mathematicians at his service. When L'Ecole Polytechnique became a permanent institution of learning in 1794, it exemplified a revolution in the teaching of mathematical knowledge. Mathematics now became used not only as a tool, but also grew as an branch of the natural sciences in its own right, and concepts such as marginal value, continuity, and integrability were analyzed and definitively proven. Paris became the centre of teaching and research, and prominent mathematicians flocked there. One of the most innovative in this tradition was precisely Augustin Louis Cauchy, the man for whom Abel was looking in 1826, and who must take the main responsibility for setting aside Abel's main treatise.

As early as his first year of studies Abel made note of tasks and probable solutions to problems that he would later work up in detail. During the last year of his life, when both acknowledgement and renown were coming to him, and his ideas were awakening interest, he continued to ponder the correlations he had seen and recognized such a long time before. It seems he discerned methods for tackling problems long before he had time to work out his theses on how they might be solved. It all had to do with finding the time to painstakingly work them through, demonstrating the propositions. It seems almost as though Abel was a visionary in his mathematical insights, that from great distance and long in advance, he saw solutions to problems he hoped one day to have the time to work out and demonstrate. Mathematical ideas frequently swept down on him, and once he had obtained a conduit to publication, via *Crelle's Journal,* he exploded into work.

"All divine emissaries must be mathematicians. Mathematics is a spiritual self-enterprise, the methodical unfolding of genius. Mathematics substitutes reason for nature. Pure mathematics is religion," wrote the poet who took the name Novalis, about 1800. A new sense of life had begun to express itself in

Germany, something that had nothing to do with cogs and wheels and mechanisms. Philosophers, poets and scientists considered that nature was animate, and that the human being was at the apex of a spiritual world. Moreover, human beings should extend their grasp, people were divinely inspired, and should one day be "..what Our Father, God, is; God wants Gods," or so Novalis, one of the foremost figures of German romantic philosophy, maintained.

It was not only in mathematics that man occupied himself in contemplation of continuity and the limits of the infinite. And whoever sought absolute truths, whether in the realm of pure mathematics, or in terms of divine reason - and however these were created - these were now concepts taken up by both poets and mathematicians.

But despite the growing fissures in old fossilized concepts, the received standard of taste continued to dominate the cultural life of the day. Art gave way only slowly to the new times: paintings should strive to attain absolute beauty, and consequently did not depart from the copying of works of Greek and Roman times. Classicism also dominated sculpture and poetry. The museums, built in classic style, were cathedrals of faith: one had to *believe* the old ideals of the eternal, and that the ideal could be realized. Only music was to escape classicism simply because nobody knew what ancient Greek and Roman music had been like. The most popular composer of the day was Joseph Hayden. Abel, however, had no feeling for music, and he scarcely had time to see any contemporary art, whether in Berlin or Paris. When he visited Dresden in March, 1826, he made the acquaintance of the painter, J. C. Dahl, who was professor at the city's art academy but his paintings seem not to have made much impression on Abel.

Several who knew Abel said he was little interested in the music that was being played before theatre performances and in concerts. Some took this as a sign of a lack of musicality. But perhaps this was more a question of discernment, of choosing what to listen to. His sister Elisabeth was very musical, and right from Niels Henrik's first year in Christiania, it was said that he used to sing ballads from his home region; and his youngest brother, Thor Henrik, who to the end of his days was a wanderer, played the flute everywhere to the enthusiastic admiration of the public.

As far as Abel was concerned, it seemed that his clear and innovative insights apparently had nothing to do with external factors such as heredity, upbringing and the contemporary milieu. Or perhaps upbringing and milieu were indeed crucial, but in a form and at a moment of time, for example in earliest childhood, when the signs were difficult to discern. He apparently had two faces. The view from his social surroundings was that of a shy and worried being who lacked the personal vibrancy one often expects to find in people of exceptional intellectual dimension. And then there was his work identity:

during those periods when, for a moment, free of everything from the outer world that hurled itself on top of him, he sank into deep thought, he became brilliant and all-conquering in his ability to make formulations, to convert the unknown to the known, to transform the obscure to the comprehensible.

It is quite certain that Abel never analyzed the causes for what can be seen as his boundless worry *about,* and *in the course of,* everyday life. His ability to find adequate expression for the human confusions that from earliest childhood he had been surrounded by, seems to have been extremely slight. But was not this gravitas, this restraint and worry perhaps useful to his working disposition? A longing to distance himself from all that lay in the depths of his private life was perhaps an extra driving force propelling his thinking. And in its turn, at least for short periods, the tangible results of this thinking became the most important anchor to his life.

PART II
Family Background

Fig. 2. Risør, 1817. Artist unknown. The Simonsen house on Strandgaten is in the middle of the picture. Private collection.

2.

The Abel Family Tree

Beginning in 1588, King Christian IV ruled for sixty years over the kingdom of Denmark and its provinces and dependencies. The king bestowed 27 visits on his dependency, Norway, and spread the view that Norway was a land rich in natural resources and labour power, a land of opportunity for new enterprises in commerce, timber, fish and minerals. The king founded the Kongsberg Silverworks in 1624 and he issued a royal command on how the city of Oslo should be moved and set up according to a quadrature gridwork plan located in the shadow of Akershus Fortress, and how it, the capital of the province, was to be called Christiania.

The king was on the road to Norway again in 1641, this time to institute another town plan in the quadrature style. This new town on the south coast also received its name from the king, as it came to be called Kristiansand.

At roughly the same time, two brothers, Mathias and Jacob, from Abild in the Deanery of Tønder in northwest Schleswig, a province under jurisdictional dispute by both Danes and Germans, voyaged northwards to seek their fortune in foreign lands. Or rather one could say that they probably both knew where they were going and were fairly certain what they would be doing. They wanted to go to Trondheim to engage in commerce and handicraft work. Presumably they had a modicum of skills and training, and a bit of money in their purses. They almost certainly travelled by the sea route; at any rate they were about twenty years old when they saw Norway's rockbound shores for the first time.

At that time Trondheim was a lively city of trade and commerce and many foreigners were among its leading citizens, merchants and officials. They came from Scotland, England, the Netherlands, Germany and last but not least, from the most southerly reaches of the realm of Christian IV.

During the Thirty Years War (1618-48), beginning with religious animosities between Protestant and Catholic princes and leaders, the struggles flared up every spring and summer, and new military forces raised armies whenever the battles thinned the ranks. In Denmark's southern Jylland, the soldiers of General Wallenstein and Christian IV rolled back and forth across the landscape; many deserted the combat. Perhaps the brothers Mathias and Jacob from Abild were among the deserters, perhaps, first and foremost, they wanted to get away from the war and the soldier's life. Norway was a secure place in which to settle, and Trondheim was particularly favourable for many reasons.

When the brothers from Abild arrived in Trondheim it was a time of overall prosperity. Copper from the mines at Røros, Kvikne and Ytterøy was being exported through that city. And there was also an upswing in the established trade and commerce activities: in the timber industry what was called the gate saw came into use, and the fine processing of logs into planks and boards created new jobs and increased trade; and in the sea, the herring had suddenly begun to swarm in toward the coast around Trondheim; an incredible fishery was quickly under way.

The brothers Mathias and Jacob were registered with the surnames "Abild" and "Abbild". Mathias always signed himself "Abell", and through the transitional form, Abelboe, the family name became Abel. They purchased citizenship papers, took part in the privileges of the city, and became inhabitants with the right to participate in trade, commerce, handicraft production and the transport of freight by sea.

Mathias Abell soon became a leading citizen of Trondheim: already by 1651, approximately thirty years of age, he had become a stipendiary magistrate in the city. He married the 17-18 year old Karen Rasmusdatter, and this union propelled him still further up the mercantile and social ladder. Karen was the daughter of Rasmus Hansen, who around 1600 had come from Kolding, not far from Abild. And this Rasmus Hansen had, after serving the country judge, Axel Urne on the island of Sjælland, Denmark, moved to Trondheim and married Elisabeth Hansdatter. They had four children but only Karen survived into adulthood. Karen's mother, Elisabeth Hansdatter came from a background of which she was especially proud: her father had been bailiff of Reinskloster in Trondheim, and on her mother's side nobility hung substantially from the family tree, with relatives among the Staurs, Benkestoks, Skankes and Rustungs.

Mathias and Karen had seven children, baptized Mathias, Elisabeth, Hans, Ester, Henrik, Rasmus and Karen. The founder of the line, Mathias, died young, in 1664; however, all in the course of five years, he had functioned as "Commissariat and Ammunition Secretary" for Northern Norway, a high military post. This was a gilded promotion from the king in Copenhagen and from Niels Trolle, the king's vice-regent in Norway. The two older sons were early on sent to the city's cathedral school - they were both to become priests - and one of them, Hans, would carry the family further in the direction of Niels Henrik Abel.

Hans, born in 1654, was sent at the age of nineteen to Copenhagen to take his student examinations. Returning to Trondheim three or four years later with a lower theological degree and a successful baccalaureate, he became a "hearer," a very poorly paid teaching position at the cathedral school in his home city.

Chagrinned at being a "hearer", and earning a mere pittance, Hans Abel left for Copenhagen at the beginning of 1690, to seek, or as was said at the time, to solicit, a post. When a position fell vacant it was best to be right there at hand in the king's city, ready to present one's application. Moreover, at any point in time one could see applicants doing the rounds of the corridors of Copenhagen, speaking politely and humbly to those nearest the king, the presidents of the chancery and the privy councillors. Some members of the king's cabinet set more store in attendance than others, and many of those soliciting a position spent years in Copenhagen. But luck was with Hans Abel. Shortly after he had arrived in Copenhagen he came to know that the priest from Tysnes near Bergen, Mr. Tyge Broch, following a judgement, had lost his post when, having ignored the protests of the bishop, he had abandoned his betrothed, and married the widow of his predecessor. Hans Abel applied, and two days later, the position was his. He travelled back from Copenhagen to Trondheim to pack up his belongings and say farewell to family and friends, except for his mother, whom he took with him to the vicarage at Tysnes, an island village south of Bergen.

A year later he married Claire Hanning, a vicar's daughter from Sogndal. Her father, Peder Olufsen Hanning, through two marriages became the pro-genitor of numerous descendants, a large number of whom, for generations, were parish preachers and preachers' wives in the region surrounding Bergen.

Hans Abel and Claire Hanning seem to have had a good life at Tysnes, and both vicar and vicar's wife were liked and respected in the village and the wider island community. They had, so to speak, a steady stream of children; the last was born in 1702. Hans Abel had gone to Copenhagen in 1700 and taken his magister degree, an examination that in addition to all the usual subjects involved defending theses put forward by a professor. Thus, in formal terms, Hans Abel could now apply for any type of position in the Church, or in the field of education. But until his death in 1723 he worked as vicar in the Tysnes parish.

Hans Abel must have been a good teacher of his own children, Mathias, Christine, Peder, Anne, Søren, Ester Marie, Hans and Jørgen Henrik. Four of the five boys attended the Bergen Cathedral School, and that they had been well-prepared is evident from the fact that each in turn began in the top form, and after three or four years at the school they were ready to tackle the *examen artium* in Copenhagen. And one after the other they began to study theology.

But the fifth son of the vicar of Tysnes chose a different calling. That was Søren Abel who was to become Niels Henrik's great grandfather. Søren dropped out of all public schooling and began immediately to orient himself in the commercial life of Bergen. While the other brothers were swotting away at the cathedral school and listening to their instructors lessons in the cathedral, Søren Abel became a "citizen of the city of Bergen" and became active in everyday commercial life. But this was during the Great Nordic War (1709-20), and commercial life was beset by prudence and stagnation. The economic blossoming and active cultural milieu, to which, for example the writer Ludvig Holberg had been part at the turn of the century, had become only a memory. A secure public servant position must also have been recommended for Søren Abel. How was it that he chose a life so different from that of his brothers?

Søren Abel saw the new peacetime period as a time for trade, commerce and new possibilities. He build his life on a careful foundation, and developed his merchant life so well that in 1732 he became executive manager of the city of Bergen. In 1737, at the age of thirty-nine and after having been a man of standing for several years, he married. But in order to marry the woman of his choice, Søren Abel had to request a special marriage license required of couples "related to one another in the second and third generation." Executive Officer Abel's bride was Margrethe Hanning, thirty years old and the daughter of a priest from Førde in Sunnfjord, a little further north along Norway's west coast. Her father was the son of Søren Abel's maternal grandfather, Peder Hanning of Sogndal.

Their first child was born a year after the marriage, a son who was christened Hans Mathias after Søren's father and paternal grandfather. And this was the Hans Mathias Abel who was to end his days as parish priest in Gjerstad. In fact he died only a couple of years after coming to know he had a grandchild called Niels Henrik.

Margrethe Hanning seems to have taken an active part in her husband's business, which, beside his city council activities, he conducted from his house right beside the council chambers, Bergen's Rådstuen. A great era for all kinds of traffic was poised to enter local history, but Chief Officer Abel only got to experience the beginning of these great changes. When he died in 1746, only forty-eight years old, the rich herring fishery of Norway's west coast had just begun; shipping and the boom years would soon give rise to an explosive upsurge in money and profits. But before Chief Officer Abel died, he managed to put aside the money for the education of his son, Hans Mathias. The same year that his father died, Hans Mathias was enrolled in Bergen's Cathedral School. The mother had been left with four children between the ages of two and eight, three boys and a girl. As was the custom, Chief Officer Abel's clerical brothers now became the children's guardians, and the two remaining brothers were dutybound to take responsibility for Chief Officer Abel's children, although both had large families of their own and could ill-afford the burden. However, it seems that Margrethe Hanning managed to care for her children herself; probably she continued with the business that she was entitled to conduct, such as the sale of old clothes and homemade linen. Housewife traders who conducted multi-branched businesses on the border of the law were, in other respects, a familiar occupational group in the city; songs and small booklets were sold; when it was time for butchering, they did the rounds of the houses, offering to sell a little meat; some of them bartered goods from the trading ships; and the beer-wives sold their brews.

But only two of Chief Officer Abel and Margrethe Hanning's children reached adulthood; in addition to Hans Mathias there was Peder who, after several years as a journeyman baker, became a "citizen of Bergen" and traded in small goods. He died a bachelor, beyond the age of seventy.

Hans Mathias Abel received unusually good instruction at the cathedral school in Bergen. When he began, Jacob Steensen was the rector, and Steensen was a pedagogical pioneer in the Danish-Norwegian school system. He pointed out that classical languages were merely a *means* of achieving erudition, not a goal in themselves. He maintained that limits had to be set on swotting and rote-learning, that the teachers' circumstances ought to be improved, and he was the first to suggest salary increments with age and experience. And when the pious Erik Pontoppidan, fired with initiative and the energy of trade and commerce, became bishop in 1748, he was able to advance Steensen's plans in many areas. In league with the reforms which prevailed at the time - as were implemented at the Seminarium Lapponicum in Trondheim and Bergseminaret in Kongsberg - Pontoppidan expanded the cathedral school in Bergen with what he called Seminarium Fredericianum. Bishop Pontoppidan also held that it took more than Greek and Latin to bring prosperity to the world. He wanted to give the pupils a taste of the sciences of mathematics and astronomy, and he felt that the contemporary languages, French and German, were necessary, in part for success in bourgeois social relations, and partly for the reading of good books. The new school plans also included a boarding school, which was to instil the pious and pedagogical characteristics highly valued at the time. In light of this goal, the school's instruction in mathematics and physics was to be disseminated to the wider public, among both the schooled and unschooled in the city's population. In the years between 1753-55, when Hans Mathias Abel was in the top "master reader" class, the Seminarium Fredericianum aroused great interest and enthusiasm, and Hans Mathias got, without doubt, a broader general education than what was common in other places. Hans Mathias Abel became a university student in Copenhagen in 1755, at the age of 17, and a year later he took the *examen philosophicum*, and decided to take the shortest route to the clerical examination; that is, one could pursue the exam by seeking a pre-selected mark; here the study programme was to an extent determined by the level of the marks selected for achievement. Hans Mathias chose the lightest course possible, the course that led to the degree awarded *non contemnendus*. And, barely 19 years old, he had his theological degree and was able to begin his ecclesiastical career at the lowest rank: personal curate under the vicar of a parish.

Hans Mathias worked for six years under his father's brother, the parish vicar and senior rector, Jørgen Henrik Abel in Bygland. After his uncle died he managed to fulfil the vicar position until the arrival of his uncle's successor. Thereafter, H. M. Abel was appointed to fill the vacant position at Oddernes in Kristiansand Diocese. Next he became the personal curate of the disreputable Erik Anker Brun, Vicar of Evje in Setesdalen, in the hinterlands north of Kristiansand. Anker Brun was accused of tricking peasants out of money and land titles; he caroused and drank, and it was said that he impregnated his own serving girls. Young H. M. Abel distinguished himself as a man who could bring about reconciliations, and undertake compromises between the ideals that men stood for, and the practices they engaged in. After three years at Evje the bishop set him an even more difficult task: Hans Mathias was assigned to be resident curate at Fyresdal in Telemark to mitigate and put in order the relationships around the spineless vicar, Henrik Berg, who lived in unsurpassed drunkenness and fornication. But before Hans Mathias left Evje, he married Elisabeth Knuth Normand; he was 28, and his bride was a year older. And according both to his own assertions and those of others, this Elisabeth came to be one of the best things to happen to H. M. Abel. She was the daughter of a merchant, Jørgen Petersen Normand, and Anne Marie Wendelboe, who in turn was the daughter of the tradition-rich Faret family of district stipendiary magistrates.

During the next forty years that Hans Mathias and Elisabeth lived together, they saw their children grow up and the wellbeing and respectability of their personal condition undergo a smooth ascent. In his ecclesiastical work, Hans Mathias Abel would serve under five bishops, all with their seat in the city of Kristiansand. These were difficult harrowing times for Church and clergy; old ties burst asunder and people's attitudes toward authority and the religious life were changing. Hans Mathias Abel soon became a vicar and his clerical work would become a lifelong struggle against what he saw as growths and defects on the old secure form of the Church and its Christian way of life. Within his own congregation he managed with equanimity and understanding to manoeuvre between intractable clerics and the popular sense of justice.

When Vicar Abel and his wife arrived in Gjerstad at New Year's, 1785, they had already worked hard, serving for twelve years in Skafså, Telemark, and six years in a parish in another corner of the district. They had two children, Margaretha Marine, who was 16, and Søren Georg, 13. Hans Mathias Abel was 46 years old and looking forward to his promotion to Gjerstad. It was said that he sang aloud with joy when he first reached that bright village with its separate vicarage. The vicar's wife, Elisabeth, was probably just as overjoyed by the light and the view, and the two children sprang ahead of them the last few metres.

The vicarage that they now took over was large and well-kept. It had enough winter fodder for thirty cows, five horses and a number of sheep. They grew winter rye in the fields and enjoyed a surfeit of potatoes. Abel's predecessor had been the first in the parish to grow potatoes, and the new people in the vicarage were soon to hear how in the autumn their predecessor had, once the farm labourers' wives had completed the day's harvest work, set large plates of boiled potatoes in front of them and urged them to eat their fill. He did this to promote potato cultivation in the district. When Abel came to the settlement, new houses were being built which, for the first time, included potato cellars.

After having lived in and helped build up three rural vicarages, the new cleric and his wife were well-equipped for the management of a farming household. At Gjerstad, no building required immediate maintenance, and the staff worked hard and conscientiously.

Gjerstad parish was a district rich in forest and mountains, with rivers, lakes and marshland. The parish also included the village of Vegårshei, and little by little the exhausting travel to the parish annex at Vegårshei became a problem in the work of the vicar, both for Hans Mathias and still more for his son, Søren Georg, who in time would take over. The riding paths led over moorland, and should one want to visit the nearest town, Risør, one could also row out across Lake Gjerstad and follow the river down to the sea. Indeed, one would have to hop out of the boat three times and drag it around rapids and over necks of land before one got down to the sea at Søndeled, but this was the usual, traffic-laden route to Risør, especially in winter when travel could be undertaken over the ice. The forest was rich in oak, which was freighted out to the coast and sold for ship timbers; earlier, eels and mountain trout had also been sought-after items of trade from the region. Ever since Egeland Ironworks had come into production in 1706, the peasant farmers of Gjerstad had standing commitments to supply charcoal to the plant. Much of this was freighted out over Lake Gjerstad and further on down the waterway; for a long time the ironwork was a good, secure and profitable enterprise.

Parish Vicar Hans Mathias Abel organised a system of relief for the poor in the community: the worthy indigents of the parish would be supervised, and tendencies to vagrancy and idleness would be thwarted. The vicar became director of the poor relief and calculated how much each household should contribute. Hans Mathias's soberness and sense of justice, his precise calculations over how much each should give, provided no reasonable basis for complaint; nor did anyone protest about how much the vicar's demands for the ambulatory school would cost each farmer in terms of monetary contributions and the labour time lost on the days that the farmer had to host the school class.

When it came to morals and behaviour among the youth of the community, the vicar, Hans Mathias Abel, quickly found something against which to take up arms. The first issue was that far too many young men associated with the trade in horses fell to drinking, boozing and dishonest practices. The second issue was that the coming generation were not looking after the cattle, and this led to the hiring of cattle-keepers from other places, often from Vestlandet (the west coast), and these strangers in the parish often ended up among the ranks of the dependent paupers. Nor did the vicar feel that the Holy Week of Easter was observed with the proper seriousness and devotion around the farms; he therefore entered into discussions with the farmers, and they decided that children, youths and servants would go out into the forest on Maunday Thursday and Good Friday when there were no regular services in the main parish church, and gather branches and shavings with which to feed the farm animals. This custom came to be called the Good Friday Regimen and was intended to contribute to "curing the desire of the youth to sin."

Abel's predecessor had managed to put an end to the copious consumption of spirits during the traditional three-day wedding celebrations by instituting contracts against what was called "over-indulgence". These were officially registered and served to set limits on such social aberrations. Abel followed this up by abolishing the tradition of holding huge engagement feasts. But gradually Vicar Abel lost faith in both the over-indulgence contracts and the Good Friday regimen. It was, after all, people's behaviour and moral comportment that had to be changed, and their deep-seated desire for intoxicants that had to be extirpated. He placed great importance on this in his sermons; he impressed it upon his confirmants, and reiterated it during catechisms in the church. The duty of temperance became a central theme of Hans Mathias Abel's work, in his preaching and promoting proper deportment. He was almost fanatical in his struggle against drunkenness and intemperance. It was said that one evening Vicar Abel went face to face with Drunkenness. On the way back from delivering logs to Egeland Ironworks, peasant farmers would stop off in drinking establishments, and it was not uncommon for them to come rolling boisterously home across the lake and past the vicarage. On the evening in question, the vicar met the farmers at the bottom of the hill. They say he staggered over to them, shouting and yelling as though he were exceedingly drunk. The farmers stopped dead and stood still, dumbstruck with horror and amazement. "The father is under the weather," one of them stammered, but the vicar retorted, slowly and clearly, "My children, look at me and see yourselves in the mirror. This is exactly how you are behaving!"

On another occasion a number of inebriates had met during a funeral procession. Abel reprimanded them severely; then one of the men followed him back to the vicarage cursing him roundly. Then Hans Mathias Abel turned around and slapped the man on the ear with the palm of his hand. The man shut up, stopped, turned around and disappeared. Hans Mathias hastened home to his Elisabeth and explained that he had perhaps committed an offence. He bid her send one of the servant girls to call the man to the vicarage. "Mollify him, give him something to eat, calm him down. He can get me into Trouble!"

To be sure, H. M. Abel's unceasing and energetic campaign against wine and strong drink led to the farmers going quietly past the vicarage, but Abel must have realized with chagrin that drinking continued on the sly. He himself drank only an occasional mug of beer. He would not tolerate liquor bottles on the table at home, and they said that whenever he was elated and happy in his state of abstemiousness, he would exclaim, "It was the Mother of my children who made a Man of me!"

Hans Mathias Abel was proud that in his clerical work he was scrupulous about attending to the dying throughout the parish. He proudly said that "Here, nobody dies without a parish visit and a blessing." And he placed great store in holy communion being taken by everyone during the two communion periods of the year, which here, as everywhere else, were at Michaelmas in the autumn, and at Easter.

The tasks and the work for the wellbeing of the people of his parish took all the vicar's time and energy. Earlier he had been interested in agriculture and animal husbandry but now, more and more, he left the supervision and control of running the farm to his dear Elisabeth. And the vicar's wife had other work as well. Elisabeth Normand Abel became highly admired and respected in the community. She read the morning and afternoon prayers to her household; she knew and cared for the servants and farmhands, and always gave them abundant food when they had been working at the farm. She went about her daily tasks as though she were the very personification of kindness and consideration. She taught people how they could make their everyday lives easier; she taught them simple household economy, and how to produce linen. She had the full approval and deference of the village people.

Long after her passing in 1817, farmers' wives curtsied to the fields of flax, in thankful memory of Elisabeth Abel, Niels Henrik's paternal grandmother.

In his time, Bishop Pontoppidan issued a call for priests to become immersed in local conditions, which had now been undertaken in many quarters, and Hans Mathias Abel followed up this call ardently. He wrote a topographical description of Gjerstad Parish and an overview of its natural history. He became Gjerstad's first local historian.

The following was later written about Hans Mathias Abel: "Mr. Abel's appearance was old-fashioned. A Pair of black Knee-breeches of Leather that he had used for Years on End, Wool Stockings and Boots - sometimes the Stocking legs had fallen in Folds around his Ankles, and a Morning cape - in this manner one saw him on a normal day, wearing an Everyday wig, and when it was Sunday or a State occasion, with his Sunday wig." Hans Mathias Abel would have had his four everyday wigs on stands in a row, with the Sunday wig on a higher stand in the middle.

The year following the Abel family's arrival in Gjerstad the bright and quick Søren Georg, aged 14, was sent to Helsingør Latin School in Denmark. That he was sent there, and not to the nearer cathedral school at Kristiansand in Norway, probably had much to do with his father's idea of what a school could and ought to be. In Kristiansand, Søren Monrad was rector. He was known as a passionate and sharp critic of the Norwegian Society in Copenhagen, but as rector he used Latin as the sole object of education. And beside this, the fact that there were no co-teachers to teach the other subjects at the Kristiansand Cathedral School, probably made the choice easier for Abel Senior.

In Helsingør Niels Treschow was rector, a man who had already distinguished himself as someone who wanted to make teaching more lively, who did not want merely to fill the disciples with dead knowledge, but also placed importance on the development of the ability to think, and maintained that the life of the feelings had their place in education as well. "Human beings can never be moved by a Confusion of mere Syllogisms," maintained Treschow, and he saw that the school's most essential goal must be the raising of children with the goal of striving for "Humanity to a higher Notability."

Just as the father, Hans Mathias Abel, had got part of the best teaching of his day in Bergen, he now, in the autumn of 1786, sent his son to school in Helsingør where he could encounter humanitarian ideas for a new era.

3.

Father's Period of Schooling and Study in Denmark

He came to Helsingør, where Prince Hamlet was said to have resided, the city that was reckoned one of the most beautiful places in Denmark. Everything that is considered typical of the Danish natural world was found there: the lovely sound, the waves breaking near the exuberant beech forests, and the Kronborg Fortress with its rich treasury of memories crowning the whole scene. There is no place more attractive than Helsingør on an early morning when the sea lies mirror-calm and sparkling, while the sun rises over the high Swedish coastline. The Danish guardship would greet the sunrise with cannonfire, and the morning bells would ring out around the many ships. Helsingør was also one of the liveliest places in the realm. About nine thousand ships sailed in and stopped at Helsingør to pay the toll for passing through the strait on their way to Baltic ports; moreover, foreign warships paused here to exchange salutes with the fortress.

White sails dominated the sound. The picture of the city was characterized by sailors and the garrison life at Kronborg. A number of foreign nations had their representatives in the city due to the toll permitting passage through the strait: English, Dutch and German families were linked by marriage to Danish families. Many German craftsmen had settled in the city, and English words and expressions from commerce and shipping quickly entered the Danish language. It was said that Helsingør had a strong English flavour both in terms of mores and mode of living. People said that English tidiness, comfort, extravagance and self-esteem had established a little England in Helsingør, and that the motto could have been "a short but merry life."

But the times seemed bright and auspicious not only in Helsingør. All over Denmark, and in Norway too, the times and the way of life were marked by great optimism; everything was awash with epoch-making reforms. Søren Georg Abel's stay in Denmark coincided with what in Danish history is called "an extremely happy period," the period of the great reforms with a strong economic upswing in commerce and agriculture. Just as the young and shrewd Abel was undergoing his own growth and development, he saw all around him a similar awakening: ideas about humanity and political and economic freedom became the basis of urban consciousness-raising and rural emancipation.

The Latin school in Helsingør was an old two-storey building; the teaching area was located on the lower floor, where the centre of a large, arched, cellar-like vaulting rested in the middle of the room on a low, bricked pillar. From this pillar a partitioning divided the space into four parts. Three of them each contained a classroom, and in the fourth stood a tiled stove intended to heat the whole room. There were two brown-painted tables in each class, with benches on either side, and between the tables was a chair for the "hearer," the teacher. School began at nine o'clock and an hour later a disciple, as the pupils were called, stood up by the pillar and proclaimed aloud, "Hora decima sonat!" The time is ten o'clock. And such stentorian reports were heard every hour on the hour for the rest of the day. Behind a wall at one end was the top form where the rector and vice-rector taught. The top form was also segregated from the other classes and it was the rector who gave the signal for the discontinuing of the lessons, by opening the door and walking out through the lower classes. In the mornings this happened at twelve o'clock.

Following the entrance examination, Disciple S. G. Abel was quickly placed in the highest class, the top form, where Niels Treschow was in command and conducted his teaching in accord with his own principles and the ideals of the period. The vice-rector was Hans Christian Hansen, who had taken both the philosophical and the theological examinations with top honours at the university, and in 1777 he had been the first to take the magister degree in philology in Copenhagen. But in the lower classes it seems that the teaching was conducted in the old, well-worn manner: Latin, Christianity and beatings. But if Treschow himself was witness to the use of corporal punishment, he consequently studied the issue, and the result became a different form of punishment for the disciple and a reprimand for the teacher. Rather than use fear of punishment as the engine driving the teaching, he wanted to awaken the individual pupil's ambitions. This was done by awarding prizes consisting of books, and by hanging up boards with the names of the school's earlier disciples who had prospered and reached public office.

The brutality among the pupils themselves could be considerable. Here, as in other places, new pupils were forced to undergo tough rites of passage. Protracted kicking and hitting sessions were reported, where the new pupil was, in the end, pulled up the stairs by the ears by two prefects as a third hit the already bruised and tender body with the flat of his hand. Later there was a brutal water baptism in the schoolyard during the mid-day break. Otherwise the usual social life of the pupils was fairly raw: corn spirits, beer, coffee rolls, and tobacco were treats at parties and gatherings; the most *risqué* verses of

Horace were sung, and the most persistent participants, under cover of mist and darkness, crept happily into the town's pubs and saloons. This kind of school gathering or symposium was referred to as "taking a swimming tour."

Otherwise, the Helsingør School had a good reputation. Jacob Baden, who was Treschow's predecessor as rector, was a prominent culture personality and one of Denmark's leading arbiters of taste. He had built up a big library for the school, laid the basis for the school's collections of natural history, and he had gathered funds for a stipend and expense money for pupils. Søren Georg Abel received about four or five marks a week in expense money during the years he was at the school. And it was certainly needed: housing and living costs in Helsingør were among the highest in the land.

The new humanistic method of teaching came into the school with Jacob Baden. The disciples should be put "in contact" with Antiquity; the classics should be studied for taste and for the sake of general education, and not for the sake of imitation. That which was popular and national had to find new expression. Baden was as well, one of the reasons why Treschow, at the age of 29, came to Helsingør in 1780. Treschow had been vice-rector in Trondheim for six years, and while there he had given a public speech, 32 pages long, entitled, "On the Need for Eloquence, precisely in Religion's Discourse and the Uplifting of Progeny." This authorial debut had awakened unbridled enthusiasm in Jacob Baden in Denmark, and he wrote an extremely flattering appreciation in his journal, *Nye Critiske Journal*. And when Baden became professor of oratory and rhetoric in Copenhagen, he made Treschow his successor at Helsingør.

Treschow had maintained for some time that rather than basing instruction on ancient languages alone, there were other means of educating the young, that being forced to learn these languages was harmful and the school system was impractical. But Treschow was unable to bring full publicity to his ideas before Crown Prince Frederik and his men took office in Copenhagen, and Minister Ove Guldberg and his regime came to power. In 1785 Treschow's paper entitled "On Public Teaching in the Sciences" was a thought-provoking contribution to the school debate. The theme of freedom ran through the debate: nobody should be forced to learn anything other than what he had the desire and freedom to learn; and Treschow argued that the ideas of freedom he championed in no way led to disturbance and rebellion: "Truth's cold and impartial Investigation is the enemy of such Rashness, and precludes it." And the point of view about the place of classical languages in secondary schooling formulated here by Treschow would come to be a red thread running through his work as school reformer, and later when he joined Norway's first cabinet, as Minister for Church Affairs. Treschow's view was that knowledge of the classics was "of the most widespread Utility," yet this literature did not need to be read in the original languages - good translations were to be found in all the better-known contemporary languages. On the same basis, it was not necessary to maintain Latin as the common European language of science. It was argued that learning Greek and Latin involved the expenditure of enormous pedagogical energy, and that the spread of good taste was more dependent upon the nurture of the mother tongue: this point of view was now being discussed in many circles.

Treschow focused a light on the living languages, and would like to have had a degree of free choice in the school's range of subject matter, except that religion and moral philosophy should continue to be basic to school instruction. Subjects such as history, geography, philosophy and mathematics should also find their place in the curriculum. In a speech that Rector Treschow gave at the school in 1787, he reasoned on the recompense of true virtue, and decided that good morals were inversely proportional to rewards. An increase in the recompense was a definite sign that morals were falling. Therefore, those studying, and not least, the teachers, must always strive for altruism! It must have made an impression on the young disciples when the rector rejected the idea that narrow self-interest should be one's motive for action. Self-interest, "which among Humanity so commonly Aspires to Riches and Power, where we only gratify our Soul's ignoble Drives," was merely "the Shady Strategem of those who Hypothesise" and could be refuted by the "Culture of the Intellect" that looks askance at anything other than inner perfection.

Treschow agreed to allow his speech to be published in the monthly journal *Minerva*. Regarding the background of the students at the scholarly schools, he wrote that they were: "from childhood familiar with Financial Difficulties, driven by Need, Exemplary, and sometimes the Opportunities of Baseness and improper Beggary." In another *Minerva* article Treschow developed an argument wherein he, unlike many other reformist voices of the time, placed great store in the fact that the home governed the family's moral experience: it was not merely the new school and new subjects that the contemporary society needed; nothing could replace the natural love of parents!

A high point in Helsingør's city life during the period Abel was a disciple, was the meeting between the monarchs of Denmark and Sweden in the winter of 1787. The mad King Christian VII came from Copenhagen. Unmoved by the changing regimes, he reigned nominally from 1766 until his death in 1808. And it was there in Helsingør that Christian VII greeted Gustav III with the words, "There's my Brother, King Quixote."

A year later, Søren Georg Abel finished school, and in the certificate he issued, Rector Treschow expressed his distress that the best pupils so quickly abandoned the school to enter university. Søren Georg Abel became a student with distinction. The examination was still taken in Latin.

Despite objections from many quarters, Latin continued to dominate school and university. But Søren Georg Abel's years in Denmark were nevertheless a time that, in many ways, freed itself from the two ancient cultures and all that that implied, both for the individual person, and for people's goals and meaning in life. Latin and Greek culture could no longer be looked upon as the sun and moon of the firmament. They became stars among other stars, where the new spirit of the times tried to map out the human realm. As well, most of what happened was understood in relation to the severance from the patrimony of classical cultures. Incidents now arose in which to some degree human features had a place. The question was, how to develop human culture further?

"The human" became the most beloved expression of the time, and to contribute to humanistic enlightenment became the lofty aim in all endeav-ours. It was a revolt against the Roman rule which through time had in a sense become, for the expression of humanism, a form fixed once and forever. There was a reaction against classical humanism which considered life as visible only when clad in historical form; "the human" had become almost a literary form.

The words of the Roman comedy writer, Terence, from the second century before Christ - "I am a human being, and nothing human is foreign to me" - formerly a citation, now became a motto.

According to the new humanistic thinking, one no longer needed to com-port oneself according to a ready-made image of life. Life itself was a veritable picture: where was the unpainted face among the classical masks? And if one were eager to experience life close up, to understand people without masks, right down to the marrow in their bones, then one also found that human feelings were an obvious starting point. Instead of imitating a given prototype, one could now speak about unfolding and developing the human self, the living soul expressed in its own tongue, the use of demotic speech. The sense of the human in all forms, a boundless human sensibility was now equated almost with originality. Just as the humanists of ancient times had tran-scended the classic view of humanity, those who would strive to be the new real human being had to transcend the more or less misguided civilisation of which he or she was a part. There was a contradiction between what was inherited from the classics, and what was from the popular culture. How was one to give the popular, with its distinctive stamp, and what would later be called the national culture, a new classical expression?

In Denmark discussion developed on, among other things, the conditions of the peasantry and how to work to improve them. The common goal now was to show oneself as a useful citizen: one respected useful work above and beyond self-centred meditation. The new ideas were not concealed in Latin and Greek; whatever was true and necessary was expressed in living languages, and therefore was, in principle, open to all social strata and both genders. Humanity and civic duty could be discussed clearly without citations from the classics. In a sense, women also came on to the agenda. At any rate they spoke about a "female philosophy" through which life was taken up not only through theoretical speculation, but also registered through experience and observation.

It was a century-long discussion that in the 1780s and 90s came to the surface everywhere, and many were permanently marked by optimism and self-assurance. Everything would soon be better for all people. The future stood in a radiant golden light. For the young Søren Georg Abel such attitudes would become central to all his later work both as a public and as a private person.

In the 1700s, thinkers and men of science undertook broad criticism of the social and intellectual foundations of European culture and found answers which they thought would unlock hidden strengths. Human reason and rationality were advanced as the more secure basis for comprehension. The times became known as the Age of Enlightenment and the pioneers of the age were called rationalists. Isaac Newton had reduced the explanation of all motion in the material world to three simple laws. With mathematical precision it became known that the same simple laws applied not only to the surface of the earth but also to the whole universe. A similar parsimony, which had been one of the ideals of the ancient Greeks, must also be found to explain the workings of human society. However, the majority of the population of European societies lived and made do under legislation and a legal system that was an incomprehensible chaos of paragraphs and decisions that, one way or another, emerged from local and temporary requirements in the Middle Ages. What were the tasks that now had to be taken up in order to get life in society on a course that would give justice and a life of human value to all?

Newton and many like him had placed the comprehensible harmony of space under the dominion of God: knowledge about Creation became evidence of God's existence. However, faith in human reason and self-worth was leading, little by little, to a detachment from the Church and its authority and wisdom.

Many of the ideas of the Age of Enlightenment were introduced to Denmark-Norway by the writer and philosopher, Ludvig Holberg, and with Holberg's lucid and worldly writings, an important division had taken place in the received tradition of Lutheran humanism. The theatre became an independent cultural organ along with Church and school. A reaction to this came in the form of a new devotion, the movement for piety.

Despite this reaction however, experiential scientific methods had, through astronomy, medicine, biology, chemistry and Newton's mechanics (one of the foremost results of mathematical thinking and experimental investigation), created a coherent pattern of thinking that again gave cause for a new attitude to life: everything could in principle be solved and explained clearly and unequivocably, everything should be capable of fulfillment, and bliss on earth for all human beings was absolutely within reach.

Faith in human ability and nature gradually made it difficult for many to accept that man's fall from grace had cast a dark shadow over this human nature. And the concealment and renunciation of man's original sin was the moral principle of the time, which united both many of the men of the Church as well as their opponents. A new profession of faith was in the air: people are naturally good, but they are badly in need of guidance, a guidance which gradually will result in bliss here on earth! The earth is no vale of tears. Quite the reverse. There was a ladder that led from the earth and the profane to the sacred and the eternal. And anyone could be led along this Jacob's ladder, moving little by little from the limitations of life toward the infinite twinkling of the stars. Human beings with all their qualities, and nature with God's evident power and wisdom and goodness played in the same ensemble. The image of Jacob's ladder was widespread in the contemporary consciousness and the contemplation of nature. In Copenhagen, Jacob's ladder was included among the designs on the porcelain cups from which ladies took their morning tea. Consequently, this visible and invisible intervention of God in nature was used both against freethinkers and as a demand in the natural science subjects in school.

In 1788 Søren Georg Abel was a fledgling student in Copenhagen. While in Helsingør he had undergone an education that left him rather open to the new currents despite the fact that school and university were still firmly rooted in the old demands of classical education. Science subjects and the scientific method had long been looked on with mistrust, and the Latin schools and universities had continued to stand as fortresses of the old culture, surrounded as they were by the natural science institutions and academies and public and private schools for the newer languages and science subjects. But precisely during this year, when Søren Georg Abel began his studies, the university opened itself to some degree to the new ideas. Through a new ordinance, extraordinary professorates were established in aesthetics and literary history, political science, natural history, and without one expending extra money, one could also take instruction in modern languages. The ordinance declared that the mother tongue would be used as the language of instruction, but Latin would be retained as the official tongue in examinations, disputations and occasions of academic ceremony.

The Duke of Augustborg, Crown Prince Frederik's brother-in-law, became the university chancellor and had the difficult task of reconciling the old humanism with the new realism. His friend, Knut Lyhne Rahbek, who was well-known and popular both as a dramatist and critic and wrote for the journal *Minerva*, received the job of lecturing in aesthetics and literature. It is quite certain that Søren Georg Abel was one of those who heard Rahbek's popular lectures. In the tension between received form and the modern attitude to life Rahbek preached reason and moderation. Rahbek never chose between Church and enlightenment, but through his articles published in the journal under pseudonyms he attacked both sides with irony.

Since 1772, The Norwegian Society had been a strong and lively opposition force in the cultural life of Copenhagen. But when the Dane, Johannes Ewald and the Norwegian, Johan Hermann Wessel, the two foremost poets in Denmark-Norway who had neither public positions nor permanent means of livelihood, died, respectively in 1781 and 1785, and political leadership simultaneous changed, the conflicts were no longer the same. Even if it seemed there was almost a command to the contrary, it was not difficult for The Norwegian Society to overcome national limitations. Rahbek, for example, despite maintaining that a literary golden age had perhaps really at this point in time ended with Ewald's literary work and The Norwegian Society's illustrious period of opposition, was now welcomed as a member of the same Norwegian Society.

As a student, Søren Georg Abel did not experience the Norwegian Society to be anything other than a social club where it was certain that literary power no longer existed. As Rahbek wrote about the Society, "Now, away with all sorts of Politics, and let us get on with our Drinking!"

In the style of the times, a radical was restless and extrovert, in whose milieu freedom and reason reigned as the house gods of philosophical optimism. The radical spoke heatedly about the American War of Independence and felt the world was developing toward greater and greater happiness. Development was like progress! The doubt and pessimism that came from Voltaire and his *Candide*, and the continuing discussions between the Enlightenment's optimism and its sceptics never really took hold in Copenhagen. There were, of course, catastrophes and setbacks, but by and large, everything was good: the problem of the suffering could not be insoluble in this the best of all possible worlds. In Copenhagen Voltaire was reckoned to be first and foremost a writer of tragedy. The French Revolution, in its first, happy phase, inspired social life and thought, where it seemed the ideas of freedom and equality for all were actually being implemented.

For the young Søren Georg Abel these were ideas and manners of speaking that he himself would later adopt in speech and action. In 1789 he had taken what was called the *Andeneksamen* [second degree exam], the philosophy-philology examination, *com laud*, and began immediately to read theology, more perhaps for reasons of subsistence than pure interest. Set in the light of his later broad field of interests it was, in any case, obvious that during his student days he took part in whatever events he encountered. Also during these years in Copenhagen he met many of the Norwegians who in later years would play important roles in the life of the young nation of Norway.

In April, 1792, Søren Georg Abel took the theological degree examination under the new protocol with its written portion being assessed by an examiner. He took both the degree and his probationary sermon with distinction. He returned to Norway and Gjerstad in the summer of 1792.

Conceptions of the world in which all appearance and all apparent powers were possessed of a wise and certain appropriateness, contributed to the creation of many enlightenment fanatics who sought a new and eternally rich consolidation of reason and freedom, and who did projects and demanded changes everywhere. But rationalism was not a concept for all. Among ordinary men and women a sense of fatalism prevailed: all creatures had their permanent place in a system of universal ranking that depended, to the end, on God's mercy. God at the highest reigned over everything and everyone, and below Him were the angels, and human beings were lower again than the angels, and animals were below humans; inanimate material lay below the living, on a scale that also descended in value. In the plan of society, women and children were under the authority of men, and public officials stood above common people; over everyone stood the king, who had the divine right. Most still believed that the monarch received his authority directly from God, and that in this way, God forced his will over each and everyone. It was the freethinkers who, rejecting both God and humanity, tried to disrupt or deny this great, eternal link between the forms of existence and value.

Candidate Søren Georg Abel returned to Norway together with the era's homage to utility and intelligence, and, by inference, with faith in the human ability to solve life's mysteries.

4.

Vicar and Curate

Søren Georg Abel probably got home to Gjerstad again in the course of the summer of 1792. Presumably he crossed the Skagerrak from Fladstrand or Løkken, in one or another freight schooner, and he was certainly greeted with great joy and celebration. He was 20 years old. He had taken his clergyman's examination, and he was the great hope and pride of the family. In all likelihood, his father had received a promise from the bishop that his son would be allowed to work with him as his personal curate. Søren Georg probably began right away to help his father, as everyone could see that age was bearing down upon the senior Abel. But still a number of formalities had to be concluded. The candidate had to undergo a catechism examination before the bishop at the head of the diocese in Kristiansand, and he had to be ordained.

Bishop Hans Henrik Tybring was delighted by young Abel, commended him *laudabilis*; he also came to Gjerstad in March, 1794, to ordain with full solemnity Søren Georg Abel into the service of the Church.

Everyone seemed to have liked young Abel. He was good-humoured and could express himself well; he possessed a personal charm that captivated everyone. Old Hans Mathias was said to have declared, "It is not difficult to fulfil the Post so long as I have this Help." He prized his son's erudition and education, and willingly went along with Søren Georg's plans to start a reading society in the parish. It was now important to spread knowledge and support practical initiatives. The idea was that the local people, by borrowing different books and writings, would be able to learn in a practical sense how to improve their lot by improving the soil, housing their animals better, growing potatoes, improving the yields of their fruit and vegetable gardens. In addition, the society's trunk of books would include volumes of Christian teachings and examples of new, contemporary thinking. The project was met with enthusiasm. Most of the fifty-six local peasant farmers quickly took part. Membership cost one mark per annum, and in the first year the reading society had 14 *riksdaler* with which to purchase books. Young Abel lent out his own books and new ones were purchased. Most were oriented to practical advice and counselling, and in terms of Christian proselytizing, the work of Christian Bastholm, the leading Danish theologian, with whom S. G. Abel was acquainted from his stay in Denmark, was richly represented. These books were the foundation of one of the first folk libraries in the country. In one place Curate Abel writes that the reading society was started in 1795, though the protocol has the foundation date written as April 9, 1796. The lending hours probably took place at the church hall or in the church's storage room (the weapon house) following the service. In the neighbouring village of Søndeled at roughly the same time, and quite certainly inspired by Curate Abel, a similar reading society was founded with books "on Religion, Childrearing and Cultivation of the Land." Here the sacristan, Gregers Hovorsen, was the motive force, but their membership consisted of only twelve of the most well-to-do in the district. Most people considered the reading society to be something heretical.

Old Vicar Abel was very happy to be relieved of the exhausting journey that stiffened the joints, through open country and over to the Vegårshei annex to advise confirmants and hold services. At the beginning, for the young Søren Georg, this was merely a pleasurable task. He loved to sparkle with knowledge and wit among both young and old, and the trips could be combined with social visits to the neighbouring parish vicar, H. C. Thestrup at Holt. Thestrup also had a son, Frantz Christian, whom Søren Georg knew from Copenhagen. Young Thestrup had taken the same second degree examination the year following Søren Georg, read theology and married a Miss Zeyer, a Catholic who a short time later converted to the Protestant faith. Young Thestrup took the theological examination in 1795. He was considered to be a fine speaker and a sociable man, and he became a private tutor in the houses of both an admiral and a professor before he was called home in the fall of 1798 by his father, to serve as curate. But now young Thestrup's overpowering desire for company and drinking soon went out of control, especially when he acquired the companionship of his contemporary, Søren Georg Abel. In any case, on a couple of occasions it happened that Søren Georg went right past Vegårshei and the waiting class of young confirmants, not stopping until he reached the sociability of the Thestrup residence. When Abel the Elder first got wind of this he would have reacted with grief and sorrow, but outwardly he would have merely glanced at this situation through his fingers and instead, emphasized his son's good qualities and capabilities. The next time that the confirmants saw the curate pass them by at Vegårshei, Abel Sr. came to know about this before his son had returned home to the vicarage from across the moorland. That evening he asked his son how the young people of Vegårshei were doing. His son replied, "They are studying well." Abel Sr. shook his head. Later the same evening he repeated his question and received the same answer. Still later, as he was mounting the stairs with a candle in hand, to his study, which was known as "old father's chamber," he repeated the question for the third time, "How was it, Søren, that you said that the Young People were studying well?" His son replied for the third time, "I said that, yes, and I don't know why you are asking me about it again."

Some time later Abel Sr. tried to ascertain which books Søren Georg had brought back with him from Copenhagen, and when among the reading society books he found several writings by the freethinker, Voltaire, he became so exasperated that he summoned the church verger and poured out his heart's innermost anguish at his son's wayward path, which he ascribed to the reading of ungodly books.

But apart from that episode it seems that there was always a good relationship between father and son, even though they were in many ways opponents: an old-fashioned, orthodox Christianity versus contemporary rationalism in a lively and talented personality. For the seven years they worked together the father praised his son's learning and his capabilities, and the son learned the practical work of being a parish priest. In 1797 the law concerning the country's conciliation boards was promulgated, a law which contributed greatly to an integrated legal procedure. In the rural communities these boards continued to be led by the parish vicar, and through arbitration between the parties many costly court cases were avoided. Parish Vicar Hans Mathias Abel had a long experience with such matters, and it would soon be evident that the young Abel had learned from his father's diplomatic practice.

But Abel the curate probably spent more time and zeal with the blossoming social life in Risør. Like other communities along the coast in southern Norway at the time, Risør experienced great prosperity with vigorous trade and commerce, export and shipping. The curate from Gjerstad, with his wit and loquaciousness, his charm and his light optimistic bearing, was a welcome guest among the town's patricians of commerce. The question among the town's *nouveau riche* was, *who* would be the personable young man's bride?

There is no evidence about when and how this occurred, but the engagement was probably celebrated in good time well before Søren Georg Abel married Anne Marie Simonsen on March 25, 1800. She was the eldest daughter of the richest man in the town, the merchant, Niels Henrik Saxild Simonsen, and his first wife, Magdalene Andrea Kraft.

5

Abel and the Simonsen Family in Risør

Public officials and their families had long been held in high esteem in Norway. Their sons travelled to Copenhagen for their education, and together with their Danish public servant brothers, they maintained the hegemonic force of their stratum of society. Ever since 1660, when the Council of the Peers of the Realm lost power, and an enlightened despotic kingdom of two equal parts was proclaimed with four prefectures, each with its sub-agencies, public officials had ruled Norway. 40 bailiffs, 50 district stipendiary magistrates, 450 priests and about 800 officers. At any one time, fewer than 1,500 public officials had the joint task of maintaining the ruling apparatus.

But now, toward the end of the 1700s and up to 1814, another stratum and class dominated society. This consisted of the flourishing houses of commerce, a merchant aristocracy, a self-confident group of commercial patricians who had now set new standards for social life.

The war for independence in North America, and from 1792, the Napoleonic Wars in Europe pressed the country's ships and raw materials into the service of the war economy, and as long as Denmark-Norway managed to remain neutral, the profits in the form of high prices for timber and iron, and marvellous rates for transporting freight, were colossal. A single ship's cargo could pay for the whole cost of building the ship, and in a year a freight boat could transport four times its own worth. Risør became one of the towns where shipping and shipbuilding were most animated. Money flowed in in immense quantities.

The public officials no longer managed to keep abreast of the lavishness and sociability of the merchant classes. Even the bishops complained that their income was too small for them to live up to the standard established by the patricians of trade and commerce in the towns along the coast.

Around the turn of the century the continuing wave of prosperity brought in great riches all along the coast, particularly a little further to the south: as Jacob Aall, who gradually began to loom large in the life of both the district and the nation, as owner of an ironmill, businessman, politician and man of letters, was to write in his memoirs forty years later, "The last year of the old century, and the first year of the new can be regarded as Norway's Golden Age."

In Risør, Niels Henrik Simonsen, as well, was one of the most prosperous. He sat on the town council, was a member of the existing clubs and societies, and was the *grand seigneur* with expensive tastes and lavish parties at his house overlooking Risør's harbour. His house on Beach Avenue (Strandgaten) was built in 1775. It was three storeys high, each composed of seventeen notched logs. The first floor contained a goods showroom, three storage halls, two masonry-finished cellars, a wash house with a chimney, and two baking ovens with walled-in iron boilers. The second storey contained six rooms above which there was a full loft. There were eight iron stoves of varying sizes in the house: one was four levels high, six were two high, and one was a single level. This was a sign of wealth. Such cast-iron stoves were expensive, and often decorated with bas-relief scenes on their sides. A good stove cost more than a cow. In a half-timbered building next door, Simonsen housed firewood, a henhouse, a barn (probably for pigs, cattle and horses). There was a high board fence along the south side of the house with a double gateway, and on the north side a corresponding fence with a single gate, while the fence continued without a gate around the back of the house.

The merchant lived here with his wife, children and servants, and he conducted his business from here with other citizens of the town, and directed his freight boats, *Hope* and *Andrea Elisabeth*, reaping the rewards of spectacular freight rates. He married three times and his third wife was the sister of his first. However it seems that he took no particular interest in his seven children; that is, as a father Simonsen gave his children only the material care that money could buy, a comfortable home, servants, and teachers who instructed them in the normal reading and writing school subjects, and in subjects like singing and piano playing which attested to the family's social standing. It was said that Simonsen had so much to do in the course of the business day that he preferred to spend his evenings at the Billiard Club.

When she was only seven, Simonsen's eldest daughter, Anne Marie, lost her mother. What had her life been like in this household, together with her siblings, servants, stepmothers and an absent father?

Mary Wollstonecraft, the English writer and one of Europe's first feminists, spent several days in Risør when she travelled in the Nordic countries in 1795. The first thing she recorded as she sailed south along the coast from Tønsberg (stopping at Larvik and Helgeroa) to Risør, was a feeling of being imprisoned. She felt herself shut in by the surrounding steep bare cliffs: "I felt my breath oppressed, though nothing could be clearer than the atmosphere [...] But I shuddered at the thought of receiving existence, and remaining here..." And she continued, "What, indeed, is to humanise these beings, who rest shut up, for they seldom even open their windows, smoking, drinking brandy, and driving bargains? I have been almost stifled by these smokers. They begin in the morning, and are rarely without their pipe till they go to bed. Nothing can be more disgusting than the rooms and men toward evening: breath, teeth, clothes and furniture, all are spoilt. It is well that the women are not very delicate, or they would only love their husbands because they were their husbands." One day she was visiting the English vice-consul at Risør: "His house being open to the sea…and the hospitality of the table pleased me, though the bottle was rather too freely pushed about. Their manner of entertaining was such as I have frequently remarked when I have been thrown in the way of people without education, who have more money than wit, that is, than they know what to do with. The women were unaffected, but had not the natural grace which was often conspicuous at Tønsberg. There was even a striking difference in their dress; these having loaded themselves with finery, in the style of the sailors' girls in Hull or Portsmouth. Taste has still not taught them to make any but an ostentatious display of wealth…"

Simonsen exhibited his wealth and status in the style of the times: the furnishings and ornamentation of Risør's 150 year-old cruciform church were paid for by gifts volunteered by the town's richest citizens. The altarpiece represented the Holy Communion and was a copy of one of Rubens' church paintings in Antwerp. It was said that the ship that carried this altarpiece was on its way from the Netherlands to a church in Riga when it was shipwrecked offshore from Risør, and the great timber merchant and founder of the city, Isaach Lauritzen Falch, and his wife, bought the altarpiece for the town church. That had been in 1667, and the Falch family had been for generations at the top of Risør society. They had a large house in the town, owned a sawmill and several farms, including the one on which the town of Risør came to be built.

But now it was Simonsen's turn. In 1794, he and his wife paid for an arched gateway for the chancel entry within the church. Between the columns, bedecked with an abundance of ornamentation and the royal crest of King Christian VII, there now stood the Simonsens' gates for all to see. And when the last Falch in Risør, the widow, Margrethe Falch, had sold off many of the family's properties, it was Simonsen who in 1789 bought the much sought-after beach idyll, Buvika.

Simonsen had been one of those who took the initiative to gather volunteer contributions for the building of an impressive bell tower in the town. Two storeys high, and with an arched roof resting on octagonal pillars, it was built in conjunction with the still more impressive new town hall, which was completed in 1798. The roof was iron-plated and surmounted by an iron spire that was topped with a gilded globe and weathervane. A clock which struck the hours, and which had a weight and two counterweights as well as clock faces on all four sides, could be seen over the whole town. Now the town could always tell the time, and the clock ran for three days between windings.

Like all the others in Risør, Simonsen paid a four percent tax that, as in all other towns, went for the most part to the poorest in the community. A system of poor relief was organised in 1797, comparatively late in comparison to the surrounding districts. Simonsen sat on the Poor Commission and evaluated who was needy and how much should be paid out each week so as to protect the town from begging in the streets. The typical pauper was a sickly widow, or a spinster in her 50s or 60s, who rented space in one or another of the town's lofts and attics, and was too worn down herself to earn her keep on a daily basis. Out of a population of 1,295 people in 1801, Risør had 38 people who, owing to the Christian commandment on charity, were assisted in this way. When a school was to be built for the children of the poor and less-fortunate, once again voluntary contributions had to be collected. Here as well Simonsen was one of those who took the initiative. His own children were in the town's other school, where admission was by paid subscription, and pupils had to be able to read from a book in an accomplished manner before they were admitted.

Niels Henrik Saxild Simonsen had inherited from his father a good economic foundation on which he could build. His father was Danish, from Saxild near Århus, and in all likelihood he had come to Norway shortly after Risør had received its royal trading charter in 1723. Presumably it was the aim of profiting from the timber trade that drew Simon Nielsen Saxild to Norway and Risør. He is registered in the town's old lists of citizens as a timber shipper and merchant and as one of the donors to the church. He married Elisabeth Marie Henriksdatter Moss in 1747. She was born in Drammen and came from the rich Moss family that, during Drammen's first boom period, had dealt heavily in timber trade and transport. It was probably through common business interests that Saxild became acquainted with the Moss family. Their son, Niels Henrik, born a year after their wedding, was christened in the Risør church, December 5, 1748. Two years later, Simon Saxild died, before the real boom period had begun. He left 1000 *riksdaler* to his only child, who was also to become the maternal grandfather of Niels Henrik Abel.

Elisabeth Marie Moss remarried, this time to one of Risør's most wealthy men, the sawmill owner and timber shipper, Claus Christophersen Winther, who was fifty-six at the time, and her little son, the young Niels Henrik Saxild Simonsen, learned well from his stepfather about how to propagate his 1000 *riksdaler* during the prosperous boom years which followed. For Saxild Simonsen, life's path led in one direction, ever upwards in terms of economic fortune and social status. At the tender age of 15 he was sent to England to study commerce, and a couple of years later he returned to Risør to begin his ascent in life.

Niels Henrik Abel's maternal father was described as an upright, honest and truth-loving man. It was said that he held firmly to his opinions, and that he was extremely conservative. He liked to keep old things around him. For example, he refused to allow the rooms in his large house to be painted or wallpapered, and he persisted in wearing an old-fashioned wig with a pigtail. Only from a sense of the social graces did he adjust himself at all to new clothing and styles. The new fashion was short-clipped hair brushed forward over the temples and forehead, and long narrow trousers called pantaloons that were shoved into half-length boots. The venerable Simonsen stopped using silverbuckled shoes, and he put away his knee-length trousers, but in speech he held fast to the old ways. He always referred to people in the polite Danish form, and never used the familiar second person singular.

Niels Henrik Simonsen entered matrimony for the first time in 1773 with Magdalena Andrea Kraft at Kragerø. Niels Henrik Abel's maternal grandmother was said to have been unusually gifted and charming. She was born at Vinje in the interior of the Telemark region, while her father, Jens Evensen Kraft, was the lieutenant-colonel posted there. This Lieutenant-Colonel Jens Evensen Kraft was the cousin of the Jens Kraft who had been born across Oslofjord, on the opposite, eastern shore, at Halden, and from 1746 had become a professor at the Sorø Academy, the writer of several brilliant works, as well as being the foremost mathematician in the Denmark-Norway of that period.

Magdalena Andrea died, leaving behind two daughters and two sons. Simonsen married Marichen Elisabeth Barth who gave birth to a son and a daughter, which cost her her life. Whereupon Saxild Simonsen fetched Christine, a younger Kraft daughter, to his house. Or was it his old mother from Drammen who demanded the marriage? Through her family in Drammen she and the two Kraft sisters were linked to the genealogical tree of the Moss family. People described the two Kraft sisters, each of whom in turn became chatelaine to shipping merchant Simonsen in Risør, as very different. Christine was described as good-natured, rather frivolous, and lacking all intellectual interests. And they said she did not take particularly good care of her house or pay attention to her own two sons or her six stepchildren.

When in March,1800, at the height of his power and prosperity, Niels Henrik Saxild Simonsen arranged for the wedding of his eldest daughter, he was probably spiritually convinced that this was a most propitious thing to be doing for his daughter. She was 19, and the bridegroom, Søren Georg Abel was 28 and had been appointed parish vicar - albeit to the Kristiansand Diocese's smallest parish, Finnøy in Ryfylke - but Simonsen was probably not the only one who saw a potential bishop in the young man, who with his well-turned phrases and witticisms enlivened both salon and dinner table. The young theologian with the best examination results from Copenhagen became a public servant right at the crest of the wave of his times. He was well-oriented both toward the secular world and the spiritual climate of the times. Despite the economic boom of the period, the civil servant still represented the hegemonic cultural values. And if Simonsen was satisfied with his son-in-law, so was the son-in-law no less satisfied with his father-in-law. Søren Georg Abel had won the hand of the daughter of the largest trading house, one of the biggest players in the town of Risør. The financial aristocrat, Simonsen, would fork out whatever necessary to help his son-in-law advance.

And what about the bride? Besides her father's much sought after wealth, she had her looks, and they said, she had learned to play the piano and to sing.

The best of bourgeois society was richly represented at that wedding in Risør. The English vice-consul was an obvious guest, and those officers of the fleet who happened to be in the district were routinely invited to the town's major social events. The town magistrate was there, the toll inspector, the pharmacist, the biggest of the shipping magnates, but scarcely any of the town's better-off craftsmen. And certainly the priest who had united the newlyweds, Christian Sørensen, was also seated at the head table. He would also play a role in the bridegroom's later ecclesiastical work. Sørensen became Bishop of Kristiansand in 1811, and in 1823, when Niels Henrik Abel was a penniless student in Christiania, Christian Sørensen reached the top of Norway's ecclesiastical hierarchy: he was appointed Bishop of Akershus Diocese, in the capital itself.

Another important person at the wedding was Henrik Carstensen. Apart from Simonsen, he was Risør's richest man. Now, in 1800, he had four freight boats under sail and in the coming year he became the owner of Egeland Ironworks. Later, as businessman, a father of the Norwegian constitution, and the nearest neighbour at Gjerstad, he was to play an important role in the bridegroom's life.

Through this marriage in March, 1800, the newly rich patricians of trade of the coastal town became linked to the old esteemed public servant stratum, and from both sides the connection was seen as the most legitimate starting point for certain success in the golden times that everyone hoped and believed would continue. And most certainly the wedding was celebrated with all the pomp and circumstance typical of a trading house of the time: the dining table decked out with polished glass and fine porcelain, and possibly with the monogram of the house embossed on the priceless East Indian porcelain that the host had obtained from a uniquely privileged Asiatic company in Copenhagen. A profusion of flowers and silver candlesticks would have decorated the dessert-laden table. The cupbearer would have struggled to see that the wine flowed richly, and the breast of duck, the trout and veal roast were borne in with the proper ceremonial mien. Asparagus, which would have been brought in piping hot and steaming in large silver salvers would have been served with good hock from the Rhine. Lively voices rose and fell, the soft light from the candelabras certainly reflected off the crystal, and the glass prisms, and of course, many speeches would have been given. With the serving of the veal, the host would have raised his glass, and with well-chosen words, delivered a toast to the bridal couple and to the whole gathering. They all would have called out, "Skål!" and drunk toasts to public servants, to the the clergy, to the fatherland, to the English people, to trade and commerce, to shipping, the ironworks, the forestry, and the working of the fields. No one would have been left out. Perhaps too, if the host owned his own private cannons he could also have saluted the newlyweds with a thundering resonance from the ships in the harbour, which in turn would have been toasted in the bridal garden. The bridegroom's father, seated beside the hostess, would have given the closing blessing; then the table would have been removed. The party would now have broken into lively talk and chat - the men with coffee, and the women, attired in silk dresses and silk shoes imported from England and bedecked with ribbons and embellishments, would have gathered in smaller groups in the drawing room. Some of the men would find their way to the host's personal *comptoir* where the clay pipes had been set out, pipe lighters prepared, the card table set up, and where the bowl of heated punch emitted curls of steam. Things became politicized and discussion ensued: where was Napoleon? What were the prospects in the freight market? And who had laid eyes on the host's brigg *Hope*, the ship that Simonsen had had depicted on the bottom of the huge punch bowl?

Suddenly the French horn could be heard. Violins and flutes tuned up for the dance, and following the lead of the bride and groom, the wedding party formed pairs for the French cotillion, for which the dancers little by little formed into two squares. At a given signal, windows and doors were opened and the watchman's specially ordered song floated in over the warm and festive guests, wafting good wishes and blessings over the whole assembly. Had the host known what was coming, he would most certainly have stopped time then and there; and who is to say, perhaps he did in his own jocular fashion. Perhaps he had the town's clocktower interrupted in order to ring out an extra peal of bells, as the final punctuation to the party.

PART III
Childhood in Gjerstad

Fig. 3. The vicarage at Gjerstad, drawn by P. M. Tuxen, 1804. Gyldendal forlag archives.

6.

Priest at Finnøy, Church Affairs in Kristiansand Diocese

Shortly after their wedding, Søren Georg Abel and his bride, Anne Marie Simonsen, took over the vicarage at Finnøy, where he began to practice his ecclesiastical calling the way he had been trained under his father at Gjerstad. From the beginning he attracted favourable attention in the local district of Ryfylke. Here too he founded a reading society with a public lending library and, given his knowledge and his practical experience, he quickly raised the standard of the schooling.

He led the Mediation Commission for Finnøy Parish, and after a two-year period, the Prefect of Stavanger, Ulrik Wilhelm Koren, wrote that Søren Georg Abel "has had such success with his reasoned Mediation that the number of given Cases referred to the regular courts has fallen to two cases."

Søren Georg Abel also wanted to improve the health of the parish children on the island of Finnøy, and he embraced the newest of the new advances in medical knowledge. Smallpox epidemics often led to death, and those who survived often went around covered in ugly pockmarks. Now, however, science had developed a real remedy for this illness. In 1798, the British doctor, Edward Jenner, published an article with his findings, that after he had injected people with cowpox they were successfully innoculated from this destructive illness. Within a short time, Jenner's method was adopted all over Europe. In many countries vaccination became compulsory - as it was from 1810 in Denmark-Norway - but parish vicar Søren Georg Abel had begun to put this new knowledge into practice ten years earlier.

The benefits of science and enlightenment also found a strong spokesperson in the new Bishop of Kristiansand Diocese, the Dane, Peder Hansen. Hansen became Bishop in 1798, and he quickly had a cooperative and like-minded vicar in the person of Søren Georg Abel. Perhaps it was also the Bishop's goodwill that contributed somewhat to the swiftness of Søren Georg Abel becoming a parish vicar, and other advances that were to come. One of the first things close to Bishop Hansen's heart, and which he had revealed to his priests, was his intention to spread the results of Dr. Edward Jenner's findings. Many of the priests saw these vaccinations as so important to the

health and wellbeing of people that they entered vaccination dates in the
Church records, along with the dates of baptisms and confirmations.

The first child of Søren Georg Abel and Anne Marie, a son, was born
November 13, 1800, and baptized on Christmas Eve in the local church. The
child was given the name Hans Mathias, after his paternal grandfather.

Following the Battle of Copenhagen on April 2, 1801, when the greater part of
the Danish-Norwegian fleet had been destroyed by the English, a coastal
defence of Norway was hastily organised, with ships and men sent from
Denmark to conduct this operation. The great military operations of Europe
were drawing nearer, and in his district, the parish vicar, Søren Georg Abel
was a driving force in this construction of defence. This was to have two
important consequences for him and his family: first, twelve years later, he
would be awarded the Red Cross of the Order of Dannebrog (the Danish Flag),
an honour that would support his political activities. Second, Abel's engagement in coastal defense in Ryfylke led to the young Danish marine lieutenant,
Peder Mandrup Tuxen, being taken into the family. P. M. Tuxen was one of
those who had come from Copenhagen to Stavanger to organise the coastal
defence, and in the course of the half-year he was in Norway, he developed
unusually strong bonds of friendship with the vicar's family at Finnøy. Ten
years later Tuxen married Mrs. Abel's sister, Elisabeth Simonsen, and ten
years later still, in a letter to his wife, P. M. Tuxen wrote about Finnoy: "That
Time, in my green Youth I bound myself to Abel and Anne Marie with a Kind
of childish Passion. I will never forget the Journey from the Vicarage down to
the Shore, where we parted company with everyone endeavouring to keep the
Tears from their Eyes. I cried for a whole Hour after coming on board. At that
Moment, they were a Pair of the most Divinely blessed Human Beings, at least
in my Eyes."

Many were quite in agreement about the young couple in the vicarage.
After a short while, Søren Georg Abel and his beautiful young wife were
inducted into the finer social life of the district. In late summer, 1802, they
awaited the birth of their second child, but this did not seem to hinder them
in the least from visiting the magistrate's estate at Nedstrand in the neighbouring parish across the fjord to the north. Parish records show that no
church services were held at Finnøy that year between the end of July and the
beginning of September. And later, people at Nedstrand said that one of Abel's
sons was born there in the village. It is likely that while they were visiting
Magistrate Marstrand in Nedstrand, Anne Marie gave birth to their second
son. Later, she herself reported that he was born three months premature and
apparently lifeless; for it was not until he had been washed in red wine that
he showed the first signs of life, and that he had been laid in soft cotton so

that the weak flame of life should not be blown out. The date was the 5th of August, and they called him Niels Henrik after his rich maternal grandfather.

But the happiness and harmony that seemed, at least from a distance, to emanate from this newly-married clergyman and his wife was perhaps not very deep. Was their conjugal life actually beginning to sicken already? Many years later when P. M. Tuxen came to hear that Søren Georg Abel had died and that his beautiful Anne Marie had become inebriated at the funeral and had comported herself in an unseemly manner with a servant lad, Tuxen wrote to his wife: "I have often wondered about Anne Marie's odd character. I have always regarded her as phlegmatic, or at least considered her a woman of little Passion. I well remember from Finnøy, there was a handsome Peasant lad that she had no trouble putting up with, and that she spoke often with him, with a special Pleasure, but she impressed me as being so cold that I could not construe from this anything but mere Kindness. So it remains somewhat puzzling for me even though I have heard more about this sort of Inclination."

Church affairs in the Diocese of Kristiansand would undergo dramatic changes during these years and have consequences for the religious life of the whole country. Bishop Peder Hansen was the driving force of this change. The main thesis of many theologians at this time was that practical enlightenment based on the knowledge of natural science should be the basis for mankind's future and for human happiness. The clearest exponent of what came to be called the spirit of enlightenment and rationalism was Bishop Peder Hansen.

When Bishop Hansen first took his seat in Kristiansand, he had already been chaplain of the royal palace, parish priest, and honorary professor in Denmark. He had been married three times and had thirteen children, nine of them living at the time he went to Norway with his third wife, with whom he had no children, but who appears to have been a respectable bishop's wife accompanying her husband on many of his pastoral visits around the diocese.

One pastoral visit that Peder Hansen took around the western part of his diocese shortly after arriving in Kristiansand made him aware that a simple teaching guide was necessary for those who taught school. In one parish he found that eighty people between the ages of seventeen and thirty could not read, and consequently were not confirmed. It was an incontestible fact that common folk usually considered the itinerant school teachers to be useless and burdensome, persons who had to be fed, housed and, in addition, given a wage. Often the door would be shut in their faces when they came to the farm to instruct the young. On the other hand, there were quite a few who became schoolmasters in order to avoid the draft, and in many places the clergymen complained about these "peculiar subjects" who sought teaching posts.

It was obvious to Bishop Hansen that measures had to be taken to deal with this sorry state of affairs. The school teachers' own knowledge and education had to be strengthened before the common people's attitude toward them could be changed. The forces of good would win if only people would discover the usefulness of practical knowledge, and the good fortune that arises from a clearer comprehension of the Divine.

Bishop Hansen moved quickly. He made it incumbent upon his priests to select well-suited young men who could come into the diocese centre to undergo teacher training. The result was that the Bishop soon had 102 students whom he divided into three groups. He gave six weeks of training to each group. Every day, including Sundays, between five and eight in the morning, he taught them to read clearly and in a declamatory style. He clarified the easiest way to teach a child to read, and how they should teach the catechism. He lectured on Christian moral philosophy and Biblical history. He intro-duced them to the main features of nature study, natural history, as well as the geography and history of Denmark and Norway. And the poor students wrote and wrote and wrote, whatever the Bishop dictated. They said that the Bishop himself was so eager that he appeared before them half-shaved and with shaving soap on his face. However, many of the students lacked the necessary schooling, and later, many droll stories circulated about these schoolmasters who came home from the Bishop's school with the conviction that they had learned all the world's knowledge. In particular, the Bishop's lessons in the art of comporting oneself "nicely and decently" contributed to turning their heads and making these schoolmasters appear ridiculous in the eyes of many ordinary people, such that some of his "peculiar subjects" were, and became precisely that.

At the same time, Peder Hansen's project reaped high praise. He was promised 100 *riksdaler* for three years from the king in Copenhagen, and plans were made in Kristiansand to found a seminary where ten apprentices, chosen by the priests from their parishes, would undergo six months of training. In order to obtain an acceptably paid teaching position in the local communities it was decided to merge the office of parish clerk or sacristan with that of the schoolmaster, and call this employee the parish clerk.

When Peder Hansen was on his pastoral tour of the eastern portion of the diocese in 1800 he found that the Parish of Gjerstad was of the finest order. The Bishop praised the active reading society that gave the common people the opportunity to borrow books on agriculture or health matters, on history and nature, and he recommended Gjerstad as a model for other parishes to emulate. Five well-qualified young men from Gjerstad parish were sent to the Bishop's teacher training programme. As well, Hans Mathias Abel had insti-tuted a pastoral journal on the ecclesiastical affairs of the parish, something that the Bishop prescribed for the vicars in the other parishes. Yes, Peder Hansen was very pleased with the parish vicar's contribution and was so certain of his conduct and achievement that, while at the dinner table at the vicarage before many guests, he ceremoniously congratulated the elderly Hans Mathias Abel as a future dean. But to the great astonishment of the Bishop, Hans Mathias Abel stood up and replied, "I thank you. However, this is a Burden that a younger Man than I ought to bear. It would be an honour, but one that I do not care about."

Three years later Hans Mathias Abel died calmly and unexpectedly. He had officiated on Sunday, taken ill on Monday, and died on Tuesday, August 2, 1803. The parish vicar, Christian Sørensen came from Risør and spoke at his funeral service. Both he and Bishop Hansen wrote glowing eulogies in verse to mark the passing of Hans Mathias Abel.

But the *young* Pastor Abel was indeed eager to advance his position in relation to Peder Hansen. When the Bishop had been on a pastoral visit to the three western divisions in the diocese in June, 1802, 11 of the 17 vicars participated in discussions and lectures with Peder Hansen in Stavanger. Søren Georg Abel took an active and enthusiastic role in the work groups the Bishop set up for the improvement and well-being of the priests and their congregations. After the Bishop had presented a list of some of the newer writings, they commented and pronounced their views, and it was decided to begin a literary society that, in conjunction with the Bishop's visits to these three western deaneries, would hold meetings in Stavanger. Also this group, in which the young Abel partici-pated, would undertake to lecture and prepare young men to undertake work as schoolmasters.

At that time there was only one body of prescribed texts in the Church, and the vicars preached from these, year in and year out from the same Biblical texts. But now Bishop Hansen proposed that the priests should to be able to request permission from the Chancellery in Copenhagen to preach from texts of their choice. Shortly thereafter, nine such requests were made from the Diocese of Kristiansand and all of them received consent. The Bishop also proposed that whenever chasubles and surplices wore out or became com-pletely obsolete, royal permission ought to be sought to obtain exemption from replacing them, and in practice abolish "this tendency in Serious Chris-tendom toward ostentatious Ornamentation." In his own diocese, Bishop P. Hansen personally abolished the use of vestments, but it seems certain that no reply came from Copenhagen in response to this request. Another real theme was the question of the psalm book: Peder Hansen sincerely expressed his desire that the new evangelical psalm book be introduced across the diocese. He strongly recommended the book's superior qualities, and maintained that it contained "a popular Dogmatics and a glowing Warmth for practical Christendom."

Erik Pontoppidan's great 1737 exegesis of Luther's catechism was still taught right through primary school to confirmation, either in the original version or in an abridged edition. But this Pontoppidan commentary had been written back during the first blush of pietism, and Peder Hansen was of the opinion that the book should be replaced by a newer and more credible exegesis. Consequently, the Bishop urged contributions to this end from his own vicars, and Søren Georg Abel felt himself particularly called to this task. He busied himself whole-heartedly with the development of a new commentary to the catechism. In 1806, Abel's book came out in Copenhagen under the title, *The Questions of Religion, with Answers, Adapted to the Comprehension of the Young*. Abel had used Pontoppidan's method and laid out the book as a series of questions and answers. Whereas Pontoppidan had used 759 questions and answers to give a complete understanding of Luther, Abel confined himself to 337. Abel's clarification of the catechism was well-received and much used, particularly in eastern regions. In the years up to 1816 it came out in six editions of 1000 copies each, the last four being printed in Christiania. During this period Abel's catechism was used at a level with that of Pontoppidan.

Bishop Peder Hansen did not spend long in Norway. In 1804 he was appointed Bishop of the Fyn Diocese in Odense, Denmark. Nevertheless, before he left, Bishop Hansen took part in two stormy controversies. The first was perhaps only an example of how a bishop routinely gets involved in criticism, while the second incident had to do with deep religious convictions and reflected one of the great schisms in Norwegian society at the time. The first involved the appointment of Søren Georg Abel as Vicar of Gjerstad to succeed his father. The second had to do with the prosecution and arrest of the lay preacher, Hans Nielsen Hauge.

It appears that the day after his father died, Søren Georg Abel was in Risør; in any case, the application to succeed his father as vicar is dated at Risør, 3 August, 1802. In the letter of application he stressed his good examination results, reminding the reader that he had earlier worked with the Gjerstad congregation, and that his family had grown beyond the limits of the slight income available in the Finnøy post; he also mentioned his contribution to the "Defense of the Fatherland," and included letters of attestation from District Administrator Koren in Stavanger, and from Bishop Hansen. Peder Hansen would certainly have done young Abel a favour, and considered it only fitting that S. G. Abel should be advancing up the ecclesiastical hierarchy. In the pastoral journal at Finnøy, the Bishop had written: "I have found, in a few words, the Enlightenment, the Order in this Parish, that bears witness to a noble Folk and their present Teacher, the Reverend Mr. Søren Georg Abel's highest, most worthy fulfillment of public office." But with regard to the available post at Gjerstad, the bishop could not avoid pointing out that the Dean of Øvre Telemark, Svend Aschenberg, was a deserving and well-qualified candidate. Thus duty-bound, Bishop Hansen wrote to Copenhagen and pointed out that Dean Aschenberg was in delicate health and deserved a less rigorous posting. But the Bishop's letter of attestation for young Abel was one of unqualified support: "That it must be one of my purest Pleasures to have found this intelligent and industrious Teacher of Religion within my blessedly devout Diocese, fills me to repletion with Happiness; to wit, to have borne Witness to the work of the vicar, Mr. Søren Georg Abel in the prosecution of the duties of his post in the Parish of Finnøy. In him I have found a pleasant, good-natured, enlightened but modest soul; an excellently educated Young Man; a Reading Society Incomparable anywhere else in the Diocese; the Psalm book of Christian Evangelism commonly accepted for public use in Divine services. A Teacher, who has charged himself with such important duties that before the Sight of God and Men of the true Faith, he has the perfect Right to my honest Testimonial, and to His Majesty's Grace, which he herewith and most humbly is recommended by P. Hansen." This recommendation was certainly written after the bishop's pastoral visit the previous year but probably he intended that every word contribute to an irresistible testament about the prospective applicant.

And thus it was Søren Georg Abel who was assigned to the vacant "Gierrestads og Weegaardsheyens Parish," and Dean Aschenberg could not fault the Bishop who was on record as having recommended him in regard to this matter. But perhaps there was a note of revenge from Aschenberg's part when he wrote in the Mo i Telemark Parish's record about Hans Mathias Abel who in his time had been that parish's first priest: Mr. Abel was "not a gifted or talented Man, nor a Learned or inspired Speaker, but to the contrary, a more useful Man for the Educating of the Youth could not be found, and this was something close to his Heart."

But the ecclesiastical work and pronouncements for which Peder Hansen was the greatest exponent began to provoke strong reactions in the country. For Hansen and his adherents, human wellbeing could be secured first and foremost through common participation in the new findings of the natural sciences and in the more rational explanation of religious concepts. But the bishop was also able to use the awakening interest in the Old Norse faith when from the pulpit of Kristiansand's "dome church" he invoked "Odin the All-good" in relation to the immortality of King and Crown Prince.

However, those who firmly believed in a more traditional Christianity now found their great spokesman in the person of Hans Nielsen Hauge. Hauge supported proselytism and faith, and recommended the old devotional books, catechisms and psalm books that Hansen and his men had discarded. Hauge criticized the priests for their frivolous lives and their appetite for money, and he warned against Godlessness, and the mind that set the profane and worldly before the sacred and the spiritual. Hauge himself was also a courageous voice for the oppressed common people struggling for better material conditions. Hauge was both evangelist and merchant. He promised his supporters both material and spiritual wellbeing, and this depended on never being either lazy or resigned. He reassured people that God blessed those of His children who were hard-working.

Bishop Hansen was a considerably bright star in the eyes of the powers that be in Copenhagen. Hansen's assessment of Hauge, and Hansen's view of Hauge's pious proclamations and activities were highly influential in relation to the decisions that led to Hauge's arrest, imprisonment and protracted trial. Bishop Hansen himself composed an article against fanaticism in general, and against that of Hans Nielsen Hauge in particular. In a sharp and trenchant letter to the Chancellery on April 24, 1804, the bishop drew attention to the audacity and boldness that Hauge and his friends had come out with. He argued that they were out to prevent the establishment of reading societies and to harm the country's economic activity. This latter point was a call to the property-owning class in the country, whom the bishop knew were watching with growing alarm the audacious and successful industrial and trading activities of Hauge and his followers. Hauge had got capable young people from the villages to travel to the cities to engage in trade and handicraft production, and following his innumerable trips around the country, Hauge had a political and economic overview and outlook, equalled by few others, on what was lacking. He suggested to young followers who were daring, and who possessed a forward-driving faith, that they establish factories and begin industrial production, and he put himself in the vanguard and helped every-where. Hauge hoped and saw himself and his flock as a Christian-based economic oppositional movement that was sanctioned with the blessing of the Lord. Only by becoming strong economically could they defeat the rich who used and plundered people, and who banished the word of God from the churches. Only by developing a national industrial wealth could they awaken admiration and understanding from the public officials and the wealthy, and only in this way could Biblical teachings call people to undertake penitential conversion of faith through good deeds.

In his letter to the Chancellery, Bishop Hansen pointed out that "these People with their fanatical Teacher and Writer, create and spread an extreme

Wretchedness among Ordinary People across the land." And he stressed "the Mistrust that these Swarming Insects spread against the State's foremost Authorities and the Teaching Profession..." and in conclusion, he wrote, "Because of this, I must, as the final Petition in my current role as Head of the Kristiansand Diocese recommend to the high Collegium, by whatever Means that it finds most useful, to stop an Evil which in one way or another would wreck otherwise amiable people physically and morally on the rocks of Perversion with tragic Consequences!"

7.

Niels Henrik's First Years

When Søren Georg Abel came back to Gjerstad with his wife and two healthy sons in the summer of 1804, the people of the local community greeted them with open arms. He was 32 years old and had come back to the village that he regarded as his home community. His beautiful wife, Anne Marie, was 23 and the people of that district probably knew her too, at least they knew her father well.

But not all of young Abel's conduct as curate in the area had fallen on equally fertile soil among the people of the community, but much had been buried and forgotten by the time he returned as the parish vicar. The local people in general hoped that the young vicar and his wife would follow in the footsteps of the old, and with time become a similar support and reassurance for their lives on earth and their faith in the Hereafter.

The hopes and expectations that things continue along the same track as before was predominant in the population, at least in this part of the country. In this region no religious fanaticism had split society into groups. People hoped that the current boom times which had in the course of a generation made life less burdensome would continue. The Battle of Copenhagen on April 2, 1801, had been a mere episode and had not destroyed the auspicious conditions for trade and commerce. Prosperity, the boom period, and the powerful optimism that followed in its wake were broken for the first time in the fall of 1807, when Denmark-Norway was drawn into the ongoing warfare on Napoleon's side.

Søren Georg Abel had come back to a landscape he knew and in which he believed he would thrive. He had come to a parish to which he was willing to devote all his labours. Duty, hard work, reason and integrity were his catch-words, as they had been for everybody in this period when enlightenment flowered and people felt that very soon things would be much better for all humankind. S. G. Abel also felt that he lived in fortunate times, when one might possibly behold a golden age on earth. His tasks and programme were clearly laid out; he possessed an abundance of drive and energy, knew he had a message to impart and the experience to convey it successfully to the people. Knowledge and enlightenment should succeed in bringing happiness and a better life for all who would follow the thoughts and methods of these new times. *Bliss*, as Pastor Abel was fond of calling it, bliss in this life was an attainable goal, a realistic possibility for whomever. In all his work, Søren Georg Abel had the wellbeing of the many in mind, and *the many* were for Abel composed of a much greater group of human beings than most public servants of the time considered deserving of happiness and a better life.

Søren Georg Abel got right to work with that which he had been most successful as curate in Gjerstad Parish: the work of the reading society, despite the praise of Bishop Hansen, had fallen off when he had moved to Finnøy. When he had started it in 1796 there were almost 60 members, but now when he returned, there were barely 20 who paid the annual subscription of one mark. Now S. G. Abel ignited new flame and enthusiasm and he opened the reading society to those he called "extraordinary members." These were monied persons from outside the local community. Among them were both factory owners from the district: Henrik Carstensen of the Egeland Ironworks, and Jacob Aall of the Nes Ironworks, and contributions to the Reading Society also came in from public officials and businessmen in Kragerø, Risør, Tvedestrand and Arendal. Pastor Abel supported the buying of books, and as before, the greater part of the books were for the guidance and enlightenment of farming folk, in the management of livestock, and the cultivation of fields and gardens. But there were also many philosophical and theological writings by Voltaire and Christian Bastholm that one might borrow from the Reading Society. Membership increased smoothly, until by 1812 it was around eighty.

The young Pastor Abel also continued his father's work of organising poor relief and managing schooling. Using what he had learned and been inspired by in the teachings of Bishop Hansen, he worked particularly closely on the education of those who would become teachers. In his own teaching, Pastor Abel placed particular weight on the way he taught catechism, that the teaching must involve the awakening of reflection, through the detection and clear understanding of words and concepts. Everything should be so clearly comprehensible that the spoken could almost literally be taken in the hands, as he put it. This seems to have been a motto for S. G. Abel, both as clergyman and father: everything must be seen and understood so clearly that one could almost feel it with one's hands!

In 1806 the first edition of his popular exegesis of the catechism came out: *The Questions of Religion, with Answers, Adapted to the Comprehension of the Young*, 337 questions and answers which outlined Christian faith and life. A second edition was published a year later. This was probably the book that Niels Henrik used as his reader when his formal education first began.

Like his father before him, Søren Georg Abel tried to wave the moral finger in front of the village people: that boys ran around visiting girls in the night, such madness had to be stopped. And when he found that speaking to the youth about right and wrong did not help to stop these indecent and obscene "night-running" activities, the pastor laid a plan that he thought would haul the evil out by the roots. He donned the clothing of a youth and went out in the dark in the company of the boys who knew the routes of the night-runners. By this means he would catch the transgressors in the act. And he implemented his plan. But the result was somewhat like the time his father had played the drunk man among the teamsters. There was an overwhelming surprise, but no appreciable improvement, and many felt that the pastor had lost a degree of authority.

When it became apparent that people found it difficult to know which rules applied to the question of public assistance to the poor in the parish, and that they had not successfully put an end to vagrancy, Pastor Abel assembled all the rules applying to these issues, and wrote them down. He got the Poverty Commission's signatures, and sent his "law protocol" to the diocesan office in Kristiansand, which approved this document and praised Pastor Abel for his juridical contribution. This paper, which also contained new, strong rules on how the parish, with the help of the village peacekeeping forces should clean up beggars and vagrants, was printed at New Year's, 1808, and many copies were disseminated in the parish. Subsequently, the arrangement with the village watchmen was implemented with good results. Beggars and vagrants were a group of people who remained in Pastor Abel's mind and his desire for a better life for all people, but they were a group of people who first had to settle down and come under the care of the Poor Relief Programme. Certain requirements had to be fulfilled before they qualified for benevolent treatment.

During that first year in Gjerstad, on August 4, 1804, Anne Marie gave birth to a boy who died two weeks later. In November of the following year she had a son who was christened Thomas Hammond. And in November, 1807, still another son came into the world. With this birth, as with the previous ones, Anne Marie sent a message to her sister, Elisabeth in Risør, and Elisabeth came, more than willingly, to the vicarage at Gjerstad. Elisabeth looked after the older boys, Hans Mathias and Niels Henrik, but the main reason she was so happy to come to her sister and brother-in-law at Gjerstad, was no doubt the freedom she found there to discuss her love for her young man, Peder Mandrup Tuxen. Her father, the rich Simonsen of Risør, did not feel that this marine lieutenant was good enough for his daughter. But in the home of her sister and brother-in-law at the Gjerstad vicarage, Elisabeth could write and receive letters, and she was even able to meet her dear marine lieutenant. And in December, 1807, Søren Georg and Anne Marie decided to give explicit expression to their sympathy and attachment to Tuxen: their fourth son, born a month earlier, was given the name Peder Mandrup Tuxen Abel.

Otherwise, things prospered at the Gjerstad vicarage with many a gallant ball held during the first years. The pastor's wife, the shipowner's daughter from Risør, well knew how to organise a successful social occasion. She loved parties and social life. And in the eyes of some, she was perhaps a little too fond of the bubbling champagne. Naturally enough, the servants at the vicarage compared her to the previous pastor's wife, and whereas Elisabeth Normand had been goodness and consideration personified, they found the new pastor's wife whimsical and vague. Anne Marie did not concern herself the same way with other people's wellbeing; she treated the servants as she had been wont to do in her father's patrician dwelling at Risør. It is probable that Anne Marie also had her party furnishings sent to Gjerstad from her childhood home in the town. The times continued to be spectacularly good for Simonsen and other members of the business class, and their view was well-founded when they said that it was the goodwill of the regime and the country's leadership that secured the established social order. Trade and commerce raked in huge earnings in the lee of the warfare that raged without, and at home the public officials put a stop to the local uprisings that brewed under the direction of the peasant leader, Christian Lofthus and preacher-business-man, Hans Nielsen Hauge.

Parties and glittering social events were held at the Gjerstad vicarage and seem to have given content to the life of the young wife. Anne Marie never seems to have been particularly involved with her children, and the servant girls certainly did most of the food preparation and housework. Anne Marie was a married woman with "empty hands." She also had her sister living with her for long periods of time, and Elisabeth helped to organise the glittering balls that in the course of an evening and a night would see up to a hundred people filling the many rooms of the vicarage. First there would be a dinner for all the married couples, public officials and merchants and their wives, who lived within manageable distance from Gjerstad, and with good weather conditions this could include Risør, Holt and Tvedestrand. Probably in these early years very few of Gjerstad's peasant farmers were guests, although in later years Pastor Abel paid less attention to the ingrained divisions of social class. When parties and social life enlivened the vicarage in the early years, the only aspect of it that the local people were able to experience was the music. From outside the well-lit building one could hear violins, clarinets, French horns and bassoons - something that was not everyday fare for the common people. To the ears of the participants however, this was probably quite normal and rather seductive dance music. And so they danced, quite happily, right through to the break of day.

However, a long time could pass between such nights and days, and right in the midst of everything, war came and put an end to it. In a letter to her dear Marine Lieutenant Tuxen, Elisabeth wrote in February, 1808, from Gjerstad vicarage: "We have not seen one strange Human Being since Christmas, but I do not miss it when I constantly have enough to do. I rise in the Morning at Seven o'clock, or a little before, and I read a little until it is light, then Abel shouts, 'Come down here, Elisabeth!' My first Task is to make Tea, and thereafter, am I School Master for the span of an hour, when I read with little Niels, and since I sew like the Wind, I usually go into the Kitchen and iron some things. In the dark winter afternoons I either play the pianoforte or, if the weather permits, Abel and I take a stroll outside. In the evenings after the whole house has gone to bed, Abel and I sit up until the clock strikes 11, then I go to my Chamber. And so go the days, one after the other, and Time rushes by."

Where is her sister, the pastor's wife, the beautiful Anne Marie, during all this? She neither gets up nor goes out walking. To be sure, Elisabeth does write "I help my Sister with whatever I can during the Time I am here and that is advantageous to us both." At this time, in February, 1808, the son Peder Mandrup was barely three months old and the vicar's wife was probably exhausted. But it also appears that without admitting it, Anne Marie more and more withdrew from her roles as pastor's wife, housekeeper and mother. Perhaps her real desire was the life of a lively society lady.

The following summer, in July, 1809, Peder Mandrup Tuxen was visiting the vicarage. Old Simonsen had still not completely come to terms with his daughter's suitor, even though now, due to the English blockade, Tuxen had been put in charge of a gunboat crew, and was bravely defending the transport of grain across the Skagerrak. But at Gjerstad, Tuxen was made very welcome by both adults and children. He entertained the children with games, told sea stories, and was well-oriented in the world of science. Some years later Tuxen lectured in mathematics and technical subjects at the naval school in Copenhagen, so perhaps could it have been that little Niels Henrik raised a little mathematical problem for attention? Be that as it may, Tuxen was very skilled at cutting out silhouettes and he probably good-heartedly cut out profiles of both children and adults. On the back of the silhouettes of himself and his wife, Søren Georg Abel wrote two small verses on July 4, 1809. On the back of his own silhouette are the words:

I preach, I love, I eat and drink and gape,
My finery a tattered hat and ditto cape,
Ask not, dear reader, what else I be,
Since this, in truth, is all I see.

And on the back of his wife's silhouette he wrote:

> Bound as I be to so odd a Patron,
> From this world I've got my Portion,
> But, dear reader, thank the Lord on High,
> If, to nothing worse than this, you fly.

This was the summer of 1809, and all that anyone could talk about were the fortunes of war.

8.

Childhood Learning, Childhood Faith

The question of a tutor for the two eldest, Hans Mathias and Niels Henrik, was not a matter for discussion at the vicarage. Father Abel himself taught his children, but presumably it was also Aunt Elisabeth who taught Niels Henrik to spell his first words and write his first sentences, and probably it was his father's catechism "comprehensible to the young" which was used as his first primer. By means of questions and answers that might arise in a father-son dialogue, Abel the father explained and interpreted his view of life and human endeavour in such logical patterns of thought that during Niels Henrik's short life it never left his mind. He would also have learned the questions and answers by rote, memorizing them the same way that the youths in confirmation classes did. The three first questions and answers in the book are:

1. What is particularly essential when one wants to acquire knowledge?
   Attention and Reflection. One must pay thorough attention to what one sees, hears, or reads, and afterwards, must reflect upon it.

2. By how many means can one come to the Knowledge of Truth?
   By three means: through Experience, Reasoning and Faith.

3. What is the most important of all forms of Knowledge?
   Knowledge about God: for without this we cannot be happy in Life, cheerful in the face of suffering, and frank in the face of Death. In a single word, this Knowledge is called Religion.

And thus began the interpretations that were repeated and repeated until they echoed through each and every head: in relation to the knowledge of God, do not pay attention only to the knowledge that He exists, but also, know what He demands of the people, and what he would do with us in time and eternity. That God really exists is easy for one to see simply by examining the world: it is just as unlikely for a house to build itself, as for the world with all its animate and inanimate beings, with its heaven with the sun and moon and innumerable stars, to have created itself. And in regarding the world, man learned to know God as Almighty, wise and good. *Almighty*, because He has arranged all things. *Wise*, because anyone can see and experience this order and regularity that rules over everything in Creation. *Good*, because He has arranged everything for the bliss of His creatures, and particularly for those of His beings who are blessed with Reason. But though we, human beings familiar to reason, can learn much about God, we cannot know Divine Providence, and nor can we know how He will treat us as sinners, nor with certainty what fate awaits us after death.

The book of Abel the father explained with such perfect logic the simplest way to acquire knowledge of God. One must first learn the Ten Commandments and their explanations. By obediently following the Ten Commandments, man demonstrated to God his love and gratitude. In return, God's fatherly love would move one to do good freely and out of love. When the Commandments were upheld only out of fear, Pastor Abel maintained, they often lost their true value.

The way one respected God was the way a child respected his father: deference, gratitude, love and obedience. To *respect,* was to esteem Him highly, and deeply feel His greatness. To show *gratitude*, was to acknowledge His goodness. To *love* was to rejoice in His perfection and beneficence.

In the course of the book's fifty pages, Søren Georg Abel, in a rhetorical style, explained the true significance of things and actions. He also explained what original sin is, who God is, and what man consists of, who Jesus is, and what Jesus has done. The book gave answers as to how God punishes trespassers, on how we can become better human beings, what prayer is, and why we practice christenings and holy communion. Pastor Abel also had clear predictions on the life to come: if one fulfils one's duties in true faith, one will find one's way into a state of bliss, both here in life, and in the Hereafter. And bliss in the Hereafter means happiness and salvation after death, a mature knowledge of God suffused with pure goodness and holiness. When one dies and one's mortal remains decay, one's soul retains its immortality. The soul was a unique vessel not composed of parts; therefore it could not be destroyed or dissolved.

In a short appendix, Abel reeled off a series of duties incumbent upon man in this, the more worldly life. There was the duty to comport oneself with modesty, decency and courtesy. There was the duty to avoid the indigence and straying from the path that might occur out of age, weakness, accident, or as a result of hailing from a numerous family which lacked the means of acquiring the necessities of life. There was the duty to pay attention to new developments in the external forms of religion that at all times undergo change in accord with the needs of people and the times they live in. There was the task of loving one's fatherland, and respecting other nations and customs. There was the task of handling animals properly and humanely, submitting them neither to blows nor excessive labour. Heartlessness toward animals could easily lead to cruelty toward people.

A year after the publication of this concordance of the catechism, Abel published a thirty-two-page devotional and prayer book that he dedicated "To Gjerrestad's and Veegaardshej's Common Folk." Here one found familiar devotions like the Lord's Prayer, and prayers written out "according to their Content and Meaning," but the greater part of the book was composed of morning and evening prayers for each of the week's seven days, and the prayers ended with reference to a psalm from the evangelical psalm book. In addition, there was a morning prayer and an evening prayer devised particularly for schoolchildren. These were certainly the texts that Niels Henrik early on had to spell his way through and, after learning them, to recite aloud. The evening prayer:

God and Father! A Day has once more come to its end, in which we have found Proof of Your Goodness and Love. We have been healthy and satisfied. We did not lack our daily bread. We have had so many Happinesses pressed upon us, and You have given us every Opportunity to learn things good and useful. O, that we have utilized this day, and that we have used all the Days of our Lives according to Your manifest Will. Never allow us to forget that Diligence and Moral Decency are the best Means to thank You and win the Respect of all good people. Let us always in this way grow not only in Knowledge, but also in Virtue, which is the road to true Glory. In this way may it be said about us as about Jesus: we grow, as we do in Age, in Grace and Favour in the eyes of God and Men. We would like one day to look back with Joy on the bygone days of Childhood and Youth; we would like to prove worthy of our Parents' Faith and Happiness, and when Death, either sooner or later, calls us to that better Life where all Goodness receives its reward. Anoint us, O God, with Your Mercy, that our love of learning never slackens, and that our love of Virtue never diminishes! And may we always have undaunted faith in You, and also, during this night, rest securely under Your beneficent Protection. Amen.

And the "morning prayer," which was of equal length:

Eternal God in all Your Goodness! Under Your care we travel the Path of Life, because You can and will do what is best for us. You called us into Life so that we could enjoy the many Pleasures with which Life blesses us. We come into the World weak, but You come to us in our Weakness and help us and free us from many Dangers. We did not know the way to Bliss, but You gave us the Opportunity that is taught through the Merciful Lessons of Your Son, Jesus. We have this night slept under your beneficent Protection, and with renewed Energy can we now take up our Daily Tasks. And what is even clearer is that we begin the Day with our hearts Uplifted to You, our Creator and Benefactor. May You always be our Protector. Bless our Work and give us the Desire to heed our Teacher's Lessons. Help us to use the Days of our Youth in this manner so that once we have grown to Adulthood we can remain useful and acceptable to You, to ourselves and to the World. Always teach us to find Pleasure by increasing the Wellbeing of Others. Let us welcome any Opportunity in which we can be of service to others. We forgive those that would trespass against us. Make us wary of, and indifferent to, Pleasures, and in everything that we undertake remind us that You are nearby, a Father who watches in Secret, and then one day, reveals Himself. Amen.

9.

Everyday Life in the Shadow of War

In addition to his explanation of the catechism, Abel had brought out a whole textbook for teaching his own children. In this hand-written volume, each subject was taken up methodically, point by point.

Under the title, *On Punctuation and its Uses*, the period, comma, colon, semi-colon, question mark, exclamation mark, dash, equation sign, ditto sign, parentheses, and paragraph sign are explained. Under *On Money, Weights and Measures*, the author goes into the relations between *riksdaler*, mark and shilling, and the different values these had in different countries, together with the correct way of writing them down. The relationship between a foot, an ell, and a fathom are laid out, as is the difference between the Danish and the Norwegian mile. Spatial measurements, casks, barrels, bushels, and quarts are gone into. In the chapter, *On the Art of Calculation*, there are four ways to calculate, "the four Species": addition, subtraction, multiplication and division, each being illustrated with the help of large, worked-out tables, six pages for each of the four species. And on the first table under Addition, incredibly enough, one finds $1 + 0 = 0$. A fantastic beginning for a mathematician. The table continues: $1 + 1 = 2$, $1 + 2 = 3$, $1 + 3 = 4$, and so on, right through the numbers from 0 to 9.

Grammar is explained in 37 points, with definitions of the verb, noun and other classes of words, under the title *On the Learning of the Danish Language*. *On the Description of the World* is a detailed description of the Nordic countries, a survey of each of the European countries, and a reeling off of a wide variety of facts about Asia, America, Africa and Australia. History (*On Denmark and Norway's History*) begins with the birth of Christ and continues to Napoleon's current war with England. That the description stops at this point in history suggests that Pastor Abel wrote his teaching book when Niels Henrik was about six or seven years of age. But it is quite certain that Niels Henrik and his older brother were bombarded from all sides with information about the actual war with England. Both news and rumours flourished in the community: the war was on everybody's lips.

The English attacked the Danish-Norwegian fleet moored in Copenhagen harbour on September 2, 1807. Admiral Nelson had earlier destroyed the French fleet at Trafalgar and now had to secure his forces against Napoleon at the point where France might replenish itself with the assistance of the many Danish and Norwegian ships. The bombardment of Copenhagen and the subsequent surrender in the autumn of 1807 penetrated the whole of Norway like an electric spark, wrote Jacob Aall in his memoirs. Aall continued: "This consequently gave rise to Relations of War with England, and naturally brought to a standstill and crippled all Trade and other Activity of the bourgeoisie."

British warships patrolled the Skagerrak and thereby blockaded the country's most important shipping route. By the summer of 1808, all timber trading had ceased. Shipping confined itself to the dangerous and not very lucrative grain trade. Earlier, most of the grain traded from Denmark was transported in Danish ships, but now to venture across the Skagerrak was a game of chance with uncertain returns, in which participants ran the risk of ending up in an English prison.

Nor was England happy that the import of Norwegian timber had been blockaded, and consequently, given the conditions, Norwegian and Danish mariners who were stopped by English warships were well-treated, whether confined to harbour or to prison. In the Agder districts of southern Norway there were more ships and mariners than there were people living ashore, and many were very familiar with the distant seas they visited between landings in Danish and Norwegian ports. This meant that there were not the same famine conditions in this region as in other parts of Norway. The two largest factory owners, Jacob Aall and Henrik Carstensen who, before the war, had dealt in the grain trade, tried at length to keep things going. And as long as their machines and equipment were running, they tried to help people by providing grain both for consumption and sowing. As well the businessmen, the smaller shipowners and other people working in relation to shipping, who thought they could manage the crossing and avoid the English patrol boats, began to try to do so. Many were successful, but many were also caught.

During the autumn of 1807 the Danish government realized that it would not be possible to rule Norway from Copenhagen. Whereupon a Governing Commission was inaugurated in Christiania, a commission composed of four men who would act as the long arm of the king in Norway. But contact with Copenhagen proved more difficult than had been supposed, and the Governing Commission had to increase the scope of its own decision-making and establish within Norway many of the institutions required by a state. Problems and their solutions looked different from Christiania than from Copenhagen, and more and more the Governing Commission realized the necessity of an increasingly independent Norwegian political system. The head of the Commission was Prince Christian August who, from 1803, had been Commander-in-Chief of southern and eastern Norway, and the three other members were all loyal public officials. Besides Christian August, who was widely praised for his gentle manner and his modest style of living, Enevold de Falsen was the main force in the Governing Commission. Ideas of equality from the French Revolution led him to eliminate the aristocratic *de* from his name. By the spring of 1808 the situation had become worse and more complicated on all fronts. King Christian VII died on March 13th. Most knew that he had suffered mental illness and had not exerted real influence on public affairs, yet he was nonetheless regarded by many as a sign and a symbol of the peace and wellbeing that now seemed to be in the process of ebbing away from the realm. From this viewpoint, the new king, Frederik VI, who in reality had ruled for almost a generation, represented the dangers and disasters that now loomed. The ruling circles in Copenhagen still believed that Napoleon was the genial and divinely gifted military leader who would never lose, but most ordinary people had already begun to experience the vicissitudes occasioned by war and the blockade.

Frederik VI abrogated the neutrality pact with England, and the day after the death of the old king, Denmark declared war on Sweden. Denmark-Norway was duty-bound by pacts signed earlier to support this move, but in Norway, Enevold Falsen and the Governing Commission tried to keep secret the Declaration of War. They wanted it to look as though Sweden had begun the hostilities and had started the war. In order to make heard the "Voice of the Common People, and their mode of Thinking" that could "express itself in its way," the Governing Commission began to publish the newspaper, *Budstikken* [The Message Stick], and here Enevold Falsen engaged in discussion on national and political problems; literary contributions were accepted, and people were aroused and stimulated by *Budstikken*, which soon reached considerable parts of the country.

Now that there was a naval war against England and a land war against Sweden, *Budstikken* managed both to represent people's privation and mobilization as "gifts to the altar of the fatherland." Most were pleased that Swedes had not managed to set foot on Norwegian soil; they also supported the army patrolling the border with Sweden. On the Skagerakk coast, they built small cannon batteries for combatting the British naval ships. Christian August was the great hero, highly regarded in all circles. After he brought about a ceasefire with the Swedish Commander-in-Chief in December, 1808, he was made a general by the king in Copenhagen.

Meanwhile, some weeks earlier, Enevold Falsen had succumbed to stress and worry. He felt the conflicts had become too great and the outlook too disastrous. Friends said that he often became melancholic and one cold November evening after he had indulged in his theatrical interest (the Dramatic Society), he went home alone. The next day he was found in the sea at Bjørvika in Oslo Harbour. Then, on December 30th, an announcement appeared in *Budstikken*: "Enevold de Falsen. He is no more, but he Lives in Memory!"

Now, the administrative officer, Count Wedel, came to the fore, and with him, a new chapter in the history of the Governing Commission began. Count Wedel became nationally known when on one of his daring expeditions to bring grain to the country by braving a full, raging storm, he had sailed right through a British transport fleet of 40 ships. And it was Count Wedel who travelled to Copenhagen with a bold banking plan that took up the politically delicate theme of an independent Norwegian bank. But the most important of the conflicts between the king in Copenhagen and the Governing Commission in Christiania came on a February day in 1809, when the Commission wrote to the king and requested authorization to conclude a ceasefire with England and Sweden, so that "the country could be saved from total ruin." In effect, this was a demand for an independent foreign policy, and the king refused categorically, although he expressed his deepest sympathy with the suffering Norwegian people, and he wrote, "The greater the Danger and the greater the Need, so must Courage multiply, and this is what I expect of my faithful and stalwart Officials." In Norway, this was received almost with scorn, and the general feeling was that the king was pursuing a policy that would bring about the complete destruction of trade and commerce, and was reducing the common people to starvation.

At the Gjerstad vicarage people were aware of the situation, and they seem to have been better stocked with food and clothing than Norwegians were on average. The farm had good houses and fields, and hardworking servants and labourers. When Søren Georg and Anne Marie arrived from Finnøy in 1804, they had a seventeen year-old farm boy, Hermann Josefson with them. When during the war years the boy was being drafted to the navy, Pastor Abel wrote to Copenhagen and requested the boy's exemption, since there was so much need for his services at Gjerstad where he was teaching the peasants better methods of cultivation. Consequently, Hermann Josefson did not have to go to war. He became the pastor's close assistant on the farm and in the community.

The year 1809 was a hungry year of the worst kind over much of Norway. Christiania sent out advice to people to use different surrogate foodstuffs. The most famous in this activity was Martin Richard Flor, botanist and teacher at Christiania Cathedral School. Among the general populace he became known as the "Moss Priest" because he travelled around demonstrating how moss could be used to replace bread grain cereals. He travelled mostly around eastern Norway, and further south, people only heard rumours about this moss priest who would have them toss anything under the sun into the cooking pot! But Gjerstad people also began to suffer want and hunger during the course of this year. In the spring, the shortage of seed grain was severe, and this had served to slow down plans to construct a building for storing grain. Pastor Abel organised the work and the people of the community joined in with will and enthusiasm: the peasants brought materials and farm workers donated their labour power. Yet, they had to ration out the consumption of the grain cereals so as to have sufficient seed grain for the new growing season. They built a potato cellar under the granary, and in the fall of 1809 there was a total of 350 casks of potatoes (1 cask = 139 litres). The following spring anyone in need could borrow seed potatoes and after the autumn harvest, the debtor could try to repay a little more than he borrowed. Otherwise, Pastor Abel thought that he had only heard one rational comment about this moss surrogate that was being discussed everywhere, and that was from an old woman in a neighbouring parish who is supposed to have said, "God only knows if it is better to starve to death or eat yourself to death!"

Pastor Abel had other ideas about how people could avoid starvation. He would get the local people to eat horse meat! The prejudices against the consumption of horse flesh were old and ingrained. In heathen times the Vikings stained red the images of their gods, with the blood of horses, the meat of which they boiled and served to the guests at the great sacrificial feasts, together with beer and small livestock. With the coming of Christianity, the eating of horse meat became permanently forbidden by law, and violations were punished with fines. Long before the year AD. 1000, the Church had decided which foodstuffs were pure, and which were impure, and among the impure foodstuffs were the meat of horses, dogs and cats. Even if the religious reasons for these prohibitions had been forgotten, there remained a presentiment that the eating of horse flesh was somewhat dangerous and unhealthy. To butcher a dead horse and cure its hide had long been considered a malediction. It was not work for an honourable peasant; it was a job for knaves. This had remained the state of affairs up to the present time, even in Gjerstad. The bodies of horses were buried in the ground, but with the changing attitudes toward old customs in these new times, people cautiously began to use horsehide for leather. But to kill his best and most faithful working comrade of many years continued to be something too abhorrent for the peasant to contemplate. But now war and starvation threatened.

Pastor Abel maintained that at the Gjerstad vicarage they all used horse meat in the household. He spoke in glowing terms about how it was the most energy-giving of all meats and that sausages made from horse meat were a pure delicacy. It was reprehensible that so much good food was simply thrown away.

One day in the autumn of this crisis-ridden year of 1809, in the presence of some of the men from the local community, Søren Georg Abel had slaughtered a "plump red pony with a White Mane, 22 years old" for which he had given 15 *riksdaler*. The following Sunday after communion, Pastor Abel invited his "parish's Most Honourable Men and Women" to a horse meat dinner at the vicarage. Everyone ate with pleasure, until they were full and content, whereupon they praised the food. From that day on, horse meat was no longer taboo in Gjerstad.

Along the coast a new source of income opened up: piracy. A pirate ship was an armed vessel that had obtained permission to plunder foreign trading ships. In Kristiansand, Chief Adminstrative Officer Thygesen was an active grain importer, who was now quick to build ships and outfit them for piracy. In this way the Diocese of Kristiansand, through plunder, acquired many kinds of goods from many hazardous journeys. The largest act of piracy was carried out in the middle of summer, 1810, when more than 200 British trading vessels were forced to land in Kristiansand, with freight valued collectively at about 5.5 million *riksdaler*. Henrik Carstensen engaged in piracy from Risør, but Simonsen and the other shipowners in Risør were cautious about engaging in this activity.

Pastor Abel taught his sons himself whenever his sister-in-law, Elisabeth, was not visiting. By the end of 1809, Hans Mathias, Niels Henrik, Thomas and Peder were respectively nine, seven, five and two years of age. The two oldest were well into Latin, and their father wrote proudly, "Hans in now a Hound in Latin and Niels is at his heels like an honest Lad. Thomas is playing these Days with Letters of the Alphabet, and about Peder it can be reported that he has his Head in the right Place. As for everything else, they are pure Children of Nature, and in this Respect, as my Mother-in-law feels, it is enough with these 4 Lads. Be that as it may, I have been until now content and do not particularly desire more; but, see! In the due course of Time a fifth will be brought forth to hang about my Neck, if I see things correctly. O, the sins of Nature! Your Blessings demand stiff rates of Interest."

The child to whom the vicar was referring was born March 16, 1810. It was to be a girl. She was christened Elisabeth Magdalene, and this little sister, Elisabeth, was to become the sibling with whom Niels Henrik had the closest affinity.

Otherwise, at the end of 1809, Pastor Abel considered the spreading war to be "a radical Cure for the dear Land of my Fathers, and of this Mercantilism, which threatens to devour everything." Pastor Abel looked at his own congregation and noted that the difficult times had actually opened the eyes of the people to the necessity for changing and improving agricultural methods. The war had forced people to take up new modes of thinking, such as the building of a granary with a potato cellar underneath; better animal breeding and better diet were possibly other examples. Perhaps also the faith of the individual who saw that through his own efforts he could improve his lot in life, had also been strengthened by the circumstances of the times.

10

Ground Swells from Offshore

Conflicts increased in the dual kingdom, problems piled up, and in these turbulent waters there were many different politicians who fished for public support. In Sweden, the old idea of a Swedish-Norwegian union arose with renewed energy. But the Swedish king, Gustav IV Adolf allowed most of his policies to be guided by mystical religious concepts through which he saw himself raised up in struggle against Napoleon as pre-ordained by the Beast of the Apocalypse in the Book of Revelations. Gustav maintained that it was necessary to wage war against the whole world. But displeasure with the king, and the desire for a quite different policy, resulted in Gustav IV being removed in March, 1809, and being replaced by his ducal brother, Karl, who was set upon the throne as King Karl XIII. One of the movers behind this radical shakeup was Count Georg Adlersparre, who, for a long time, had militated for a Swedish-Norwegian union. Already in 1790, on commission from King Gustav III, Adlersparre had travelled widely in Norway, sounding the waters and testing the terrain for union with Sweden, and now, in 1809, the opportunity had arisen again. Discontent with the Danish regime had spread through the Norwegian population, which in addition had its own powerful spokesmen for a different political union. First and foremost, Count Adlersparre contacted Count Wedel, who for his part could count on the support of Jacob Aall and several other powerful men of the business community. Adlersparre and those of his way of thinking believed that if only Sweden had a new king whom the Norwegian population could accept, then the game would be won. The first point then was to choose a royal successor who had the potential to unite people. Karl XIII was old, weak and childless and without opposition he gave way to Adlersparre's bold proposal to choose the Norwegian Commander-in-chief, the young Christian August, as heir to the Swedish throne. Although probably ridden with doubt and misgivings, the Swedish aristocracy and the Gustavians also supported Adlersparre's plan.

In July, 1809, with two days' warning, Christian August received word from Stockholm that he had been chosen successor to the Swedish throne; immediately thereafter he was ordered by King Frederik VI in Copenhagen to seize

Swedish territory as far as Gothenburg. Christian August chose a double play. He disappointed Count Wedel by not giving a determined yes to the Swedish offer, and he disappointed Commission Member Kaas (who above all desired to keep the established contact with Copenhagen) by not complying with King Frederik's order to attack. The others in the Governing Commission supported Christian August, and in their answer to the king in Copenhagen they maintained it was "physically impossible" to carry out a military incursion into Sweden. At the same time the Commission maintained that the only means by which the country could find salvation was "an Armistice of the Land and the Sea, united with the ability to Trade as freely as possible." If this were not done, famine, misery and despair would find even the best and most faithful of the citizenry breaking the laws. Kaas went to Copenhagen to lay the matter before His Majesty, but by then Frederik VI had lost a great deal of ground. His grand political loyalty and agreements had not achieved much in the way of favourable results, and his dreams of a triple union of the Nordic countries, with himself as king, had come to nought. But he made one move that came to have great significance: he entered into an agreement with England to prohibit piracy and open a legally licenced trade across the British cordon. In the course of a few months the trade blockade and standstill had been replaced by a degree of timber export and a boom in employment that surpassed all others. During the years when the delivery of timber from Norway had been stopped, a great demand for Norwegian timber developed in England, and the prices reached heights never seen before. Risør and the other towns along the coast experienced a new period of rapid economic growth.

Officially, Denmark-Norway and England were still at war and there was much wrangling involved in manoeuvring the shipments into the English ports. For the most part the shipments officially went to French ports and the shippers had the requisite papers in the event that they were stopped by French or Dutch pirates. But at the same time, they carried British licences securing against confiscation by English warships, and Danish letters of safe conduct against confiscation by the Danish navy. After delivering their timber, the ships loaded up with salt, coal, foodstuffs from the English colonies, and manufactured goods from the English factories. The shortage of goods from the first year of the war was replaced by a flood of commodities. All this created contentment and public optimism with the state of things. The need for drastic changes, as would result from union with Sweden, was felt to be no longer necessary. When in the course of autumn, 1809, King Frederik VI also had concluded a peace treaty with Sweden, signed at Jönköpping, Sweden, on December 10th, everyone hoped that things would return to what they had been before the war.

At Gjerstad, Søren Georg Abel followed the developments keenly, and it is quite certain that he explained both to his sons and his congregation what he knew. One could read the official news in *Budstikken*, and first-hand information also came from Peder Mandrup Tuxen.

In February, 1810, Niels Henrik and the whole family were invited to a wedding at Risør. Father Simonsen had finally acceded to his daughter's desire to marry P. M. Tuxen. What had in the end changed Simonsen's mind was Tuxen's exploit in August, 1809, when disguised and using a Swedish fishboat, he had boldly conducted Prince Frederik of Hessen-Kassel safely and soundly through a large British convoy, and landed at Moss in Norway. Prince Frederik of Hessen-Kassel was the king's brother-in-law, and the new general who was to be commander-in-chief of southern and southeastern Norway after Christian August, who had by then accepted the offer of becoming the successor to the Swedish throne, despite the fact that King Frederik in Copenhagen had showered him with favours and made him Vice-Regent and Field-Marshall of Norway. Merchant Simonsen in Risør understood well the way the wind was blowing, and saw the honour accruing to Tuxen for the daring exploit he had carried off, and the value that Tuxen's boldness now possessed. It was reported that old Simonsen embraced Tuxen with full reconciliation and the following words: "He shall become my Son-in-law. God bless him and my Elisabeth!" And so they celebrated the wedding joyfully with all the sense of occasion, sparkle and glitter that the merchant patricians of Risør were still able to muster. And this was not insignificant now that the suddenly lucrative trade of the licence period had made things seem just like the good old days.

On December 29, 1809, there was also a large celebration in the assembly hall of Christiania Cathedral School. Three hundred people had gathered there to celebrate the peace that had been signed with Sweden at Jönköpping, and to honour and say farewell to the popular Prince Christian August, who was now moving to Sweden to be the Crown Prince of the neighbouring nation. Christian August, who had been the country's leader and Commander-in-Chief since 1803, had together with his successor, Prince Frederik of Hessen-Kassel, circulated through the ranks of the guests and joined in the jubilation as the ovation swelled through the rooms and through the crowd gathered outside. Here, as well, most people hoped things would now be as before, while others considered Christian August the man capable of uniting the Nordic countries into a more stable and stronger region, grounded in their common history, language and aspirations. But all of them were able to gather together in the desire to work for Norway's wellbeing in all matters and in all fields. Within this jubilation, amid toasting and shouting of "bravo!" and in an atmosphere of wistfulness, The Society for Norway's Wellbeing had been founded.

The invitation to join the Society was penned by the senior teacher of the Cathedral School, Ludvig Stoud Platou, and was signed by Frederik of Hessen, Count Wedel, the rector of the Cathedral School, Jacob Rosted, and Cathedral School teachers Søren Rasmussen and Martin Richard Flor, and 22 other prominent men of business and public life. Shortly thereafter, District Commissioners were instituted in most of the larger towns around the country, and soon the Society would become a comprehensive organisation with central functions. When, in the course of 1810, the Governing Commission gradually scaled down its mandate, and was finally dissolved, much of the country's activity, enthusiasm and dynamic thinking went into The Society for the Wellbeing of Norway.

At Gjerstad Parish, Vicar Abel received the reports from Christiania with a mixture of happiness and sorrow: sorrow because the capable Christian August was forsaking the country, and happiness because the atmosphere of the times had found robust expression in The Society for the Wellbeing of Norway. It seems that in many ways Pastor Abel had already worked hard within his congregation to promote many of the ideas and undertakings that the Society now wanted to institute: out of love for the country "Enlightenment should be spread, Agriculture developed, the Forest Industry improved, Handicraft expanded, Factories constructed, and each Branch of Industry assisted and improved."

The year 1810 would be a turning point, but not back to the good old days that many had hoped for. The central power of Denmark had renewed its control over Norway, the dissolution of the Governing Commission in Christiania was evidence of this. And in Sweden, Christian August, the new Crown Prince, now named Karl August, fell dead from his horse during a peaceful military parade. Evil tongues had it that he had been poisoned by the Gustavians, something that led to riots and the murder of one of the Gustavian leaders who had wanted one of the younger sons of the deposed King Gustav IV Adolf as successor to the throne. Sweden now went through a new struggle of royal succession, where many candidates were under discussion. Frederik VI proposed himself as a candidate, the Crown Prince of Denmark-Norway, Christian Frederik, was proposed, and Count Wedel had his supporters as well. But the Swedish government backed the brother of the deceased crown prince, Frederik Christian, Duke of Augustenborg, known among other things for his school reforms in Denmark, and the matter seemed to be concluded. But when the Swedish Parliament finally undertook to hold the vote, the Prince of Ponte Corvo and Marshall to Napoloen, Jean Baptiste Bernadotte, was chosen as the successor to the throne. He arrived in Sweden in October, 1810, under the name, Crown Prince Karl Johan. He had sacrificed his Catholic faith for the throne of Sweden, and his new position was to wage war against his former homeland. Adlersparre's plan to conquer Norway now also became Karl Johan's goal.

At Gjerstad, Abel the father was impatient and not completely at ease with the way things were going. Shortly before Christmas, 1810, he wrote to his dean in Arendal: "Today I have visited several Sickrooms, I dissipated myself with a quantity of Tobacco and Brandy. This is my Salarium, but also my Palliative."

11.

New Sounds in City and Hamlet

The Society for the Wellbeing of Norway rapidly received an overwhelmingly positive response. In the course of its first year, the Society had 1,500 members, who each paid 10 *riksdaler* by way of annual subscription. In addition, many gave private gifts that swelled the income for the year 1810 to 23,000 *riksdaler*. The members were spread over the whole country, but the greater part were from eastern and southern Norway, and the relatively high subscription rate limited membership considerably. However, the District Commissions that had been instituted around the country worked to set up local branches of The Society, and one found increasing amounts of information in *Budstikken* about different subgroups; after a year there were around the land about thirty of these organisations, popularly known as Parish Societies.

In February, 1811, Vicar Abel was one of the driving forces behind the establishment of one such Parish Society in his district, which was called the "Gjerstad and Sønneløv Parish Society." Another of the initiators was Gregers Hovorsen, the sexton of the neighbouring Parish of Søndeled. The leadership of this Society also included the sexton of Gjerstad, the sheriffs of the two communities, two peasant farmers, the owner of Egeland Ironworks, Henrik Carstensen, and Pastor Støren of Risør. The District Commission with which Pastor Abel was in correspondence, met at Arendal and was led by Jacob Aall and the district doctor, A. C. Møller.

Søren Georg Abel's many-branched labours for the wellbeing of the common people were now spread more broadly through the local Parish Society. By the spring of 1811, 173 people had signed up as members of the society. The subscription had been set at one *riksdaler* per annum, but voluntary initial contributions gave the Society a start-up capital of 700 *riksdaler*. This sum was rapidly used for interest-free loans to farming folk who wanted to improve the fertility of their soil, the planting of useful herbs and the establishment of fruit trees, and cash premiums were awarded for spinning, needlework, weaving, the making of wooden shoes, brassware, knives and other useful tools like rakes, shovels, trowels, and axes. As well, pamphlets were purchased and circulated with information about improvements to agriculture and horticulture, housekeeping and animal husbandry, and they took out two subscriptions to *Budstikken* that were to be circulated among the members.

When The Parish Society's leadership gathered at the Gjerstad vicarage in late October, 1811, Søren Georg Abel managed to award the gardening premium of 10 *riksdaler* to a man who in 200 square metres had grown 1 cask [139 litres] of turnips and 68 cabbages. In handicrafts a premium of 4 *riksdaler* went to a man who had made 60 dozen pewter buttons, and it was announced that 16 farmers had received loans of approximately 40 *riksdaler*. The Parish Society also gave a premium to bridal couples and confirmants who appeared before the altar in clothing they had made themselves, and to those who used these clothes as their Sunday best all year round. And there was a prize promised for whoever produced the most linen and hemp, and to whoever had the best hop garden. To the first peasant to demonstrate that he had trained an ox to work as a beast of burden there was a premium of 20 *riksdaler*, and the offer that: "Pastor Abel at Gjerstad gives Instructions to the Bullock's Taming, Shoeing and Handling." They had a bullock at the vicarage that could be used as a draught animal.

Every farmhand who could show that he had prepared a new potato field received, after the autumn harvest, half a *riksdaler* "for each cask harvested over and above what had been Planted." A prize was also given to farming people who in the course of a winter had taught their six- to twelve-year old children to read and write.

It is highly probable that Niels Henrik and his brothers frequently went around with their father and the hired hand, Hermann, to the various farms to inspect the fields and check up on the progress toward goals and premiums. But any independent cultivation work on their part was naturally not encouraged. None of The Parish Society's leaders or their dependent children were, according to the rules, eligible to compete for prizes or premiums. For them, whether adult or child, gratification came in a different form. It was: "Payment enough from the Honourable Calling to work for the Guidance and Encouragement of others."

There were many bears and wolves in Gjerstad District at this time, and it was not unusual for a bear to break into the summer livestock pens and kill the animals. Consequently, the Parish Society promised a premium of 50 *riksdaler* for anyone who felled a bear apprehended in a livestock pen, 30 *riksdaler* for he who killed a bear within the farm, and 20 *riksdaler* for felling a bear wherever. Ironworks-owner Henrik Carstensen, who had a special interest in people feeling safe enough to fall trees and make charcoal for use at Egeland, contributed an extra 50 *riksdaler* to the bear premium.

To rid themselves of the wolves, as proposed by Pastor Abel, six portions of wolf poison were to have been purchased that the vicar, together with the sexton, should have set out in appropriate places. Due to problems of liquidity, The Parish Society was only able to purchase four portions of poison, two for each parish.

There was also discussion in The Parish Society about purchasing a Spanish ram for breeding purposes; they also concentrated on getting "a Hat-maker to settle down in the Parishes, and to at first assist him with free Lodgings and a Monetary Loan for Materials." This person was to make hats for inhabitants of the parishes and sell them "in Relation to the most Moderate Prices of the Times." When this market became satiated, the hat-maker could work for strangers. Shortly thereafter, Hatmaker Blomberg settled in the neighbouring parish of Søndeled. He received 200 *riksdaler* in the form of an interest-free loan and, during his first year, made 300 hats for the men of Gjerstad and Sønderled.

Another priority task was to establish a tannery, but when they could find no one with the appropriate training, the Parish Society wanted to send a man to Christiania to train him as a tanner. Pastor Abel energetically sought out the right young man, and wrote several letters to Christiania to the Society's "Class for Household, Manufacture and Artistic Industry," seeking elucidation on what things and expenses were involved in the apprenticeship that could take up to five years at a master tanner's in Christiania. And when Abel finally found the lad, Knud Larsen, he kept him for over a month at the vicarage where he could "practice his Reading and Writing." In November, 1813, Knud Larsen, provided with new shoes, adequate clothes and pocket money, travelled to Christiania. And in his rucksack he carried a number of homilies that Pastor Abel had written down for him in his own hand:

1. Think often of God and picture yourself in His presence. Nothing directs us toward the Good as strongly as this does. He watches in Secret, according to Jesus, and rewards the Open-hearted.
2. Be utterly True to your Master and Superior. Never take what is not yours, even the smallest iota of what belongs to Others. Remorse, Disgrace or Punishment are the Rewards of Infidelity.
3. If you transgress, which is most Human, then confess your transgression immediately. Sincere Confession is halfway to one's Betterment, and finds Favour with both God and Men.
4. Avoid Swindling and Deception, Drink, Card-playing, Fornication and whatever other names are given to the Vices that disgrace Mankind. Guard yourself against Mendacity and Choose the path of Righteousness. My Son, consider the Word of God, whenever you are tempted and beguiled by Sin, do not follow.

I would like you to read through these few Rules often, and remember me, your cautionary Friend. When you can no longer happily contemplate God and your sincere Friends, then, dear Knud, clear your Virtue and Peace of Mind.

Now be vigilant and pray that you fall not into Temptation.

Pastor Abel noted down in the Parish Society records what Knud Larsen had taken with him: Two Pairs of Shoes, one old. A good new Hat. Two Cravats. Three Shirts, one of them new. Three Undervests, one of them good. Two Vests, both of them second-rate. Three Pairs of Trousers, one of them quite good. Three Pairs of Hose, two being old."

Once in Christiania the boy was taken in hand by senior teacher, Flor, and everything went according to plan. Knud Larsen returned three years later to Gjerstad, a tanner.

The district commissioner in Arendal commended Pastor Abel and the Society for "a patriotic sense of Patrimony and Reality, which we confess exceeds what finds its Place in the other Parts of the District." In Christiania as well, the activities of the Gjerstad and Søndeled Parish Society were favourably noticed.

Another project that The Parish Society was eager to undertake was to build a brick and tile kiln but this seems to have led to contradictions among the members. Pastor Abel put himself in charge of the work and formed a stock company wherein twelve peasants and he himself committed themselves for 200 *riksdaler*; in addition, he obtained capital support from the rich Peder Cappelen in Drammen, who also sent an expert to lead the practical work. The tileworks was almost completed, except for work on the oven when construction on the site was stopped. The reason seems to have been that Ironworks-owner Carstensen seems to have feared that the tileworks would require so much firewood that this venture would compete with the Egeland Ironworks for the same finite resource; Egeland was a great consumer of charcoal. Carstensen therefore bought up all the shares he could get his hands on, and at a significantly high price, so as not to advance the work of the tileworks, but rather, to stop it. The result was that the tile plant never really came into production. Only some rough bricks were made, and sun-dried. Pastor Abel used them to build an oven for heating tar. He wanted to improve and make simpler the process of burning tar, and he employed a new manufacturing process, but after some years had passed even this tar oven was not in use.

Now in the immediate surroundings several things began to crumble around Søren Georg Abel, and the tide began to turn against him. The tileworks project was aborted, its lumber rotted, or under the cover of darkness, was used illicitly for firewood. All the iron also disappeared without permission, and was melted down. The remnants of the big water wheel sat for some years like a reproachful memorial for grandiose plans and the "tar oven" gradually became no more than a village landmark.

In Christiania there were those in the leadership of The Society for the Well-being of Norway who felt that the Society's projects emphasized to too great a degree practical and utilitarian issues. Count Wedel seems to have been among those who had hoped that the Society would play a greater political role, become a Trojan horse in the struggle against the Danish regime. But conditions had changed. Already in January, 1810, a peace agreement had come into effect between Sweden and Napoleon, such that any Norwegian-Swedish union seemed now out of the question. Napoleon is said to have stated, "The thought of letting it (Denmark) lose Norway was to me like something out of a novel." On the other hand, it also seemed that in Norway there was no longer the widespread dissatisfaction with the Danish king and the associated concept of Absolute Monarchy. But a strong Norwegian demand for rights and equality persisted in large parts of the populace. One of the greatest demands was for Norway to have its own university, and in this matter The Society would play a decisive role. In the summer of 1811, all the inhabitants of Norway were sent an invitation to contribute economically so that "Norway got a University on its own Soil," and the results were overwhelming. Thanks to the persistent and tactical conduct of The Society for the Well-being of Norway, King Frederik saw himself forced to accede to the Norwegian demands, and The Royal Frederik's University of Christiania saw the light of day in record time. Abel at Gjerstad was also astonished at the speed at which the university plans developed. Driven by idealism, and certainly also by concern for his sons'education, he became, in his district, one of the issue's most ardent collaborators. He thought that his father-in-law Simonsen, in subscribing 500 *riksdaler* a year for four years, was contributing very little, while Henrik Carstensen gave 1,500 *riksdaler* a year for four years, and Jacob Aall gave "once and for all time" the luxuriant sum of 20,000 *riksdaler*. Pastor Abel himself, in excess of what was prudent for a man of his means, pledged 100 *riksdaler* in cash and 10 *riksdaler* annually plus a half cask of grain annually from the harvest at the Lunde farm. That half-cask of grain would also become a duty that Niels Henrik would later have to deal with.

12.

Home and Abroad, Crises and Turning Points

At three o'clock in the morning on February 27, 1812, Pastor Abel rose from his bed and woke up his children in "the Expectation of seeing an Eclipse of the Moon," as he expressed it. But clouds hung thickly over the village.

The year 1812 was a new turning point in the affairs of war and in ordinary people's life circumstances. Frederik VI linked his Kingdom ever more closely to Napoleon, at the very time when Napoleon's great army was facing near annihilation in its march against Russia. In Norway, trade tariffs ceased, the English blockaded Norwegian ports once more, and British imports were stringently prohibited. In the cities, commissioners were set in place to conduct domestic investigations to root out prohibited goods. Blockades, food shortages and times of need had returned, and in addition, nature showed its uglier side. 1812 was one of the years of great hunger; the whole country suffered crop failure.

In Gjerstad, the spring came abnormally late. They managed to begin ploughing the fields only in late May; summer was cold with a great deal of rain; there were frosty nights during the full moons of summer, the grain crops froze, and many did not even manage to get in their meagre harvest. And the potato harvest was just as disastrous. Many were unable to harvest more than they had planted, and some took in considerably less. Prices rose monstrously: up to 144 *riksdaler* for a cask of rye, 70-80 *riksdaler* for a cask of barley, and well over 30 riksdaler for a cask of potatoes. The forests were full of worms and larvae that ate almost all of the branch tips and gnawed away most of the bark. It was said that the insects worked with military precision. Bears were also common that year; in several places they broke into the pens and killed the livestock. In Søndeled there was also an outbreak of an infectious intestinal sickness that caused bleeding and took the lives of many.

The 1812 fish runs failed as well. Discontent with the state of things travelled all the way to the central powers in Copenhagen. Ordinary people were up in arms, not primarily from political consciousness, but rather from hunger, destitution and desperation. Peasants stormed the granaries and the warehouses of the merchants in several towns, threatened the bailiffs and the priests, and abused the merchants. Once again Swedish agents were spying and reporting back to Count Adlersparre and Crown Prince Karl Johan about the degree of hatred in Norway toward Denmark.

It seems that at Gjerstad the shortage of grain was not as severe during this crisis-ridden year as it had been in many other places. In a contribution to *Budstikken*, Pastor Abel admitted that their relatively good condition "is particularly due to the Works Owner and Merchant Hr. Carstensen's zealous Efforts."

Apart from this, what both adults and children discussed that autumn was something that had happened nearby on the coast at Lyngør. The story was about the newly built frigate, *Najaden*, with its 42 cannons, that had been pursued and totally destroyed by the English. This story was told with full details, and no doubt in many different versions. The four British naval ships (*Dictator* with 68 cannon, *Calypso* with 20 cannon, *Feamer* and *Podargus* with 14 guns each) were pirates of the worst type, and the Chief Officer of the British detachment, the 24 year-old James Stewart, must have been drunk out of his mind in order to have mounted this Damn Fool Operation. The cannons were heard all the way to Gjerstad, it was said, and even one Gjerstadian was killed in the attack at Lyngør.

On Saturday, July 5th, one day before the cannons rang out over Lyngør, Pastor Abel assembled 25 men from the village at the vicarage. They had been chosen "by all the people", and together with the priest they undertook a contract that would reform local feasts. There was an old ordinance against "Sumptuousness" which Pastor Abel wanted to renew. This was a Church ordinance instigated more than forty years earlier, but had been less strictly observed by Hans Mathias Abel who, after all had maintained that external commands and means of pressure neither put the fear of God into people, nor led to good morals. But now Søren Georg Abel felt that the difficulty of the times required that something else should be done about people's use of resources. It was decided that at weddings, people would not use more than one cask of malt, 18 jugs of spirits and 2 pounds of coffee (1 jug = 0.97 litre). Other foreign varieties of spirits were forbidden and the bringing of spirits by guests would not be tolerated. Those guilty of violating this order would face a fine of 20 *riksdaler*, to be paid into the parish's poor- and school fund. At funerals no more than half a cask of malt, 8 jugs of spirits and 2 pounds of coffee were to be used. The coffin had to be brought to the church before three o'clock, and only relatives from other villages and closest friends should be invited. The fine for violating these agreed terms was set at 10 *riksdaler*. There was also a proviso added to the contract: "Incidentally, there shall be for all of us, when Funerals Occur, a sacred Duty toward Moderation and Moral

Conduct that Delivers up Evidence of the higher Feelings that such Occasions ought to Inspire in Men of Reason and in Christians."

These men chosen from the village also decided, together with the priest, that from New Year, 1813, each bridal couple on the day of the nuptials, and each confirmant on Confirmation Day, *should* wear homemade clothes. It was legitimate, however, to purchase kerchiefs and other head-coverings. If this regulation were to be violated, each bridal couple would have to pay 20 *riksdaler*, and each confirmant, 10 *riksdaler*.

Pastor Abel arranged a similar contract for Vegårshei. There, 20 of the village men, representing the whole community, were in agreement, and the agreement was read out from the pulpit twice a year. The contracts were approved by the higher authorities in the diocese, with the comment that this initiative by the priest and congregation was "a shining Example of Living interest in the Good."

A heavier atmosphere also seems to have prevailed at the vicarage at this time. The enthusiastic encouragment and the incentives of the earlier days were now replaced by regulations and punishment. But the priest himself did not stand in the fore as a good example for others. People's respect and opinion of Pastor Abel was falling. One factor was the abortive tileworks; more serious were the relations around the Vegårshei annex. Right from his curate days, Pastor Abel seems to have had trouble with the congregation at Vegårshei, but at this time it became worse. Certainly the geographical situation had something to do with this. It was impossible to work for the good of the local congregation at Vegårshei to the same extent as he did at Gjerstad, and perhaps in order to set this right, Pastor Abel had given extra books to the reading society branch at Vegårshei, and had put himself in charge of, and contributed to, the building of a new church there in the hamlet. And when on August 19, 1810, the new church building was consecrated by Dean Krog from Arendal, Pastor Abel gave a speech that was rewarded with such applause that he was later persuaded to allow it to be published.

But despite this, Søren Georg Abel was blamed for doing less for the Vegårshei congregation than he ought. The fact was that the poor relief and school arrangements did not function as well at Vegårshei as at Gjerstad, and this was used against him. But there were also other reasons for discontent. People had not forgotten that he, as curate, had travelled right through the village and, instead of instructing the confirmants, had ignored them and gone on to enjoy social events at Holt and Tvedestrand. And almost the same thing had happened again in the first years after he had become the parish vicar. Instead of coming to Vegårshei on the Saturday to spend the night in the little cottage as the late Hans Mathias had been happy to do, Søren Georg Abel drove through the village and stayed with his boyhood friend, young Curate

Thestrup. But when Pastor Abel had gone back to Vegårshei on Sunday morning to conduct the service, some of the village folk had hidden the church key from him. He was unable to get into the church and there was no service.

Now in 1812, the first parish vicar of Holt, Hans Christian Thestrup died, and his son, Curate Thestrup was defrocked due to drinking problems and worse. And in the midst of these circumstances, Pastor Abel proposed that he be quit of Vegårshei annex. Right enough, there had been talk that Holt Parish should be reorganised after Thestrup's death, and in his written proposal, Abel argued that Vegårshei annex ought to be part of Holt because the road between Vegårshei and Holt was easier and better since people took that road with their loads of logs, bound for Jacob Aall and his ironworks. Abel's plan was the subsuming of Søndeled within Gjerstad Parish. Perhaps he had reasonable grounds to argue that this could be easily brought about. He himself had been the driving force in The Parish Society for Gjerstad and Søndeled and in this respect he saw the interdependence between the two communities. However, Søndeled came under the jurisdiction of the main parish at Risør, and Parish Vicar Støren at Risør resented the proposal and bound Søndeled more closely to Risør Parish. When Støren died a year later, and Catechist Schanche, upon his appointment to Risør, declared that he would not under any conditions lose Søndeled, Pastor Abel understood that the proposal was lost.

And with that, Abel was left with only the Gjerstad main parish and considerably less income. He wrote to Dean Krog in Arendal: "I am quite satisfied in spite of all this, for the very essence of Art is being able to live on almost Nothing." But in the coming year Abel applied for several different pastoral positions, and he applied for half the vicar income from Vegårshei, but the new vicar at Holt had naturally enough not acceded to this, and there was a long and harrowing struggle, where both the Mediation Commission, among whose members was Jacob Aall, together with the authorities, had difficulty determining how much each party should have. The hullabaloo ended with Abel being awarded one-seventh of the Vegårshei annex income.

A versified obituary for a thief is ascribed to Abel, and possibly reflected his state of mind during this period:

O so rare a Man, brighter than the sun,
While others slept, he'd grab and run.
What others lost, was what he won.

Abel received great encouragement when on January 28, 1813, he was awarded the Red Cross of the Order of the Danish Flag, for the efforts he had made on behalf of the coastal defense while he was vicar at Finnøy. Søren Georg Abel had not been forgotten in Copenhagen. Bishop Peder Hansen had gone away in 1810 but Abel's energetic contribution to the well-being of ordinary people was known to many, and his former teacher and rector, Niels Treschow, was still professor in Copenhagen. A new attraction to, and involvement with, public affairs seems to have gripped Abel. He ardently wanted to work for a larger constituency than Gjerstad Parish, and he would seize the chance when it came.

During the spring of 1813, many Norwegians thought that the double monarchy was about to explode, and the future seemed uncertain. Some believed that an independent Norway under the protection of England was a possibility. Others, and first and foremost among them, the business community of southern and eastern Norway, thought that the time was ripe for union with Sweden. Count Wedel continued to be the leading voice for this position, despite the fact that he rather feared that Karl Johan, in one way or another, was in league with Napoleon. Again others placed their hope in Crown Prince Christian Frederik and the continuing union with Denmark. On May 21, 1813, Christian Frederik came to Norway as the new Viceroy after Frederik of Hessen, and in the course of the next seventeen months, Christian Frederik played the role, in quick succession, of Viceroy, Regent and King.

Christian Frederik tried to link himself to the patricians of trade, and proved to have some success. Besides Carsten Anker, who had become the owner of the Eidsvoll Ironworks, after having been director for many years of The Asiatic Company, and then later in Copenhagen became well-known to Christian Frederik, there was also Carsten Tank of Fredrikshald, who was one of the prince's most trusted men. In Copenhagen, King Frederik had begun to understand how desperate conditions had become in Norway, and at the end of April, 1813, he had given orders that ships and boats from the west coast of Jylland should set an "uninterrupted Course for Norway" to bring grain and goods to the suffering fraternal folk. A premium was offered to the six first ships that, despite the danger of the British ships, managed to reach Norway.

To help alleviate the lack of food this time, Pastor Abel had proposed neither horse meat nor moss; but rather, kale (*Brassica campestris*). He experimented in his own household with the use of this plant which grew freely everywhere, but which everyone considered to be a weed. "During the current Year the fields are bursting with Kale," Abel wrote in *Budstikken*, and gave detailed recipes to his readers: The kale plant's leaves, stems, tops and flowers should be chopped up finely and boiled at length, stirring them until they become a smooth mass. Thereupon one may add flour to one's taste; then, with more boiling and a steady churning, the gruel is ready. With the addition of a requisite amount of flour, he guaranteed that boiled kale had the customary taste of porridge, and only the green colour betrayed the substitution. He had tested this on both his wife and his children, he wrote, and as well, "the Parish's most Reasonable Men, and Tomorrow I am gathering together the Wives of the most Needy Farm Labourers to instruct them in the Preparation and Possibilities thereof." The recipe could also be extended to the making of flatbread. One had only to boil the kale longer until it became thicker still, then, once it was cold, knead in flour. The vicar urged everyone to pick supplies of kale, dry them in a shady place, and store them for future use; seedpods and stems could also be used.

As the Governing Commission had earlier wanted, Christian Frederik now approached England to obtain some alleviation of the impediments to the traffic in grains. He proposed several times that Carsten Anker should travel on a special mission to London but the king in Copenhagen refused. Instead, King Frederik VI gave the order to renew the war against Sweden, but Christian Frederik refused, arguing that they lacked supplies, money and equipment for the soldiers and "the Swedish Nation has been given new Power and increased Ability to defend itself from an Attack from our Side." Despite the fact that Denmark-Norway was officially at war again from September 3, 1813, peace and tolerance prevailed to a large extent all along the Norwegian-Swedish border. Christian Frederik, and with him, many others, understood that Norway's fate should be fought out on other battlefields. It is quite probable that Søren Georg Abel in Gjerstad also interpreted the news and signals in this way, and he hoped to be able to participate in the coming struggle.

13

S. G. Abel Steps into Politics

Stories were told in the surrounding communities about how the sons of Vicar Abel at Gjerstad, along with their like-minded comrades played cards outside the church on Sunday mornings while their father was performing the holy service inside. And if it had not been for the extra cold winter that year, perhaps the boys would have been sitting there playing cards on March 11, 1814, as well, when a national day of solemn prayer was to be held in the Gjerstad church. But perhaps the boys had other reasons to be at church on that day? At any rate, the vicar's sons would certainly have heard enough about what was about to happen to make them curious. Perhaps they followed their father to the church door and watched everybody going in with an air of worry and gravitas. Perhaps they also heard their father's imploring voice coming from within: "Look graciously down on our beloved Fatherland and give us leave to Unity, Power and Citizenship ever linked more firmly in defence of our Independence, Freedom and Honour."

Like all the other priests in the country, Søren Georg Abel had received a solemn call from Prince Christian Frederik to hold a national day of solemn prayer. On that day, March 11th, no unconfirmed persons or servant girls were to be admitted into the church, and females were urged to keep themselves in the background. An oath would be taken that day in church, and some male representatives should be chosen from among the civil, clerical and military officials, manufacturers and farmers over the age of twenty-five. Most people knew very little about this; not even the civil officials really knew what was going to happen. Christian Frederik himself had urged the bishops and priests to remain silent until the solemn moment in church.

In the Gjerstad church, Pastor Abel gave his congregation an historical and political orientation. He pointed to the hundreds of years that Denmark and Norway had been united under the same sceptre, when common benefits linked the peoples together like bands of sisterhood and the gentle angel of peace so often graced the united fatherland. But the year 1807, he said, had been redolent with the horrors of war. Denmark had become involved in an all-consuming strife, prosperity had fallen away, and above all, people were forced to endure hunger. But now the fortunes of war had also turned against our unhappy ally, Emperor Napoleon. A large portion of Germany had almost annihilated his power, and enemy armies were overrunning Danish territories. And Sweden, which had never ceased undermining Denmark and Norway's good fortune with intrigue and violence, now seized the chance and saw its extension into the territory of Norway as the only terms for peace. Pastor Abel also maintained that Frederik VI had been forced to surrender to these terms in order to save his remaining territories. Without any expression of sympathy for Sweden, Pastor Abel read out King Frederik's statement on the Kiel Peace Accord and his notification that Norway had been relinquished to Sweden, before regretting the actions of the unfortunate King Frederik whom he believed really did desire peace, but who had been robbed of peace by covetous neighbours. Pastor Abel presented the situation with the words: "In this way the Ties binding Denmark and Norway have been dissolved and we have been set free from our oath of Fealty to our former King. How dark would be the Forecast for Norway if in this Position we were left to ourselves. Consequently, we must neither submit with Serenity to the Swedish Yoke, nor watch our Fatherland split in Spirit, in Parties, and in Commotions. We must hold firm to the belief that Providence has willed us something better. It has happened that last year, by way of our own Leadership we have been sent the realm's Crown Prince, Christian Frederik. This noble Prince, who, up to this point, has manifested fatherly care towards us, and warm feelings towards Norway, has accepted to head a Government of our own, that takes charge of our Citizens and Wars. Order is maintained, the Laws are obeyed, and Old Norway's Independence shall be asserted and stand as firm as a Mountain Peak."

And in this manner the announcement from Prince Christian Frederik was made known: that Norway did not agree to give itself to Sweden, and that right here and now, immediately following the holy service, a selection was to be made for representatives to attend a Constituent Assembly to be held on April 10th at Eidsvoll. Norway would become an independent country! But, in order to be crowned with success this required a unified nation standing behind it, and therefore people were now invited to swear the oath of independence: "Do you swear to support and uphold Norway's Independence and risk Life and Blood for the beloved Fatherland?" And with hands uplifted, the congregation answered in a clear voice: "Yes, we so solemnly swear, so help us God and on His Holy Word!"

Abel was glowing with enthusiasm, contemplating what happened and what would yet unfold, and was naturally one of the two electors chosen from Gjerstad Parish. As an elector he would be among those who could be chosen as representatives from Nedenes County, to participate in the Constituent Assembly at Eidsvoll; he most certainly had an ardent desire to be chosen.

Among the electors with whom Søren Georg Abel would eventually have been competing was the priest, Hans Jacob Grøgaard. Theologically they took the same line. Grøgaard, like Abel, warmly supported the new psalm book and in general criticised "The out-of-date, Close-fisted Adherence to the Old." But it was Grøgaard who was chosen, along with Jacob Aall, who had unanimous support. The district surgeon, Alexander Christian Møller, chosen by the town of Arendal, was also sent to Eidsvoll, along with iron works-owner Henrik Carstensen from the town of Risør. Pastor Abel had to wait some months longer before he himself could seriously enter the grand game.

People around the Gjerstad vicarage scarcely knew about, nor discussed most of what happened at Eidsvoll between April 10 and May 17, 1814. People were preoccupied mostly with the fact that it had been an uncommonly slow spring replete with its many everyday problems. The national assembly at Eidsvoll had little impact on the everyday life of most. Otherwise, an average summer came to pass, with tolerable harvests.

Pastor Abel was probably more aware of events than most. He spoke enthusiastically during the spring and summer of 1814 about Norway having become an independent country again. That so many had participated in, and been given the right to vote by the Eidsvoll constitution, and could participate in government and administration, was a principle he had always championed, and which he had managed to bring into effect in his own district. Abel felt he had a covenant with the new times: *equality* in rights, and *social obligations* for adult men. This, he maintained, would secure the best for ordinary people and give the best possibilities for common wellbeing. He most certainly understood in July that relations of war between Norway and Sweden were unavoidable, but he also greeted as acceptable Karl Johan's offer of peace, as it became known through the Moss Convention of August 14th. The most important thing was that the Constitution was secure. Whether or not Christian Frederik really had to leave the country was still unclear. And when it did become clear that representatives would be chosed to an "Extraordinary Parliament," Pastor Abel put himself forward and was chosen the first representative for Nedenes County.

Father Abel left Gjerstad in a state of optimism, and he had the greatest faith that he would move with, and make his mark upon, the new Norway. The country would be firmly established on the basis of reason and freedom. He still had a happy faith in human perfection and in the people's ability to reach it.

Niels Henrik and Hans Mathias were twelve and fourteen years old when their father went to Christiania. They knew what their father stood for and where his sympathies lay. And the national events also led to changes in the everyday lives of the boys. Before their father left, he found a new tutor for his children, the best teacher in the district. It was this lad from Gjerstad, Lars Thorsen Vævestad, who in his turn had been confirmed and taught by Pastor Abel, and had been made the parish's itinerant teacher. Lars Thorsen was 24 years old when he became tutor at the vicarage and the boys naturally knew him from earlier interactions; but in recent years, Lars Thorsen had been in service to Mayor and War Commissioner, Henrik Georg Tønder of Kragerø. H. G. Tønder with his luxurious manor, Taatø, was extremely prosperous, and generous in helping the district's peasants. Søren Georg Abel had secured Tønder's support both for the reading society and the parish society. It was probably also at Abel's initiative that his most promising pupil, Lars Thorsen, "according to the royal Resolution" in 1811 was removed from his teacher training and sent to Tønder so that this prosperous gentleman could support and subsidize Thorsen's "studies." But this was during the difficult years of the war, and instead of studying, Thorsen was forced to take up a backlog of tasks at Taatø. In Tønder's luxurious house with its elegant ballroom there were many fine furnishings and works of art; the garden was full of rare plants and the greenhouse replete with strange blooms; a fishpond had been built and there was a game park with donkeys for guests to ride. In addition to all this was both a fruit orchard and berry garden with a multitude of apple, pear, cherry and plum trees; there were also peach, apricot and walnut trees, and grapevines, as well as innumerable blackcurrant and brambleberry bushes. Lars Thorsen's intellectual studies went no further, and Søren Georg Abel called him back to Gjerstad now that he himself would be away. Abel summed up the fate of Thorsen's study plans as follows: "The Times confounded the Conjunctures and several Circumstances were to Blame for the fact that he had to give up this Plan, no matter how ardently he wished that it be carried out."

As a teacher for Niels Henrik and the other children at the vicarage, Lars Thorsen did a superb job. He seems to have been well-liked by the children and now while he was tutor, he continued with his own studies as well. He read German, kept a diary in French, was interested in history, and studied the Bible. Father Abel found him so "competent in the Field of Teaching" that he allowed Thorsen to continue as tutor even after he himself returned home from his first period in Parliament.

Pastor Abel sent several letters home from Christiania to Gjerstad. He had gone to the capital some days before the Parliament opened on October 10th, and was astonished by the atmosphere in the city. Everyone was talking about the union with Sweden as a necessity; the question was only about the conditions for such a union. Abel's old rector, Professor Niels Treschow, was one of those who most plainly doubted the wisdom of the spring's mood of independence, and who now maintained that the Parliament must undergo a rapid period of reflection, approve the union, and undertake the selection of a king before all other matters were debated. Count Wedel held the same view. He had reason to remind people that things had gone just as he had predicted, and now he could point back to the stigma of treachery and the derisive words he had been forced to tolerate at Eidsvoll. Many now thought that Count Wedel would become the foremost man in the country. However, there were others who thought that Ponte Corvo, as Karl Johan was known, would be a congenial and accommodating head of state. But rather it was Wilhelm Frimann Koren Christie, who at Eidsvoll had been Secretary of the National Assembly, who now became the new "extraordinary" Parliament's first President and the dominant character in the debates.

The most intense discussion occurred around the decisive question of whether or not Parliament should *first* select the king and *then* change the constitution, or should there first be a contract between the king and the people. Formulated differently, should the people themselves ratify the constitution, or should it be a contract between the king and the people?

The realities of this question were debated for many days, and in the debate Søren Georg Abel made a crucial contribution against his old rector and in support of the line advanced by Christie. In a deep, serious voice that some called sepulchral (others said that Pastor Abel had "a hollow Voice, as though he spoke from the bottom of a Barrel"), Søren Georg Abel impressed upon his listeners in the Parliamentary chamber, the following argument. "We are, God be praised, still a free People, and as such, we ought by all means to act..." Sweden, Abel pointed out, had no right to expect that Norway should compromise its basic principles for a possible union: "it is we, in this country, who should submit the terms by which free Norwegians decide to call the Swedes their brothers!" He would not hear of any supreme Swedish rule, even though the Kiel Peace Accord had agreed to this, and the Moss Convention seemed to support this understanding. "Therefore, with calm and mature reflection befitting a free people, we should decide the basic points and conditions that the union between Sweden and Norway should be based upon, if the men of Norway find this course of benefit for the country." Abel was impressive with his well-formulated, almost mathematical precision: "When, in these Determinations we have taken suitable Consideration of our national Honour, our Freedom, and our Rights of Citizenship; when, on the other hand, we have seen to it that whatever type of Oppression possible, by any Means, becomes impossible under whatever Regent; then, let us be the First to stretch out to the Swedish People a sincere hand of Brotherhood; then, allow us as a free Nation, to offer Carl the Thirteenth the Sceptre, which until now has not become his Lot to hold!" In conclusion, Abel argued that if this constitution, which nobody was more entitled to sanction than the people who should be bound by it, if this constitution were to be abrogated by an unmistakenly tyrannical Regent, yes, then, would Norway's full power be preserved, and with "...this, we could triumph, or facing this, we could die, and in both Eventualities we would have the capacity to reclaim our Honour."

After a series of day-long meetings and negotiations, Parliamentary President Christie formulated three questions that were presented for the vote, and a decision was to be taken by October 21, for then Karl Johan was within his right to end the cease fire, and war could break out again with only one week's notice. The first of Christie's questions was that on the occasion of Christian Frederik's abdication, was it necessary to choose a new king, and this was rapidly decided by a voice vote, the outcome being yes. The next two questions were as follows: Should Norway as a country, the terms of whose independence are recognized, unite with Sweden under one king? Should that king now quickly be chosen?

At this point speeches were made with a use of language never again heard in the history of the Norwegian Parliament. Perhaps the most lengthy was given by Abel's fellow representative from Nedenes County, Pastor Reiersen, who declared in cascading waterfalls of words, that if the Swedish King would not give the people their freedom, swords would hiss out of scabbards, and with a death-defying rage of despair, the people would liberate their violated honour and the shipwreck of freedom. At that moment, the despairing wail of the widow and the fatherless child will be the victory song at the funeral pyre of the fallen warrior! And the death rattle would blend with the sobbing voices of the mourners, and like a rasping sigh from the abyss, a cry of woe should fall over Sweden's King, a cry of woe over his lust for conquest!

Dean Niels Hertzberg declared that the Norwegian who, in the event that the Constitution was violated, did not fire his crops, grab his weapons and go on a berserk rampage of yore, should be hanged from the nearest tree. But the majority of the independence men were more sober, and when the vote was taken, only five voted against union with Sweden. Treschow and Count Wedel now wanted a quick vote on the king, but, by a vote of 47 to 30, the Parliament approved Christie's position: the vote on the king would be stood over until the Constitution had been revised and approved.

The Swedes were very happy with the approval of the union, and even though Karl Johan expressed his disappointment about the vote on the king as head of state, he began to send his troops home.

At this time, Father Abel wrote the following in a letter home: "Yesterday, October 22nd, Parliament convened at 9 o'clock and remained in session without a break until 6 o'clock in the Evening. The grand Question was the following: whether or not Norway, as an independent Country, should unite with Sweden, and whether or not this Union would be considered beneficial to Norway. These questions were debated vehemently. 17 speeches were made and finally it was decided with a vote of 74 in favour, and 5 against. The Union was considered beneficial in the customary Terms. It now follows of itself that the Swedish King must be voted on, but the Conditions are now such that people are gradually coming around to the position of being in favour of it. Thank the Lord we have come so Far already, for the Country could not endure the renewal of War. I Believe that the Parliamentary session will go very well. I shall soon be recounting several Intelligences. May whoever count themselves Christians and good Citizens remain Serene."

On October 24th, Parliament began the work of revising the Constitution of May 17th. This became an intense tug-of-war between Parliament, led by Christie, and the Swedish delegation which several times threatened to break off the proceedings. But with the help of supple and ambiguous formulations, Christie managed to present the issues and set up motions upon which to vote, and to do so in a way that gave the politically unavoidable a veneer of voluntarism, a factor important to the electoral process, and it was at the same time a format that the Swedish negotiators were easily able to overlook.

The most difficult paragraphs dealt with the king's control of military power and whether or not he had the power of veto over the question of naturalising foreigners. Pastor Abel made an outstanding contribution to this latter question. It was commonly felt that the naturalisation of foreigners would have considerable implications for the recruiting of future Norwegian civil servants, and although the Swedish Commissioners had withdrawn from discussing the admittance of Swedes to civil service posts in Norway, they argued that the king should have a veto in such matters, and it was indicated that their demand would have the character of an ultimatum. In a straw vote there were 39 for and 39 against. But the day before the final vote, Pastor Abel gathered together five farmer-Parliamentarians and plied them with strong words and strong punch to get them to vote against the right of veto. The following day the demand for the king's veto fell by a vote of 43 to 34!

Pastor Abel's punch squad was criticized from many quarters, and some were almost saying it was a scandal, but many were very happy with the result. The vote was accepted by the Swedish negotiators only with the gravest of misgivings. But time was short and the same day, according the the agenda, the vote on the king was to be held, and so as not to postpone this delicate matter, no more was said about the earlier issue. In the afternoon of November 4th, the big vote took place. The Union was finally in effect, with a majority *voting* Karl XIII as Norway's king, and a minority declaring that they *recognized* him as king.

Søren Georg Abel went home to Gjerstad at the beginning of December. By
then he had also met Karl Johan and his son, Prince Oscar. On November 9th,
the royal family conducted their ceremonial procession into Christiania. In
the dark afternoon, when the air was heavy with rain and sleet, and accompanied by a brass band and mounted cavalry bearing flaming torches, they had
come down into Christiania from Ekeberg, a promontory from which one can
view the whole city. Cannon fire was detonated: nine rounds when the party
set off from Ekeberg, nine rounds when Karl Johan entered the square, and
nine rounds when he arrived at the Paléet. Here the grand event was celebrated
by the Parliamentarians and invited guests: a splendid banquet table with
speeches and toasts, followed by promenading in the illuminated palace
gardens. Karl Johan charmed them all. The whole population was on its feet
and they said the hordes of people made lively throngs in the illuminated
streets. Others said that the procession from Ekeberg ressembled more a
funeral cortège than a splendid ceremonial entry, and that a kind of scream
could be heard when the Swedish carriages rolled through the streets. In any
case, grain was given out to the poor, and money given to the marching
soldiers. Ponte Corvo was perhaps not quite as bad as some would have it. On
November 20, he and Prince Oscar went out with the fire brigade to extinguish
a fire in the capital, and stories were told that Karl Johan had with his own
hands taken hold of the fire hose and directed the water where the flames were
most dangerous.

During the whole time, Parliamentarian Abel, despite his vote for the
Union, had remained a patriotic independence man. The episode with the
punch had done him no harm, rather the opposite. At the end of the parliamentary session he had the honour of being chosen as part of a seven-man
grand deputation which, with Christie as its spokesman, would be travelling
to Stockholm to present Karl XIII a formal address to mark the occasion of
the Union between the two countries, and Norway's ratification of the king.
They also took with them a copy of the Constitution signed by each and every
parliamentarian. This was an undertaking of great prestige, and it was with
great disappointment that Abel had been forced to send his regrets. After
having slipped and fallen against the raised edge of the pavement "in one of
the City's ill-lit Streets" he had injured his foot. "I neglected to Take Action
and the Wound grew worse," he reported.

However, he managed to get back home to Gjerstad. Father Abel took a
broad view of things now. He would probably not have known much about
the sense of resignation that had spread out across the country in the fall of
1814. He kept his faith in the ability of the individual human being to use
reason and insight to do everything for the better. Great things could be
achieved if human needs and human reason were in harmony with the reason

that governed the world, what Pastor Abel called God's foresight and meaning. Perhaps his children, exposed as they were to his interpretations, learned that both inspiration and pathos were natural and self-evident forms of expression.

On December 20, 1814, yet another son was born to Abel at Gjerstad vicarage. The birth went well, but a week later Anne Marie began to haemorrhage. A message about her condition was sent by horseback to the doctor from Kragerø, Dr. Homann, who at that moment happened to be on a series of rounds in Risør. At the beginning of January, Abel wrote to his dean in Arendal, "Heaven be Praised! My wife is out of Danger."

14.

The Times during which Niels Henrik Grew Up

There was an expression that was common at the time when Niels Henrik was growing up: "The children of the Priest amounted to very Little, the Bishop's, to absolutely nothing."

Most people thought that this was so because theological students so often found a wife in Copenhagen and that such women did not adapt well to Norway and its customs, particularly in the countryside. Yet they did their best to cope with the household and children. This interpretation was reflected as well in the more or less official view of the situation. In the struggle through the years to establish a Norwegian university there was a constantly recurring argument which concluded that young Norwegian students ought to be saved from the many big city temptations of Copenhagen. And not the least of the considerations were the erotic temptations that impressionable youths carried around in their minds.

When Niels Henrik was growing up in Gjerstad, the first generation of Abels had set a standard for the village. In the eyes of the common people, Niels Henrik's grandparents stood as an example of what the good life, devoted to the service of people and God, entailed - an ideal that embodied both human goodness and the security of Christianity.

"The unforgettable Hans Mathias Abel!" was the way he was referred to in people's conversations. The sexton had named his own son after the old priest. The sense of confidence and compassion that Elisabeth Normand Abel always seemed to have is something Niels Henrik would also have experienced. She had always been obligingly everywhere and involved in everything. She taught village people to grow flax and produce linen, and Niels Henrik on many occasions saw farm wives in Gjerstad curtsey in front of the flax fields in honour of Grandmother Abel. Niels Henrik's grandparents became a yardstick for measuring what an Abel was and should be, an ideal that descendants would find impossible to live up to. The card playing in front of the church by the vicar's children was a most unambiguous example of this.

Niels Henrik's father was solemnly preoccupied with the wellbeing of the common people, and in his vocation, Søren Georg Abel mobilized all the knowledge and education at his disposal for achieving a better life for the greatest number possible, both here in this world and in the Hereafter. He proclaimed and impressed upon people a set of rules by which to live, and patterns of thinking that could create quite strong expectations about the possibility of the world becoming a good place to be, provided only that one acted and thought correctly. From the standpoint of Niels Henrik and his siblings, this must have functioned as though all life's questions were answered by, and within the scope of their father's encyclopedic explanations.

Niels Henrik had a mother who often seemed absent, and her missing sense of contentment must have been something that, from an early age in the eyes of her son, did not square with the absolute certainty of his father's systematic worldview. She was distant and unsympathetic, and possibly created a pattern of reaction in her son which later often found him rigidly on guard against possible rejection. But his grandmother lived at Lunde, a couple of kilometres from the vicarage, and she most certainly, and often, took her grandchildren home with her, and gave them the love and care that they seemed to lack in everyday life. Once in a while Grandmother Abel travelled to Risør as well, where her only daughter lived. This daughter, Margarethe Marine, was a spinster and worked as a seamstress. Perhaps Niels Henrik went along sometimes to Risør; in any case, he had good enough relations with his father's sister that he wrote her letters later while he was a disciple at the Cathedral School in Christiania.

Niels Henrik and his brothers seem to have been rather lively boys, "pure Children of Nature," as their father called them, the sort of boisterous youngsters who are capable of ferment in the most close-knit of families. It was said that Niels Henrik was very plucky and he distinguished himself at an early age with his physical prowess: he swam faster than the others, a veritable eel in the water. In winter he succeeded in skiing down slopes so steep others did not dare try them. Was this particular behaviour an expression of a child's excessive energy, combined with an inability to visualize the dangers that people normally avoid, or could the reason have been quite the reverse: a great appreciation of all threats and menaces, that challenging these dangers and extending the boundaries, simply became essential in order that he could find himself a certain personal breathing space?

Gjerstad is an inland community with moorland, forests and lakes, that toward the north and east is bounded by rather inaccessible, rough and steep hills and crags. Nevertheless what distinguises the landscape, particularly around the church and vicarage, are the rounded fields sloping down to their boundary with the long expanse of Gjerstad Lake, a body of water that lies in the landscape like a ladle.

It was said that back in the very old days, *Nøkken*, a troll figure, lived in Gjerstad Lake, and that he had come there after having been imprisoned within the Sundsfossen waterfall at the end of the lake. While imprisoned in the waterfall, *Nøkken* was said to have demanded one human life a year, through drowning, and would achieve this by upsetting a pram being rowed near the waterfall on the way between Gjerstad and Egeland Works. But a theological graduate had arrived one day long before the era of Niels Henrik's grandfather. He was called "the Green Student." The Green Student had taken four fellows with him and they rowed with all their might against the current and into the waterfall, and there the graduate had grabbed hold of, and hauled out of the water, a black creature, a bit like a small, wet dog. Without uttering a single word the student had bidden the men to row back across Gjerstad Lake while he held this creature fast between his knees. Offshore from the glacial rockfall at Tvetsuren, *Nøkken* had been consigned to the deep and anchored fast. Since that time there had been no boating accidents around the waterfall, however, sometimes people had been drowned offshore, opposite the rocks deposited by receding glaciers, and many reported hearing laments coming from there, as though from people in peril.

Since Gjerstad Lake was so dominant in the landscape, *Nøkken* remained a clear preoccupation in the folk culture. However, in Pastor Abel's systematic litany of how things were articulated with one another, and how one should comport oneself, such folk beliefs were considered simply to be superstitions. In the context of the open opposition between Søren Georg Abel's rock-solid faith and this vague folk belief, the landscape itself must have encouraged the people to talk about the supernatural. Perhaps the great robust swimmer, Niels Henrik, had been dared by the other boys to swim the waters offshore from the Tvetsuren rocks?

There was a slope near the vicarage at Gjerstad called Ronne-Hellen, where of yore people laid out deformed newly-born infants so that they might be exchanged for the healthy children of the trolls. The trolls had many healthy human children because they seized human infants as soon as parents left them unattended in the fields or meadows, and in their place, they left their own troll children. But at Ronne-Hellen, people tried to lure the trolls back again. And the trolls, not being exceedingly bright, willingly made the exchange on that slope, and took the deformed child in return. People said that even though the trolls were rather dim, this had to be done at night, around midnight. But were there people who still did this?

There was also talk of *hulder* (small, attractive but sway-backed young women with tails) and small grey gnomes. Bommen Farm, or Solbakken, was called Trollkjerringbakken (Hill of the Troll Hags) because it was here that the old magical women were said to assemble before they rode their broomsticks to Blokksberg, the realm of the witches. On another farm, Bortigarden, everybody possessed unusual strength. Another farm, Hallen, was known to have the most abnormally green hills in the region. And Jens Beintsen from yet another farm down by the lake, was found one day on top of the church spire, spinning round and round, horizontally with the church spire acting as a fulcrum against his navel.

As well, the boys from Tangen Farm, which was also down beside Gjerstad Lake, were particularly good shots. The boys from this farm had shooting contests using lighted candles, the aim being not to put out the flame, but merely to graze it. The Tangen boys were also excellent skaters, particularly one Ola Tallaksen, who was the same age as Niels Henrik. Ola competed on skates with the fastest horses, and when such a horse was galloping at top speed, would place his hand on its back, and vault over it. Once he was not so lucky and came to sheer off most of the horse's mane with his skate blades.

Besides this, people also told a story that had happened some forty years earlier about the vagabond called Jonas Kruse. Every summer, Jonas, together with his wife, Ingeborg, were allowed to live in a little cabin in the woods on the way to Vegårshei, a cabin the farmers had built to use in winter when they were cutting wood far from home. And this Jonas had the ability to shift his identity! He could transform himself into either a werewolf or a brown bear. One day when he and Ingeborg were peacefully cutting hay around the cabin, Jonas could feel a transformation coming on, brewing up in his body. He tossed his wife on to the top of the haystack, handed her a stout staff, and told her to defend herself the best she could. Jonas himself disappeared into the forest but returned a short time later as a huge bear which savagely attacked Ingeborg and threatened her life. Only by the skin of her teeth was she able to defend her position atop the haystack, and she had grasped her staff with such determination that blood ran out from beneath her fingernails. In the end, the bear abandoned her, and a short time later, Jonas, once more in human form, came shuffling out of the woods. But a man from the village, who had happened to come by immediately afterwards, noticed that Jonas had threads of fabric stuck between his teeth. That place in the woods on the way to Vegårshei, even when the cabin sat empty, was thereafter known as Krusen-hus (the Kruse House).

What had happened to the seven year-old girl, Anne, at Egeland Works, must also have left a harrowing impression in the district. In this story, the person most discussed was the girl's father, Christian Amundsen, who in his youth had been confirmed by Hans Mathias Abel in the Gjerstad church. This Christian, known as Long Kristian, was arrested in the spring of 1810, and charged with murdering his own daughter. Many people showed up, and even small children were lifted onto the shoulders of adults so that they could see the man who was being conducted to the place of execution. "O-ho," he was reported to have said just before his head was hacked from his body and stuck on a long stake, "a great crowd has turned out for my funeral today!" The body was buried in the churchyard without the normal ceremony, and when that grisly head with its long black hair and beard had become a dry, empty skull, and had evidently been paraded around Egeland Works, it was immured in the stone wall of the roadway. But what had happened, and why? In the stories that people told, Kristian was a widower who wanted very much to remarry, but his intended was said to have insisted that he first get rid of the child of his previous marriage. So one day, probably a Sunday, when he was collecting firewood in the forest, he took his daughter along. And on a cliff above the western shore of a small pond, he beat her to death and threw the remains in the water. Later, when questions had been raised about the where-abouts of little Anne, her father fell under suspicion, and after long interrogation, he confessed. The descriptions of the girl when she was finally found decomposing in the water, were grusome: her lips, apparently, were un-touched, but her head had been bloodied from a deep hole on the right side of the cranium, a powerful blow, probably with a rock, had smashed her nose between her eyes, and there were wounds on her chin and bruises on her cheek below the left eye.

15

Niels Henrik Leaves Home

Abel the father had evidently had an exciting time in Christiania, and distinguished himself during the autumn session of the Extraordinary Parliament of 1814. It is quite certain that he also spoke with Rector Jacob Rosted about places for his sons at the Christiania Cathedral School when the time was appropriate. For the eldest son, that would most likely have meant the autumn of 1815, and quite probably Niels Henrik hoped that he too would soon follow, probably very pleased with the idea of getting away from the well-known landscape - and away from his discontented mother. Perhaps one day he would be able to do things better than his parents had?

There were probably several reasons Pastor Abel wanted to have his sons enter the Christiania Cathedral School. First of all, as a parliamentarian for a period of two or three months he had become familiar with the school premises: the school's assembly hall had functioned as the first Parliament. He had come to know the teachers at the school, and certainly knew that there was a stipend from which his sons could well benefit, since his own means were insufficient to pay for the schooling. And perhaps the most important point of all, Pastor Abel had been swept up in the great political awakening that had washed across the country, and he thought that the capital and the assembly hall at the Cathedral School would be his own main arena in the coming years. The drive and enthusiasm that earlier had resulted in his Reading Society and Parish Society activities, and the establishment of regulations and new initiatives, seemed no longer to have been directed toward the congregation and local people in the same way. Parish Vicar Abel wanted to take up other tasks, with a wider field of activity and a greater public.

It is likely that he had also met his relatives during the autumn of 1814 in Christiania, the brothers, Jonas Anton Hielm and Hans Abel Hielm, who had at that point just returned from Copenhagen. Their mother was called Hulleborg Abel, and she had been the grandchild of Jørgen Henrik Abel, the Dean of Bygland, under whom Hans Mathias, in his day, had worked. The Brothers Hielm were also keenly preoccupied by the welfare of the new Norway, and they quickly got to work in public affairs, the former, as a high court lawyer, and the latter as a bookseller, and later, acting publisher of the newspaper, *Det Norske Nationalblad* [The Norwegian National News], an organ that would come to represent the new Norway in all ways.

Under these conditions, Pastor Abel had neither the time nor the desire to teach his sons himself in Gjerstad. The boys had to go to the Cathedral School. Lars Thorsen, who was continuing as private tutor at the vicarage until further notice, lacked the necessary education, first of all in Latin, required to prepare the boys for the *examen artium*.

To Søren Georg Abel the following were important facts: On November 25, 1814, he himself had been part of the deliberations which decided that the first official parliamentary session would be convened at the beginning of the month of July, 1815. Abel also had good reasons to believe that he himself would once again be selected to represent his district, Nedenes County. He had demonstrated his talents and ability to work in the nation's best interests. With exaggerated modesty and humility he wrote to Dean Krog in Arendal on January 8, 1815: "I am not really opposed to participating in one more Parliamentary Session. Still, I am rather insignificant."

First there was the congregation vote, and there, once more, it was quite certain that the vicar would become the representative to the District Assembly that in its next session would choose the county's Members of Parliament. Abel was one of the nineteen representatives who met on January 12, 1815, to select the two parliamentarians for Nedenes County. Five of these nineteen stood as candidates. The two with the highest number of votes would be sent to the Parliament, and the next two would be reserve candidates. According to the customs of the day, to vote for oneself was considered a form of election fraud. In fact, no one ought to be on record as having judged his own worth. Besides S. G. Abel, the two Eidsvoll representatives who stood for election were Jacob Aall and Hans Jacob Grøgaard. The farmer, Christen Andersen Neersten was also a candidate. Neersten was sheriff of Øyestad, a man of about sixty, known for his amazing strength. It was said that once on the Neersten wharf he carried ashore three one-cask sacks of rye in one load, one sack under each arm, and one in his teeth, and that this image of him had frightened away several potential thieves. Sheriff Neersten had also been one of the main people behind the arrest in 1787 of Christian Lofthus, the Nedenes peasant leader, and he had been presented with a beautiful silver pot in appreciation for this deed. Neersten got the most votes in the election in January, 1815. Thirteen of the representatives wanted Neersten as their Member of Parliament, and Jacob Aall received ten votes. Only eight voted for Vicar Abel. Thus the battle was lost; there would be no trip to Christiania for the first reserve Member of Parliament. Grøgaard received only two votes.

Fig. 4. Gjerstad church and vicarage drawn by Vicar John Aas, 1826. Aas took over the Gjerstad parish after Søren Georg Abel, and became an important contact person for Niels Henrik Abel. John Aas came from Røros and was a collateral descendant (at the sixth remove) of the peasant, Hans Olsen Aasen, who had been the first to discover the copper ore for which the Norwegian town of Røros was long famous. In 1813, Johan Aas was in the first cohort of students to take the *examen artium* at the new Norwegian university. Five years later he took his theological public service examination with distinction. In 1820 he arrived in Gjerstad and lived out his life there - in all, 47 years: 20 as dean of the Nedenes pastoral district. He was chairman of the town council several times, and Member of Parliament four times. Collection of Anders Mo, Gjerstad. Photo: Dannevig, Arendal.

However, Pastor Abel held firm to the plans for his sons' education, even though he complained about his insufficient income in this time of inflation, saying that he himself had to "live like a Peasant in order just to survive." He complained to the dean about his health, but what was worse was that Anne Marie, following the birth of her last child, remained supine and exhausted, and more and more frequently S. G. Abel referred to his wife's ill-health as "her Indisposition" and "her Troubles."

But in July, 1815, a state of joy descended over Gjerstad vicarage as Peder Mandrup Tuxen and his Elisabeth came from Copenhagen for a visit. At the beginning of August, when Tuxen was about to embark on a ten-day visit to Christiania, Pastor Abel decided to accompany him. And on this occasion

Fig. 5 a, b. Paintings in Gjerstad church that portray Parish Vicar Hans Mathias Abel and his wife, Elisabeth Knuth Abel, née Normand - Niels Henrik's paternal grandparents. These unsigned portraits were painted in 1788, three years after the Abel family came to the community. Hans Mathias Abel died when Niels Henrik was two; his paternal grandmother lived until he was fifteen, and seems that he loved her dearly, as indeed, did all the servants and village folk. Manuscript Collections, National Library of Norway, Oslo Division. Photos:Dannevig, Arendal.

Pastor Abel seems to have enjoyed himself in the capital city. He had dinner with the Swedish governor, von Essen, who, Abel commented, "is popular with all, and to my Thinking, a particularly worthy Man." He paid his respects to his longtime acquaintance, Niels Treschow, and commented that, "it is a shame to see one of Europe's greatest Philosophers dressed in a State uniform and sporting Spurs on his Heels." Abel also visited Parliament which was now in session, and reported, "even now there is Much that has not been Straightened out." Later on, in trying to deal with the contrasts and differences between city and countryside, this Parliament decided to lift all restrictions on the distilling of alcohol, an action that later would have great consequences in Søren Georg Abel's life.

And while he was in Christiania, Pastor Abel also arranged for room and board for his son Hans Mathias in the house of a merchant near the school.

At the end of September it was time for Hans Mathias to ready himself for the move from the farm to schooldays in the capital. At the vicarage of

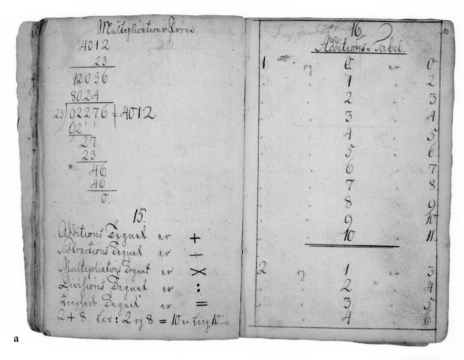

a

Fig. 6 a, b.

b

a Pages from the teaching manual that Søren Georg Abel wrote and used for the first instruction of his children. The first line in the first addition table is thus: 1 + 0 = 0. Manuscript Collections, National Library of Norway, Oslo Division.

b Silhouettes were the "family photos" of the day, and it is quite certain that these of Niels Henrik's father and mother - Søren Georg Abel and Anne Marie Simonsen Abel - were made by family friend P. M. Tuxen, who later married Anne Marie's sister, Elisabeth Simonsen. Now in a private collection, these silhouettes were probably found at Peter Collet's hunting property, Markerud, at Nittedal.

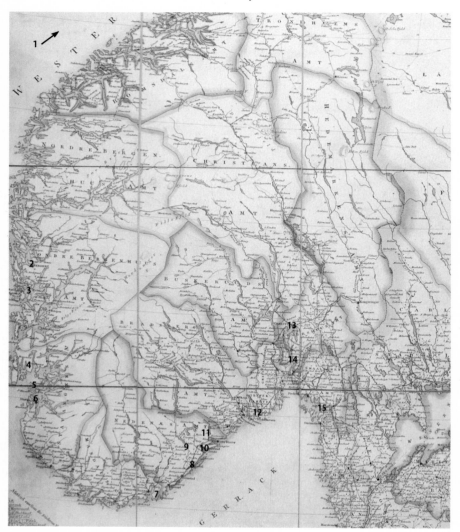

Fig. 7. Map from the 1820s. The ancestors of Mathias Abel came from Abild in Schleswig during boom times, to Trondheim during the 1640s. The next two generations of Abels lived on Norway's west coast (Vestlandet), dealing in timber, iron-smelting, and shipping before they took the serious decision to move to southern and eastern Norway in search of wealth and prosperity. *1* Trondheim (north, in Trondheim's Amt), *2* Bergen, *3* Tysnes, *4* Nedstrand, *5* Finnøy, *6* Stavanger, *7* Kristiansand, *8* Arendal, *9* Froland, *10* Risør, *11* Gjerstad, *12* Fredriksvern (Stavern), *13* Christiania (Oslo), *14* Son, *15* Fredrikshald (Halden) on the Norwegian-Swedish border. [Identif. of Original Map: Map Collections, National Library of Norway, Oslo Division, Nr. 1571]

Gjerstad the grain had been harvested and stored away, and when the potato crop had been ready to take in, the priest urged fifty young people to attend "a little country Celebration," and in this manner they took in two hundred casks of potatoes in one day. But shortly before Hans Mathias was to have journeyed to Christiania he fell ill. His father referred to it as "the Drooping sickness," and he was ailing for the whole month of October, and as well, because he was so "completely withdrawn and quiet," that Abel the father did not dare send his eldest son away from the farm. Instead, he sent his 13 year-old son, Niels Henrik.

On October 31st, Niels Henrik sailed from Risør, and his father wrote, "May God protect him! but it is without Anxiety that I send him out into this depraved World." To his dean in Arendal, Abel continued the account, telling him that he had heard Niels Henrik had arrived safely and was installed at the school in the capital: "God watch over him. I can do no more, but I shall pay with the Shirt off my back."

But what was it that Niels Henrik had so abruptly and unexpectedly left behind? His big brother, Hans Mathias, soon recovered. The "drooping sickness" was a local expression for an undefined condition, probably closest to a depression. But their mother was sick. Anne Marie could barely move due to pain in her legs, and Abel the father complained to the dean: "Everything is going wrong. Does even God know how our luck stands?" A horse for which he had been promised 3,500 *riksdaler*, died after two days of sickness, and eleven piglets, for each of which he could have realized 60 *riksdaler*, died of convulsions in the course of one night. And: "...my wife's ill-health continues, and I am living a truly sad Life."

PART IV

Disciple in Christiania, 1815–1821

Fig. 8. Christiania, showing the harbour of Bjørvika, painted by an unknown artist before the fire of 1819 that razed the lots on which timber was stored for shipping. *Paleet* with its garden and seaside pavilion is on the far right-hand side. [City of Oslo Museum]

16.

Meeting A New Era

Niels Henrik arrived in Christiania and became a disciple at the cathedral school precisely at the beginning of a new era for the city, the learned school and the country. From being a regional administrative centre, Christiania had become Norway's new capital city with a university and a national assembly.

Five days after he left Risør, Niels Henrik found himself ensconced in life in the capital. Christiania was encircled by several suburbs, villages and abundant fields. The city itself, King Christian's "quadrature" back behind the Akershus Fortress, consisted of no more than about 400 houses of one and two storeys. The streets had names, and most of them were paved, either with cobblestones, timbers or the bottoms of bottles, although here and there they consisted of hard-packed earth and solid rock. Most of the streets had pedestrian pavements of flagstones and cobblestones interrupted by deep gutters in which floated sewage of both the human and animal varieties. In the sparse light of evening the risk of stumbling into these gutters was considerable. This was precisely how Abel the father had injured his foot and been prevented from participating in the esteemed deputation to Stockholm. One found the occasional garden between houses, and on most street corners there was a water pump protected with a wooden box. Even though the streets had names, addresses were not yet in use. However, this did not hinder Niels Henrik from finding the house of the merchant with whom his father had arranged lodgings. Niels Henrik was well satisfied with what he called his "Quarters," no doubt only a little room with bed, table, stool, and perhaps a window overlooking the stable yard where horses and wagons came and went with packing crates and various goods.

In order to be admitted as a disciple at the Christiania Cathedral School one had to be at least ten years old, demonstrate that one had some insight into history and geography, and "could read Danish and Latin Scripts with Proficiency," and that with regard to mathematics one was "acquainted with the four Species." Niels Henrik must have fulfilled these demands brilliantly, and been pleased with himself about that state of affairs. Some days after his arrival in Christiania and his admittance to the school, he reported home to Gjerstad that he had found himself "right in his Element."

The Cathedral School property lay at the corner of Dronningensgate [Queen Street] and Tollbodgaten [Toll House Street]. The establishment was old, having been inaugurated by the king in 1719, but rebuilt in 1800. There was now a separate room for each class and the school's large auditorium was accordingly the city's most prestigious meeting hall. The first regular Parliament had been sitting since July, and was continuing, and, as during the previous year, the people's parliamentary representatives occupied the school premises. For this reason the teachers and pupils were moved out to Merchant Henriksen's property on Tollbodgaten, Thomsegården, as it was called. The new school year had begun on October 1st. Niels Henrik was probably the last of the eleven new boys to be placed in the first class that year. The total number of pupils was about eighty.

A large ballroom at Thomsegård had been made over into classrooms with the help of plank walls. But it was noisy, such that everyone could simultaneously hear all the other classes. The teachers' shouts and reprimands rang out from all quarters, and for a newcomer it must have been difficult to know which voice he should listen to. This most certainly resulted in dereliction, disorder and noise. The teachers complained about "this unpleasant place for the School" that they now were forced to put up with during the time the Parliament was in session. The classrooms were too small, they said, and the atmosphere was "stinking and dirty."

The older boys complained most about how at Thomsegård it was so difficult to "christen" the new pupils properly. This baptism and christening at the water pump, which in normal conditions was a usual ordeal for new pupils, now had to be conducted with buckets of water. This water baptism was a one-time phenomenon, but it was understood that by virtue of being the youngest in the schoolyard, be he prince or pauper, meant that the newest pupil had no contact with the older disciples other than by means of earpullings, kicks or demands that he run one or another errand, such as being sent to the bakery to purchase buttercakes.

The manner of dress in the capital city was unlike anywhere else in the country. Boys from the countryside stood out in their homespun clothing, and were often made fun of, and given nicknames. Niels Henrik's first period at school was surely, at least on the surface, like the lives of all the other disciples: he wanted new clothes and ardently desired to fit into the environment.

Christiania had changed from a city of merchant patricians to the main seat of public officials. It was also the centre of education, the brightest radiance of which emanated from the new university. But in the same way that the Parliament had seized the premises of the Cathedral School, the university had taken all the best teachers. Søren Rasmussen had been a brilliant head teacher in mathematics; he taught physics and had been a central man in the leadership of the School. But now he was Professor of Theoretical Mathematics. Ludvig Stoud Platou had been responsible for the teaching of history and geography at the Cathedral School before being appointed Professor of History and Statistics. In 1814 as well, Stener Johannes Stenersen, who had been a sessional teacher, was employed as "Lektor" of Theology at the University, and the natural sciences teacher, Martin Richard Flor, "the Moss Priest," became university lecturer in botany, but continued simultaneously as the school's teacher of the natural sciences.

The backbone of the Cathedral School was Rector Jacob Rosted and his new teaching staff who had to come to terms with the fact that the school no longer had the same standing as when it had been the cultural centre of the city. In the ensuing years, during Niels Henrik's schooldays, the teaching staff was noticeable for its discontent and for being lured away by other institutions.

Even though Niels Henrik began his schooldays in Christiania in premises not unlike those experienced by his father and his father's father, the school's ambitions were quite different. In its teaching resources and organisation the Christiania Cathedral School had undergone comprehensive reforms. New teaching methods had been instituted, at least in theory and principle. The direction of the School had been separated from the Church. The old class-room teacher system, where the "invigilators," as well as the vice-principals and the principal/rector, each taught his class all subjects, had been replaced by a system wherein teachers had been trained in, and now taught, one specialized subject. A more humane way of treating the disciples and a stronger emphasis on the school's educative activity had come to the fore. Where before corporal punishment, for the most part, was used to enforce industriousness and inculcate knowledge, now this was to be accomplished through appeals to the pupils's sense of honour and reason. Emphasis was being placed on the awakening of interest. Only the gravest of offences were to be punished physically. Perspicuity was demanded of the teaching; black-boards came into more frequent use; and the perpetual rote-learning of the past had been replaced by a gradual and easily-understood progressive form of teaching suited to the pupil's age and ability. *Speaking* Latin, for example, was not required until the final year.

On paper at least, the new regulations required also that living languages and the sciences be included in the range of subjects taught. The permanent teachers, each with responsibility for his own subject, were appointed as public servants and their salaries were increased. A specialized teacher was required to teach only one or two subjects, and he should follow his pupils right through to their final examination. The specialized teacher with the highest responsibility for his subject was called the head or senior teacher in that subject, and sometimes he had an assistant, who was called an adjunct.

But compromises quickly snuck in; for example, the teaching of modern foreign languages suffered from one and the same teacher having to teach all subjects, that is, German, French and English. And these language teachers were not permanent employees like the head teachers and adjuncts; they were sessional appointees. Nor was there teacher training for the sciences, and there were no textbooks. Instruction was totally reliant upon what the teacher found useful, often knowledge that he had managed to acquire more or less on his own. When Rasmussen left to teach at the University, physics was dropped from the syllabus. Flor was now the only one left in the natural sciences, and after his departure in 1820, his subject was also dropped from the curriculum.

Even though the head teachers were now royal appointments, and the Cathedral School was directed by a council composed of the rector and four senior teachers, the rector had the determining word when it came to the hiring and letting go of teachers. It was still the rector's personality, capacity and method of operating that, to a great degree, characterized the school, its reputation and prestige. The rector of the Cathedral School was an important person in the public life of the city, and for the School's disciples he was certainly a dominating presence.

In the eyes of the general public, Rector Jacob Rosted was seen as a man of strong character, an upright colleague, "an excellent Man in every Respect," a true, noble and steadfast man who could not be faulted as "one who trims his sails to every breeze," in other words, a decent *gentleman*. He was said to have been thoroughly educated, and also, zealous about his pupils' intellectual and moral well-being. But by now some considered him to be an extremely old man who was not really capable of maintaining the established discipline, and they thought he was much too credulous to be a good teacher. They were particularly afraid that beginners would get the impression that they could achieve good exam results merely by dabbling in their studies. Rosted's subjects were "the Mother Tongue" and Greek. Niels Henrik had the rector only for Danish-Norwegian in his first year, three or four hours a week. Apart from this, the disciples recounted that he could be fairly peppery in his teaching, but he had such weak eyesight that he was not always able to detect the disorder when he came to take his customary tours between the rows of desks. One of his foibles was to fiddle with his vest while he spoke to the class, and one of his favourite expressions was, "By the Faith of God, I am going to teach you!" and when he got more excited: "For the Faith of God, I am going to brush you down!" This had resulted in him gaining the name, "The Brush."

Old Thura was another nickname, until in the end he was simply known as "Fosen" (The Torrent).

Jacob Rosted had succeeded Niels Treschow as rector of the Christiania Cathedral School in 1803, precisely when the introduction of the new, sweeping reforms should seriously have been implemented in practice. And even though publicly Rosted had hailed the new ordinances, he certainly sighed in private about all that was new, and how it meant double work for both rector, teachers and pupils. Rosted was conservative, but he had a sense of the pedagogical winds of his day and he protested against those who, in the current theme of the proper position of classical languages, too strongly argued in favour of learning, understanding, and becoming competent in the classical languages. Nevertheless it was Rector Rosted who formulated the declaration that for many decades became almost a doctrine in the country's debate over schooling and the condition of the common people; namely, those who studied the classical languages would, under roughly similar circumstances as far as mental abilities were concerned, have a greater capacity to learn, and in general, greater ability even in other sciences or fields of studies that demanded some larger application of the powers of thought and judgement, than would those who had not diligently applied themselves to the ancient, learned languages.

In spite of all the reforms, Niels Henrik had come to a school where a disciple's natural abilities were to be enhanced and perfected, first and foremost by applying his intellect and powers of discrimination to the acquisition of the classical languages. And when later Niels Henrik was to show a tendency toward a one-sided brilliance in mathematics, the rector, and most of those around him, reacted with great alarm and scepticism. But almost three years would pass before Niels Henrik would seem to have discovered that *some* school subjects were anything other than the duty-ridden, time-consuming requirements of a teacher's individual will.

17.

School Life, Money Worries
and the Joys of the Theatre

Being a disciple at the Cathedral School was a fulltime occupation. School began at nine o'clock with four hours of morning lessons, and as a rule, three hours in the late afternoon. The day ended at six in the evening. In addition there was homework and preparation, day in and day out, six days a week without variation, and almost without any free time. There was a monthly holiday, one day every fourth week, as well as Christmas holidays from December 24th until January 2nd; Easter Week was free of classes, and summer holidays were of two weeks' duration.

With the decree established around the turn of the century, making mandatory the payment of school fees, the practice of having the pupils serve in church and choir at Christiania's main Domkirke, the large Church of Our Saviour, was discontinued. The Cathedral School's duties toward and subsidiary income from these practices were now taken over by the city's orphanage. But the customary practice, that the school's disciples were to attend Sunday service, had not been revoked: half were to attend the morning service, and half, evensong, under the supervision of a teacher. But Dean Lumholtz and Curate Garmann were considered to be "almost Ridiculousness itself, poor and unintelligent Preachers," so neither pupils nor teachers were zealous in upholding the tradition. The biological sciences teacher, Flor, had even permitted himself and his pupils to go off on botanical tours on Sundays. He had no doubt been strongly reprimanded for this, but the custom of going to church every Sunday became impossible to uphold in practice, and when Niels Henrik came to the school, not even Rector Rosted insisted on the disciples' dutiful church attendance.

In spite of the new regulations and new subjects of study, Latin remained supreme. Everyone in the first class took 13 hours of Latin every week. The new Schools Act of 1809 stated that: "Among the Subjects for Instructions, a thorough and cultivated Study of the ancient Languages and the Classics shall always occupy the Place of Honour." But permeating the expressions of theoretical superstructure about schooling there was the desire to give each and every power of the mind valuable employment; the educational plans stated: "The Wealth of History, the Thoroughness of Mathematics and Philosophy must be linked to each other if One-sidedness in Mode of Thought is to be avoided." A written semi-annual examination was held halfway through the school year, and the main examination of advancement, in both written and oral forms, was held at the end of the school year. An invitation was issued in relation to this annual examination session to "all the Benefactors of scientific and scholarly Teaching and esteemed Compatriots," about "cooperating with the now forthcoming Advancement Examinations at the Cathedral School by Joining us to assess the Progress of the Pupils." What was called the *School Programme*, composed of reports written by each head teacher, giving information on his subject's goals and content, were sent out together with these polite invitations. Beyond this, the school's subjects fell into three categories: Languages, Science and Scholarship, and the Arts.

The first category began with the Mother Tongue, the lowest level of which consisted of exercises in correct usage and grammar, essay-writing and rhetoric. The grammar was based on the Latin scheme with Danish examples, cases and syntax. In rhetoric the schemes and examples were also taken from the ancient classical writers. The only items read from mother-tongue literature were examples assembled to explicate lessons in grammar and rhetoric. It was the classic languages, Latin and Greek, that in every way composed the main subject; Latin had an approved right of precedence in regard to occupying the time and efforts of the disciples.

In the *Languages* category, Hebrew was also offered for the prospective theologians; French was taught in four classes, and German in two. English was set down as the practical language of trade and commerce, and lessons were as a rule given only in the senior class; English was not on the syllabus for the *examen artium*.

Somewhat over half of the 40–44 hours a week that the disciples had to spend seated on the school benches were devoted to language-learning. Danish-Norwegian and modern foreign languages received the same emphasis, but altogether fewer hours than were devoted to the ancient tongues.

The *Science and Scholarship* category included the following subjects: religion and morals, physics, the natural sciences, anthropology, history, geography and mathematics. In physics and natural science the disciples were to obtain the point of view of the Enlightenment relating to nature and historical development. And this empirical knowledge of things ought to be followed up as well with pupils reflecting upon them, "solving" their higher connections and considering their "Perfection." But when appropriate specialist teachers were not available, the lessons were happily omitted and would disappear from the syllabus.

Mathematics consisted of arithmetic and geometry, with exams and marks in both branches. There were no textbooks to be found in these subjects in Niels Henrik's time. The teacher took them through the material and dictated, but teaching plans are to be found which indicate there were broader perspectives: "Geometry is read with Euclidean Rigour, and the little Opportunities, whenever given, are to be utilized practically in the study of Logic as well." Niels Henrik's mathematics teacher during the first two or three years was Adjunct Hans Peter Bader. Bader's instruction did not venture much beyond the demand that the pupils copy things down from the blackboard, and Niels Henrik seems not to have had any particular feel for mathematics as long a Bader was the teacher, except that even in these years he was the one who most often got the best marks.

The third subject category, *Arts*, included calligraphy, drawing, vocal music and gymnastics. None of these subjects were included in the syllabus for the *examen artium*. There was no teacher for gymnastics, and the subject was presumably restricted to the free utilization of some of the apparatus in the school's Place of Recreation. Only occasionally were there teachers in the field of vocal music and song, and Niels Henrik seems never to have taken part in any such classes. The teacher of writing and drawing for the school was one Galschjøtt, but he neglected to hold many of the classes, and after Niels Henrik had had him for a couple of years, and had achieved the worst possible marks, Galschjøtt abandoned both the school and Christiania without notice, marching away, pockets filled with a large advance on his salary. He went to ground in the town of Sandefjord and refused to return to Christiania. In the end, the school's account had to be written off, and so Galschjøtt "in addition to Laziness, wherewith he has been for a long Time afflicted, now as well has fallen to Drink, such that he is not capable of Employment."

The great breaks from everyday life at the school were the visits to the theatre, to the Comedy House in Grænsehaven, which was outside the Quatrature[4], on what is today the major Oslo street called Akersgaten. There the Dramatic Society had developed into a semi-professional theatre with frequent performances. The school regarded the theatre as an innocent amusement, and a certain number of tickets were given to the school's pupils, and to the children of the city's better families, so that they might attend the dress rehearsal performances. However, active participation by disciples from the

---

[4] Following a major urban fire in the early 1600s, a new city centre for Christiania was planned and implemented under the Danish-Norwegian king, Christian IV. This was laid out on an innovative grid plan with relatively wide streets running at right-angles to one another, and was, and is, located on the northerly, or inland side of Akershus Fortress. From its gridwork plan, it has been given the name, the Quadrature.

School in the theatre's productions was frowned upon. The School could not forbid pupils from taking roles in the Comedy House performances, but it was considered less than seemly for a School disciple to appear in front of a "tolerably large and mixed Public." The School's duty was first and foremost to watch over the disciples' "Diligence and Moral Order."

It probably did not take long for Niels Henrik to discover the delights of the theatre. The theatre and its actors became his great recreational interest, perhaps to a greater degree than was common at that time. Going to "the comedy" was his greatest ways of disconnecting from his work and worries.

The autumn of 1815 saw six performances in Grænsehaven, and if Niels Henrik missed the November performances, he probably had a ticket to the one in December. The front rows at the dress rehearsals were usually occupied by girls, and the interval provided a chance for the boys to prove their courage and venture forward to try to chat with them. Theatre performances functioned as an exciting meeting place for young people, somewhat like the social-literary salon of the adults. The theatre milieu was a place where the weaker sex could hold forth as much as the men, and that applied to women both acting on stage and assessing and discussing the performances afterwards. The words "For Taste and Spirit" were inscribed on a blue board over the Dramatic Society's proscenium at Grænsehaven.

The first theatrical performance attended by Niels Henrik was almost certainly at the beginning of December, 1815, when there were two items on the placard; one was a three-act comedy, *Dog Eat Dog, or Out and Out War*, by a French playwright named A. J. Dumaniant. This play had been showed four times in previous years, and many were eager to see how new actors would interpret the familiar roles. But the greatest excitement had to do with the second play. This was Holberg's *Master Gert Westphaler, or The Grandiloquent Barber*, that was being played for the first time in Christiania. The garrulous Gert plagues everybody with his unstoppable muddle-headed chatter about every possible subject under the sun, but above all, most on Cromwell and English politics, although all the while his actual aim is to propose to Leonora, the pharmacist's daughter, who however is taken over by the sympathetic Leonard, while Gert continues his perpetual chatter. And while Leonora and Leonard celebrate their wedding, Master Gert Westphaler raises the town's dust with his feet and continues his flood of words about how he no longer wants to live among these dull-witted Philistines, but rather to "travel to other places, where erudition receives more respect."

Holberg had said, "This comedy has always been my favourite child." That particular evening at the theatre must have been the experience of a lifetime for Niels Henrik, who now for the first time saw these well-known foibles of human nature rendered in such a cultivated form, and for the first time saw a person characterized with such broad strokes. The braggart Gert Westphaler was no doubt quickly stirred into the growing brew of personalities who surrounded Niels Henrik's schooldays. Among the teachers there were those who could *talk* holes, and others who could *knock* holes, in a person's head. The school had teachers who stood at the desk in front of the class, day after day, conveying information that was predetermined, repetitious and one-dimensional in both form and content.

In the coming spring term seven plays were presented at Grænsehaven; three of them were by the German, August Kotzebue, but Holberg's *The Amateur Politician* was also played, and Ludvig Holberg seems to have become a favourite writer of the young pupil. Disciple N. H. Abel borrowed Holberg's *Peder Paars* several times and in two editions, both from the school library and the city's Deichmann's Library, which for a time shared the same premises. As well, Holberg's *Metamorphosis or Transformations* was one of the few literary works among the books borrowed by N. H. Abel. Could Holberg's epic parodies on academic erudition have acted as a sort of antidote to everyday life at the school?

From the relatively exalted social position of a vicar's son who had grown up in a country village, Niels Henrik was probably not used to looking up to all adults by virtue of their age, but *the teachers* at that learned school would nevertheless have functioned as personal examples and ideals for the disciples. In addition to the rector, Niels Henrik was familiar, from his father's conversations, with two of his teachers. These were Christian Døderlein and "the Moss Priest", Martin Richard Flor, both of whom had been active in The Society for the Well-being of Norway.

Adjunct Christian Døderlein had been connected to the school as a permanent teacher of religion and Hebrew for nearly ten years. It was known as a fact that, as a joint-editor of the newspaper *Tiden* in 1814, and possibly also for other reasons, he had caught Christian Frederik's particular attention, had received frequent perks from the king, and had been appointed to the position of senior teacher. Now, in 1815, Døderlein had gone over to edit *Den Norske Rigstidende*, the newspaper which was the official organ of the new government and the new king. The students referred to Døderlein simply as "The Book's Word." As a teacher his only demand seems to have been that the textbook be repeated word for word. If his pupils were to replicate a concept in their own words, this did not suffice for Døderlein, who would always retort that "this was still not the Book's Word" and "what is it that our Writer in this Passage says so beautifully?" The least deviation was met with: "Book's Word, Peasant! You cannot make it any better!" But because Døderlein never left the teacher's desk at the front of the class, since he was both near- and weak-sighted, it was not difficult to fool the teacher: the pupils let themselves be seen sitting in rows with their books open. Døderlein had his suspicions and it did once happen that he confronted a pupil who had been leafing through a newspaper, two pages at a time, in the classroom. He had remarked coldly, "This Time you have been turning improper pages." The book that was constantly referred to was *Religion and Morals*, by Professor Niemeyer of Halle, translated by Knud Lyhne Rahbek. This was a rationalistic interpretation and explanation of the Bible's teachings that Niels Henrik was very familiar with from home. At first Niels Henrik received the mark of 2, and later, only 3, in religion, that is, *very good* and *good*, on a scale in which 1 was the best and 6 was the worst mark.

The natural sciences teacher, Martin Richard Flor glowed with interest in his subject, and was highly educated. But in and among the Latin-schooled times and teachers, the natural sciences did not have a very high standing in the curriculum. Flor received less pay than the other head teachers, and he was never a member of the school council, only its recording secretary. Most were rather dismissive of the fact that the school even offered a post in the natural sciences, and many doubted whether a teacher of natural science could really be a man of cultivation. Nonetheless, these were also the years, 1800-1820, when due to the efforts of teachers Flor and Rasmussen, natural science instruction was instituted at the Christiania Cathedral School - a unique event of its time. Despite this, physics was stricken from the syllabus when Rasmussen was made professor at the university in 1813, but Flor seems to have been tireless in his ardour, and spared no effort to obtain a decent natural science collection for the School. In order to overcome his colleagues' contempt for his subject he was always trying to formulate the correct function that the natural sciences ought to play in the school. His activities can perhaps be best explained by his School Programme from 1810, where he maintained that natural history "serves to Inspire in the Breasts of the Youth, Warmth, the exercise of moral Feelings, teaches them to know the Moral Law, provides a good, great and dignified Conception of the highest Being, works on the soft Heart such that, with Yearning, Aspiration and Sensibility, it moves one more closely toward the state of Perfection."

Botany was said to be Flor's favourite field of study, and in his classes he loved to gather the pupils around the plants that were being examined and demonstrated. But the wildest of stories were told about his teaching. He quite certainly never managed to keep his pupils between the traces; those who were not interested in the subject would kick up a fuss in another corner of the room, or busy themselves preparing for the next class. The pupils also told stories about how Flor, among his various experiments, had allowed the boys to slaughter, grill and eat a cat. Perhaps in reality he had only been dissecting

Fig. 9. The corner of Dronningensgate and Tolbodgaten. A watercolourby Anna Diriks from a sketch made in 1820. The building on the right is the Cathedral School. The main post office is located on the site today. On the other side of Tolbodgaten is Kanslergården, the military school that is still standing. A university student can be seen coming down Dronningens gate in the new student uniform, that had just been sanctioned by the king. The building to the far left became Hotel du Nord in 1829, the capital city's first hotel. [Photo: Rune Aakvik, City of Oslo Museum]

the cat. It is true, however, that Flor used a child's skeleton in his teaching, which the school had obtained from a doctor. It was also said that Flor's lectures could be rather incoherent, lacking a beginning and an end, and when he was examining them, the students noticed that it did not matter what they said so long as it poured out of their mouths. Flor was radiant so long as they could maintain a coherent torrent of words. It did not bother Teacher Flor in the least that they might be repeating the same sentence or talking utter nonsense, but if one's flow of words came to a stop, or one stammered or searched for a word, Flor was immediately on guard, and one's mark was reduced. But Flor also submitted to the boys interjecting "Herr Flor" between every second or third word; for example, "The ass, Herr Flor! is known for its Laziness, Herr Flor! has long Ears, Herr Flor! or the Ox, Herr Flor! is used as a Beast of Burden, Herr Flor!" and so on and on. It was also said, with rather vicious enjoyment, that the bachelor Flor had been absorbed by, and eagerly in pursuit of the entrails of a bear, and to his "Heart's Delight" had sawn away, flailed, torn out and dug into the inner organs of a bear that had been shot on the property of Chamberlain Anker, just outside the city, at Bogstad.

Fig. 10 a, b.
a The grand auditorium and ballroom of the Cathedral School that Parliament occupied for its sessions, and which it took over completely in 1823 and used until 1854. *Stortingssalle*, the Parliamentary meeting room, was drawn by I. L. Losting during the 1830s. [Photo: Rune Aakvik, City of Oslo Museum].
b Here, Member of Parliament Søren Georg Abel, Fig. 10b., (adorned in the silhouette [most likely cut out by P. M. Tuxen] with his Red Cross of the Danish Flag) held his fiery speeches during the fall of 1814, and here he made his famous plea in April, 1818. This room is now in the Norwegian Folk Museum at Bygdøy, Oslo.

Head Teacher Flor was perhaps in some ways rather like Gert Westphaler in that he talked and talked, and just barely, in the midst of this, paid heed to what was really going on around him. But on other occasions, when he became furiously worked up, and he caught disciples reading other subjects instead of following his lesson, he could rip their books into little pieces and toss them out the window. There is much to indicate that Niels Henrik, at least during the first year, was among those who respected Flor's professional insights. At the end of the first year he received the mark of 2 in natural history, and for

Fig. 11. Watercolour by B. M. Keilhau, from the summer of 1820, when he and his friend C. P. Boeck, with their mountain expeditions, put Jotunheimen on the map: "Snowcapped and ice-trammelled Alps in Indre Sogn at the border with Akershus Diocese. (From a Peak in Koldedalen, d. 14 July, 1820.)" This panorama seems to have been from Snøggeknosi looking out over Fleskedalen toward Hurrungane. From Keilhau's memoir album: "Recollections of the Mountain Tour in 1820". [Picture Collections, National Library of Norway, Oslo Division].

his general performance at school he received as a prize for his diligence, *Helmuth's Book of Nature*, a book with which he was probably already very familiar from the bookcase at home in Gjerstad.

In addition to awarding marks in each subject, the school carried an evaluation of each individual disciple in the Examination Protocols under the categories: *Natural Gifts, Scholastic Industry, Domestic Industry, Progress* and *Morals*. After the name of Disciple N. H. Abel, at the end of his first school year, Rector Rosted wrote: "Natural Gifts, Good; Scholastic Industry, Remarkably Good; Domestic Industry, Very Good; Progress, Very Good," and under Morals: "Very Good, orderly and devout." Rosted had Niels Henrik in "Mother Tongue," in which for the first year's oral work in the language he awarded him the mark of 2.

The other teachers who assessed Niels Henrik's input and abilities did not give the same high certification as had Rector Rosted. Only Albert Lassen, who taught history and geography, felt that Niels Henrik's natural gifts were *very good,* and in addition, gave him the mark of 1, *remarkably good,* in both history and geography. The lessons in these subjects continued to consist of copying down Lassen's copious lectures in which anecdotes and dramatic sketches were used to put a little life into the past. Teacher Lassen had been one of Rector Rosted's favourite students and had been hand-picked for a position at the school. Lassen became a university student in 1808, and studied theology until he was hired as a teacher in 1813 upon Stoud Platou's having been made professor at the University. And no doubt in his first years Lassen relied in his teaching on the historical compendium put together by Stoud Platou, and which came to be copied down, disciple by disciple. Later, Lassen himself wrote textbooks, and his lessons had been so comprehensive and longwinded that his pedagogical work in history swelled to five thick volumes. Early on, Lassen seems to have had a comprehension of history inspired by the German romantics; he saw history as a progressive advance toward an ideal that was quite certainly unattainable because it changed and developed along with changes and developments in humanity. But nonetheless, he maintained, the pursuit of the ideal grows in strength to the extent that human freedom becomes victorious. Greek history showed "an unbroken Pursuit of Ideas," and showed "the role of Intellect in Matter." As well, the period of the late Roman Empire showed that revolution could be "a Means by which Providence brings progress to Humanity." The protagonists in the French Revolution were freedom's spiritual heroes; Robespierre was quite certainly guilty of atrocious cruelty, but Lassen, the history teacher, still pointed to his strength of character. There were two requirements from history demanded by the *examen artium*, one from ancient history, which was to be answered in Latin, and one from modern history in Danish-Norwegian. Niels Henrik's mark in Lassen's class rapidly fell to *good*, and then to *middling*; at the completion of his studies in history he stood with a mark of 4, *tolerably good*.

It was said about Lassen that he was lively and convivial and that he had in spades what were known as "social Talents." He performed at The Comedy House and had considerable success as an actor. But, in this period when people took so much enjoyment from the glow of the glass, the desire for amusing distractions and his own disposition toward the joys of cup took over more and more. In the period that Niels Henrik was a disciple, the results of this were seen in the occasional day of heavy, dull lecturing; later there were frequent absences, and in 1828, before he had reached the age of forty, he was forced to resign.

The language teacher, Poul Christian Melbye also assessed Niels Henrik's first year abilities and efforts as *good*. Melbye came from Ebeltoft in Jylland, Denmark, and following his student examination in 1806, spent several years in Paris. He came to Norway in 1814 and shortly thereafter became a teacher at Christiania Cathedral School. He was the teacher of French, German and

Fig. 12. Old taproom on Maridalsveien. Oil painting by Johannes Flintoe, from the period when Abel came to the city. Gamle Aker Church can be seen in the background. Private Collection. Photo: O. Væring.

English. He was a francophile, revelling in all things French - art, poetry and everything grand and noble, and clearly he prided himself in being the epitome of good taste, breeding and refinement. However, totally lacking in self-control and self-criticism, and advancing step by step with his glass raised, Language Teacher Melbye met his downfall. He, too, was forced to resign and only managed to survive by tutoring in a household at Eidsvoll.

It seems that Niels Henrik did not have any special friends during his first year at the school. But to those at home in Gjerstad he maintained steadfastly that he was in his element living in Christiania. Maybe he was also expecting his big brother to arrive in the city. Toward the end of November, 1815, Vicar Abel reported to his dean in Arendal that Hans was well again and that he was ready to leave. Perhaps Hans Mathias did journey to Christiania at some point before Christmas, 1815, and possibly received some sort of teaching; there are documented arrangements that suggest this. But what is most certain is that Niels Henrik had the company of his big brother when the school year began the following autumn. In any case, Hans Mathias did not receive confirmation that he had secured a place at Christiania Cathedral School until August, 1816. At that time Pastor Abel reported that both sons had places at the school and would be living at the same good house in Christiania, and that he had to pay 180 *speciedaler* per annum for this, in addition to the cost of books, clothes, et cetera. As he drew this sum from his income, Pastor Abel wrote to his superior in Arendal to say: "There is nothing left but Milk and Gruel, Gruel and Milk for my remaining Family." Father Abel stated in his rather frequent reports to the dean that he was satisfied that Niels Henrik had received honourable reports from the teachers in Christiania, but otherwise, Pastor Abel seemed to have been more preoccupied with his eldest son, Hans Mathias.

With the lengthy travel time involved between Gjerstad and Christiania, over land and sea, it was almost impossible for Niels Henrik to get home for either Christmas or Easter, and it is likely that he did not go home for the summer holidays in 1816 either. At least we know that it was only to Hans Mathias, on his way to take the Cathedral School's entrance exam, whom Father Abel waved goodbye from the quayside in Risør in September. A short time later, Vicar Abel wrote that Hans Mathias "distinguished himself in the Examination...admitted to the 2nd class...Both Rosted and Flor declared him to be exceptionally competent. Niels also advanced to the 2nd Class...has received a Commendation for Diligence. Heaven be Praised! One Portion of the Grief of the Father has now been lightened." And a little later in the autumn he wrote, "Now only my Desires and my Purse can accompany Them."

Having both his sons at school was also to a great degree a question of economy. School fees were set at 20 *daler*, plus 3 *daler* for light and fuel. But the Christiania Cathedral School was well endowed with legacies and stipends; a third of the students received free tuition, and the Legacy of the Anker Trust helped many. From the second half of 1816, Niels Henrik consistently received 20 *speciedaler* from this legacy at six-month intervals, and each time the reason was because his father "sat in a feeble Calling and had many Children." From 1816 on, Pastor Abel stopped paying school fees for Niels Henrik. The next year, Hans Mathias also received free tuition from the School and approximately the same amount of legacy support.

But there were always expenses, and the cost of lodgings, some clothing and books, and from New Year's, 1817, the landlord requested 25% more for the two brothers. There was a shortage of housing in Christiania, and prices continued to rise. Pastor Abel wrung his hands at home in Gjerstad, but he stuck firmly to his purpose: "I will pay with the Shirt off my back. My children are my All."

18.

Everyday Life in Christiania
S.G. Abel Enters the Public Spotlight

At the end of September, 1816, big brother Hans Mathias Abel also came by boat to Christiania, and everything points to the fact that Niels Henrik was there to meet him. Christiania with its 11,000 inhabitants was not a big city. But nonetheless a panorama of the main square would show street crowds surging around the stalls, with horses, wagons and animals everywhere. On an ordinary promenade outside the ramparts of Akershus Fortress, the city's beauties showed themselves off to advantage in the latest fashions, while young men engaged them in conversation and conducted "Undisguised Flirtation". The scene was splendid on this occasion as well: the blue fjord toward the south; to the west one looked down on Piperviken (where Oslo City Hall is located) and the long valley of Bislett Brook; from there, which today is part of the heart of the city, the countryside began, beyond the vacant lots and scattered buildings. Around the elegant stone-walled house with fishponds at Tullins Ruseløkke were a number of small wooden dwellings, some of them under construction with piles of plank ends and wooden blocks cast around helter-skelter. Row upon row of steep hilltops were all that was to be seen in the area in which the Royal Palace later would be built. Rosenkrantzhaven was located on the south side of Drammensveien, with greenhouses and an old-fashioned main building that had galleries and a mansard roof. Further south there lay a single enclosure with a crofter's cottage, a log cabin, two large barns and a dwelling for smaller livestock. To the north, north of Grensen Street, Storgaten was lined with houses all the way to Vaterland which sprawled outside the city proper. Prisoners serving long sentences were housed at Akershus Fortress. They could be seen in the city streets in their prison garb and chains attached to their arms and legs, which clinked as they moved to or from their day's work, or as they were engaged in sweeping, ditch-digging or repairing roads. Might not Niels Henrik have told his brother about the dog that, once upon a time, was immured in the thick stone walls of Akershus, said to be ever poised to reveal its terrifying spectral form to any enemy intruder.

The Paléet and the lovely garden behind it had already been described to them by Father Abel. It was there that he had met Karl Johan and Prince Oscar. The government now held its meetings in that beautiful building standing on its white-washed foundation block, red-brick walls, blue shutters and dark blue glazed roof tiles; the building was all one-storey and had been given to the state by Lord Chamberlain Anker, the richest man of both the city and the country around the turn of the century. The Head of State would now be in residence at the Paléet if, it being September, he was not still at his summer residence on Ladegård Island in the fjord near the city. At the harbour, over by Bjørvika, the brothers could see huge piles of planks and boards. Perhaps they met some of the timber carters with the chalk marks on their backs, rushing to get their pay for the day's work? That is to say it was common practice for the carters, after they had delivered their consignments of timber, to have these chalk marks made on their backs to show how much they had delivered. The cashier at the delivery office paid for the notated transport, and brushed off the marks.

If they had the time, the brothers may also have hiked up to Tøyen to look at Gardener Siebke's greenhouses and his monkey. They had to be back to their quarters in the merchant's establishment before dark. It was the common practice to go to bed early; moreover the two boys had to save on the tallow candles. They were perhaps asleep before the night watch came around the streets and announced in his monotonous voice, "Ho, Watch Ho! The clock has struck ten," and then would either light or extinguish the streetlamps that burned cod liver oil; the times at which this was done were relative to the moon's progress: was it "Off, or Rising"?

The Abel brothers were now in the same class at school. They had as their Latin teacher the young Hans Riddervold, another of Rector Rosted's former brilliant pupils. Riddervold was assigned to the beginner classes and had been Niels Henrik's teacher in the first year, which consisted of only spoken Latin, and in which Niels Henrik had received a mark of 2 on the year-end report. It seems to have become more difficult now that essays in Latin regularly had to be turned in. Riddervold, who later became both President of Parliament, Bishop, and Minister of Church Affairs for 24 years, used to look back and tell a story about his famous pupil, Niels Henrik Abel. It happened that one day Riddervold found a note on Abel's desk that said, "Now Riddervold thinks I have been writing my Latin Essay, but in that he has made a *real* Mistake. Abel." Who the note was addressed to, and what the reaction had been, Riddervold did not recount, but in the end-of-term examination in 1817 he awarded Niels Henrik 3 and 4, respectively for spoken and written Latin.

After three years of teaching, Riddervold left the school to take up his theological studies. Ole Kynsberg now took over as Latin teacher for the beginning classes. Kynsberg as well had been a disciple at the Cathedral School, and had distinguished himself as a particularly talented linguist and historian. He was the son of a businessman at Elverum, had become a university student in 1811, and graduated *cum laud* in law at the University of Copenhagen four years later. But in June, 1817, when the first defense of a doctoral thesis was held at the new Norwegian university, Kynsberg seems to have impressed Rector Rosted *ex auditorio* with the beauty and fluidity of his Latin, such that the young man was quickly offered a position at the school. Kynsberg was treated with respect and reserve by both teachers and disciples. He was bright, possessed a sense of justice, and was known for his uncommon kind-heartedness. Some people said he spoke a brilliant and superior Latin like a native of Rome. Kynsberg did not need to exert force to maintain discipline in his classes. A sharp glance, or a word, was enough to disarm even the most audacious. In any case, even Niels Henrik must have learned a certain amount of Latin from Kynsberg. In one year-end examination, Kynsberg gave him a mark of 2 in both written and spoken Latin. In his *examen artium* results, Niels Henrik received a 3 as his final mark, and seems throughout his schooldays to have been below average in Latin.

Like so many others of the period, Kynsberg also loved to drink. Even if Adjunct Ole Kynsberg's alcohol problems were not very noticeable while Niels Henrik was a disciple, there were lessons that went up in smoke due to drink and the revels of the night. It later happened that Kynsberg came to school roaring drunk, but it was said that the disciples formed a ring around the teacher, led him down the stairs and shielded him from the gaze of Rector Rosted, or at least created a situation whereby the rector could close his eyes to the event.

Adjunct Hans Peter Bader had become the successor to the clever Søren Rasmussen in January, 1814, in the field of mathematics. As a scholar, Bader was certainly also well-qualified. In any case, Rector Rosted said that it was not easy to find a person of his qualifications. But his reputation was not of the best. It was said that he, while celebrating with other teachers, had been observed in "a Place in which it was not respectable for Them to come," and with respect to reputation as well, Bader "allowed his irresponsible Opinions about God and Religion to spread to the Disciples."

The disciples gave Bader credit for his learning and his talents, but he was ill-tempered and considerably harder in his beatings than the other teachers were. Corporal punishment and castigation were in theory no longer among the means used to increase respect and hardwork, but continued to be used by the majority of teachers. Moreover, Adjunct Bader maintained that for those who did not understand mathematics, there was only one thing to do: pour on the scorn and the beatings, and then more beatings and more scorn. During Bader's classes Rector Rosted frequently had disciples sent in to him with bloody noses, swollen cheeks and aching heads. And the invariable explanation was that Bader had beaten them. In order to maintain stability, the rector had to get involved and remind Bader that thrashing was no longer the way to go. For a long time Rector Rosted believed that the pupils' complaints were exaggerated, and were due to the fact that Bader had awakened indignation and hatred in them, first and foremost due to his unfortunate disposition, but also due to his rude and tactless habit of deriding and making fun of those who did not understand mathematics. Rosted himself never administered beatings; the strongest punishment that Rector Rosted gave a disciple was to call in the offender for a counselling session where, with the fullest seriousness and solemnity, he admonished the miscreant.

Niels Henrik did well in mathematics right from the beginning; from Bader he received *exceptionally good*, but he did not win the mathematics prize that was given out annually to the best pupil. The lessons continued to consist largely of copying things down from the blackboard and cramming tremendously. The requirement in arithmetic on the *examen artium* of 1814 was to explain and give self-chosen examples of how fractions are multiplied. The question in geometry was to demonstrate Pythagoras's theorem. It was later said that Niels Henrik had been tormented by the mathematics teacher, Bader, and that for this reason he had been away from school for some weeks in the autumn of 1816. There is no sign of this in the school records. Because Niels Henrik was receiving a stipend, such an absence would no doubt have been noted in the records. But nevertheless was there not perhaps something that made Abel the father worry about his sons in the big city? In any case, Pastor Abel now began to make plans to travel to Christiania. But he also had other reasons to make the journey. In January, 1817, he wrote to Dean Krog in Arendal that he wanted to see his children, and he would like to get Vicar Schanche at Risør as a temporary replacement for the two-week round trip. In practice this had all probably been arranged, as earlier S. G. Abel had covered for Schanke in Risør, so that this was a simple reciprocity. Getting the bishop's approval took more time. Christian Sørensen occupied the episcopal residence of Kristiansand. He had been the colleague of old Hans Mathias Abel, had ordained Søren Georg and also approved his desire to separate Vegårshei from the Gjerstad parish. Christian Sørensen knew the prevailing conditions well, and was probably extremely familiar with the current nuances of Church politics in the capital. Perhaps the bishop had misgivings about what would happen if Søren Georg Abel journeyed to Christiania at this time. Perhaps the rumours had already reached the highest ecclesiastical circles that there were some in the University's Faculty of Theology who had found fault with Vicar Abel's activities?

The Christianity and the type of ministry that Søren Georg Abel had so ardently championed and given such clear expression to in his concordance to the catechism, was precisely the target. The Theological Faculty in Christiania was led by Professor Svend Borchmann Hersleb and Lektor Stener Johannes Stenersen and they both regarded with intense distrust the rationalism that had long been in vogue in the Norwegian churches. Both were admirers of the looming theological personality of the day in the Nordic countries, N. F. S. Grundtvig in Denmark. Lektor Stener Stenersen in particular considered it was his holy duty to lead the Norwegian people back into the fold again, and in his eyes, Søren Georg Abel's concordance of the catechism was the most obvious example of the worldly delusions that had snuck into ecclesiastical teachings. Vicar Abel's catechism was frequently used, particularly in districts around the capital, for confirmation preparations. In fact, the confirmants were divided into Abelists, Pontoppidanists and the readers of other catechist trends, depending on which book was used as their text. And for Niels Henrik Abel, who was to be confirmed this year at the Church of Our Redeemer, and who had Curate Jens Skanke Garmann as instructor, the choice was simple. Garmann willingly used Pastor Abel's book; Niels Henrik was thus an Abelist.

But there were quite a few people at the University who were aware of Lektor Stenersen's plans to annihilate the Gjerstad priest's conception of Christianity. In January, 1817, Stenersen published three articles totalling twenty-five pages in which he went through Søren Georg Abel's book point by point. His conclusion was clear: the Gjerstad priest was one of the greatest heathens in the country; so heinously had he strayed from the truth that he now had to take responsibility for the souls of at least 10,000 who were now on the way to eternal damnation. The figure of 10,000 was arrived at by calculating that each of the five editions of 1,000 copies had been read by at least two people. The articles were published anonymously, as was the custom of the day, but most people rather quickly came to know who the author was. Already on January 30th, Curate Garmann himself had tried to defend S. G. Abel. Garmann maintained that even if one admitted that Abel's book contained substantial shortcomings, the consequences were not nearly as grave as Stenersen would have it; the shortcomings could easily be remedied by any competent teacher of religion who understood and had a feeling for the holy teachings and who allowed Luther's catechism to be the foundation for building the pedagogical edifice.

From his farm in Gjerstad on February 2nd, Pastor Abel had sent his first reaction to what he called "the quite notorious Assessment" of a book that earlier, "enlightened Teachers" had greeted with acclaim. Abel wrote that because he hated the darkness of ignorance, he had made this critique available to many members of his congregation, and that everyone had "declared the Review to be from someone who either was Malicious or a Religious Fanatic," but that he himself had considered the writer, who still had not publicly declared himself the author, to be simply mad. The only thing that Abel wanted to thank the writer for was his "precise Calculation of the 10,000 Souls that my Book has put on that Road which will lose for them their Salvation. Be so Kind, Mr. Reviewer, if You indeed have any *lucida intervalla*, to indicate how many souls Pontoppidan's Concordance has saved from Perdition, and how many You yourself have taken on to lead to Salvation."

On February 23rd Søren Georg Abel finally set out for Christiania himself. Maybe he had already written out the expanded reasons for the madness of the reviewer which, on March 6th was published in *Nationalbladet*. Because the allegation of madness might, "with the absolute exception of the Reviewer," be considered somewhat excessive to the readers, Abel explained in twelve points why he considered the reviewer to be crazy.

On this trip to Christiania, in order to save expenses, Abel the father took his own horse and a manservant; on February 26th he arrived at his sons' quarters in the capital. On the previous day his article dated February 2nd was published and Pastor Abel was burning to know who was saying what about whom. He later reported to his dean in Arendal on his visit to the boys: "My Children are flourishing. Hans is quiet, and Niels is excessively cheerful, but the Teachers are particularly pleased with both of them."

Why was Niels Henrik so "cheerful"? Why not? The whole town was going around making rhyming couplets about Abel. *Nationalbladet*, the nation's most popular publication had, in a burst of wit, begun on February 20th to carry verses that rhymed with the name Abel, and this set off a near avalanche of related contributions; the raw materials for these endeavours were immediate and easily comprehensible: *sabel* (sabre), *snabel* (elephant's trunk), *parabel* (parable), *fabel* (fable) and many others of like character, plus a series of French adjectives - *aimable, honorable, capable, incroyable, admirable, miserable, detestable* - which in Norwegian rhyme perfectly with "Abel." In addition, they ridiculed his forenames, using the peasant variant of Georg, which is "Jørgen", and thus providing a hint about where the rhyme-smith believed Abel belonged. It seems that Pastor Abel never knew who had written these verses. Presumably he tried to find out from his relative at the *Nationalbladet*, the publisher, Hans Abel Hielm, who had to be aware of who, in relation to potential litigation, actually wrote contributions printed in the paper. But ordinarily, *Nationalbladets* editorial policy was to function as a mere postbox for contributions from the public, or rather as it maintained,

Nationalbladet strove to be a public forum in which anyone, however ephemerally, could enter; and those who did not have "sufficient faith in their own Abilities to write for the Public," were promised by the editor "the most inviolable of Discrete Silences, within the limits afforded by the Law." However, it frequently happened that those in charge felt that the language and concepts of public contributions were not suitable for a state "that thrived on the Freedom of the Press and a Free Constitution." And if matters progressed toward litigation over the freedom of speech and publication, it was to the benefit of *Nationalbladet* that the editor and publisher had the expert assistance of his brother, the high court lawyer, Jonas Anton Hielm, who with great persuasiveness steered cases through the procedures of the courts.

Niels Henrik was thus "excessively cheerful" during this flustered and turbulent time for his father. Almost certainly both the accusations about Abel's fundamentally flawed faith, and the rhyming doggerel attacks resonated in the schoolyard: "Abel spandabel/Hva koster din Sabel?" (Generous Abel, How much does your sabre cost?) Big brother Hans Mathias was very silent and melancholy. The 15 year-old Niels Henrik, cheerful and apparently indifferent, was probably also insecure about what really had taken place.

In effect, Pastor Abel was under attack from two directions. The background to the doggerel attack on the Abel name arose from his engagement in the day's most burning issue, namely the discussion on Norway's relationship with Denmark, in the past and the present, a discussion which also revealed relations to the country's new rulers, Karl Johan and Sweden. The "Man of Historical Truth" derided by Abel in rhyming couplets throughout the land, was undoubtedly Nicolai Wergeland, who in the fall of 1816 had published his great work, *A True Account of Denmark's Political Transgressions against the Kingdom of Norway*. This book, which came out anonymously, and without even a place of publication (simply "Norway, 1816") had led to a violent increase in the level of stress found within the delicate relations between "Dana, Nor and Svea," as the three countries were personified. The book was brought out by The Society for the Well-being of Norway, and in Denmark was consequently viewed as a representative expression of Norwegian opinion. The anonymous writer received violent public reactions from all sides. Vicar Søren Georg Abel had also picked up the pen. The same day that he got his hands on the book he wrote to *Den Norske Rigstidende*, and on December 11, 1816, his contribution came out as a "Paid Contribution":

Where is the Wretch who has written: A True Account of Denmark's Transgressions against the Kingdom of Norway? He is hereby challenged, and if he has any Sense of Honour left, to emerge from his Lair, and account for himself in public, because as a justice-loving Dane, Swede or Norwegian, and our former and present equally wise and humane Governments must recall: hicce niger, hunce caveto, which is construed as: He is a blackguard, shun him.

Gjerestad's Vicarage, near Brevig, the 28th Nov'r, 1816.

Søren Georg Abel, Vicar of Gjerestad,

Diocese of Christiansand, Knight of the Order of Dannebrog.

In regard to the payment for this, Abel stressed that this would rapidly be discharged if the honour of the country did not demand that it be printed for free. And the reason that he had undersigned it, "Gjerestad's Vicarage, near Brevig" was probably to remind the readers that the book on Denmark's Political Transgressions had been pilloried in the town of Brevik.

For his engagement in this polemic, which was soon to be found in all quarters, in *Rigstidende, Nationalbladet,* and *Intellegenz-Sedlene,* Pastor Abel was closely associated with "the mock Norwegians" together with the likes of Curate Garmann and Professor Hersleb, who both made strong contributions to the issue. Professor Hersleb, who was among the first to come out in vehement condemnation of the book, which everyone soon knew had been written by Nicolai Wergeland, who at the same time had sought and obtained the posting as parish priest to Eidsvoll. In *Rigstidende* of November 23, 1816, Hersleb had declared that the author of *A True Account of Denmark's Transgressions against the Kingdom of Norway*, which a short time earlier had come out for sale in the bookshop of Hans Abel Hielm, had abused the name of Norway by setting it on his title page, and taking the liberty of being the nation's spokesman. This text was an outbreak of "the most foolish National Arrogance and the basest Hatred", written by a writer who, "with Frivolous Licentiousness seeks to ridicule the Danish Nation," and does not even have the courage "to acknowledge his own Monster bestamped with the Mark of Cain."

But in the early months of 1817 the Vicar of Gjerstad was one of the most controversial figures in the capital's press. His pronouncements on Nicolai Wergeland's book inevitably entered the wave of sympathy that opposed the anti-Danish sentiments that had broken out among the population after the Kiel Accord and the difficult arrangements for monetary settlement, feelings that only grew in intensity as, later, the old Danish-Norwegian national debt was to be apportioned. But because Abel was simultaneously criticized by Stenersen, the Danish Grundtvig's great spokesman in Norway, for the most horrendous heresy, in the eyes of many people, and not only Abel's sons at the Cathedral School, this must have blurred into an incomprehensible rhetorical bruhaha. Immediately on the heels of Stenersen's long devastating attack in *Nationalbladet*, a poet also came forward to cast Abel in a miserable light.

Abel probably tried to be balanced and at ease toward his sons during this time. He had arrived in the city on a Wednesday and stayed over the weekend. Perhaps he also paid a visit to his tanner's apprentice, Knud Larsen, who had almost finished his training in the capital. On Monday, March 3rd, Abel the father left Christiania. It was the same day that a new round of rhyming couplets came out in *Nationalbladet*, both for and against Abel and his critics. The 6th of March saw the publication of Søren Georg Abel's commentary on his reviewer's madness, and this stood as a preliminary conclusion to the discussion on his ecclesiastical book.

On his journey back to Gjerstad, Søren Georg Abel gave himself time to visit friends and acquaintances. He spent one night in Drammen, one in Holmestrand, a jovial dinner at the home of the merchant and Eidsvoll man, Carl Peter Stoltenberg, in Tønsberg, three nights with Magistrate Bryhn in Larvik, and he paid visits in Porsgrunn and Skien as well. When he eventually got home to Gjerstad he reported to his dean in Arendal that he had spent 400 *riksbankdaler*, more than half of which was used during four days in Christiania.

Pastor Abel also described some of the other experiences from his travels: that of all the cabinet ministers, he liked Christian Krogh the best, which was a little dig at Treschow whom he constantly considered to have sacrificed philosophy for politics. But how in reality should Vicar Abel's observations be interpreted, when he wrote that "Lasciviousness" increased as he travelled eastwards, and his conclusion that "these Times are Depraved"? He mentioned the demented rhyming ditties about Abel, but also wrote that many supported him, and that shame lay with those who desired to disgrace him, and that he was not the only one who now considered the university to have become the "Father of Falsehood."

19

Confirmation and Direction in Life

The two brothers went home to Gjerstad for the two-week summer holidays. They were back in Christiania and at school again by August 13th. Only a few days later the message came that their beloved grandmother, Elisabeth Abel, had died. On August 20, 1817, Vicar Abel wrote to his dean, on the eve of his mother's funeral: "The times have become stark raving mad, and for the remainder, the Elements seem to have united purely to lead us into Misery. Now I can certainly say that Poverty is becoming my Lot in life."

Since Niels Henrik had left to become a disciple at the Cathedral School without first being confirmed, he now therefore was to be confirmed in the Church of Our Redeemer, where the curate, Jens Skanke Garmann, was resident. The preparations for the confirmation were under way right in the midst of the agonizing debate about Abel the father's book on religion. At the end of September, at Michaelmas, 1817, Niels Henrik became a confirmant, and Garmann wrote meticulously in the protocol: "Niels Henrich's [sic] Knowledge and Performance are remarkably good."

But despite the fact that Niels Henrik followed Garmann's lessons and repeated his father's catechism commentary as well as he could, it seems that even for such an Abelist as Niels Henrik, these were uncertain and stressful times. His academic results at school grew worse. At the end of his second year he was advanced to the next class only *on probation*. The discussion about his father's book presumably had a stronger effect on Niels Henrik than he was willing to admit at the beginning. At first he was merely cheerful, but little by little he seems to have become infected by his big brother's suffering and melancholy. In reality it must have been a shock for the Abel brothers to experience their father's whole system of thought being demolished and derided right in front of their eyes. That thousands of souls would be heading for hell and damnation as a consequence of their father's persuasive teachings must have been interpreted by the sons as almost an attack on their own rationality.

Niels Henrik had been in Christiania for two years, watching and learning, and even if he did not go around behaving as the Word of God fresh from the countryside, he nonetheless was a 15 year-old boy who most certainly had been imprinted by his home conditions and his father's rules of conduct and patterns of thought: so long as one acted correctly, so long as one thought correctly, then the world would be a good place to be. But if one's father's thinking was not *really* right? Under that excessively cheerful surface of this period perhaps there smoldered a desire to improve his father's work, to make it better, more precise and so fundamental that it could be assailed by no one: find an inherent orderliness that on the whole could not be thrown in doubt.

In his critical examination of Søren Georg Abel's book, Lektor Stenersen had expected the reader to have Abel's book in hand, and he had referred to the question numbers in Abel's book as he painstakingly made his comments, page by page and point by point. Readers were also expected to have the Bible at hand; as well, scholarly readers ought to have the basic text clearly in mind. After having indicated that "Religion is once more deemed a National Matter," and the estimation of the ten thousand souls, he began with concrete attacks on Abel's Question 26: "What is the value in particular of the Old Testament?" And the answer, which the Abelist Niels Henrik also had to repeat to Curate Garmann at the Church of Our Redeemer, was "It teaches us to celebrate the World's Creation, Mankind's first Condition, Religion's future, and presents many important Examples." The objections here were that all predictions about Jesus Christ were left out of the answer, and in ignoring *that*, Pastor Abel was denying Jesus. And where Abel wrote that the Church's traditions were the clothing of religion, Stenersen argued that "clothing" was an extremely blasphemous expression. For just as the writer, for all his wisdom and education, would be considered a jester and a fool were he to dress himself as a harlequin, so would improper Church traditions awaken contempt for the religion, no matter how divine these traditions might be.

Where Abel wrote that the duties a human being owes to God are the same as a child owes to its father, he ought to have added, "but to an infinitely greater degree." And to use the word "deference" instead of the old "Fear of God," to refer to one's relationship with God, was deficient in the extreme. Deference was something that one owed to any superior; Fear of God exceeded all deference that one might pay to even the most upright of gentlemen.

Abel had written that to love God, in the knowledge of his perfection and beneficence, was to rejoice in Him, eagerly to contemplate Him, and to strive to please Him. Stenersen, the Grundtvigian, raised the question of why the desire and yearning for heartfelt union with God was omitted.

"Must we not love the Earthly?" was one of Abel's questions, and the answer had been, "Yes to be sure, but we must not entangle our Hearts in so doing." Stenersen scoffed over the answer, doubting with amazement that the writer had ever been able to love without his heart being entangled in the object of his affection; if that were the case, then he must have an ice-cold temperament. Besides, Abel's answer was un-Biblical, and in opposition to John 2:15.

There were also various Biblical texts that ought to have been cited, according to Stensersen, and many of Abel's questions dealt with completely irrelevant issues, were mere padding, and often wrong and misleading. Moreover, many assertions were made without explanation. For example, it was certainly meaningless to do what was *true* without first explaining what *truth* was.

Stensersen ridiculed Abel on the question of explaining superstition: *If* what Abel wrote were true, that superstition was to belief, what to act on that which others said and did without investigating what the situation really was, *then* certainly by the same logic it was superstition to belief that America lay to the west of Norway so long as one had not, oneself, been to America, or to believe that rat poison was deadly without first having tried it out! Right enough, ghosts do not exist, but did Abel not believe that God might allow certain revelations to occur? And to maintain that a natural cause can always be found when something unusual happens, that this is merely taking for granted something that does not exist.

Abel had included "Agriculture" among the useful forms of knowledge. Stensersen replied sarcastically, "Agriculture benefits only the Agriculturalist." Abel had defined thrift as a virtue that lay between extravagance and avarice. Stensersen maintained that "a Virtue that lies between" was a peculiar expression, and for common people, quite incomprehensible.

Abel's question on whether we ought to judge the unfortunates who took their own lives, should never have been posed. Stensersen was firm that the person who takes her or his own life is not within the realm of our judgement, and offenses such as suicide, can never be sufficiently abhorred by the living.

Lektor Stensersen was also afraid that Abel gave young people too much freedom of choice in whom they should marry when he said that parents should advise and not coerce their children to choose or reject one or another relationship.

In reference to prostitution and the lack of chastity, Abel stated that these acts damaged and defiled both the body and soul and did irreparable harm to civil society. For his part, Stensersen wanted venereal disease, which he called "whoring disease" to be designated as "that illness which killed in the most painful manner and was evidence of God's punishment." In addition, this professorial critic maintained that the expression, "civil society" was certainly not easy for children of the common people to understand. On this point, Pontoppidan's explanation of the Sixth Commandment was considerably clearer, better, and more concrete.

Abel had written that even if one was aware of wickedness in one's neighbour, one should say nothing about it, unless one could improve him or caution another person. Stenersen could not understand that it would be wrong to expose the evil that one knew about and could substantiate in the next person. This would not harm the transgressor, and it would teach other people to be on guard against the ungodly. However, one ought certainly to be sure, to judge leniently, and not ascribe to someone the possession of an evil heart.

Under the assertion that in particular it is a person's *desires* which tempt one to stray into evil, Abel also added that the examples and persuasions of other people were more tempting. Stenersen interpreted this as a clear example of "our elucidating Times" not being able to abide the mention of the Devil's name. The Devil had quite rightly been removed from the evangelical psalm book, but was the Devil's ability *to tempt* thereby diminished? How could Abel be so certain that there was no devil to be found, that the opinions of Noah and Abraham were deluded? Where Abel preached that a person's sensual desires in themselves were not sinful, that they only became such when they degenerated into passions, there, Stenersen protested. He was of the opinion that Abel was speaking wholly against Jesus Christ. Evil desires should be rooted out, killed off, and not merely dampened down by one of Abel's expressions of "the enlighted Reason of Jesus' teachings."

Abel's vision of original sin was also completely on the wrong road: to maintain that a person was not born in a state of sin, but rather with a capacity for both good and evil, and merely that evil characteristics frequently became strongest in children who had poor upbringing and bad examples, was something not sanctioned by the teachings of the Bible. To say that original sin was in opposition to the justice of the Lord, to judge in this manner the right and wrong laid down by God, was an abomination created by a man of short-sighted vision. The source of human life and human wretchedness lay in the newborn child, and a disposition toward evil would inevitably lead one into sin and perdition if common sense, illuminated by the holy teachings, were not to lead the way, and religion's power through Jesus, were not to eradicate all evil. Abel was too mild as well on the question of God's punishment. Stenersen reminded the reader of the transgressor who went into the eternal fires of damnation, the wages of infidelity and disobedience toward God.

Stenersen indicated in various formulations and with different points that in reality Abel, in one way or another, was perhaps contemptuous of God's mercy and the atonement of Jesus Christ. In any case, Abel explained so obscurely and indistinctly the atonement of Christ - that was really the only consolation for the sinner - that ninety-nine out of one hundred confirmants would not understand a word of what he said.

To maintain, as Abel had, that in the great hiatus between man and God, it was natural to contemplate invisible beings, good and evil angels - this was nothing other than having both truck and trade with a pantheistic outlook on life.

And when Abel, on the question of human improvement, in Question 249, wrote that one point of departure for improvement was that man seriously use his reason - this gave Stenersen the opportunity for a long deduction in which he went on to use a named person with whom, by way of example, he warned and threatened people: there was the peasant, Halvor Hoel, about whom it was known publicly that he lived with two women, did not believe in marriage, and maintained that it was no sin to sire children out of wedlock, but nevertheless maintained that he was a true Christian. The man did not lack *reason*, Stenersen wrote, but he had applied it in completely erroneous ways. "With Speculation you can arrive at anything, Oh misguided Man! but never at the Knowledge of Salvation." First, when one lived according to the teachings of Jesus, as the proper use of the Bible proclaimed, then first one's spirit began to function, and first one's faith was verified and embodied in goodness!

All the things we cherish we owe to our Father, and to Him alone. On this subject, Stenersen protested, Abel's book had nothing to say. God's mercy had been passed over in silence. The sufferings of Jesus were much too superficially dealt with. A false significance had been attributed to the betrayal by Judas. That Jesus Christ endured his sufferings for the sins committed by human beings was either completely ignored or not examined in appropriate detail. In the same way, in Abel's book, Jesus descended to hell, and his seat on the right side of his Father was omitted. The Church was also completely ignored; likewise, the Resurrection and Judgement Day.

"Do we obtain something by means of our Prayers alone?" was Abel's Question 267. The answer was: "No, we must ourselves work seriously to achieve what we want." Here Lektor Stenersen felt he had to point out as well that Pastor Abel would have managed to get what he prayed for, if only he prayed, *in Jesus' name*, and in a deeply heartfelt manner, imbued with faith and patience.

And Abel's treatment of The Lord's Prayer was also imprecise, full of gross and serious errors, and extremely misconceived. "Give us each day what we need for daily sustenance," Abel had written. By employing several references to the original Greek Biblical texts, Lektor Stenersen was able to show that the translation "each day" was wrong. The meaning of the Greek words was clearly "this day." Stenersen indicated that Abel, in his translation, could, in a way, pray all at once for several days' running, and thereby render daily prayers superfluous. Abel's formulation, "Preserve us from every Seduction," was also directly wrong. The original text used the word "Temptation" and not, as Abel had written, "Seduction."

Abel's choice of words to explain baptism and communion was reprehensible and revealed his unbiblical attitudes. That the christening should be a solemn ceremonial activity whereby the child was admitted into Christian society, yet wherein the baptism was of utterly no help with salvation unless one lived according to the covenant of the baptism, caused Stenersen to groan: Good God! are christenings then to be completely useless? Why be baptized in that case? And is society the same as Church and congregation, and are "ceremonial" and "holy" the same thing? And perhaps the word "holy" is no longer useful? And how can the sacrament, holy communion, be a means of improvement if people are not granted absolution from sin, but merely took part in a ceremonial activity of consuming the bread and wine to commemorate both the good deeds and the death of Jesus? And why would Abel make the youth believe that it was only St. Paul who taught that all Christians ought to celebrate the sacrament?

Abel had written that man can give the impression of being Christian by attending communion, and Stenersen asked why he used the Danish-German word *bivåne* for "attend", which not once indicates participation, but only presence. The essential thing that ought to move one to take the sacrament was the acknowledgement of one's sins and faith in Jesus Christ. On the whole, and above all, Abel took little account of the fundamental nature of *absolution from sin*. With regard to the afterlife, Stenersen would have the reader believe that Abel maintained that in the fulfillment of one's *duties*, one could secure eternal bliss.

When man died and his body rotted, his soul remained immortal because the soul "is a single Entity, that does not exist in Parts, and therefore cannot be dissolved." Thus was the Abelist's vision of the soul! Stenersen pointed out that this was completely different from Christ's teachings, but was a philosophical idea that was difficult to understand for those not initiated into metaphysics. How otherwise could Abel know that the idea was true, and that nothing could be eternal without being indivisible?

According to Abel, the fact that in this earthly life one finds depraved good and virtuous evil, is proof of a life beyond, a just and blissful life. Stenersen felt that this was a particularly fragile proof for those who were familiar with the joys of virtue "extinguishing all the red-hot Arrows of wickedness" here, in *this* life.

Stenersen's purpose and conclusion were clear: all the errors in the book had been attacked heavily because he believed that they more or less led toward a heathen godlessness. The impression that young Abelists would be left with was that Jesus was only a great teacher and philosopher, and not a real saviour.

In the debate that ensued there were certainly others besides Pastor Abel and Niels Henrik's confirmation priest, Curate Garmann, who made rejoinders. An anonymous contributor to *Nationalbladet* argued that the intensity of this spiritually arrogant attack was pervaded by, and quite opposed to, the teachings of Jesus and the Christian spirit. Wrath, indignation and ill-will were the reaction of someone who saw himself as infallible, who sat himself down on God's throne of judgement, and mercilessly dealt out a sentence of damnation. Halvor Hoel called Stenersen "Per the Hypocrite," and several contributions could be interpreted as sympathetic to Abel. But in the end, it was Stenersen's sentence which persisted. Besides, many felt that Abel had completely disgraced himself by calling his critic insane.

Otherwise, the year 1817 was a great anniversary year for The Reformation. It was three hundred years since Luther had formulated his theses, and this was celebrated wherever his teachings had become the state religion. In Norway, where for the first time the country, in its own right, was to celebrate a Church festival of commemoration, the jubilee was a particularly big event. The actual festival, during the days around October 31st, were celebrated with cannon-fire salutes, heraldry, bassoons, trumpets and cantatas, the peeling of bells, and processions. In Christiania three days completely without work were observed. And Lektor Stenersen's lectures on the history of the Lutheran Reformation were held to mark the jubilee. These Stenersen lectures became so popular that they had to be moved to the ballroom of the Cathedral School. Three or four hundred people came two evenings a week to listen to Stenersen; the governor, Count Mörner and his wife were in the audience. There was no longer any doubt about who represented the correct teachings. In fact, with his lectures during that autumn, Stenersen instituted popular university lectures as a feature of the city's cultural life. Two years later, the lectures came out in book form, in two volumes.

Niels Henrik was subsequently confirmed during that autumn of 1817. At school he was only just advanced to the next class; his marks that year were the very worst. Big brother Hans Mathias only became more and more silent, while in terms of his schoolwork he still managed to be perhaps one notch better than Niels Henrik. The opportunities afforded the two Abel brothers to sustain their will and their lust for life were probably extremely thin. The dissolution of the lives of their father and their family, which began with the attacks on the ecclesiastical book, and later when their father's political failure came into the light of day, probably affected Hans Mathias most strongly. It became more and more difficult for him to do his lessons, motivate himself and concentrate. Perhaps this did not affect Niels Henrik's motivation and joy of living quite so much. But then, precisely at that time, something happened at school which would lead seriously to the beginning of Niels Henrik's ascent of the spheres in the field of mathematics.

The fateful event occurred in November, 1817. One of the pupils at the school, Henrik Stoltenberg, the son of the Eidsvoll constitutionalist, Carl Peter Stoltenberg from Tønsberg, whom father Abel had visited on his way home from Christiania, died after having been treated violently by the mathematics teacher, Bader. In any case, that is the way the incident was seen from the perspective of the pupils. The naked facts, which gradually came to light, were that on November 16th, Henrik Stoltenberg had suddenly become ill, and after being bedridden for a week, had died. It was ascertained that Disciple Stoltenberg had received, shortly before, a violent beating by the mathematics teacher, Bader. Bader himself admitted that he had hit Stoltenberg "who had fallen from the Stool," but he denied having kicked the boy while he lay on the floor, something that several pupils alleged. Other disciples explained that the basis of Bader's violent rage had been that Stoltenberg, after receiving a rebuke from Bader, had allowed his lips to curl into a smile, and thus Bader had knocked him to the floor. When Disciple Stoltenberg got right back up again, Bader had pushed him into a corner of the room and beat him repeatedly with clenched fists. Eight classmates signed statements that they had witnessed Bader cruelly mistreating Henrik Stoltenberg. Beyond this, it was said that Bader was hard of hearing, and the pupils often got his attention by bowing or nodding, or by asking silly or obscure questions that Bader answered with a perfunctory "yes," as he anticipated the question being a request to leave the room. But if Bader discovered a slight smile or any laughter then he would attack and beat the offender.

The incident with Stoltenberg now came on top of the pupils' longstanding indignation toward, and hatred of, Bader, and they resolutely refused to show up for his classes. The first strike of school pupils had become a reality. Henrik Stoltenberg had received board and lodging from a house near the School, and his host, Merchant Lidemark met with Rector Rosted to protest verbally; thereafter he went to Bishop Bech, who was a member of what was called the Board of Trustees, with a written complaint in which he explained that in the last days before his death, the boy was heard to "frequently invoke the name of Mr. Bader, and to do so in the greatest state of Anxiety." This had not been eased when Bader himself visited Stoltenberg's sickbed in tears, and to be sure, had gone down on his knees and begged forgiveness.

The corpse was said to have been markedly black and blue on one side, as well as on the cheek. State Physician and Professor of Medicine, Nils Berner

Sørensen, was unable to determine the cause of the Stoltenberg's illness, but ascertained that the illness, at some point in time, had developed into nervous fever, and consequently typhus accompanied by headache, nausea, high fever, diarrhoea, and intestinal hæmorrhage, was the cause of death.

Bishop Bech quickly called in the eight disciples who all confirmed that Merchant Lidemark's written statement was true, and they most certainly gave the bishop a lively impression of why they did not dare to go to school so long as the mathematics teacher, Bader, was going to show up the coming Monday. This was on Saturday, December 6, 1817. Bishop Bech promptly wrote, on behalf of the Trustees, to Treschow's department wherein he requested that Bader be dismissed. The bishop delivered the text personally, and received a reply the same day, to the effect that Bader should for the time being be suspended and the matter investigated. Presumably Rector Rosted could not avoid mentioning in his report the earlier complaints made by the disciples against Bader. But what was more important was the statement of Professor Sørensen, who had attended Henrik Stoltenberg in the days *before* he died. Professor Sørensen stated categorically that as, "during the whole Illness there was no Sign of any medical Affliction in him, I have therefore no Grounds to believe that a Mental or Post-mortem examination would have given any Illumination on the Cause of his death." Following this statement, no other responsible authority raised the possibility that Bader was in any way guilty in the matter of Henrik Stoltenberg's death. But the episode uncovered a grave state of relations at the school, and in particular, in the activities of the mathematics teacher, such that Adjunct Bader could not be allowed to continue to teach. On being dismissed, he was awarded half of the yearly salary, a sum that the government swiftly reduced by half. Compassionate friends, among whom was most certainly Lektor Stenersen, would assist Bader, and they encouraged him to sit the public service examination. Bader received the assistance thankfully, and used all his powers to struggle against becoming a greater burden on his benefactors, but he had been living a dissolute life, and this great strain certainly drained him of all his strength. A violent nerve fever also ended his life, barely two years later.

The school's reputation had fallen into disrepute. The collegium of teachers most surely included some with skills and gifts, but in general they continued to be "the Flower of Incompetence," a gathering of half-disabled subjects who, as was said, consoled themselves with the bottle. As well, it was now in the autumn of 1817 that the boozer, Galschjøtt, had abandonned the school. Conditions at the Christiania Cathedral School were commented upon in many circles. Thus, in the midst of the critical situation in which the school found itself right before Christmas, 1817, Rector Rosted was forced to hire two more of his bright former students: Bernt Michael Holmboe and Johan Aubert.

It was the twenty-three year-old Holmboe who now succeeded Bader in the teaching of mathematics. The eighteen year-old Aubert began teaching Latin, a position that had become vacant when the catechist quit in order to marry.

The two young teachers, Aubert and Holmboe, quickly became an inspirational feature of the school. For Niels Henrik, Holmboe's teaching represented a turning point and the beginning of a life-long comradeship. B. M. Holmboe and J. Aubert were friends. Johan Aubert went on to become a leading figure in the gang of young people who, calling themselves *The Colony*, gathered outside Akershus Fortress and ardently discussed the ideas of the day. Johan Aubert in particular was a pioneer in the field of pedagogy. He brought teachers' pedagogical training into focus, and advocated healthy human relationships. And in the event that Niels Henrik did not have Aubert as teacher, he most certainly heard his appeals to be self-reliant in everything, and in all situations, to be proud of one's own gifts. Johan Aubert could not dissuade his friends and disciples strongly enough from "falling under the sway of Others in Matters, when a man with Justice on his side could by his own Assessment, make a choice without having to consult with anyone." The motto was: "One should only consult the Advice of Others according to one's own Faith and Conscience." Perhaps this sounded in Niels Henrik's ears as an echo of his father's ideas?

Maybe the Abel brothers had set their hearts on going home for Christmas, but in the days before the holidays southern Norway suffered the worst snowfall in memory. The roads were impassible for a long period of time.

20

The Fateful Year, 1818

The twenty-three year-old Bernt Michael Holmboe was now responsible for all the mathematical teaching at Christiania Cathedral School. Holmboe had become a university student in the summer of 1814, and had been a zealous spokesman for the voluntary student corps, which had been set up when the Swedish forces had advanced into Norway. He had taken the university's Second Examination with distinction, had followed Søren Rasmussen's lectures in mathematics, both the obligatory ones and those Rasmussen gave on topics of special interest. B. M. Holmboe had decided to go on with mathematics, even though it was not a course of studies that led to any public service examination: there were no public service exams in the sciences. By 1815 he had already become an assistant to Lektor Hansteen who that year had been made professor and most certainly needed the services of a good mathematician for his astronomical calculations. Holmboe studied mathematics on his own, had received a little teaching experience at the newly-opened Institute of Commerce, and above all, he had a well-thought-out view of what mathematics really was, and how it could best be learned.

Mathematics Teacher Holmboe maintained that the basic reason so many, both young and old, complained that mathematics was some "soul-destroying, and tedious Stuff" resulted in people not taking enough time to become familiar with the use of mathematical signs. It was the unique symbols that divided mathematics from all the other sciences. Therefore it was essential that these symbols be learned through drills, and that the teacher, through constant repetition, train the disciples to see the significance of these signs. Such an approach to mathematics required systematic instruction on the part of the teacher, and when so many simply felt an aversion to mathematics, and were not able to see in a mathematical formula "the necessary relationship between causes and effects," this was due to unsystematic instruction. Before Holmboe undertook anything, he therefore got the disciples to repeat in words what the mathematical statement expressed; and vice versa: when the pupil heard a proposition expressed in words, he had to learn to write it down in mathematical signs. When, for example, the pupil saw the following formula:

(a + b) - c = (a - c) + b, he had to be able to say quickly: "instead of subtracting a number from the sum of two other numbers, you can subtract this quantity from one of the numbers in the addition sequence, and to the outcome you then add the other number to the out come." Similarly, when the pupil was given the following sentence: "when one factor in an outcome is the sum of two numbers in an addition sequence, then the outcome is equal to the sum of the two outcomes that result when each of these factors in the addition sequence is multiplied by the other factor" - at that point the pupil had quickly to write down: (a + b)m = am + bm. Such proficiency was most easily achieved, according to Holmboe, when the teacher, no matter what the proposition was, for example: (a + b)m = am + bm, gave the variables a number value and allowed the pupil to calculate the proposition's numerical value. Thereafter one continued with compound expressions, for example, x = (a + b)(c - d), or

$$x = \frac{(a + b - c)(a - b + c)}{(a + b + c)}$$

and so forth, when the pupil calculated the value of x, where the number values for all the remaining variables were given. By means of such exercises the disciples were forced to pay attention to the meaning of the mathematical signs, which for the first time became comprehensible, such that the proposition's proofs and systematic order would never be a big problem. Holmboe advanced from such exercises to the solving of equations of the first order that contained one or several unknowns. By using examples from everyday life he argued that both the pupil's power of discernment and passion for science would be sharpened. Holmboe stressed that the teacher had to prepare the foundations by which the beginner could easily learn without too great an effort, by not moving forward too quickly, not working at too abstract a level, and going lightly over the most difficult propositions until everything contained therein had been learned. But then he got a pupil who quickly understood everything, and exceeded all the propositions that Holmboe had come up with. With regard to Niels Henrik it was never a matter of having to repeat difficult parts; with him it was only a question of replying quickly enough to correlations and propositions that he felt were obvious. Holmboe moved from simple numerical examples, to their general expression; and, in addition, he tended to express algebra rhetorically, and Niels Henrik must immediately have grasped more than Holmboe formulated in this manner:

In whatever science one decides to take up, one begins with *simple concepts*, basic concepts, that cannot be broken down into a number of others; for example, whole, part, space and time. Many simple concepts are combinations, considered to be one; they are called *compound concepts*. The expression of a relationship between two concepts is a *proposition*. The first idea in combination concepts is called *the subject*, the other, *the predicate*. Propositions are categorized as either *direct* or *indirect*. A proposition is called *direct* when the subject's relationship to the predicate is evident without regard to other propositions. A proposition is called *indirect* when the subject's relationship to the predicate cannot be seen without recourse to other propositions, from which the foregoing indirect proposition is deduced as a consequence. To deduce a proposition from the implications of other propositions is called *inferring*, and thus, this deduced proposition, is an *inference* of the propositions that it is deduced from. The interpretation of the inferences in order to help one demonstrate the subject's connection to the predicate in an indirect proposition, is called the proposition's *proof*. Any indirect proposition must consequently be proven.

After having stressed that the words that one uses in order to signify numbers and distinguish them from one another, are called *numerals*, while the written signs that are used to express numbers are called *ciphers* or *digits*, Holmboe began to teach arithmetic. He explained the position system whereby a cipher can signify different quantities depending on where it is placed in the number. Addition, subtraction, and the signs, plus (+) and minus (-), as well as multiplication and division were things that all the disciples had mastered before they came to the Cathedral School, but Holmboe's review of this material gave many of them a new understanding. How does one solve any number into simple factors? What is a multiple? And why was there no one greatest primary number? Long ago, Euclid had shown that assuming that one finds a finite, quantity of primary numbers, and calls its largest value p, then we can write all the primary numbers in an ascending series, such as 2, 3, 5, 7, 11, 13…p. And if we thus examine the number $q = (2 \cdot 3 \cdot 5 \cdot 7 \cdot 11 \cdot 13 \cdots p) + 1$, this would not be a primary number, for q is certainly greater than p, which was indeed the largest primary number. Thus q must be divisible by one of the primary numbers, but if we try to divide by 2, 3, 5, 7, 11, 13…p, then we will always get 1 left over. Thus we have gotten a self-contradiction, and the hypothesis that there is a finite multitude of primary numbers, must also be wrong.

Apart from this, Holmboe stood securely anchored in Euclidean forms of geometric interpretation.

Niels Henrik advanced quickly but he could hardly prove that there were infinitely many primary numbers, when at the end of January, 1818, his father came back once again to Christiania. There was a new Parliament, and this time, Vicar Abel was chosen as the front-ranking representative from his district. From the beginning of January, until late September, 1818, Abel the father lived in a block of houses right outside Akershus Fortress. And once again, the school was forced to move out to other premises, this time to Chamberlain Glückstad's nearby estate.

Once again, Pastor Abel had been able to fulfil his desire to become a parliamentarian. On December 10, 1817, twenty-eight electors from Nedenes and Raabyggelaget District had chosen Vicar Abel as their first representative. According to rumour, Søren Georg Abel had conducted a sort of election campaign, and had actually defeated the rich and well-known Jacob Aall. It was probably the case that Aall was not particularly interested in running for that Parliament, in any case, that is what he said. In a letter to his friend, the magistrate, G. P. Blom, in Drammen, dated January 3, 1818, Aall gave the assurance that he was "intensely happy" at not being chosen. He was still completing work which his absence at the last Parliament had inflicted upon him, and if he should now have been away again at a time so vital to the future of the ironworks he would "indeed have been inconsolable." Aall continued, saying, "Thus it came as a great relief that Priest Abel had such a great Desire to go to Parliament that he put all Powers in Motion to achieve his Goal. He goes in Peace, but if he employs the same disorderly comportment at the Parliament, as at home, then he will shame the Constituency and the People he represents. Otherwise perhaps an intoxicated Representative like him is better suited to the Parliament, than a sober Person, like me, who does not deal in my Countrymen's exalted Feelings for the Fatherland's brilliant Constitution?"

In the middle of January, 1818, the main road from Arendal, all the way to Christiania, was hand-cleared by people from each local parish; which incidentally was the same way the roads were built. After the enormous snowfall, the worst in human memory, that had begun before Christmas, the snow had been so deep that it could no longer be ploughed with horses. Thus, at the end of January, Abel the father journeyed to Christiania with horse, sleigh and driver, in all likelihood with the other representatives from Nedenes.

Seventy-eight Members met for that second regular Parliament, opened with great ceremony by the Viceroy, Count Mörner, on February 6, 1818. Of the seventy-eight representatives, twenty-seven came from commercial centres, and fifty-one from rural districts. Pastor Abel was a member of nine committees, including the important Procuration Committee, together with, among others, Professor Georg Sverdrup and President W. F. K. Christie. Apart from this, he was on the Editorial Committee, which was a permanent committee, and three others under the jurisdiction of the Department of Church and Education, but was not a member of the committee that he probably deemed the most important, the committee that was to draw up new laws governing academic schools, such as the Christiania Cathedral School, and as well, to discuss the educational institutions positioned above and below the academic schools, namely the university and the common school.

Many difficult issues had to be dealt with in the months that followed. And when on September 15, 1818, the same Count Mörner disbanded the assembly, among the other things that had happened, Pastor Søren Georg Abel had, in everyone's estimation, disgraced himself so fundamentally that his career as a public figure could be definitively considered over. And around the same time, on September 1, 1818, Lektor Stenersen, who in the meanwhile had been appointed professor, got in the deathblow against Abel's catechism book. Pastor Abel did not then take the trouble to answer it.

The basis for this situation had arisen before Parliament had convened in February. On January 31st, *Nationalbladet* had published a piece entitled "Fragment of a Fable," that was quickly perceived as a very coarse and scandalous libel of King and Government. This fable was a subject of conversation in all circles, this fox fable, as it was called. Some were scandalized, others apprehensive of unpleasant consequences, and many public officials in Christiania urged Parliament to express its displeasure with what had come out in the press. It was a question of saving the nation's honour. But who was the author? Many names were mentioned, but it was "the deranged Pastor Abel" who in the end was pointed out, and from more or less responsible quarters, as the writer of the fable. But what was it in this animal fable that could set tempers boiling to such a degree? The choice of words was limited strictly to the animal world, but parentheses were read into the text:

In a remote mountainous region there stood a beehive (= Norway), and in the neighbourhood there was a fox burrow (= Sweden). The hardworking bees lived in happiness and contentment with their lot, but the inhabitants of the fox burrow wanted to conquer the peaceful hive. The bees consequently saw their little Shangri-la attacked by their crafty neighbour, but the bees sorely stung the fox on the snout such that the fox, shame-faced, drew back. The one time that the losing bees were afraid for their dwelling was when, during the superior enemy's plunder, the enemy even set fire to the hive (= the Fredrik-shald [the present-day city of Halden] fire, 1716), and colonized the old property once more. The Fox King (= Karl XII) was killed in the course of these attacks. Meanwhile, his successor renewed the struggle, without much success. But the conquest of the hive was ended once and for all. The foxes thereupon got the support of powerful allies, and a wolf, a donkey, a bear and a seal now arrived as ambassadors to the hive with the ruler's command to give up (= the envoys of the Great Powers, Russia, Austria, Prussia and England sent to Norway in the summer of 1814). The law of might triumphed, the bees' steadfastness disappeared, and their chosen headman (= Christian

Fredrik) had to abandon the hive, where faintheartedness, apathy, sluggish-
ness, treachery, adulation and low trickery began to reign (= the atmosphere
in Norway).

The hive came under the burrow's control, and the cunning foxes proved
ingenious in obtaining for themselves, by underhanded means, the hive's wax
and honey. The fox king and his council employed sly and clever trickery: they
instituted a blinding luxury in the anaemic hive, and one saw bees going
around in gilded embroidery, and others, with painted wings, and yes, indeed,
even royal medals were dealt out with gay abandon, medals that were mocking
symbols of the hive's bondage encapsulated in a little chain wrought from
silver stolen from the peasants (= Sweden's Royal Order of the North Star).
The patriotic among the bees wrung their hands and sighed, and above the
entrance to the hive some wit had written: *Splendida miseria* (= Splendid
misery)!

And so things went for a time, the resentful bees stuck to their ignominious
lot, but they realized that the foxes were not powerful enough to stop animal
states of a higher rank (= Denmark with its legitimate royal house) from
restricting them in their endeavours in the nearby blossoming meadows. The
fog enveloped them, spread and deepened, and they could only wait for a more
fortunate moment (= for Karl XIII, who was at death's door, to die). While
the fox king ate himself to death on a poisoned carcass, a wild scramble broke
out over the succession to the throne. And it so happened, as it does in the
real world, that political gratitude turned into human ingratitude (= speculations about what would happen in Sweden: the new royal house, the Bernadot-
tes, was not very strong, and Karl XIII was not the last in the line of the Vasas).
Therefore, perhaps, an absent fox youth (= Gustav, born in 1799, the son of
dismissed Gustav IV Adolf) of royal descent would come home, and under
the protection of the mighty polar bear, make good his claim to the throne,
and after many upheavals, the result would be that...

The fable ended here, that is, with suspicions about a state *coup d'état*,
perhaps the fable's most dangerous point.

The fox fable engendered great discussions and speculations about censor-
ship, in any case, about legal action being taken against the publisher of
Nationalbladet. Viceroy Mörner reported the matter to Stockholm: he could
only with difficulty contain his anger over this fable, the contents of which
were of such audacious and punishable character that they could not be
allowed to go unchallenged. Mörner therefore called together the government
in Christiania to discuss the matter which, in his opinion, was an attack on
His Majesty's inviolability. Naturally enough, the cabinet shared Mörner's
wrath and indignation, but felt that, vis-à-vis the law, they could not attack
Nationalbladet, something only the king could do. But in royal circles at that

moment another event preoccupied people's minds: as predicted in the fable, King Karl XIII died in Stockholm on February 6th. The Regent, Karl Johan, therefore put off the matter of *Nationalbladet*. He was soon to be crowned in both countries and wanted very much to believe that the Norwegians could be won over by goodness. The government had already proposed on February 12th a charge against *Nationalbladet*, but it later dismissed the matter. The fox fable went on to be one of reasons for Karl Johan proposing before Parliament in Christiania in June, 1818, a Royal Proclamation limiting the freedom of the press. The Swedish *Riksdagen* had been the target of a similar Proclamation that came out prohibiting the publication of anything that could offend or damage the Fraternal Kingdoms, the Fraternal Peoples, the Union, and the Royal House. The Norwegian Parliament replied very formally that to declare that the freedom of the press could be trifled with was the same as tampering with the Constitution, Paragraph 100, and this could only be done by an extraordinary Parliament.

In the debate on the establishment of a chivalrous order for Norway, that came before Parliament only a few weeks after the session had begun, Vicar Abel expressed a view on this gilded order that ought well to have disqualified him as the author of the fox fable. But nobody showed any interest in refuting "Abel capabel til fabel" (the proposition that Abel was capable of writing the fable) and who knows what else. In the matter of a Norwegian chivalrous order, it had been decided during the previous parliamentary session by a vote of 70 votes to 3, to agree to ask His Majesty, provided he agreed to honour worthy persons with orders and titles, to set up a Norwegian chivalrous order. Now the matter had come up again and a large majority wanted to follow up the proposal. A counter-proposal declaring orders, ranks and titles ought once and for all to be banished from the land, was defeated. Parliamentarian Abel, who indeed himself had received the medal of the Danish Order of Dannebrog, voted with the majority that a Norwegian honorary order be instituted, even though he in general was "of the daring Opinion, that Orders, like the Nobility, are completely unnecessary in a well-organized State." But orders had existed for some time, and the Constitution referred to them and gave the king the legal right to award them. And if one, at that time in 1814, had been thinking of Norway as an independent state with its *own* king, then, declared Abel, those who served Norway had to be honoured with Norwegian orders. This was a declaration that would be difficult to reconcile with the fox fable's mocking words about all gilded Swedish medals and orders.

From the Swedish side, the Norwegian National Assembly was seen as a rather unified mass voice, where everyone continued to be in agreement that it was the *government* and above all, *Sweden*, which were to be blamed for everything that went wrong. But apart from this prevailing opposition, there was also a smaller, minority group that the Swedes called "the rabble opposition". The leader of this rabble opposition, the Farmers' Club, as it was also contemptuously called, was Wincents Lassen Sebbelow. It was rumoured that among other prominent members were the fanciful and loquacious parish priest, Pastor Abel, Major Pierre Poumeau Flor, who shouted himself white in the face when he spoke, and Lieutenant Hoel, a small, pale man who looked like a thief, and just like his brother, Halvor Hoel, hailed from so-called learned peasant stock and had an extremely bad reputation.

Moreover, the Swedes considered it unfortunate that the parliamentary debates did not take place behind closed doors. That was to say, "the mob's heroes" allowed themselves to perform for the approval of the crowd who had come to listen, so much so that the parliamentary gallery often ressembled a theatre gallery. And it was said that Pastor Abel never seized a single word without casting a beseeching glance at the gallery as if to say, "What do you think about it?" On one occasion he was the butt of laughter when he promptly sat down in response to a hiss from the gallery. Abel had thought that the hiss had come from the President.

Parliamentarian Abel was otherwise very active both in committee work and plenary discussions, but some of the things he did were considered mere hair-splitting, and many of his motions were defeated. When he proposed that every representative should, in writing, give reasons for his absence every time he ran an errand outside the House, and Abel himself was consequently proposed to be the supervisor of these endeavours, the matter - quite probably with a certain amount of laughter - came to nothing. His motion that the existing election committee, that chose the members of various committees, ought to be made permanent "so as not to waste Time on Elections," was also unanimously defeated. His motion that all Members of Parliament ought to be addressed as "Repræsentant" (Representative) before their names, was defeated by 43 votes to 27. And when Easter was approaching, at the beginning of March, Pastor Abel presented a motion that Parliament ought to break its proceedings from Saturday before Holy Week until Tuesday after Easter. This provoked a debate on whether or not the many pressing issues before Parliament would be offensive to the sacredness of Easter Week, and should Parliament lead by good example, or was it only going to show its religiosity in superficial forms? Abel's motion fell. Parliament continued with its meetings up to and including Wednesday of Easter Week.

Parliament's first large topic for discussion was the monetary system, monetary laws and statutes, The Bank of Norway (Norges Bank), and the silver tax. The work concerning a voluntary bank had run completely aground, and decrees concerning an obligatory bank still had not come into effect. State banknotes that ought to have been called in before the end of 1817, were still valid legal tender. Complaints about a lack of money came from all quarters, but a demand for an extension of the banknote mission was rejected. Parliament was clear that stabilizing the speciedaler, a stable monetary system, was in itself the basis for Norwegian political independence. Suspicions, that in order to succeed in his efforts to build a closer union, Karl Johan would actually want to take the banking system in hand, made many parliamentarians extra vigilant. Pastor Abel was not among those who were most active in this first great debate, but, like most public officials, he was interested to have a stable unit of currency. As well, he supported W. L. Sebbelow's finance plan, that, as it were, would put an end to all the monetary worries with one blow, a plan that most characterized as the phantasm of an overwrought brain and a "tolerable wonder" that was totally unrealistic, and that in the end would also be defeated.

But when the question of the law on naturalization came up once again, Pastor Abel had his arguments ready. This, indeed, was the matter that in the autumn of 1814 he had struggled through with the help of hot punch and persuasion: "The Constitution ought to, and must be, sacred and inviolable," and then it was quickly adopted that the law on naturalization be left absolutely as a parliamentary matter, requiring neither royal sanction nor any committee to settle such questions. When he who applied to become Norwegian fulfilled the requirements laid down by the Constitution, and had, in addition, virtue, capacity and a warm feeling for Norway, together with certain assets, then Members of Parliament, without any contribution from committes could easily deal with such matters.

Søren Georg Abel acted like a seasoned politician, and seems to have felt himself at home in every milieu of the capital city. How much time he managed to spend with his sons at the Cathedral School is uncertain, but perhaps they were together on the evening of April 1st when there was a grand dress rehearsal at the theatre on Grænsehaven. Kozebue's show piece, *The Crusaders*, was being staged for the first time, and sixty actors decked out in magnificent costumes were taking part. There was a complete mob of armour-clad knights, Turks, pilgrims, nuns, monasteries, churches, seiges, violated maidens, and wars; this was supplemented by the despair and salvation of fathers and innocent mothers, by songs and speeches ringing out between riffs of Turkish janissary music and calls to Allah.

Beginning with the extraordinary Parliament in the fall of 1814, and up to the present in 1818, it became customary for Members of Parliament to go to the theatre together, half of the Members attending the dress rehearsal, and the others going to the premiere on the following day. As well, on April 3rd, there was a special performance of *The Crusaders* in support of the city's poor and homeless, at a cost of five *riksbankdaler* per ticket.

As for April 2nd, it proved to be a fateful day for Søren Georg Abel. Namely, that was the day he went up to the podium in Parliament and read out a complaint he had received from a former constable and overseer at the Eidsvoll Ironworks, a text that contained serious accusations against two Members of Parliament, and in addition, against the venerable cabinet minister, Carsten Anker, owner of the Eidsvoll Ironworks, where the drawing up of the Constitution had been completed on May 17, 1814, the date that was to become Norway's national day. Abel's conduct on that day would have severe consequences, and in the end, finish him off as a politician. It seems as well that the episode was the beginning of the ultimate downfall that broke even his will to live, and his untimely death which would weld Niels Henrik permanently to the family's poverty.

Abel had met this overseer, Johan Hjort, some days earlier; that is, it is quite likely that Hjort sought out Parliamentarian Abel, whom he convinced of the justice of his complaints. As argued by Hjort, the issue was that at some point in 1805 his former employer, Carsten Anker, had promised an annual pension, but instead of receiving this pension, Hjort had been arrested and thrown in jail in an illegal and mortifying manner and had been submitted to "Bolts and Iron bars" for ten months on the premises of the sheriff, then two months at the Rådstue (Town Hall premises) in Christiania. Hjort had been accused of having falsified the document his pension claim rested upon. The case against him had been conducted by Prosecutor Knud Carl Krogh, then appointed to the circuit court of Øvre Ringerike, and now, a Member of Parliament. The sentence in the case had been passed by stipendary magistrate Laurentius Borchsenius, who had also been chosen Member of Parliament in 1818, but who, due to illness, was absent for the whole session. In the autumn of 1816, Borchsenius had sentenced Hjort to five years hard labour for having falsified a letter from Anker. The higher court reduced the sentence to three years, and following an appeal to the Supreme Court in July, 1817, Hjort was acquitted of further indictment on the condition that he pay the court costs. But Hjort considered this to be exceptionally hard on him and his family, who had received no compensation for the tyrannical and violent treatment of which he had been victim. Hjort had thereupon decided to complain to the National Assembly, and he had written an elaborate document that he managed to publish in the *Norske Intelligenz-Seddeler*. But in order to get his complaints and demands more directly into public view he managed to convince Parliamentarian Abel to read his letter of complaint into the record in Parliament. The whole matter was thus repeated from the podium of the Parliament; Hjort's story included the allegation that he, on the day of his arrest, sent his horse, and an open letter, home to his children, but both horse and letter had been stolen from the public royal highway, and according to rumour, this had been done by Carsten Anker's men on instructions from Prosecutor Krogh. And it was added that Magistrate Borchsenius, as both judge and probation officer, had also contributed to the total ruin of Hjort and his family. Thereupon the following question issued from Abel's mouth on the podium of Parliament: under the provisions of the Constitution and the country's legislation, could such barbaric and inhuman behaviour exist? Had these two men, Krogh and Borchsenius, incurred the responsibility to prosecute? How should the said person, who possessed no means of life's sustenance, and certainly no means with which to litigate, seek compensation for the injury he had suffered?

In his answer five days later, Krogh declared that by the very fact of Pastor Abel's recitation he himself had now been charged with behaviour that would render him unworthy of being a parliamentary representative. That everything in this matter could come before Parliament for deliberation was indeed an outrage to which he had been exposed; in addition, he argued, Parliament itself had been outraged by Pastor Abel's reading. Prosecutor Krogh went through the said case against Hjort in detail, and acquitted himself of every suspicion that he might have committed an offense. Hjort's complaint involved only claims already known to, and evaluated by, the courts, and besides, they were claims relevant to the jurisdiction of the courts, and not of the Parliament.

For his part, Pastor Abel clarified that he had never had a doubt but that every Norwegian had the right to have his signed complaints taken up by Parliament, and that he therefore did not hesitate to read the complaint into the parliamentary record, even though Hjort was a person completely unknown to him. Abel also maintained that he, prior to mounting the rostrum on April 2nd, had declared before Parliament that he himself in no way supported the contents of this complaint, and that therefore it had never been in his mind to offend anyone. Abel tried to separate the principle from the personal, and added that he had since come to know that Hjort had an equivocal reputation, both as a private person and as a pettyfogging solicitor at Eidsvoll, and that he, on the other hand, had heard only good words about both Krogh and Borchsenius as public servants. Indeed, Abel went so far as to say that he thought the complaint in the moral respect perhaps been unfounded, and he hoped the matter could be dismissed, and that any suspicion that had possibly incurred in Parliament, could be annulled.

A committee that had been set up to look into this matter was of the same opinion: Hjort's complaint had to be rejected, but the committee nonetheless considered its duty was to ascertain whether or not Parliament or Prosecutor Krogh had been insulted by the manner in which this complaint had been made public in Parliament. An offense against an individual Representative had, according to the Law, to be regarded as an offense against the whole Parliament, therefore it was also "the National Assembly's unalterable Duty, in the most scrupulous manner, to risk Parliament's juridical Sanctity to protect any Representative against unmerited defamatory Attack from the Midst of the Assembly." The majority of the committee felt that Pastor Abel had completely failed to do this, and allowed himself to be used as an organ of malicious vindictiveness. The regulations expressly said that matters that came before Parliament and had not come from the government, Members of Parliament, legitimate committees, or selection committees, could not be raised without the approval of the majority; rather, they should be reported to the President who as quickly as possible should make the contents known to the Assembly. Since no representative could argue that he did not know either the clauses of the Constitution or of these regulations, it was not unreasonable to expect that Pastor Abel, since precisely he himself had read out the complaint, had known what he was doing, and therefore was responsible to Parliament for the scandal that occurred with the reading "of yon Lampoon within these Hallowed Walls."

The committee found that it had to say it was in agreement with Krogh when he stated that he had been offended and insulted both by the terms of abuse and the claims that would make him unsuited to be a parliamentary representative. Abel ought to have known that the assertions contained in the complaint could simply be handled by the courts or the government, and not by Parliament, and Abel could just as well have looked into Hjort's reputation sooner rather than later. It was thus for the committee to find out how Parliament and Representative Krogh could best be assuaged.

The committee finished its work on August 12th, and only one of the members felt that Abel had already apologized enough for his conduct. This was Gabriel Jonassen, citizen of Stavanger, who made his living in shipping and salting herring on his many estates. He was also known as the "Lord of Jæder" and he was known and feared for his apt witticisms, both those of a high-minded nature, and those that were decidedly low. Jonassen's view was that Pastor Abel, in relation to this matter, had said in advance that he himself was not vouching for a single word in the complaint, and therefore he, Jonassen, as a parliamentary representative did not feel at all offended by Abel's recitation. Moreover, Jonassen argued, the issue had gotten such an airing in the "Public Tidings" that the reading in Parliament could not have harmed Prosecutor Krogh, who could so easily defend himself from the accusations. Yet, in Parliament's session of August 15th the President was authorized, by a vote of 41 to 21 votes, to call upon Pastor Abel to submit an explanation, an "Apology to Parliament," about how he, misled by a sense of compassion toward an unknown man, had committed an affront against

Parliament and Representative Krogh. But Abel refused to do this, and in the end, on August 28th, the matter was submitted to Odelstinget (the largest administrative body of the Norwegian Parliament), with a request that a determination be made as to whether "the Nature of the Matter occasions that Pastor Abel be placed under Charge according to the Law of Impeachment." Odelstinget did not find the issue to be one worth wasting energy on, and sent it back as a matter to be dealt with by the full Parliament, where it was unanimously dismissed.

During these last days of August Vicar Abel announced in *Nationalbladet* and as an announcement in *Rigstidende*, under "Paid Contributions," that he would reply later with his views on the issue: those who did not know him personally, or the matter in its entirety would, by reading the committee's findings, get the "most unfavourable Impressions" about him; impressions that would not only reflect on his public office but also lead to the mortification of the whole district he represented. In order to deal with such unpleasant impressions it was therefore his duty to offer the public a decent answer to the request that was made of him by the President of Parliament. Abel's answer was published first in *Nationalbladet*, one week after Parliament, with due and solemn ceremony, had ended.

But before things had reached that stage, Pastor Abel had also had his victories in Parliament. Abel's motion to establish a veterinary school in Norway was an additional basis for Parliament concluding: "The Norwegian Government is requested to allow the establishment in Christiania of a Veterinary School, wherefore a sum of 1000 Spd. per annum, is granted against the next Parliament's budget." Even though Abel did not take part in the committee for school affairs, this was the area in which he stood out as the boldest spokesman for advanced ideals. Surely Pastor Abel had heard about Niels Henrik's fabulous advancement in Holmboe's class, but it is also quite certain that he had other reasons to argue committedly against the massive parliamentary majority who wanted to go back to the old-fashioned classroom teacher system, a system that had no place for subject-specialized teachers, as for example, in mathematics.

Ever since a desire for a distinctly Norwegian law concerning schooling was advanced at Eidsvoll in 1814, work had advanced on such a law by means of committees and rounds of hearings, carried out by, among others, the rectors of the academic schools. Professor Georg Sverdrup had been a central man in this whole process, and he submitted a draft "Law regarding the learned Schools" to the Parliament of 1818. Many felt that these school matters were the main reason that Georg Sverdrup had once again willingly engaged himself in politics and become a Member of Parliament. Some felt as well that Sverdrup, coming to know about the disgraceful conditions at the Christiania

Cathedral School, had felt appalled, and somehow responsible, and wanted to involve himself in straightening things out. Senior Latin Teacher Lorentz Wittrup seems to have been a particular thorn in the side for Sverdrup. With the establishment of the old classroom system again, wherein one teacher had responsibility for one whole class in all subjects, Sverdrup, the Professor of Greek, maintained that this would be the way to put a stop to such incompetences and drunken carousing, of which there was far too much at that particular academic school. To abandon the specialized-subject teaching system would, as well, make more room for the old major subjects, the classical languages. The parliamentary committee led by Professor Sverdrup now pushed the issue forward in a detailed motion which included the various subjects' allocation of instruction hours for each class. Georg Sverdrup certainly had a sufficient portion of the majority with him right from the start, but Pastor Abel argued strenuously against the motion, and he had a couple of other speakers on his side, among them, the honorable Messrs Flor and Koren. At the time, most of the farming representatives were unsure about the issues in this, a field with which they had had so little experience. The matter was dealt with four times by the largest body, Odeltinget - the first time on April 10th, and three times in Lagtinget (the smaller parliamentary body) before, in front of the assembled Parliament on June 11th, the vote on the bill was held. The bill accordingly included the liquidation of the subject-teacher system that dated from 1800, and the limiting of instruction in the living languages and the sciences. It was maintained by the majority that sound teaching presupposed a solid foundation, and that a child's thinking would become confused from hearing the lessons of too many men on too many topics. The youth had been led into superficial knowledge about *everything* without a basis in *anything*. This was the new humanism's critique of the Enlightenment, now expressed as *not many, but much*, or for those knowledgeable in Latin, *non multa, sed multum*. Another argument advanced by the majority was, as well, that the scholarly schools should be preparatory institutions for university, and not schools for the general citizenry that opened their doors for instruction in subjects "which to any well-educated Member of Society were indispensable."

Abel had made several contributions to the issue. He brought to attention the basic preparations that underlaid the present school system, and that the rector of the Christiania Cathedral School, the "worthy Rosted, about whose Understanding and Heart there is only one Meaning, and who has been Teacher and Helmsman of the School under the former System," had publicly declared himself against the new proposed parliamentary law. Beside it had certainly not been proven that one learned more Greek and Latin in the old system, and the fact that now one's native language would not be taught at the lower levels was completely upside down. Likewise, it was incomprehensible that most of what were called the sciences were about to be omitted to make more room for Greek and Latin. Otherwise, there was his ironic addendum: "As is evident in the Motion, in the final analysis, there is the desire to create Greeks and Romans; splendidly appropriate for our beloved Norway!" Latin and Greek were subjects one could go deeply into at university, Abel maintained, and felt it most curious that the subjects of greatest interest to the young were being done away with in the academic schools, subjects precisely like "the Natural Sciences or Information on Nature." But after a certain amount of debate the parliamentary majority expanded its view, saying that also *Hebrew* was useful, indeed necessary to he who would become "a Man of Science in the Word's broadest Sense." Abel replied that boredom and passivity, confusion and negligence would herewith come to reign over the teaching; and what was to happen, not infrequently, where "there is no Changing of Teachers, and of a sudden one of these is grumpy, or one cannot abide him, as the Englishman says, *I don't like the face* - what then? Disorder and uproar, of which the chronicle of schooling is full, will once more be the result. And was it not remarkable, Abel stressed, that Sweden was now setting store more and more in teaching the natural sciences, and that "even in Russia" in recent years "Nature Cabinets" had been arranged from which to supply the schools? In Denmark, the mineral collection "from the prospective great museum" was sent around to the schools. "But in Norway people wanted to rule out these Sciences. It seems that we want to distinguish ourselves, and Praise be to they who have the Gifts and the Heart to do so; I myself have neither of these Parts, but I stand upon Experience, and the Guidance of sound Human Understanding, from which I have until now been well-served. I thus propose that the Motion be defeated."

Abel was clear that the motion put forward by the majority had its models in other countries, and he indicated that "the Motion itself seems to have arisen from a great Desire for the Germans' *bessermachen*"[improvement policy]. Abel felt that the school proposals were inspired by Grundtvigist currents, and could not restrain from getting in a few digs against his opponents (they being Stenersen and Hersleb) at the University: "Grundtvig's system, whether completely or only half mad, seems to have become *à la Mode*, at least in the Faculty of Theology. May Heaven be wary of these Gentlemen!"

Parliamentarian Abel's motion to reject the proposal received three votes.

But the matter dealing with the character of the school in the countryside, the law on general schooling, and the question of the education and salaries for these teachers, *that* was defeated. By two votes. In effect, Parliament felt that the law of July 1, 1816 contained provisions on how school teachers' education and salaries should be dealt with. In actuality, this law was an extremely ramshackle affair, an agreement governed by fear of the economic burden that permanent comprehensive schooling would entail. Ambulatory schools would continue to exist.

On August 13th, the day after the strong criticism was lodged against him in the Hjort affair, Pastor Abel made a speech that clearly characterizes his relations to school and to his own work:

> Mr. President!
> That by itself Upbringing makes a Person into a Man, is something that Nobody really can deny, and indeed the Means to bring this about is still certainly Reasoned Enlightenment. One can easily see that by Reasoned Enlightenment I do not refer to that which is delivered by dark Dogma, which can very well silence the Human Mind, but is never capable of raising it to experience its own Worth. By means of reasoned enlightenment, as I understand it, one can lead Human Beings to the Knowledge of what they owe to God, to themselves, and to others, knowledge drawn from pure Moral Principles, as well as from Christian Teachings. One can perceive that I speak particularly about common people on the land. Some Classes in the Nation require more basic Knowledge, such, to be sure, are the above-mentioned. One grieves about so many Things; one awards sums of money to bring forth many and much; I pose a question, can anything be of a more important Influence than the improvement of the character of schools in the countryside? Much zealous work has gone into the task of changing the organisation of the academic schools, and in this I believe Everything, as experienced school teachers have born witness to, was in the most desirable order; but only see, when it is a matter of the Plan for a better and more adequate Education for the Youth of the General Population, this proposal was quickly shelved. [...] I still allow myself to take note that all the enormous concern that has been paid to the Higher Schools, and all the Wisdom that has been learned therein, will be of little benefit until the Effects of School Improvement are spread to the lesser Classes. These High Schools are, after all, only Rainbows that delight with their Colours, but contribute nothing to the Earth's Fertility.

In any case, the day after he made this speech, Father Abel had what might have been experienced as a victory when the parliamentary majority's "Law concerning the learned Schools", was denied royal sanction. Cabinet Minister Treschow, who in his time had also worked to support the specialized-teacher system, was in deepest disagreement with the view given, by his professorial colleague, Sverdrup, on what the times needed. For Treschow, enlightenment and learning were most possible in all fields when they contained practical and theoretical philosophy that provided the moral and scientific starting point for the great development that humanity saw in store for itself.

Søren Georg Abel had been an active parliamentarian, and was in many ways also before his time. But he had miscalculated. He had lost by supporting the split that would come later in the culture of public servants, between a class-conscious cultivation of classical European tradition, and a patriotic will towards the popular character and national fellowship.

The parliamentary debates came to an end; it was September, 1818. For Niels Henrik and his big brother, a new school year waited expectantly just around the corner. They might have been home to Gjerstad during the weeks that were the summer holidays, but it is more likely that they stayed in Christiania, following their father's work, and waiting for Parliament to close so they could travel home together. Niels Henrik had become more and more obsessed by mathematics, his father had his own affairs to attend to, and big brother Hans Mathias seems to have distanced himself more than ever from everything. It is uncertain how much time they had together in the capital. But they probably at least went to the theatre together, saw *The Crusaders*, and later perhaps, Holberg's *Masquerade* and *Ulysses of Ithaca*. The latter play had awakened considerable attention because the Greek goddesses, played by men, had so boldly displayed their bodily charms. On May 10th Parliamentarian Abel and his sons might have gone to Rådhusgaten 11 to witness the opening of Norway's first art exhibition, an exhibition that was also to become an economic success with a profit of 250 speciedaler, that was quickly put toward the establishment of a "provisional Drawing School," that later became the State School of Applied Arts and Handicrafts.

And on one or another Sunday in the course of the spring, that was unusually late that year, or in the course of the summer, which had been unusually dry, Father Abel and his sons had probably strolled up to Ekeberg. Nobody could spend time in Christiania without admiring the view from there; it was unique and magnificent, painted by both Norwegian and international artists, and depicted both by foreign travellers and Norwegian writers. From Ekeberg one could see the blue fjord with its forest-clad islands, the white sails spectacular against the water, and in the valley, at the foot of the hills, the city with the silhouetted towers of Akershus fortress encircled by a mass of farmsteads and estates with manicured trees, and farms green with fields of grain crops rippling in the breeze, and which, like the steps of an amphitheatre, ascended toward the dark and mighty hills.

But one can only guess at what Father Abel did otherwise, and beyond this, in his spare time. It is not improbable that his sons had seen their father in situations he would rather not have been seen. Peder Mandrup Tuxen, who had not seen Søren Georg Abel since the summer of 1815 when they had been in Christiania together, later wrote: "When he arrived in the City, he found Life and Gaiety; here he took Compensation perhaps for what he missed at home, and the unfortunate Passion - Drink - which rather than diminishing, always Rages."

But in September, 1818, Father Abel longed considerably to be home, or at least away from the capital. Niels Henrik in all likelihood was so taken with new mathematical propositions, proofs and assignments that he would ask Holmboe for private tutoring and not think any more about his father's downfall.

On September 1st the first part of Professor Stenersen's final sentence on Abel's catechism book was published in *Nationalbladet*. The next part came out on the 8th. A week later Parliament was dissolved and probably Vicar Abel left Christiania a few days later. The last part of Stenersen's "extermination" of Abel's book came out on the 22nd; the three parts together were about thirteen pages long. The same issue of *Nationalbladet* contained Abel's explanation of his behaviour in Parliament, written on August 28th. The support that Abel had received from the "Lord of Jæder", Gabriel Jonassen, was published in the same issue. On behalf of himself and the nation, Representative Abel regretted sincerely "the Existence of this Controversy," regretted that Parliament's valuable time had been used to "gratify a private Man's Ill-will or Lust for Revenge," but felt that he himself had equally been offended by Prosecutor Krogh in Krogh's explanation of Hjort's written complaint, and the principle stood glaringly evident: that he, following his conscience and best convictions had acted in conformity with the constitutional freedom of the representatives, that he, both before and after the explanation of the complaint, had assured Parliament that he did not in any way vouch for the contents or the truth therein, and that therefore no one ought to have felt himself offended.

This contribution would prove to be his last public utterance. It was signed "Yours sincerely, Søren Georg Abel, without reference to either "Vicar" or his royal Danish title.

Home at Gjerstad again, finished as a politician, compromised as a person, the priest sought what was by now familiar solace: it was clear to many that the clergyman had become inordinately fond of strong drink. The village people said that the vicar and his wife drank, and they each drank separately. Servants received their pay in potatoes, most of which was now used as raw material for home distilling. The potato crop was otherwise miserable that autumn. Vicar Abel wrote to his dean in Arendal: "All that I ask is that I be relieved from bitter, baseless Attacks, when I am as well beaten down by Illness, Pecuniary Worries and Tiresome Troubles."

21.

Christiania, Fall, 1818

The school year began on October 1st. In the exam that was now immediately held, and that determined the composition of the new classes, Niels Henrik received the overall mark of *good*. In all subjects with the exception of mathematics, in which he had the best marks, he was below the class average. In history, geography, calligraphy (or copybook writing), Latin composition, spoken French, and Norwegian-Danish composition he received the mark, 4: *tolerably good*. With such results, particularly in Latin, it once again meant that the only way Niels Henrik could advance to the next class was by means of probation. The rector expressed his concern that Niels Henrik would develop in a one-sided manner; other teachers felt that, all in all, Niels Henrik was only mediocre. Lassen, the history and geography teacher, had noticed that Niels Henrik had achieved a special standing in the class, in all probability as a result of his facility with mathematics, and Lassen thought that the boy suffered due to this.

The conditions at the Christiania Cathedral School which were exposed to the light of day following the dismissal of the mathematics teacher, Bader, and exposed to parliamentary debates on the academic schools, these conditions now led, among other things, to the following developments: immediately after the school year began, an enquiry came from Treschow's department, asking if Senior Teacher Wittrup and several other teachers at the School "did not only lead a disorderly Life which was both an Affront to the School's Disciples, and prevented these teachers themselves from the requisite Preparation and Mental Good Cheer that they took with them into their work at School, but also that they even exhibited a Condition that for the Most Part deprives the Disciples of the Benefits for which their Teaching ought to be the basis." Rector Rosted had to consent to the removal of Wittrup; the other teachers, who were not cited by name, were allowed to continue. The young Latin teacher, Kynsberg, took over Wittrup's position, not as senior teacher for he was too young, but he received a salary augmented to 400 speciedaler (senior teachers received 500). Johan Aubert took over for Kynsberg as well, and became a deputed adjunct with a wage of 330 speciedaler. These changes had little or no meaning for Niels Henrik; he was so completely enveloped in the mathematical world that he devoted even less time that before to his Latin compositions. The previous year he had had Wittrup for Latin: Wittrup had found Niels Henrik's ability only barely "tolerably good in general." Now Niels Henrik got Kynsberg back again, and even though Kynsberg too could have great difficulty in mobilising the correct Mental Good Cheer in the classroom, he was always well-liked and continued to be the teacher the disciples protected. But now, for Niels Henrik, it was the mathematics teacher, Holmboe, who mattered most. As well, Holmboe was eager to stimulate Niels Henrik, who with such astonishing ease, understood and oriented himself toward all the mathematical material that Holmboe came up with. The library's record of borrowings also reveals that Niels Henrik had seriously begun to borrow mathematical literature during the autumn of 1818. Earlier he had borrowed only literary works, and if one can draw any conclusions from the lending records, Niels Henrik had concentrated his reading on famous and respected writers. In addition to Holberg, he had read Johan Hermann Wessel and Johannes Ewald; as well, seven of the twenty-one books that Niels Henrik borrowed during these three first years at the Cathedral School, were written by Adam Oehlenschläger. The last of Niels Henrik's borrowings of a literary nature from the Cathedral School library was in November, 1818, when he borrowed Jens Baggesen's *The Spectre*. Young Niels Henrik seems thus to have wanted to familiarize himself with the mighty poetic battle that was raging between Oehlenschläger and Baggesen. His next borrowed book was Newton's *Arithmetica universalis*, and thereafter everything he borrowed had to do with mathematics.

And it seems again that the Abel brothers were not even able to journey home to Gjerstad for Christmas. This was perhaps not a matter of importance to Niels Henrik. Together with his classmates, Holmboe and Holmboe's friends, it seems as if Niels Henrik had begun to find a social life in which he felt at home. But as for Hans Mathias, he did not thrive, and homesickness seems seriously to have darkened his life. The school's record of marks shows that from now on there was a rapid decline in Hans Mathias' results. All he managed was *tolerably good* and *middling good* in most subjects and *tolerably good* as his average mark was the school's dismal summary of his performance. But he was to suffer for a whole year before the rector made it clear that Hans Mathias would never be able to pass the *examen artium*.

22.

Into Mathematical History

Holmboe wanted to bring Abel up to date with the mathematical developments of the day. In the way in which he did this, Holmboe was inspired by the French mathematician, Joseph-Louis Lagrange, "the proud pyramid of the mathematical sciences," as Napoleon had called him. It is probable that Holmboe had read the biography that had already come out in Paris about Lagrange, and written by J.-B.-J. Delambre in *Moniteur officiel*; be that as it may, Holmboe had thoroughly read the version of the biography that was reproduced in *Zeitschrift für Astronomie und verwandte Wissenschaften* in 1816. Lagrange's father, a close relative of Descartes, had been a prosperous army paymaster in Sardinia, had married the daughter of a rich doctor in Turin, and of their eleven children, only the youngest, Joseph-Louis, reached adulthood. Lagrange senior seems to have been a passionate speculator as well, and by the time his son had grown up, the family fortune had been reduced to zero. "If I had had a fortune to inherit, I most certainly would not have thrown myself into mathematics." This is the way the great mathematician later commented on his home conditions. In his earliest schooldays he had been more interested in the literary works of Cicero and Virgil than in the geometric works of Euclid and Archimedes. But then one day he had come across a piece written by the astronomer Edmond Halley, the comet's namesake and friend of Newton, on the superiority of differential- and integral calculus in relation to the Greeks' synthetic geometric methods, and the young Lagrange became converted to mathematics. It was not long before he was involved in what at the time was known as analysis, and as a sixteen year-old, he taught mathematics at the artillery school in Turin. Together with some of his older pupils, Lagrange next founded an association that later would develop into Turin's Academy of Sciences, and from this society came the first volume of communications in 1759. After having shown the inadequacy of the old formulas in this field, Lagrange developed new methods for calculating maxima and minima, what was called variation calculus, that ever since the days of the legendary Queen Dido and her founding of Carthage, had been an actual problem of mathematics. And in addition to variation calculus, Lagrange wrote that he later wanted to develop a whole mechanics, both for solids and liquids, from the same principles.

At the age of twenty-three, Lagrange already had the plan worked out for what was to be his major work, *Mécanique analytique*, published in 1788. Among those who read this work was Leonard Euler, at that time leader of the Academy of Sciences in Berlin, and esteemed as the giant among the mathematicians of the day. And in Paris the young mathematician from Turin quickly came to the attention of the secretary of the Academy of Sciences there, Jean d'Alembert. Lagrange received encouraging and flattering letters from both Berlin and Paris. In 1764 the learned prize of the Paris Academy was offered to anyone who could calculate the moon's small vibrations, what are called librations, which allow one to see the rotating lunar circumference, always with the same side toward Earth. Lagrange calculated these phenomena, deduced them analytically from the principles of universal gravitation, and his essay, *Librations of the Moon*, won the Academy's prize. Two years later he won the prize again, this time for work on Jupiter's four moons, and on the forces that mutually influence each other around the planet. The determination of Jupiter's moons and other heavenly bodies was also an important tool for determining positions at sea. And to secure dominion over the seas was a great practical problem that caused many rulers eagerly to support mathematicians. The King of Sardinia now financed a trip for Lagrange to Paris and London together with a friend who would serve as Sardinia's ambassador to London. In Paris, Lagrange was welcomed by d'Alembert and other prominent mathematicians, but after a lavish dinner held for their guest of honour, Lagrange became so ill that he returned to Turin as soon as he was strong enough. Perhaps it was the gall bladder problem that was later to plague him that on that occasion appeared for the first time? In the course of his life Lagrange was bled twenty-nine times, most frequently for his gall bladder disorder.

The Swiss Leonard Euler had been the leading mathematician in Berlin for twenty-five years when in 1766 he decided to accept an invitation from Czarina Catherine II to go to St. Petersburg. King Frederik II of Prussia, Frederik the Great, as he was called, very much wanted d'Alembert to take Euler's place in Berlin, but d'Alembert preferred to remain in Paris and he therefore suggested Lagrange as Euler's successor. Lagrange was enraptured by the offer and quickly agreed to go to Berlin, but the authorities in Turin would not release him from his responsibilities without prior notice. King Frederik II had personally to intervene in order to sort out the matter. Thus, in June, 1766, Euler, with his household of eighteen, left Berlin and travelled to St. Petersburg, where a large furnished house awaited, and where one of Catherine's own cooks was put at his disposition. Then in November, Lagrange arrived in

Berlin and he was cordially received by the king. Lagrange was appointed director of the physics and mathematics section, and for the next twenty years he was to fill the Academy's publications with a series of his great works.

A short time after Lagrange had arrived in Berlin he wrote to a cousin in Turin and asked if she would marry him. She agreed, and travelled to Berlin. When d'Alembert came to hear of this event in the autumn of 1767, he wrote, "My Dear and Esteemed Friend, one has written to me from Berlin to say that You have done what among us other philosophers is designated a somersault, *le saut périlleux*, and that You have married one of Your relatives, whom You have gotten to come from Italy; please receive my compliments regarding this event, for I calculate that above everything, a great mathematician ought to be able to reckon his Happiness, and after having engaged in said calculus, found that marriage is the solution." Lagrange thanked him for the compliment and replied, "I do not know whether I have calculated well or badly, or closer to the point, that I have calculated at all, for in that event I would have done as Leibniz, who, because he had to contemplate, was never able to decide." Lagrange maintained that he had never had the proper sense for marriage. It had been practical details that had caused him to ask his cousin to come and share his fate, and look after him. He was in a foreign land, had poor health and was plagued by melancholy, and that he knew his cousin from before, and that he could get along with her: "This is the precise story of my marriage. If I have forgotten to communicate any of it, it is because the matter has been for me so mundane as to be inconsequential and not worthy of communication to You."

After having passed such a sentence on the conjugal state, Lagrange did have a happy life together with his cousin. When after they had been together several years, she became ill, Lagrange did not want anyone else to help look after her. He was no doubt tireless in his exertions, but she died in 1783, the same year as d'Alembert in Paris and Euler in St. Petersburg.

It is quite probable that there were those in Berlin who did not like the fact that King Frederik had obtained another expatriate, Lagrange, as the director of science and scholarly research. But in addition to superior mathematical skills and knowledge Lagrange also revealed an amiable disposition. He had a strong aversion to demonstrating power, to engaging in intrigues and controversies. "I don't know" was his motto in all meetings about everything that did not concern his own interests.. "I have a deep aversion to disputes," he is said to have declared. This was a quite different behaviour from Euler who constantly engaged in philosophical and religious discussions. In his popular book, *Letters to a Princess in Germany on the Differentiation between Physics and Philosophy*, translated to Danish in 1782 from the French version of 1770, Euler had thrown himself into the struggle of the day about the ability

of bodies to function in relation to one another in a vacuum. Euler championed the view that inanimate objects must have inanimate properties. Space
had to contain a medium, and everything that happened in the physical world
must be the consequence of each individual object intrinsically asserting the
power of its own basic properties: extension, impenetrability and inertia.
Euler's hypothesis was that bodies self-assert themselves, that is, that bodies
function exclusively where they are located, in opposition to the principle
involved in the doctrine of distant effect: where bodies are seen to assert
themselves far from their actual location. In the midst of the Enlightenment's
rationalism, Euler had taken it upon himself to defend the Christian religion.
The story was probably well-known in St. Petersburg about how Euler had put
to silence the atheist mouth of Denis Diderot. Catherine the Great was as well
the high protectress of Diderot, but may perhaps have suspected that Diderot
was trying to turn her courtiers toward atheism, and may have given Euler the
task of silencing the loquacious philosopher. Euler, who had lost his right eye
in earlier life, and was now about to lose the sight in his left, was said to have
met Diderot with the news that a learned mathematician had come up with
an algebraic proof of God's existence, and that he would demonstrate it before
the whole court if Diderot would be so kind as to be there to respond. In a
sepulcral and convincing tone of voice, Euler was said to have intoned:

"Gentlemen, $\dfrac{a + b^n}{n} = x$, therefore, God exists. Answer!"

Diderot, who was rumoured to have little understanding of mathematics, grew
embarassed and completely silent, and as his silence was greeted with roars
of laughter by the others in attendance, Diderot shortly thereafter requested
permission from Catherine to return to France, a petition that was subsequently granted. Diderot and his circle of Encyclopedists were otherwise
favourably disposed toward mathematics. They maintained that it ought to be
the first subject in the schools, and that it was a superb means of developing
intellectual abilities.

In Berlin, Lagrange staunchly maintained his "I don't know" position, and
he came to be almost an agnostic. Despite this, he was eagerly cited precisely
for his statement of faith: "I have noticed that people's pretensions always
stand in inverse proportion to their merits; this is one of my moral axioms."
Among Lagrange's mathematical works in Berlin were his treatises on partial
differential equations and higher arithmetic problems in connection to what,
in French mathematics, was considered among the most honourable: Fermat's
unproven theorems about the connections between whole numbers. During
his time in Berlin, Lagrange also took up the question of when it was possible
to solve algebraic equations with root indicators, and he showed that with

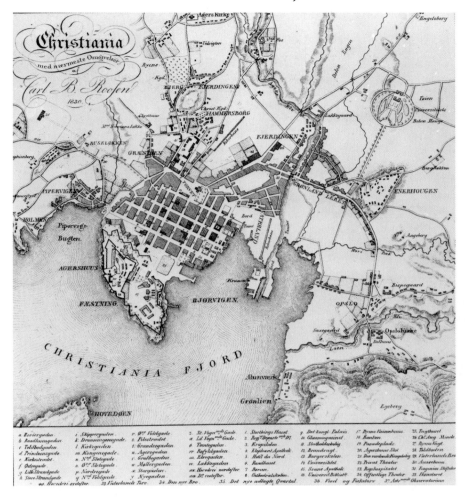

Fig. 13. Christiania and its close surroundings, by Carl B. Roosen, 1830. [City of Oslo Museum]

cubic and bi-quadratic equations there were linear relationships with the roots which could be determined with equations of a lower order. In relation to equations of the fifth magnitude and higher, Lagrange *thought* that such relationships could not exist. It was Niels Henrik who would be able to *prove* this supposition.

A general ill-will toward non-Prussians broke out in Berlin following the death of Frederik the Great in 1786. Lagrange, now fifty years of age, received permission to leave the city on the condition that he continued to submit his

Fig. 14 a–d. Abel's friends and associates, from top left: Bernt Michael Holmboe (**a**): drawing by Johan Andreas Aubert (before 1832), photo: O. Væring; Balthazar Mathias Keilhau (**b**): drawing by Johan Gørbitz in 1835; Aschehoug forlag archives; Carl Gustav Maschmann (**c**): from Elephant Apotek. *Elephant Apotheket gjennem to Hundre og femti Aar*, Kristiania, 1922; and Nikolaj Benjamin Møller (**d**): painted by Carl Peter Lehmann. [Norwegian Portrait Archives]

Fig. 15a–d. From top left: Christian Peter Boeck (**a**): a portrait from later in life, painted by P. N. Arbo. City of Oslo Museum. Christopher Hansteen, with one of his instruments in the background (**b**): copper engraving, by E. C. W. Eckersberg, 1828. Aschehoug forlag archives. Søren Rasmussen (**c**): oil painting. City of Oslo Museum. Niels Treschow (**d**): copper engraving by F. Fleischmann, of a pastel made by Christian Horneman. [Picture Collections, National Library of Norway, Oslo Division]

Fig. 16. Market days in Christiania, the annual popular festival that was opened with the pealing of bells from the church tower on the first Tuesday in February and turned the city on its head. Schools closed and the University readjusted its classes. The Supreme Court stopped its deliberations so that the lawyers would be accessible to their out-of-town clients. This is the oldest known picture of Christiania's market, at the square still called Stortorvet, by an unknown artist, circa 1830. [City of Oslo Museum]

treatises to the Academy's publications. And this was a promise he kept, even after he had come to Paris in 1787 at the invitation of Louis XVI and found living quarters in the Louvre itself. But in spite of all the glory and the honours with which Lagrange was met, he felt himself to be old, exhausted and lethargic. He had a nervous breakdown and lost all interest in mathematics. Queen Marie-Antoinette considered Lagrange a special favourite. The fact that she had heard of him from her home city of Vienna and considered him German, also was a reason that she assisted him. The leading men of science in Paris were now generally considered to be the chemist and man-of-many-parts, Antoine Laurent Lavoisier, and the mathematician and astronomer, Pierre Simon Laplace.

Laplace had shocked many with his "nebula theory": a system of creation that had no room for the personal guidance of God, but rather, advanced the idea that this world and the neighbouring planets had been cast into space from their source, the sun, and from a state of expanded, superheated gas, they had shrunk into small stable bodies. Lavoisier, "the father of chemistry," had won his first prize in science for a treatise on how one might organize street-lighting in a large city; he had given advice on experimental management in agriculture, and determined the paper that banknotes ought to be printed on in order to deter counterfeiters. In addition, he had written on magnetism, water supplies, chairs for invalids, and divining rods; he had demonstrated the completely erroneous nature of the common notion that water, after repeated distillations, could be reduced to earth. Lavoisier introduced the name for oxygen, and explained how combustion takes place; moreover, he asked the age-old question: what is it that ensures that the world does not perish, why does it not wear out? Plants require air and water for their survival; animals sustain themselves from plants or plant-eating animals, and in turn, they give back to the earth, by means of processes of fermentation, decomposition and metabolism, that which had been taken from it. Life goes in cycles, dust into dust. When there were scientific gatherings at Lavoisier's place, Lagrange planted himself at the window with his back to the others, staring out, distraught and melancholy.

However, the great revolution of 1789 put an end to Lagrange's apathy and brought him back to life. At the beginning he followed events with sympathy. When peace and order were restored in 1791 he expressed his admiration for the great nation that had gotten itself a new government, not by force of arms, but by means of the word and the opinions of the public. When a short time later, events began to run out of control and slipped into the unforeseen Reign of Terror, Lagrange could have gone back to Berlin. However, he wanted to remain in Paris where he could see "the experiment through to its conclusion," as he himself put it. Lagrange sympathized with the people who had suffered so much injury, and had lived through poverty and injustice. On the other hand, in the plans of the populist leaders for the renewal of humanity, it seems he did not find much proof of human greatness: "If you want to see the human mind in its true greatness, then you should walk into the workroom where Newton analyzes white light through refraction and otherwise lays bare the system of the world," said Lagrange. And when he came to hear that Lavoisier had been guillotined, he was said to have uttered, "It did not take more than a second for his head to fall, but perhaps hundreds of years will not be long enough to recreate another one like it." During the case against Lavoisier, the court stated that the people had no need of science. The Academy was abolished, but a commission would soon come forward with a draft that retained the metric system of weights and measures. Lagrange was chairman of this commission, and it was presumably his reputation and his ability to keep his silence that saved him. It was also thanks to Lagrange that the number ten, and not twelve, was chosen as the basis of the metric system.

Many of the others who were back on the commission, after Lavoisier had been executed and Laplace banished, had supported the number twelve. They knew that ten had only two divisors apart from one and the number itself, but twelve had four divisors! But Lagrange held fast to ten, and in order to get his opponents to give in, he suggested the primary number eleven as a compromise. The commission grasped the irony and gave in.

When l'École normale supérieure was founded in 1794, Lagrange was appointed professor of mathematics. And when, two years later, the school became the legendary École Polytechnique, it was Lagrange who formulated the syllabus and became the first professor. The expectations were great and the outlook was propitious for those who passed the entrance examinations. Lagrange taught class after class of those prospectively tumultuous military technicians who, under Napoleon's command, would roll across Europe in conquest. Because these students found it difficult to grasp the concepts of "great and small infinities," which were the infinity concepts Newton and Leibniz had used, Lagrange now developed differential calculus, without making use of Leibniz's infinitesimals, and without Newton's concept of marginal value. Despite the fact that Lagrange's method would not be used in later teaching, the achievement he made by developing the method was of great significance for mathematical analysis.

But there was still another chapter to be added to Lagrange's life story. Despite all the successes in his work, he still felt worn out and depressed, yet, even though his patron, Marie-Antoinette, had been sent to the guillotine, there were others around him who showed their love and care for the famous mathematician. The sixteen year-old Adélaide, daughter of an astronomer called Lemonnier, insisted on marrying the 56 year-old Lagrange. And this would prove to be a happy union. She awakened in him a new lust for life. He followed her everywhere, including into ballrooms, and in the end became so dependent on her that even if she were briefly absent he grew unhappy. It was said that Lagrange loved music. Be that as it may, he was always eager to find a way to divert people's attention away from conversations during the interlude of a concert. When Lagrange was once asked what he actually felt about music, he was said to have replied, "I like music because it isolates me. I listen to the first three bars, and by the fourth I cannot distinguish one thing from another. I give myself over to my reflections, nothing disturbs me, I am not interrupted, and in this way I have solved many difficult problems."

Napoleon held Lagrange in high regard and it is quite certain that they had many private discussions on philosophical questions and on the role that mathematics played in society. Lagrange was showered with honours and medals; he was made a high officer of the Légion d'Honneur, and he was made a count. And yet he consistently refused to have his portrait painted. His health began to fail during the spring of 1813. By the end of March he was seriously ill, and he himself clearly understood that his end was nigh. As though he were the dispassionate observer at yet another great and rare enquiry, he now carefully studied what happened to the body as it neared death. On April 8th, the mathematician Gaspard Monge, whom Napoleon had also made a count, together with two other friends paid him a visit. There was a general agreement regarding what occurred during that visit, and Lagrange was said to have uttered, "Yesterday and the day before I was extremely ill; little by little my body grew weaker, my physical and intellectual abilities were reduced to an imperceptibly low ebb. I observed with interest and amusement the fine calibrations of the diminution of my powers. I am approaching the end of life without remorse, without regret, as though descending a soft and gentle hillside. It is a final function that is neither strenuous nor unpleasant." Next, after having advanced some opinions on life, which he felt was seated in each organ, he continued his train of thought: "In a little while there will be no activity anywhere; death will be all-pervasive; death is nothing other than the body's absolute repose. I am willing to die, and indeed I want to die, and I even find pleasure in it. But this is not what my wife desires. I wish now that I had chosen a less good wife, someone exhibiting less effort in trying to resuscitate my energies, who would have let go of me gently. My career has come to an end; I have acquired a modicum of renown in mathematics. I have not hated anyone, nor have I done ill by anyone; it is good to come to the end." Lagrange died two days later, on April 10th.

Holmboe, the mathematics teacher, seems to have stressed particularly that Lagrange had been self-taught, and therefore had been cautious about pointing out to others the best way to learn. But Lagrange had said *something* about the study of mathematics, and Holmboe had taken the advice, which was published in French in *Zeitschrift für Astronomie und verwandte Wissenschaften*, and incorporated it into his notes: those who are really ardent, ought to read Euler because everything in Euler is written with clarity, is well expressed and well calculated, with floods of apt examples, and because one always ought to study the sources! Study Euler! Lagrange had said, and continued: you should take great pains to solve all the problems you encounter, for when you study someone else's solution, you do not learn the bases upon which he chose this, and not some other method of advance, nor will you detect the difficulties he has met *en route*. Lagrange regarded the reading of great works of pure analysis, to be rather useless, as one saw far too many methods flashing by. In his view one ought to spend one's time and effort on application. The great works of analysis ought only be consulted when one encounters methods that are unfamiliar or peculiar from an analytical standpoint.

Holmboe seems to have expanded Lagrange's personal principles, advocating that one never study more than one work at a time, and if it is good, read it thoroughly. Do not give up in the face of difficulties, but rather, let them lie, and go back to them twenty times if that should prove necessary. If, after such endeavours, something remains unclear, only then should one investigate how another mathematician has treated this point. Never abandon the book you have chosen without working through it; skip over those parts that are no challenge, in order to get at what is new to you. Reflect on the reading, particularly about what could have led the writer to this or that transformation or substitution, and above all, what it leads to. Then, in order to gain experience of the assistance that this analysis provides, investigate whether or not another transformation or substitution would have solved it in a still better manner. Always read with pen in hand so that you can work out all the calculations and yourself practice all the questions you encounter. When a work is important, it is an excellent habit to draw up an analysis of the methods, including a summary of the results. Beyond this, draw up lists of certain subjects to give you a chance to develop your own theories. Coming back again to geometric reflections can be a suitable way of strengthening and securing one's judgement.

From Lagrange Holmboe further noted: "to conclude, I have never neglected to give myself homework for the next day. The mind is an idler, but one must take it by stealth, in its natural state of relaxation, and keep it busy developing all its powers, and hold them in readiness for when they are needed; this is a question of practice. It is also an excellent habit, to do, as far as possible, the same things at the same times every day, so long as one saves the most difficult tasks for the morning. I learned this from the King of Prussia, and I have found that little by little this regularity makes the work easier and more pleasant."

23.

School Days and Many More Stories

The Christiania Cathedral School was a handsome building. In addition to the library with its six thousand volumes, the auditorium, and the room with the natural history collection of plants, insects and fossils, the school building had five extremely well-suited, well-appointed reading rooms, as well as a room where the school archives and a collection of antiquities, belonging to The Society for the Well-being of Norway, were kept. The School had received two new fine Swedish globes of the world in 1816, and the administrators of the Bernt Anker Trust had donated a large English chart of the world's oceans. As well as this, the School had a collection of 84 land maps, and something that was called a mathematical and physical apparatus. This was located on the second floor, overlooking both the street and courtyard. Rector Rosted and his wife lived in the rooms downstairs together with a niece, and the porter also lived at the school. Due to dry rot, the school's library had sunk somewhat on one side. The roof, it was said, had too slight a gradient, and this was the reason that the library hall, the auditorium, the rector's kitchen, and the dining room had earlier been exposed to vigorous dripping when the snow and ice melted on the roof in the spring.

In spite of the fact that the School possessed interest-bearing capital, income from goods bequeathed, from various legacies, "per diems" from Akershus Diocese and fees from the City of Christiania, the School also received considerable support from the state coffers.

Even though physical punishment was no longer commonly employed, disciples who had committed an offense could, in addition to receiving a verbal reprimand, be incarcerated in "the Jug." And the Jug was described as a room without windows, where the only light came from a crack between boards that boys had managed to wiggle loose in order to get enough light with which to read their novels about thieves and pirates. But this was in the time before Rosted's rectorship, that period when disciples could choose between sixteen blows with a length of rope, or incarceration in "the nick." Nor was it uncommon at that time for pupils to treat the porter, who administered the blows, to strong spirits, so that they might succeed in their efforts to insulate their backsides with exercise books shoved inside their clothing. During Niels Henrik's schooldays, the Jug was located under the gallery of the school's ballroom/assembly hall; moreover, the room connected to the school's library was quite probably used in one way or another as a prison before this method of punishment gradually went out of use.

It was now that Niels Henrik's outstanding talents and interests in mathematics were seriously taken note of at the Cathedral School, both by disciples and teachers. Rector Rosted remained steadfast in his vigilance against the one-sidedness he considered to be a dangerous development, and in this anxious concern the rector probably received support from most quarters. In his book, *Rhetoric*, on the art of speaking, published in 1810, diligently used in the popular election for the founding of the constitution at Eidsvoll in 1814, and which was becoming a standard school text, Rosted convincingly explained how reason could be developed and expressed. The art of speaking had not only to do with giving a lecture to elucidate a given topic, *speech* also had "great Influence on human Perfection and Bliss. With it are developed, and made useful, our own Mental Powers; by these means we become, in many respects, charitable toward our Fellows. In the previous year's examination Rosted had given Niels Henrik *tolerably good*, or a mark of 4, in written Norwegian-Danish. In spoken Norwegian-Danish, and in Greek, Rosted gave Niels Henrik the mark of 2, but this was about to go down. Quite probably the rector reminded his pupils of the utility of the art of speaking: "By studying Eloquence we cultivate Reason itself. True Rhetoric and healthy Logic stand in an extremely close Relationship with one another. He who studies how he shall properly order and express his Thoughts, learns to think as well as to speak, with accuracy. The Relationship between Thoughts and Expression are so strong that the Improvement of Expression cannot be contemplated without the Improvement as well of the Ideas therein."

It was this sort of view of oral expression that lay to some extent behind Holmboe's rhetorical exposition of the basic axioms of mathematics. But for Niels Henrik the approach to the problem was in all probability already something quite different. Behind all the authorities' well-formulated, useful and successful expositions on thought and expression, Niels Henrik must have asked himself where had all such talk gotten his father. What had all the world's arts of speaking given to Abel senior? Perhaps Niels Henrik now felt that only mathematics would lead him out into a world where words, themselves a source of concepts, ceased to exist, where a way of thinking lay so far beyond everyday experiences that it could not possibly be identified with talk about human purpose, perfection or bliss.

Yet there was one man who wholeheartedly supported Niels Henrik. Holmboe unashamedly boasted about his disciple, and with the conclusion of the school year in the summer of 1819, he wrote in the school register about Niels Henrik: "a remarkable mathematical Genius." After rapidly completing his instruction in elementary mathematics, Holmboe had given Abel private lessons in higher mathematics, pointing him to further readings and going through the classics; in other words, Euler, Lacroix, Francoeur, Poisson, Gauss, Garnier and above all, Lagrange.

One of the first mathematical themes that Niels Henrik was particularly preoccupied by, and one which also had its genesis in elementary mathematics, was namely the solving of algebraic equations. To solve equations of the first degree, with one unknown, and equations with several unknowns, was easy enough for most. But Holmboe went further, teaching his disciples to solve quadratic equations; that is, second degree equations of the type: $x^2 + ax + b = 0$. And Holmboe informed his disciples that there were also normal ways of solving equations of the third and fourth degree, but that a solution to an equation of the fifth degree had not been found using the five calculating operations of addition, subtraction, multiplication, division and square root. A general fifth degree (quintic) equation can be written in the form: $x^5 + ax^4 + bx^3 + cx^2 + dx = e$, where a, b, c, d, e are given numerical values. Trying to solve such an equation had long been one of the most popular mathematical problems in Europe. There were many from both recreational and learned circles who sought the solution. Indeed, there were myriad attempts to find the solution, but in every instance it was revealed that something had been done wrong, or that the starting point had not been inclusive of the whole problem in its general scope.

Holmboe must certainly have shown his prize student how the general third and fourth degree equations could be solved, and Niels Henrik also had no doubt obtained the knowledge of the dramatic history behind it, the history of equations up to the point where Lagrange had advanced it further.

To solve a general equation of the second degree, which was something that every pupil now had to learn, was ancient knowledge, known by both the Babylonians and the ancient Greeks. But neither the Babylonians, the Greeks, nor later, the Arabs who all tried to solve the problem, were successful in finding the solution to the next challenge: the third degree equation, the cubic equation, as it was also called. The learned Persian poet and mathematician, Omar Khayyam, who lived in Bagdad around the year 1100, wrote in his works that the third degree equation could almost be considered a mathematical mystery. Omar Khayyam solved certain third degree equations using geometric considerations over the classic conical sections, but a real solution to the third degree equation did not come into being until the Renaissance in northern Italy, where it arose in a great, bloody and dramatic row, with the mathematicians Gerolamo Cardano, Niccolo Tartaglio and Ludovico Ferrari playing the main roles in front of a large and curious public.

An itinerant Franciscan monk, Luca Pacioli, had published a book in Venice in 1492, a collection of mathematical facts that for the most part consisted of old, well-known material. Yet nevertheless Pacioli's book became highly important, and the author himself travelled from place to place giving folksy lectures on mathematics in the Italian universities. Leonardo da Vinci was among his good friends. The book gave good instructions in computation and in algebra, and was the first that clarified the rules for double-entry bookkeeping. It was also Pacioli's book that put the word "million" in use for the first time. In the algebraic section he solved both first and second degree equations, and stated flatly that there was no way to solve equations beyond the level of the second degree. However, it was not many years before a solution was found for solving the third degree equation of the special form: $x^3 + ax = b$, where a and b have numerical values. The discoverer was Scipione del Ferro, a professor at the University of Bologna. When he died in 1526, his papers fell into the hands of his son-in-law, Annibale della Nave, who was also appointed del Ferro's successor at the University. Another who was let in on the means of solving the third degree equation was Ferro's student, Antonio Maria Fiore from Venice. None of them wanted to publicize their knowledge, as they knew that they could earn money by keeping it a secret. At that time, public spectacles involving challenges and debates between men of science were common, and members of the general public placed bets on their favourites. It was almost like the old medieval jousting tournaments with large monetary prizes. Thus, secrecy about an effective means of solving known mathematical problems could be an effective weapon in the struggle for survival in this learned society.

Fiore felt that he was free to utilize the ideas of his old professor, and in 1535 he challenged Tartaglia to a problem-solving *disputas*. Each would pose his opponent 30 problems to solve, and the loser would pay the cost of arranging 30 feasts for the winner and his friends. All Fiore's questions were built up around del Ferro's method of solving the equation; that is, all the questions came down to an equation of the form $x^3 + ax = b$. Tartaglia naturally ran into problems, but he worked frenetically in the time at his disposal, and during one sleepless night right before the deadline, he was overcome with inspiration; he discovered the method and solved all the problems in short order. For his part, Fiore was strong in calculation and weak in theory; he had trouble with the problems Tartaglia had set for him. Fiore clearly became the loser, although Tartaglia relinquished the demand for 30 banquets. The honour itself was adequate: he was now the leading mathematician in Venice.

Niccolo Tartaglia was born in Brescia, the son of a postal messenger, from a poor family. In his memoirs, Tartaglia tells about his hometown being overrun by the French army in 1512, under the leadership of Gaston de Foix, with the majority of countryfolk being massacred. The thirteen year-old Niccolo escaped with his head bleeding from sabre wounds and his mouth so bandaged that he could scarcely speak. thus, as an adult, he always wore a long beard to hide the disfiguring scars, and a speech impediment had given him the nickname Tartaglia or Tartalea, The Stutterer.

The great mathematician of the time was Gerolamo Cardano, a learned player, and also known as Europe's most sought after medical doctor. Gerolamo was born in Milan where his father, Fazio Cardano was a lawyer, and otherwise was known as a man of learning that Leonardo de Vinci, among others, consulted on questions of geometry. Fazio lived together a long time with Gerolamo's mother, who came from more modest circumstances, without marrying her. He did not marry her until just before he died. The good bourgeoisie of the city therefore maintained that Gerolamo Cardano was an illegitimate child. It was even whispered that he had been born after innumerable unsuccessful attempts at abortion. As a young boy he helped his father, but he wanted to get more education, and thus, at the age of twenty, he was in the midst of his education as a medical student in Padua. Employing superior knowledge, wit, a good memory and a sharp tongue he took part in the public debates wherever there was a cash prize involved. All through his student days he was invincible in such tournaments. Gambling was not only his favourite pursuit, but also his main source of income following the death of his father, as he himself admitted in his memoirs. He would scrupulously calculate his chances of winning, and had begun to formulate propositions on "games theory," ideas that later came out as *Liber de Ludo Alea*, the world's first book on the calculus of probability. Cardano became a doctor of medicine at the age of twenty-five, but because he had been born out of wedlock, he was not allowed to practice within the City of Milan, where his mother was still living. Consequently, Cardano became a country doctor at Sacco, a short distance outside Padua. He got married, and by his own admission, the six years in Sacco were the happiest in his life. In addition to conducting his medical practice, he began to publish books, not only on medicine, but also on astronomy, palmistry, mathematics and a whole series of other topics, all written in a manner that was easily understood by ordinary people. Local people began to see that he wrote about practical medicine, and about what was wrong with ordinary medical treatment; the medical faculties of the universities, however, criticized him. Gradually Cardano swept aside all the ill-will and opposition; he became Milan's foremost doctor, both popes and kings were aware that no one could provide them better health care than Cardano in Milan.

Cardano certainly became aware of the competition between Fiore and Tartaglio in Venice, and he grew curious. Even though Cardano was critical of Pacioli's book and could detect obvious errors in it, he had been in agreement that only equations of the second degree could be solved. Cardano now tried to pry the secret out of Tartaglio but Tartaglio brushed aside all of Cardano's overtures. Cardano was in the process of writing a book on algebra and wanted very much to include the latest developments, and he offered to publicize the solution under Tartaglia's name and give the full honour to the mathematician from Venice. Tartaglia, however, refused; he wanted to write his own book. Cardano offered a considerable sum of money, 100 *scudi*, for a mathematical dispute in Milan, but to no avail. And finally when Tartaglia's book came out, it did not contain the much sought-after solution to third degree equations, but rather a study of projectiles and an enquiry into falling bodies. This was half a century before Galileo carried out his experiments at Pisa.

But Cardano, who now had good contacts everywhere, and in particular, in Milan, wrote to Tartaglia and said that there were great prospects for him, in terms of money and honour, in the military in Milan. Thus lured, Tartaglia moved to Milan, and during one of his discussions with Cardano, probably on March 25, 1539, he revealed, in a coded form, the means of solving the third degree equation, on the proviso that Cardano keep it secret and never write about it. Cardano swore on the Holy Gospel, and as "a gentleman,"and promised as well to note down the solution in code so that after his death no one would be in a position to understand it. But it now happened that Cardano had a student from Bologna, Lodovico Ferrari, who had come to Cardano's house as a poverty-stricken boy in search of a position as servant.

Cardano discovered the boy's extraordinary abilities, and Ferrari became one of his closest collaborators. Ferrari was teaching mathematics before he reached the age of twenty, and in a short time he had such a reputation that he was invited to teach the emperor's son. Ferrari graciously refused this offer, saying he preferred to remain in Cardano's company.

After having seen Tartaglia's solution to the special third degree equation ($x^3 + ax = b$), Cardano began to do more work on it. He soon found how a *general* third degree equation could be solved. And not only that: when, in the course of the solutions, he encountered what were called irreducible cases - the highly paradoxical situation wherein the formula resulted in expressions where the roots could not be calculated as ordinary numbers - Cardano clearly saw that there must also be three completely calculable solutions to be found.

He understood that there must always exist three roots, or solutions, in third degree equations, and consequently he was forced to attack the calculations by making use of these imaginary numbers, which normally had no meaning. He expected that they had to behave like other numbers, and rise to correct and intelligible results expressible in ordinary numbers. This treatment of complex numbers, as they were later to be called, must have appeared as magical as the occult studies that were still being undertaken in that period. Cardano devised approaches for solving such equations, and he had some ideas about the connection between the roots and the coefficients in the equation. He had approached the field that would later be called higher algebra, and the achievement becomes greater when we realize that negative numbers were not fully understood before Cardano's time.

Cardano discussed mathematical problems every day with his student, Ferrari. Then, after having seen Cardano's solution to third degree equations, Ferrari solved with great ingenuity and cunning dodges, fourth degree equations. This method of solving the problem continues to bear Ferrari's name. Master and student now had a great abundance of new and important mathematical knowledge, and they had to find a way of making it public. What held Cardano back was the oath he had made to Tartaglia. Indeed, Tartaglia's solution had been the starting point for these fantastic discoveries. But no help came from Tartaglia; quite the reverse. He had promised to make public his solution, but he had still not done so, and it was fruitless to try to find out what his real plans were.

In 1543 Cardano and Ferrari journeyed to Bologna to visit Annibale della Nave, and here they got to see a number of Ferro's papers, and came to see with their own eyes that he too had found the solution to equations of the type $x^3 + ax = b$, thus precisely the same thing that Tartaglia was now holding secret. Whereupon Cardano maintained that Tartaglia was not the first to have found the solution, and no oath could hinder him in making public a discovery made many years before Tartaglia's. Cardano felt himself free to publish both del Ferro's solution to the special third degree equation, and his own epoch-making solutions to the general third degree equation, and in addition, Ferrari's work on the fourth degree equation. Thus, Cardano's book, *Ars Magna*, was published, in Nürnberg in 1545, and was earth-shaking for equation theory, and rapidly won praise from among all mathematicians with the exception of Tartaglia of course. Despite the fact that Cardano had honoured Tartaglia, Tartaglia became livid, and put out a challenge for a meeting and a competition with a large cash prize. Cardano now allowed Ferrari to take over the relations and the controversy with Tartaglia, something that annoyed Tartaglia intensely, who desired to do battle with the most famous mathematician in Europe, and not with the young, unknown Ferrari. The dispute about who had copyright to the solution, and the debate about whether or not a scientific discovery in relation to fundamental principles ought to be kept secret, or allowed to circulate freely washed forward and back, in front of a large and avid public.Tartaglia's point of view was that of the Middle Ages: the solution was his, and he had been personally deprived of ownership!

After many "ifs, ands and buts" and discussion about money, and the questions that should be posed, the *disputas* between Tartaglia and Ferrari was arranged to take place in Venice on August 10, 1548, under the direction of the Franciscans, and with the Governor of Milan as the highest judge. Despite the fact of the existence of only Tartaglia's version of what had occurred, and that he pooh-poohed the whole incident, it was clear that Ferrari won overwhelmingly on all points. Some professed to know that the defeat had affected Tartaglia so strongly that in the end he took his own life, or at any rate, they said that he did so in order to show that he could predict the day of his own death.

Since that period - during the 1550s - mathematicians over the whole of Europe had been on the hunt for the solution to the next natural challenge: the fifth degree equation. Over the years there were many who claimed that they had found the solution but they had accordingly always revealed that their starting point for approaching the solution was either a special case of the general equation, or that they had made mistakes in the process.

It was to be the student, Niels Henrik Abel, on the farthest fringe of learned Europe, who would finally demonstrate that this was impossible, and that the solution which so many had sought, simply was not to be found. And it was now, as a disciple at the Christiania Cathedral School that he began to interest himself in the problem. Niels Henrik studied Lagrange, the works in which the great French mathematician had managed to demonstrate that the root expressions that occurred in the solution of equations of the third and fourth degree could be written as polynomial expressions of the equations' roots. Lagrange used this angle of attack in his own efforts to solve fifth degree equations, but he had to give up. Niels Henrik would succeed in demonstrating that the formula of a solution to the fifth degree equation necessarily led to a self-contradiction. This, for which he had made his own calculations, was what he published in 1824, and which he thought would be his passport to Europe's learned societies.

At the beginning of July, 1819, Niels Henrik was advanced to the senior class. This time he received *very good* as his main mark. Hans Mathias was once again left with a *tolerably good*.

What was being most discussed during the early summer season that year was the violent fire that had destroyed the old trading houses of Anker, Collet and Thrane. One day in May, 1819, all the highly valuable but uninsured timber of the lumber yards in Bjørvika had gone up in flames. Despite wholehearted efforts to extinguish the blaze, all the stored timber, which had been ready for shipping, was reduced to ashes within a period of twenty-four hours.

For the disciples at the Cathedral School there was a one-month's summer holiday and in all likelihood the Abel brothers went home to Gjerstad. It was an unusually good summer, at least with regard to weather and crop yields. As writings from the Gjerstad region attest, the crops exceeded themselves, particularly in July after the good rain that fell at Midsummer's Eve, followed by heat that was to remain memorable across the region. But around the vicarage conditions were not nearly so prosperous. Everyone could see that Vicar Abel drank too much. Spirits were sold even on the grounds of the church. It is likely that Niels Henrik did nothing to improve conditions. He quite certainly continued to be an avid participant in the card games in front of the church, presumably he was a constantly good player who could calculate the winning odds. But what he did not know and could not calculate, on the day when he waved goodbye to his parents at the end of July, 1819, was that he would never see his father again. On August 19, 1819, some weeks after Niels Henrik and Hans Mathias had journeyed back to the capital, Vicar Abel wrote to his dean in Arendal: "Constriction in the Chest, coupled with intermittent Blood-spitting do not seem to promise a very cheerful Future."

24.

To the Examen artium
and Further Studies

The school year began again in Christiania on August 1st. This was a new arrangement in 1819: formerly school had started on October 1st. Niels Henrik was now in the senior class, a prefect, and had two years to go until he took the *examen artium*. According to custom he was now probably baptizing the newcomers, the wretched "plebs," in the school courtyard. One of those who was now beginning his school career at the Cathedral School was Henrik Wergeland, who was to become one of Norway's leading national poets and intellectuals. There was also a new man among the teachers: Lieutenant Erik Christian Busch was hired by the hour to teach French, and would also be responsible for instruction in writing and drawing. It was reported that his most cherished command was, "Shut up! numbskull, when you're talking to me!" but during the writing and drawing classes he sat glued to the teacher's desk, engaged in his own affairs. Niels Henrik had the lieutenant for French. In English, in which he was now instructed for the first time, he had his old language teacher, Adjunct Melbye. That the party-loving francophile, Melbye, who was also to have taken a journey abroad at this time, had now been removed from the teaching of French, was perhaps a first reaction to his growing inability to concentrate. Nine years later he was forced to give up his post.

Otherwise, the outer contours of the school session were as before. But now it seemed that for Niels Henrik everyday life had taken a new turn. Not only were his unusual mathematical talents noticed and remarked upon, but also Niels Henrik himself seems to have taken a more active role in the events of his surroundings, and through his association with Holmboe and his friends he found new circles in which to move. It was now that the merriment friends later remembered him for saw the light of day. Perhaps there was a little of the same "excessive cheerfulness" that his father had seen, now revealing itself in social gatherings. But Niels Henrik's practical proficiency in mathematics

was not rated very highly; later his friends recounted jokingly that they never
once suspected he was even capable of keeping score in card games.

Cardplaying aside, and more importantly, it was precisely now that Niels
Henrik began to associate with students interested in the natural sciences. This
circle consisted of Baltazar Mathias Keilhau, Christian Peter Bianco Boeck,
Jens Johan Hjort and Bernt Wilhelm Schenck, who all became university
students in 1816, and it also included Christen Heiberg, who had taken the
examen artium the following year. Heiberg came from the Bergen Cathedral
School, Boeck was privately tutored for university entrance, and the others
had been comrades at the Christiania Cathedral School. Keilhau studied
mineralogy, but the others had chosen medicine. To be sure, medicine was
traditionally the only subject in which natural sciences were taught; the choice
of courses of study for those proceeding to public service positions were
theology, law and medicine. In 1821 Keilhau became the first to take the
mineralogy exam at the University in Christiania. That these students in 1819
invited *Disciple* Niels Henrik to their weekly meetings must mean that young
Abel was already considered to be a legitimate man of science. These students,
following the pattern of other countries, had eagerly formed a kind of scientific society associated with the university milieu in Christiania; they "read out
small scientific Compilations" to one another and discussed the future tasks
in the natural sciences.

Niels Henrik took part in new, exciting get-togethers. One black cloud
hanging over his pleasure was perhaps the fact that his big brother continued
to sit at home in an ever-darkening disposition. But what could Niels Henrik
have done? They went to the same classes: did he help his brother with
homework or skip out to play cards, or visit Holmboe and his university
student friends? Did he take Hans Mathias along when as a rule the disciples
of the Cathedral School were invited to dress rehearsals at Grænsehaven?

That autumn the first item that the theatre presented on its bill was a family
drama, *Journey to the City*, by August Wilhelm Iffland, starring one Ole Rein
Holm. Later in the season a lively English comedy was staged. But the perhaps
most important performance opened at the beginning of November. For the
first time a real event was dramatized on stage, and with brilliant acting it was
said. The piece was called *The Slave Girl of Surinam*, written by an F. Kratter,
and translated by Knud Lyhne Rahbek, a guarantor of bourgeois enlightenment and healthy reason. The play depicted conditions that in reality could
have existed in Surinam, and was so genuine that people's real names were
used. The Dutch plantation owner, van der Lyde was played by Ole Rein Holm,
and she who played the slave girl, Assessor Peckel's wife, had never before
performed this well. The only negative feature of the play, according to some,
was that the sets from the plantation were so beautifully made, that the

dreadful social and working conditions of the negro slaves lost some of their gruesome reality.

Cathedral School pupils would certainly not have wanted to miss the December bill at Grænsehaven either: Enevold Falsen's play, *The Jolly Cousins* and J. C. Brande's *Ariadne of Naxos*. As well, *The Crusaders* was mounted again, more or less the same performance as in April of the previous year. The final presentation before Christmas, 1819, was Holberg's comedy, *Don Ranudo de Colibrados - or Poverty and Haughtiness*, about the proud aristocratic couple who preferred to starve rather than accept help from their daughter's lover, who came from a less venerable and less august family.

In the December exams, Niels Henrik did worse than in the advancement exams of the previous July. He went down in all the language subjects, standing with a *tolerably good* for Greek, French, German and the new subject, English. But things were even worse for big brother Hans Mathias. In several subjects, *failed*, was used on Hans Mathias, and he got six *indifferent*s, the lowest mark it was possible to receive. Only in Latin and French did he do slightly better. But this would be the last time that anyone would try to evaluate Hans Mathias' learning. Right after the Christmas holiday, which the boys probably spent in Christiania, Rector Rosted noted, "It seems his Mental abilities have become so weakened, and together with this, his Diligence and Desire, that there is no Hope that he will ever be ready to be admitted to the University." When father Abel received this news he wrote back to say that it was best that Hans Mathias be sent home immediately. This was done in March, 1820.

Conditions at home at the vicarage were more dismal than ever. In November, 1819, Vicar Abel had once again written to the bishop requesting a transfer. He assembled many letters of attestation and reminded the bishop that Cabinet Minister Treschow "had prepared me for university, affirming that should I ever have need thereof, Transcripts could be sent." Father Abel was in despair. He wrote, "I, with my wife and 6 children have been brought to the Penury of the Beggar's staff and we can only be saved by the more favourable Prospects that a rapid Transfer would afford." In a letter that Pastor Abel wrote to Dean Krog in Arendal in February, 1820, he defended himself against claims and rumours that were advanced in the parish about the priest being a liar, and he urged at the end that a disagreeable matter could be postponed until the summer when both bishop and dean would be at the vicarage. On March 9th Abel wrote again to Bishop Sørensen in Kristiansand, or rather, he had to get his twelve year-old son, Peder, to write the letter, as an "arthritic Ache in my right Hand makes it difficult for me to hold the Pen." The letter continued, "Penury, Sickness and domestic Bother have, in this recent period, ground me down, and now in addition I must even take my eldest son, Hans Mathias, who has weakened in both Body and Soul, out of the Latin School in Christiania. Nothing can save me other than Advancement to another Ministry, be it even to endure, for here most of the Peasants are owned by Carstensen..."[5] After explaining that he was so poverty-stricken that neither he nor his children could manage to obtain necessary clothing, Vicar Abel continued: "The sole happiness I have is that Niels Henrich [sic] is doing well and will become one of the foremost mathematical Geniuses." He urged the bishop to recommend his two "Solicitations, the one to Halden's and the other to Eger's Parish ministry," particularly the position in Halden.

He ought certainly to have gotten this position, but Søren Georg Abel was seriously ill; he had only a few months left to live.

Niels Henrik's reaction to his brother being sent home probably was one of relief; he must have agreed that it was the best solution. But everyday life in that room in the garrett must now have become somewhat different. He was quite certainly delighted by the theatrical offerings in the capital that February. A play by Oehlengschläger was performed in Christiania for the first time. *Håkon Jarl* was the name on the marquée. The play had originally been planned for May of the previous year, but had been postponed until autumn, quite probably because of illness, and then postponed again. The Dramatic Society shrank away from tackling such a huge enterprise. At the beginning of January it was decided that they would perform *Håkon Jarl* during the annual market days in February and, as was the custom, the proceeds would go to the city's poor. Many had read the play, and many had seen it performed in Copenhagen, where it had been first staged twelve years earlier. But the production in Christiania was one of the big ones in the theatre's history. *Håkon Jarl* dramatized the struggle between the champion of Christianity, Olav Tryggvason and the old Viking, the Earl of Håkon, defender of the old pagan culture. The performance was vivid and it gripped the audience. Håkon, the tragic earl, with his obstinate, manly Nordic spirit (played by a Lieutenant Juel) was the main character of the play, and his thrall, the murderous Kark, was played by Ole Rein Holm. What was most gripping was the final and concluding scene: after Olav Tryggvason had been acclaimed and made king, and the multitudes followed him to a huge feast in "Trondheim's cheerful grove," there came the following scene: in an underground vault by burning torchlight two men entered bearing a black coffin. They exited and the Earl's

---

[5] Henrik Carstensen (1753–1835), prominent in business and politics, was not someone that Abel the father saw eye to eye with at this point in time, probably because Carstensen had survived the economic downturn that followed on the heels of the Napoleonic wars; and to add insult to competitive injury, Carstensen had bought out the bankrupted Merchant Simonsen of Risør, Søren Georg Abel's father-in-law.

beloved, Thora, played by the lovely Kaja Lasson, slowly appeared. She advanced with a drawn sword and a wreath of spruce boughs which she lay upon the coffin, and among other things, declaimed to the dead Earl:

You were true Nordic, our rarest hero;
A flower stifled by a Winter frost.
One day will Northland's saga coldly tell
When Time's hand's erased the Colour out,
And only the Exploit's contour stands behind:
He was an evil spirit, cruelly did worship Idols!
Stammering men may dare to mouth your Name.
I stutter not; for You I knew.
The best of Powers and a mighty Soul
Sacrificed to his Time's Delusions.

Thora stood by the coffin, wishing her beloved good night, and said with a finality that chilled the audience:

Now I go and lock myself away.
And later when this Door be opened,
Bring Thora's servant to fetch her Corpse,
And set it by her Håkon's side.

The lights faded away on the scene, and the curtain came slowly down. In the solemn silence that ensued the players left the stage and quietly closed the doors behind them.

At the same time that Hans Mathias was preparing himself to leave Christiania, something else happened at the Cathedral School that would change everyday life for Niels Henrik. On February 25th, the natural science teacher, Flor, did not appear for his classes, and the rumours that flew were soon confirmed. The unfortunate Flor had thrown himself into the harbour at Bjørvika and drowned the day before.

An uncommonly large cortège, composed of people from all walks of life, followed Flor to the grave. Many held in high regard this man, who earlier had been ridiculed as the "moss priest". Two of his former students, who had now become well-known poets, Maurits Hansen and Conrad Nicolai Schwach, wrote memorial verses to M. R. Flor who had given so much for the well-being of both disciples and the general public. Schwach wrote:

Greatly have you struggled, much endured;
To you the world accorded oft cold favour;
Sleep soft in your sheltered room inured;
Till morn doth bring a better World to Savour.

At the School, Flor's subject "natural history" was taken off the curriculum. In the previous year's examination Niels Henrik had been given a *very good* in this subject, but nevertheless, perhaps it was not difficult for him to find a way of using Flor's two weekly lesson hours for something else? As well, the few crumbs of instruction of a somewhat sexual nature that Flor was to have given his senior class had now disappeared. Bachelor Flor's view on this type of instruction was that an experienced teacher could provide better preparation for such "Things of Nature" than that which the disciples learned from "smutty Books and uninformed People." But the "fathers of academic life" in their lessons to the senior class, gave precious little ethical guidance regarding the temptations that the disciples faced. And first and foremost these temptations were a lust for drinking and whoremongering, "Drunkenness and Lust". All the same, during this post-Christmas winter and spring season two chapbooks came out and were most avidly gleaned for their intimations of sexual enlightenment. These two long poems which could be sung as "The water that brings evil and the wine that brings good," had recently been published in spite of the fact that Justice Minister Diriks had wanted to prohibit this. The poems opposed "the Laws of Modesty," as it was phrased, but because they had been published privately, the Minister of Justice could do nothing. Niels Henrik's classmate, Niels Berg Nielsen, spread the message. Berg Nielsen was a cousin of the poet Schwach, who had written the poems, and in addition had lived in the house of Schwach's father and had been tutored by him before he entered the senior class at the School. University students and School disciples were now eagerly reading that "There is no loss of Paradise" which should be located:

At the foot of a hill, that divides in two,
A smiling Valley unfolds its jests;
And round its Sides do Thickets grow,
And small birds flit and build their nests.
Midway in the Dale, a Spring of Power,
Like the Source from which the Nile doth grow,
This water, as the Nile, bursts from its bower,
And swift does its Fountain flow.

And back there at the Fountain's core
There opens of itself a deep and splendid Cave
Where Adam can, when the Devil cries for War,
Hide safe away and put to use his stave.
Its Portal narrow, and driven thus by Need
Did Adam oft break through;
Though t'was never the trouble he had foreseen
So sweet the Balm, that to his senses flew.

But just as Adam, the moment of his Fall,
Was cast out from Eden's fruit and flower,
So now must our Adam leave his Valley's thrall
Amid the swampy wastes of this desolate hour.
Though the Fallen is condemned to loss forever,
The Fallen can rise himself again,
And go back in and enjoy once more the pleasure
That one ought never construe as pain.

There were several more verses, and the poem was entitled "Kennst Du das Land?" This was taken from Goethe himself. The Goethe poem was well-known at the time, and the first line was "Kennst du das Land, wo de Citronen blühn?" (Do you know the Land where the Lemons bloom?). The other poem circulating among University students and Cathedral School disciples that season was called "The Tree of Life" with a citation from Oehlenschläger under the title. This poem was enigmatically both the tree of the Bible, and a tree of fecundity and sensual possibility. The poem ended with an unambiguous statement that declared, indeed, while many learned writers considered this to be the Biblical tree, the knowledge of whose fruit could cause death, it was more correctly:

...that from South to North
Wherever human souls are Rife,
'Every day it bears new Fruit on Earth
And acts as the very Source of *Life*.

From his cousin, Schwach, senior classman Berg Nielsen also had exciting stories to tell about this Ole Rein Holm, whom people more and more talked about as the city's leading male actor. Holm and Cousin Schwach were best friends, and Berg Nielsen would tell about the great glittering star, Holm, who, right in his own house, would have his ears boxed by his angry, ill-tempered wife, one Helene Thoresen, whom Holm had chosen out of pure love, but who, in her turn, had married him for his money. Ole Rein Holm had been one of the city's richest bachelors, good for more than a casket of gold pieces when, a couple of years earlier, he had married shopkeeper's daughter Helene. The actor's father, Wholesaler Holm, had taken his own life, and his mother was the sister of the poet, Jonas Rein.

Could it have been that Niels Henrik's interest in the theatre, and his taste for the rumours and tidbits of information that Berg Nielsen could contribute about various actors' lives, both on stage and off, may had something to do with a hunger in Niels Henrik to see relationships, and understand what it really was that drove people together, and according to which rules?

From Risør the news now came that his mother's father, Niels Henrik Saxild Simonsen had passed away on March 10th.

By Easter time, Father Abel had become so ill that he was in bed most of the time, and he was growing weaker by the day. Upon his homecoming, perhaps the twenty year-old Hans Mathias had some days respite in which to find out what was happening and to prepare himself to receive his father's admonitions. The other brothers, Thomas Hammond and Peder Mandrup, were fourteen and twelve years old, and Father Abel called these boys to his sickbed. He would have admonished them, and among other things pointed to their mother and warned them against the steady decline in her moral conduct and rectitude. He would also have bid them take care of their youngest siblings, Elisabeth who was ten, and Thor Henrik who was about five. But it would be only the absent Niels Henrik who seems to have taken his father's last wishes seriously. As for Hans Mathias, it would not be long before neither sorrow nor happiness would be capable of making the least impression on him.

It was getting on for spring and spring farm work in the Gjerstad region. After a hard winter the mild weather had arrived and it was a quite normal spring. But it was reported that snow and slush covered the landscape on May 4th and in the morning the surface of the snow was strewn with earthworms, so thick they looked like sown fields in many places. This was the day that Pastor Abel died at the Gjerstad vicarage. The next day was mild again and the spring planting was in full swing.

From whom and how quickly Niels Henrik received the news of his father's death is uncertain. In any case, it is quite certain that he had no chance to come home for the funeral. But news of what happened at the funeral probably reached him quickly enough: the paternal aunt, Margaretha Marine, Miss Abel, came from her sewing room in Risør to help out at the vicarage. Vicar Schanche of Risør came in beautiful weather to conduct the obsequies. There were many people on the church grounds the day of the funeral, and quite a few more than usual at the vicarage afterwards. Perhaps Miss Abel had arranged a fitting conclusion to events here in her brother's house, the very place that had also been her childhood home: there would be no potato hootch with the fusil oil still in it at this public servant's funeral, but rather real gin, *genever* from the Netherlands, and Jamaica rum! But what everyone talked about was the behaviour of Mother Abel, the priest's wife. On that day, the day of her husband's funeral, she got openly drunk, and just as openly, beckoned her servant boy to her, and with him at her side, she went to bed. When Pastor Schanke and Miss Abel greeted her with reproaches the next morning, she only replied that Jørgen L. had promised never to leave her.

In Christiania Niels Henrik threw himself into the mathematics books that he had borrowed from his teacher, Holmboe, and those that he could himself obtain from the library. These consisted of Newton, Euler, Lagrange and d'Alembert, and books that in addition to mathematics, contained studies in mechanics and astronomy. At the weekly gatherings of the students, Keilhau, Boeck, Hjort and others, it seems that Niels Henrik presented his first "Compilations." As to what these compilations dealt with we know nothing, but in any case, by now Teacher Holmboe stated that Niels Henrik had made colossal progress in mathematics. In the school records following the year-end exam in June, 1820, Holmboe did not bother with the usual categories of Natural Gifts, Academic Diligence, Craftsmanship and Future. He wrote right across the page: "With the most incredible Genius he unites Ardour for and Interest in Mathematics such that he quite probably, if he lives, shall become one of the great Mathematicians." Beyond this, in front of the words "one of the great Mathematicians" several words have been stroked out, which can almost be detected as "the World's greatest Mathematician", and the addition of "if he lives" might indicate that his health, even now, was not of the best.

In the course of the four weeks of the summer holiday in 1820, Niels Henrik was home at Gjerstad. It was the first time since his father's death. Maybe he had had a greater desire to accompany his comrades, Keilhau and Boeck, when in June that year they went on a hiking tour in the Norwegian mountains, a hiking tour that would leave great tracks and would be much talked about in the coming years. Niels Henrik had no doubt taken part in the meetings where the aims of the journey were discussed and planned. Norway the Fatherland had to be mapped, and Keilhau, who earlier had done geological research in the mountains of the Valdres region of Norway, had long had plans to push deeper into the mountain wilds, and Boeck, who during the summer of 1819 had been making botanical collections at Fillefjell, was an obvious travelling companion. On the map then in use, drawn up by J. Pontoppidan in 1785, there was only one name between Lom and Valdres: Skagen Fjell with Fillefjell to the south. Neither Lake Tyin nor Lake Bygdin had been marked in. Boeck and Keilhau had accordingly decided to mount an expedition to this "Plant-rich Chain of Mountains," as Christen Smith called it.

The botanist Christen Smith had long been a sort of model for young researchers in the natural sciences. He had in the course of his many long and exhausting travels across large parts of southern Norway studied and cata-logued Norwegian plant life and made it known abroad. He had been made professor at the University, but he had hardly begun his lecturing before, while on an English expedition to the interior of Africa, he had died of fever, at the tender age of thirty-one. On the Congo River at a place called "The Tall Trees," he was lowered into the ground in the fall of 1816. During the summer of 1812, the natural science teacher, Flor, had joined Smith in an expedition in Tele-mark, westward from Bolkesjø, Gausta, and Møsvatn and over the open moorland to Sørfjord and Bergen, and onward to Jostedalen and back via Fillefjell. 1812 was the year of the great famine, the summer was uncommonly cold, the grain crops froze, and people were forced to mix bark with meagre amounts of grain in order to make their bread. Teacher Flor had not managed to stay with Smith beyond Møsvatn, but before they parted company at a point below Gaustatoppen, one of the highest peaks in Norway, the two friends had carved their names in the bark of a large spruce tree. Subsequently, during a winter storm in 1816, the tree had been hit and splintered by lightning; it was like a warning of the demise of the two friends, one in the Congo and the other in Bjørvika. In memorial verses written to commemorate Flor's passing there was an allusion to this coincidence: "The Spruce fell to the Force of the Winter's Storm,/ While there upon a far-off Shore your brother died;/ Tired, you too fell to Death's cold form;/Go now, and sit by your brother's side."

In the course of their summer trekking in 1820, Keilhau and Boeck set Jotunheimen on the map for the first time. (This is the high mountain region in the touristic "fjord country" of western Norway, south of Trondheim and north of Bergen.) And for this feat the two young men would harvest great honour. Jotunheimen was an area beyond the normal realm of reindeer hunters, shepherds and fisherfolk; no one before had shown the world this expansive alpine world that now received the name Jotunheimen ["Home of the Giant Trolls"] and which everybody at that time expected to contain the highest peaks in the fatherland. The young students had calculated peaks of 7,100 Rhineland feet[6] above sea level and still thought there were higher

---

[6] 1 Rhineland foot = 31.55 cm.

mountains in the region. Earlier Professor Jens Esmark had rather categorically declared that Snøhetta was the country's highest mountain.

When Boeck and Keilhau returned to Christiania in August, Keilhau wrote about the discovery in *Morgenbladet*. This sensational exploit made them well-known, and in the eyes of many, their little circle of young students and their weekly gatherings were now regarded as something more than just a private society. Inspired by all this benevolent fame, the young researchers now wanted to form a physiographical association "where the Aim should be with Respect principally to the physical and in part geographical investigation of the Fatherland."

It is not know to what degree Niels Henrik took place in these plans and formulations, but it is quite certain that he listened to his friends' unreserved joy and astonishment, following their experiences with the mountains and glaciers. When they stood on the shore of Lake Bygdin surrounded by four or five hundred cattle, they probably felt that "The tract is the uttermost Wilderness and Appalling," but the longing to come into this unknown and alluring region was overwhelming. Keilhau could lose himself in descriptions of how deep undertones would well up through the sonorous ice when water flowed down mighty fissures, together with rocks and chunks of ice and things would chink against one another in the deepest tones: an organ played by unseen hands! It aroused feelings and thoughts that indeed there was life in all inanimate objects; it was the chaotic motion and the very struggle of the elements. One was thrown back into primal time, before the beginning of organisms, to the childhood of the earth. On many occasions they stood in admiration and watched the competition of stones in rockfalls that danced and thundered down from the peaks and came softly to a common stop in a snowdrift below. The rocks passed over them "as though stuck together at a certain point in the air." Another story was that Boeck had almost been killed by such a rock that the snow had *not* stopped. The barometer on his back had been smashed and he himself had been slung to the side by the flying boulder. However, dizzy and with great effort, he got up again and continued, and after a couple of weeks he said he no longer noticed his wounds. They were young researchers on an expedition of discovery and they boldly defied all dangers. They paid no attention to the treacherous crevasses in the glaciers, or to the snow up to their armpits as they struggled up toward the peaks, all the while, avalanches and rock falls tumbling around them in the warm afternoon sun, or during the violence of thunderstorms.

But on one occasion they had both been in extreme danger. They were on the ascent of Koldedalstinden [Cold Valley Peak] that, due to its unusual shape, the writer, A. O. Vinje, later called Falketinden [Falcon Peak]. The local porter that they had hired from Øystre Slidre raised strong objections when they stood at the foot of the perpendicular pinnacle, and Keilhau as well understood that the assault on the peak was foolhardy, at least he said so afterwards, "but it was as though a Mania had seized my Travelling Companion." Boeck absolutely wanted to conquer the unclimbed peak, and he ordered the porter to accompany him. Keilhau climbed a lesser peak instead, but three hours later when he saw the others on top of Koldedalstinden, he also had the urge to be there with them, and put on a burst of speed in order to meet the others on top before they started down again. But every time he grasped the rock face, water poured down his arms and soaked his chest, and sharp stones continued to bounce over him; since the stones "frequently dealt out painful blows and wounds," he clambered up over the precipitous rock faces and again and again sensed he was on the very edge of nothingness. He made a sacred pledge to himself that *if* he made it to the peak he would never go back the same way. He reached the top, was reunited with the other two, and after having knocked some rocks over the near-perpendicular precipice of the north face, and over the slightly gentler south side, they began to discuss their descent. The farmer from Slidre maintained that they had to go down the way they had come up; Keilhau felt that the southwest side looked a bit easier, and under no circumstances would he go back the same way. Thereupon they divided in two once more, with the peasant farmer going back down the same way they had come up. Keilhau and Boeck went toward the west, and came upon steep glacial ice, encountered unstable, sliding stones and gravel that almost sent them over the edge; moreover they came up under an overhang and were drenched with ice water. It was dangerous everywhere, bad weather was moving in, the roll of thunder could be heard in the distance, and it was getting on toward evening. Keilhau said that he saw a possible way down further to the west, but that entailed climbing up again for a stretch, and Boeck regretted bitterly that he had not followed the advice of the peasant. "For a Second we stood silently on the Rim of the Glacier, and with just a glance at one another, we parted company. This frozen landscape of ice and snow had finally seized the very Heart with its cold; the Dangers of the day and the Physical Exhaustion slayed and defeated Sensibility. All right then, let him, if he wants to, go off and smash himself to bits, the Blind Fool!"

In this manner the two friends parted company at the brink of a precipice on Koldedalstinden. Keilhau carried on over a thin layer of snow that obscured "the Abyss that with each step was ready to welcome the Foolhardy." He worked his way down laboriously and after four or five hours came to Tyinboden, a primitive stone hut that they had left that morning and where they had agreed to meet again in the evening. But no one was there. It was dark and raining hard. Finally, after a while the peasant rode up on horseback. He had been out looking for his horse. They managed to light a fire and prepare a meagre meal. Then the farmer immediately fell into a deep sleep, but nightmarish visions flashed in front of Keilhau: Boeck, the friend that he had forsaken: "appeared there with his Head smashed in and broken, blood-stained Limbs - or I saw how he had tumbled from the high Stonewall Peak (Muurtinde), he had plunged over the Cliff and was flung to annihilation in the horrible Chasm…A fierce thunderclap crashed as rain shot down from a Cloudburst. Then a Shout was heard, a voice came down through the Smoke-hole in the Roof. It was his!"

This mortally dangerous event on Koldedalstinden was a topic that the two continually returned to; they said it was their most cherished memory of the whole expedition, a journey that continued westward, over Nordre Skagestølt-ind and further toward the sea at Nordfjord, to Lodalskåpa that no one had climbed before, and that *they* now had a great desire to climb, particularly due to the stark warnings by local people. It was near there that the rock fall had smashed the barometer and almost killed Boeck, and they had given up the attempt, continuing on to Lom and Skjåk, Vågå and Sjodalen on the east side of Jotunheimen, and back to Slidre which the peasant and his horse called home. Jotunheimen was now on the map.

Beyond this we do not know much more about this "physiographical association" that Niels Henrik took part in together with Keilhau and Boeck, Heiberg and Hjort (the student Schenck died that year). But it is certain that the mapmaking of the country's geographical features was a task much talked about. In this new, nationalistic Norway it seemed that everyone was eager to know what was contained and involved in the country. Keilhau's geographical research and his discovery of granite between younger types of rock provided material for real discussion on what the earth's core was actually composed of. And the medical men, Heiberg and Hjort, went out eagerly looking for new treatments for the commonest illnesses; to this end they had begun to dissect human bodies. What could Niels Henrik contribute to this society? He certainly worked with equation theory, tried to locate new properties with different functions and problems soluble with the help of definite integrals. But perhaps one might suspect that, in good-humour, he might have taken up problems related to the needs of these geographical discoveries?

Everyone knew Jonathan Swift's book, *Gulliver's Travels,* and these young men of science would certainly have remembered Captain Lemuel Gulliver's visit to the flying island kingdom of Laputa. To be sure, this long Laputa episode was considered as one of the weakest sections of Swift's book. Men of learning also felt that much of it was a flat imitation of the classical Greek writer, Lucian, the lusty satirist from the second century A.D., who had written *A True Story.* Swift also wanted to poke fun at the natural sciences and research: the inhabitants of Laputa lived in constant fear that the sun and the comets would come down and destroy the world. For since the world was governed by mechanical clockwork as Newton had established, many were afflicted by the fear that this clockwork would run down, and the great comet that Halley had talked about, might come and destroy the known world. Mathematicians and astronomers were not afraid that the stability of the earth depended on a fine balance between the velocity of the earth being pulled toward the sun, and the angle of the earth's tangential velocity from the direction of the fall line, but for most others it sounded alarming.

But what might have been a starting point for a mathematical treatment among the students of Christiania, was Swift's humorous description of the courses of a dinner at Laputa's Academy of the Natural Sciences: the three courses were, first, shoulder of mutton cut in the shape of an isosceles triangle, then a joint of beef in the shape of a rhombus, and finally a pudding in the form of a cycloid. The cycloid was an interesting curve that quite probably Niels Henrik considered in the course of his mathematical reading. In several instances Lagrange had engaged in geometrical reflections as well, and apart from Archimedes' spiral, the cycloid was one of extremely few examples of what were called transcendent curves. The cycloid had even been called the Geometric Helen, after the most beautiful of all Greek women. The curve had been studied zealously and admired passionately, and had caused strife and discord among European mathematicians since the 1500s. The Geometric Helen can be explained as the curve that a point on a wheel describes when the wheel rolls across a straight plane base. In more exact terms: when a circle with radius "a" rolls, without slipping, along a straight line, then a point "P" on the circle describes a *cycloid*. The cycloid is accordingly an extended arc. Galileo Galilei was the first who had managed to utilize this arc line in the construction of bridges, and it soon became evident that this arc was far superior to others, without anyone being able to explain why. In the same vein, the English architect, Christopher Wren studied the cycloid with zeal in the course of designing St. Paul's Cathedral in London. But it was Pascal who discovered two fascinating mathematical properties of the cycloid; namely, that the cycloid's arc line is precisely eight times as long as the radius of its circle, the rolling wheel, and that the area within the arc line is precisely three times as great as the area of the rolling circle. Moreover, these considerations of the cycloid were said to be Pascal's only backsliding into mathematical speculation after he had converted to Jansenism at Port Royale in Paris. This recidivism can quite likely be attributed to the night in 1658 when, suffering from a strong toothache, Pascal had noticed that when he lost himself in considerations of the Geometric Helen, his toothache went away, and he took this as a sign from above that it was permissible to think about the cycloid.

But the results of these eight pain-free days he nonetheless published under a pseudonym.

About the same time, the Dutch scientist, Christian Huygens, made a fantastic discovery about the cycloid, a discovery that put him in a position to design the first functional pendulum clock, the Huygens pendulum, Anno 1656. Huygens proved that the cycloid was a tautochrone; that is, he demonstrated that this cycloid curve, when it is turned upside down, like a bowl, has the property that, if, for example, pearls are dropped down its sides, they will all reach the bottom in precisely the same amount of time. Whether beginning high or low, a pearl (Huygens used pearls in order to reduce the intrinsic friction in the movement to a minimum), would reach the bottom in the same quantity of time, "tautochronically." Huygens had thereby demonstrated that a point without friction, functioning only with the power of its weight will roll back and forth in a bowl-shaped form, made as an upside down cycloid arc, will always use the same time for a swing no matter what size that swing happens to be.

In 1696, the mathematicians G. W. Leibniz and Johann Bernouilli posed two challenges to Europe's mathematicians. The first problem was as follows: assume that you choose two points at random in a vertical plane, and ask the question, what form is the curve along which a particle, without friction, must glide, such that under the functioning of the power of its weight, it passes from the upper to the lower point in the shortest possible time? This was "the shortest time" problem, the so-called brachistochrone problem. Newton, who by that point in his career had for the most part shelved his scientific challenges and enjoyed the honour and prosperity of his position as head of the British Mint, had certainly taken up the problem after a long workday and a good dinner. In any case, he solved the problem in a couple of hours and sent the solution anonymously to The Royal Society in London. However, the method used to solve the problem betrayed the old master's identity. Newton had shown that the brachistochrone they sought was a tautochrone and therefore exactly a cycloid.

It was all this talk about the cycloid that had led Swift, in 1726, to write about "cycloid pudding" in his satire about the milieu of the natural sciences. But perhaps the youthful society in Christiania in the autumn of 1820 also liked to have cycloid pudding when now and then after a gathering they went out to one of the city's inns. Pudding was a cheap meal.

It does seem that Niels Henrik was now more at ease and more self-assured. In a letter at the end of September from his paternal aunt, Margaretha Marine Abel in Risør to Elisabeth Tuxen in Copenhagen, we find: "I had a letter from Niels Henrik Abel that really in all Aspects was well written. He was home for a Visit during the Holidays. He becomes a student in a Year and hopes then to be able to help his Brother Peder who has an extremely good Head and has a desire to study."

And on one occasion that fall when Holmboe was ill, and when Professor Jens Jacob Keyser was summoned into the class, the following occurred: J. J. Keyser, Professor of Physics and Chemistry, had formerly used his mother's name of Krumm, and under this name he had published a couple of mathematical textbooks. Many had jested that this was the reason he changed his name from Krumm to Keyser, and that these textbooks were little more than adaptations and translations of Danish and German originals. The books were used in the schools of commerce but scarcely at all in mathematics teaching at the Cathedral School. In any case, Niels Henrik had familiarized himself with Keyser's mathematical contributions. He must have read the books, and he must have had them with him, or in any case brought Krumm's textbooks to light in the pure mathematics that Keyser was now to be teaching the class. One of Niels Henrik's classmates, Fredrik Bonnevie, recounted later that Niels Henrik had shown him a fundamental mistake in Krumm's book, and said, before Keyser entered the room, that now they were going to have fun! Niels Henrik had then quickly gone up to the professor at the teacher's desk, with the page open in the Krumm textbook of pure mathematics. He asked about the meaning of the passage in question, and did there happen to be any meaning to be found in it at all? Keyser had grudgingly agreed that both the author and the calculations had been wrong. And following this admission young Abel had expressed himself with great candour and in rather dramatic terms, asking how it could happen that a writer could present such nonsense for the consumption of young minds. Among the boys in the class this afforded great amusement and satisfaction; they knew that the two names referred to the same person, and that the professor in front of them was the guilty party.

25.

Student and Homeless

Niels Henrik now thought that he had found a formal solution to the fifth degree (quintic) equation. He laid out his work before Holmboe, who also considered the solution to be correct. Holmboe showed Abel's work to Christopher Hansteen and Søren Rasmussen, professors respectively in applied and pure mathematics. Neither of them could find errors or shortcomings in Abel's conclusions and understood probably that this would make quite a sensation if indeed it were a solution to a problem that had stumped mathematicians all over Europe for close to three hundred years. They wanted Abel's work to be published but still there was no channel of publication in Christiania for such scientific material. Hansteen had good contacts abroad to Scandinavia's reputedly most gifted mathematician, Professor Ferdinand Degen in Copenhagen. Degen was unable to detect errors in Abel's reasoning but he looked at the work with obvious scepticism, feeling it was ridiculous that a disciple at the Christiania Cathedral School could solve this problem when so many great mathematicians had been unable to do so. In his reply to Hansteen on May 21, 1821, Degen wrote that the work "exhibits, even if the Goal has not been proven, an uncommon Mind, and uncommon Insights, particular for his Age." Degen wanted very much, with the thought of publication, to present Abel's treatment to The Royal Academy of the Sciences but desired, for two reasons to wait a little while. First, he urged that Abel send a more detailed treatment of the deductions in his results, and a numerical example, for instance, the solution of an equation of the form $x^5 + 2x^4 + 3x^2 - 4x + 5 = 0$. And second, Degen urged a new transcript of the final part of the treatment. Such as it was now, it would be unreadable for most of the members of the Academy of Sciences. But Degen also had an important additional observation: "In this Context I can scarcely repress the Wish, that the Time and the Mental Powers of the Head which Mr. A. blesses us with, not be utilized, from my Perspective, on sterile Subject Matter, but ought rather be applied to a Theme, whose Edification would have the most important Consequences for the whole Analysis and its Application to dynamic Explorations: I am thinking of *the elliptical transcendents*. With the proper Approach to

Investigations of this Type, serious Scrutiny by no means becomes static, nor do the highest and most remarkable Functions, with their many and handsome Properties, become something in and for themselves, but rather, it is going to reveal the Magellan-Voyages to great Regions of one and the same immense analytic Ocean."

The expression "Magellan-Voyages to analytic Oceans" had deep significance in the milieu of discovery that Niels Henrik frequented. By elliptical transcendents, Degen meant what was also called elliptical functions and integrals, called such because the ellipse's arc length could be expressed by this type of integral. And precisely these elliptical integrals, that have the common property of not being expressible by means of elementary functions, would be the next field in which Abel made fundamental new discoveries. Niels Henrik found himself on real Magellanistic voyages to a new analytic ocean. But Niels Henrik was not in the least spooked by Degan's words about the quintic equation being some "sterile Subject Matter." He now proved the formula of his solution with several numerical examples, and he himself found an error in his earlier reasoning. But he continued, and read Lagrange, who was one of the people who had analyzed equations most thoroughly.

Professor Christopher Hansteen was the nation's brightest light in the field of the natural sciences. He enjoyed great respect in all circles; he had precisely calculated the exact location of Norway's capital city, and brought the punctilious spirit of the times to the city. Since 1815, the almanac, calculated to the latitudes of Trondheim and Christiania, had been published in Norway under Hansteen's direction. He had the support of the academic collegium, and a royal appropriation to build a small, spartan observatory outside the south wall of Akershus Fortress, and as often as he had occasion, he utilized his sextant and chronometer to scrutinize and improve the old maps and charts. But abroad, first and foremost, it was his research on magnetism that drew attention. His great work, *Untersuchungen über den Magnetismus der Erde*, published in 1819, awakened international attention, and his plan to follow up with a treatment of "magnetic light phenomena," and in particular, the northern lights, was being followed with excitement in Europe's centres of learning. That Hansteen was now made aware of Abel's remarkable mathematical talents signified a turning point in Niels Henrik's life, not for his scientific track record, even though Hansteen made several attempts to make Abel known abroad, but rather that Hansteen opened his home to the gifted boy from the countryside. For it was in this "home" that he was consequently to meet the woman he referred to as his "second mother," Mrs. Hansteen. More and more she would become his adviser and helper in personal problems to which he himself saw no solution. Mrs. Hansteen seems to have been the person who gained the deepest glimpse into Niels Henrik's feelings and sensibilities.

There was another professor, the many-sided, interesting man of medicine, Michael Skjelderup, who opened the doors of his home to Niels Henrik. The professor's son, Jacob Worm Skjelderup, was Niels Henrik's classmate; they played cards together and would now graduate together. It is also quite certain that Niels Henrik would have heard about Skjelderup the father from his student friends' discussions. The professor was a very popular teacher, and was also one of those in public life who had been set many tasks in relation to bettering the health and well-being of the population. In addition to teaching anatomy, physiology and forensic medicine, he also lectured on medical history and a broad spectrum of the public came to his many popular lectures where he laid out new discoveries in medicine. Even though as a youth he had been seriously afflicted with a stutter, and was still plagued by this, his lectures were both clear and lively, his wit and his anecdotes, often with a melancholy tinge, were talked about in many quarters.

Niels Henrik now had many he could visit and talk to when he was not concentrating on mathematics, and he was very fond of social company, such that without it, he said he became melancholy. And everyone would meet at the theatre on Grænsehaven. Holberg's *The Restless* was put on" as a national artistic undertaking" in January, and during the market days in February, the city's great musician, Waldemar Thrane, appeared in a violin concert, playing the music of Viotti, before Holberg's *Jacob von Tyboe* was performed to packed houses, with admission fees going to the city's poor. Market days in Christiania were the city's annual popular festival which was opened with a pealing of the bells from the church tower on the first Tuesday in February. This was the time that it was generally the easiest to get around in the streets and environs. The city was now stood on its head for three days. The university suspended its lectures, the schools had a holiday, and the high court stopped its deliberations so that the lawyers could be available to meet their out-of-town clients. The city swarmed with people. Peasants in their Sunday best came to receive their pay from the timber traders. Out-of-town traders entered into new business relationships. Families of consequence from the surrounding areas, and public officials from the countryside streamed into the city and the market to meet friends and acquaintances whom they did not see the rest of the year, and the citizens of Christiania opened their doors to them. The more well-off held parties every day to which anyone who had the remotest relationship to these families could count on being invited. The square was the great meeting place. There were stalls here with honey cakes and candy, gold and silver goods, all kinds of finery and toys. At the back there was a tent where jugglers performed, and a tent where one could gamble for real stakes.

Clouds of smoke came from low stalls where hot punch was served. Peasants dressed in wolf- and bear-pelts stood with bunches of ptarmigan and wild hare, and others with wooden tubs, sleighs and harnesses. And of course the shouting horse traders were everywhere. People swarmed everywhere, walking, talking, and flirting. Horse-drawn sledges were driven between the rows of stalls, and finer carriages were mounted with runners. In the evenings when the booths were illuminated with coloured lights, people milled around, happy, shouting, and drunk.

During market days there were also driving races over the fjord ice in front of Akershus Fortress before a huge and expert crowd. Everyone would be discussing who had the fastest horses and who were the best drivers. Ice racing was a sport that both peasants and townsmen engaged in with passion. Skiing was little developed, but tobogganing was something with which even adults could amuse themselves. The boys would go skating on the ice at Bjørvika and on the many reservoirs in the area.

That year, on February 1st, shortly before Market Days, the new Parliament opened. This meant that once again the Cathedral School had to move its teaching, and for Niels Henrik this might well have brought back the good memories of the Parliament three years earlier when his father, Vicar Abel had participated, bold and unafraid.

Again this spring the greatest event at the theatre was once again an Oehlenschläger play. On April 13th, the love tragedy *Axel and Valborg* was presented. For the first time the Christiania public was confronted by a dramatic work set in Norway's Middle Ages. The audiences were moved by Axel, who had to journey to Rome to beseech the Pope for permission to marry his half-sister, Valborg, and on the way back through the forests of Germany, they cut the substantially symbolic letters A and V in the bark of many a tree, from longing for the Nordic love marked by fidelity and "High Esteem for Women." As long as incestous relations were portrayed as having occurred in a dim and distant past, they were perceived as romantic and piquant. Nevertheless, the play contained a wealth of information about Norwegian power struggles in the 1100s, the wretched life of the monks, the kinship between Goths and Germans; and this osculant amorousness was recognizable and engrossing for everyone. Valborg was played by Miss Lasson, whom everyone thought so beautiful, and the action extended out from Christchurch, Nidaros, in Trondheim: " a high hollow vaulted Dovrefjell."[7] An old greybeard sat there as well,

---

[7] Dovrefjell is the high, mountainous, alpine region north of Gudbrandsdalen that separates the mountains of southern and northern Norway.

played by O. R. Holm, and longed for the good old days when the cup was not *sipped*, but when great drinking horns were *drained*, and you slept on the bare earth, with your helmet over your temples and your shield on your breast.

To be one of those who was graduating, that is, to take the *examen artium* at the University meant that one had a different conclusion to the school year than in earlier years. What was called the "candidate holiday" began in April, a forerunner to what later has become "ragging," the period of festivity that school pupils engage in before they graduate from school. From this point to the end of the school year at the beginning of July, the candidates met at school only at certain hours used for review and repetition, and for the assessment of their homework projects. For Niels Henrik this reduction of school obligations quite certainly had an extra appeal. It seems that for a long time he had very much wanted to plan his own time, as his mathematical studies had long since jumped the confines of the school's curriculum and teaching. In relation to the other subjects too, it seems as if he had a reasonable idea of how he stood in relation to the requirements for the exams. In the School's final examinations for the year, marks were given in fifteen subjects. Niels Henrik received one mark lower in religion, Greek and German style, but otherwise his marks were as usual, apart from the remarkable fact that Rector Rosted on this occasion gave Niels Henrik his best results, *excellent*, in Norwegian composition! What could he have written on? Previously Niels Henrik had only received *good* in "the mother tongue" or Norwegian-Danish, which they now began to call a subject. *Good* was also Niels Henrik's mark in the *examen artium*. There he answered the essay question: "What is the Reason that almost all, even the most unrefined of the nation, love their Land of Birth?"

Through his university student friends, and probably also from Hansteen himself, Niels Henrik was at this time initiated into the professor's plans for a foot tour from Christiania to Bergen, a hiking tour that Hansteen later wrote about in detail in *Budstikken*. On June 22, 1821, he started out from the capital and walked westward via Kongsberg, over Bolkesjø, then Rjukan, over Gaustatoppen and further west; thus, at the beginning he followed Christen Smith's route. But Hansteen's commission was geographical surveying. With the assistance of sextant, chronometer, barometer and thermometer, he determined such facts as the geographic coordinates of width, length and height above sea level. In addition he worked on his magnetic investigations; he had already begun to plan his Siberian expedition and wanted to acquire experience in making observations. On the way northwest over part of the high alpine plateau of Hardangervidda, he "discovered" Vøring Waterfall, and with the help of his assistants he threw a stone out from the top of the falls and measuring the time of its fall with his chronometer, calculated the height of the falls to be 933 Rhineland feet. By means of observing the arc of the stone and the opposing wind velocity, he established that the height was somewhere between 850 and 900 feet. (Today the waterfall's height is said to be 182 meters.)

Hansteen and his followers continued down Måbødalen, 1,500 paces over the steepest talus slope to tidewater at Eidfjord, where the grass, he wrote, had turned brown, but the grain crops had been saved by means of artful watering. And thus "the Almanac Chief from Christiania" paddled down Hardanger-fjord to the well-known dean and man of science, Nils Hertzberg at Ullens-vang. Hansteen calculated the settlement's precise location, and together with the dean, determined the distance from the fjord of the venerable farmstead, called Aga, to be 5,586 *alner*[8] or approximately two English miles. The two discussed the legend wherein a whole village, because the priest was ungodly, was punished by entombment beneath the ice mass of Folgefonna Glacier. It was said that the streams that gushed out of the glacier had borne with them water wheels, hewn logs, barrels, buckets and other domestic utensils. Chris-ten Smith had, on both physical and historical grounds, declared the legend to be irrational, and Hansteen was in agreement, although he was astonished that a real glacier could form on so low a mountain. Could electro-chemical reactions of the metal-bearing mountainside have the ability to impede heat? Dean Hertzberg was about to journey by boat and foot, to his brother, Dean Christian Hertzberg on the island of Bømlo to attend a visit by the bishop, and Hansteen gratefully agreed to accompany him. They rowed down the fjord in the afternoon and through the whole light summer night, past Utne, and further along the shore, and by midday Hansteen was able to go ashore and take a reading at 60° 7' 15", and to the south he caught a glimpse of the Barony of Rosendal. At eleven o'clock that night they reached the magistrate's estate at Kårevik on the island of Stord. The next day Hansteen worked out the location of the flagpole to be 59° 25' 22". By midday they reached Bishop Pavels' and ate together before they all journeyed onward as a group, to Dean Hertzberg of Finnås on Bømlo Island; that Sunday they held a solemn visita-tion to the main church on Mosterøy, Norway's oldest church. But about the meeting at the magistrate's estate, Bishop Pavels wrote in his diary for July 12, 1821: "Hansteen told us all about a Son of Priest Abel of Gjerstad, who goes to Christiania School, and is becoming one of the greatest mathematical Gen-iuses imaginable. The other day he made an algebraic Undertaking that both Rasmussen and Hansteen genuinely believe to be the Solution to an hitherto unsolved Problem. *He ferreted out the Computation's Incorrectness,* and is

---

[8] One alen was equivalent to two feet or 0.627 m.

now studying its rectification. There ought to be, once he has become a university Student, a Subscription formed that he may Travel abroad, and expect in him to see one of the Earth's premier *mathematici*."

But what everyone was talking about in the capital in July, 1821, was Karl Johan's displeasure with the Norwegian Parliament and the King's threat to assemble the troops right outside the city at Etterstad. From the Swedish side it was called a "recreational camp," but the Parliament and population were alarmed by the assembling of troops. Two things had aroused Karl Johan's ire: the final settlement of debt with Denmark and the question of a nobility for Norway. Many considered it to be outrageously unjust that the country had to pay three million speciedaler for four hundred years of oppression. Parliament wanted to reject the matter, it being said that the conditions of the Treaty of Kiel did not concern Norway. Karl Johan was implacable in his demand that Norway participate in the discharging of the debt obligations toward Denmark, and in addition he demanded that Parliament abandon its efforts to abolish Norwegian nobility. Were Parliament to go ahead, it would for the first time in history have a law which lacked royal sanction; and the King from his side threatened to abolish the whole Constitution and give Norway a completely new one. Now, at first three hundred hussars in their beautiful uniforms, and then four hundred dragoons, all heavily armed, arrived outside the Norwegian capital, at Etterstad. Moreover, a Swedish naval brig bristling with cannons, suddenly appeared one morning in the fjord. It later became evident that it had only come to deliver an enormous tent that later would shelter the king's banquet at Etterstad. Rumours went around that Karl Johan, when he came to the city, did not want to stay at the royal residence, Paléet, but would go directly to the camp at Etterstad. Both the Parliament and most of the people were apprehensive about what would happen next. Six thousand men were gradually assembled at Etterstad, the tents could be seen from the city, and at night, the flames of hundreds of bonfires. People strolled out to the camp to take a look, to sell their wares, and gradually to take part in the fraternization into which things would eventually dissolve. For when Karl Johan came to the city on July 29th, in pouring rain, Parliament had voted by a large majority to support the payment of the debt obligations. The strongest counterplea had come from Vicar Abel's old fellow partisan, Major Flor from Drammen, who was not diverted by the prospect of a better post, and maintained that at least Sweden should pay half. But on the question of a nobility, Parliament had not backed down, even when it came into the open that the proposal for a new nobility might be put forward as a constitutional proposal. Following the visit of a Swedish naval division and meetings of the Cabinet, it seemed as though all the difficulties had been cleared aside. The King was in a good mood, seven hundred guests were invited to a grand farewell ball at Etterstad on August 19th. The camp had become a pleasure garden; in the evenings much of Christiania's population streamed up there. Dancing continued without interruption until late into the night, the flames leapt high from the bonfires; rounds of cannon fire were set off at Akershus Fortress. The ladies of Christiania were enormously delighted, and not only those from fashionable Church Street, but perhaps even more those from the less-fashionable Vaterland and Piperviken.

According to plan, the University held its annual *examen artium*. Candidates came from the cathedral schools of Trondheim, Bergen, and Kristiansand, to the capital city. Moreover, fourteen of the forty boys who took the examinations that year had been privately tutored to the level of graduation from a cathedral school. The written examinations were spread over four days, with two four-hour tests each day. Apart from Norwegian composition, Niels Henrik received roughly the same marks as he had in the year-end exams at school. He shared the overall mark of *good* with twenty-two others; eight got *very good*; the rest, *tolerably good*. Classmate Skjelderup was in the top group and Berg Nielsen in the lowest. Before the formal graduation from the Cathedral School the candidates normally submitted to a special examination in the School's grand auditorium in the subjects not included in the *examen artium*. For Niels Henrik's cohort this meant only English, as natural history and physics had been dropped from the curriculum. In any case, Niels Henrik got a *very good* in English. But five years later, on one occasion in Trieste, he was unable to communicate to an English seaman his desire to obtain a boat. English no longer stood very high on the course of studies in the capital. Only a generation earlier Christiania had been called "Londonified" by virtue of its citizens' facility with the English language.

Niels Henrik was probably not much interested in these examination results. It goes without saying that he got his best results in arithmetic and geometry. With the *examen artium* he would now begin at the University, but there was no one who could teach him more of what he really wanted to study. Niels Henrik was probably already, as a neophyte student, more skilled in pure mathematics than anyone else in the country.

But to have become a student was to inspire respect, and that constituted a notable step up the social ladder. A student was one who had passed through the eye of the needle and was on a clear road toward a post in public life. Quite certainly now in the weeks before the university began its teaching year, Niels Henrik expended money on a trip to Gjerstad to see his family, and experience the new conditions under which they were living. His mother and siblings had moved out of the vicarage; Widow Abel had the right to use the Lunde Farm. The new priest at Gjerstad parish, John Aas, had been appointed on October 28, 1820, and John Aas was not a nobody. He was among the first cohort to

graduate from the new Royal Frederik's University in 1813, and Aas stood at
the top of the list in terms of academic achievements. He was called the first
student of the new Norwegian University. Aas had then begun to study
theology under the teachers, Hersleb and Stenersen; he was among those who
founded the Student Society, and became its leader in 1816. In January, 1818,
he took his theological public service examination with distinction. John Aas
was thus Lektor Stenersen's top student exactly at the time when Stenersen
was flogging and condemning to eternal perdition, Pastor Abel and his religious instruction. It is probable that John Aas had intended to pursue a
university career; he remained on at the university for another year to continue his studies, and acted as the curate to the Akershus Castle congregation
before he subsequently chose to go to Gjerstad. It is highly likely that Niels
Henrik had met the still-unmarried John Aas in Christiania before this new
Gjerstad pastor moved to his vicarage around Christmas time, 1820. The two
would come to have some correspondence by letter, and some years later,
when it dawned on John Aas what an extraordinary scientist Pastor Abel's son
was, he clearly showed his respect whenever Niels Henrik came home on short
visits.

But in August of 1821, there was little, if any, respect for the Abel family in
Gjerstad. Widow Abel had certainly moved to Lunde Farm, and it was known
that she received 16 *daler* per annum by way of widow's pension. Pastor Abel
had secured three widow pensions for her, one each from Copenhagen,
Christiania and Kristiansand. But when she had to leave the vicarage she dug
in her hind legs and demanded twenty-six head of livestock under the pretext
of helping Niels Henrik in Christiania. In addition, when she had grudgingly
agreed to pack up and change her dwelling, Dean Krog had had to interfere
and remind her that it would be thievery against her own children were she
to keep the twelve silver spoons she had in her possession. Dean Krog wrote
about her: "Bacchus and Venus are diabolically pursued by the wretched - in
relation to both Brain and Heart intolerable, and in morality, below all
Critique." There is no doubt that she wanted to marry her servant and
horse-wrangler, Jørgen L., who now ran Lunde Farm, but when she found out
that in so doing the widow's pension monies would come to an end, she gave
up this idea. Dean Krog could also recount that Pastor Abel, from his sickbed,
had written a letter to the bishop and the dean, where he described in very
strong terms not only his miserable marriage but also the reasons for his own
plunge from Grace, his anguish and despair, and an eccentricity that he
bitterly regretted. But Mrs. Abel and her "Champion" were said to have
destroyed this letter immediately upon Abel's death. During the last period of
Vicar Abel's life, Mrs. Abel was reported scarcely ever to have appeared at her
husband's bedside, or indeed if she were to give him a cool drink or any such

administration, she did so without the least compassion. If it had not been for the sacristan and two or three others, Vicar Abel would have died unkempt and in his own excrement. And what Niels Henrik's brothers and sisters might have said, and beyond them, the dean himself said, "A dog's life is much better than that of these poor unfortunate down-and-outers - may God have pity on them."

Pastor John Aas shared the dean's worries and wrote to Bishop Sørensen in Kristiansand on February 28, 1821, that he hoped "that a Sense of Religion and Devotion will speedily return to this Parish." Aas learned "with a Feeling of Sadness…the Condition in which the Vicar's Widow, once the housewife of a virtuous and honorable Man, finds herself," and he felt it was worrisome that those who earlier had been the priestly couple's friends now had withdrawn in disgust, and most tragically, "with Regard to her lack of upbringing of the Children", who were badly in need of a strict hand.

It was probably in order to educate *all* the children of the parish that Pastor Aas now wrote on the lone wooden cross on Father Abel's grave:

> Hark, Wanderer, whoever you be! May this Grave recall
> That Time doth from Happy Smile to Anguish go.
> Life's beauty blossoms in the Sun's sweet thrall,
> Until Sighs and Tears drown Joy in Woe.

Niels Henrik could read these words himself when he visited his father's grave in August, 1821. The wooden cross remained in place until it rotted away, and the grave was almost forgotten; then, in 1902, in relation to the one hundredth anniversary of the birth of his second son, a tombstone was placed over Father Abel's grave.

At the end of August, 1821, Niels Henrik was back in Christiania. As a student without means he could now apply for free lodgings at the student residence, Regentsen. In a letter dated September 3, 1821, and signed "Most faithfully yours," he wrote: "In the month of August of this year, the Undersigned, who passed the Examen artium, petitions the Collegium as to the possibility of obtaining one of the University Institution's vacant Places. The reasons for this Request are the death of my Father, and that my Mother, in addition to me, has five Children unprovided for, and finds herself in such Circumstances that she, in her difficulties, can offer me no Help."

PART V

Student Life, 1821–1825

Fig. 17. Drammensveien, from Arbien's field and villa during the 1820s. Aschehoug forlag archives.

26.

Free Room and Board at Regentsen

Niels Henrik was assigned space at the student residence, Regentsen, in September, 1821, and he swiftly moved into an attic room that he was to share with Jens Schmidt, a boy who had begun his student life a year earlier. About twenty of the most needy students lived at Regentsen, and it was here that Abel would stay, in various rooms, until his grand tour of Europe in the fall of 1825.

The University's Regentsen establishment was located in the Mariboe Building that the University had purchased a couple of years earlier; and since then, as it were, most of the teaching was amalgamated under one roof. The Mariboe Building was considered one of the most handsome establishments in Christiania, one of the first three-storey buildings, with its façade overlooking three streets: Kongens gate (King Street), Prinsens gate (Prince Street) and Nedre Slottsgate (Lower Palace Street) known at the time as Northern Street. The estate was built by artillery captain and wholesaler, Ludvig Mariboe in 1810, and was the harbinger of the empire style in the city. Halfway down its length, on Prinsens gate, the entry area was marked with high pilasters; along Kongens gate there was space that opened both to the street and the courtyard, but otherwise it was simply rows of rooms with closed entrance portals to the courtyard.

The University building was composed of thirty-four rooms. The first floor contained two rooms for the University Collegium, four rooms and a kitchen were let out to the University Association's inspector, Gregers Fougner Lundh. Lundh had a many-sided background; he had been a medical student in Copenhagen when the war broke out in 1807, had become second lieutenant and captain in the Norwegian defense forces, and head of division before, in the autumn of 1814, being appointed Lektor of Technology and University Secretary. Now, as the "house father" of Regentsen he was to become jack-of-all-trades to the students and part of their everyday life. The remainder of the first floor housed the laundry and three rooms used by the porter and the waiters. Here the students could obtain hot water for their customary cup of tea in the mornings at six o'clock before they began the day. The second floor contained six rooms that held the natural sciences museum, or "the cabinet" as it was called; the cabinet of medals and coins was in another room. On the floor above these rooms again, in addition to the auditorium, there was "the Great Auditorium," or examination hall, also used for matriculations and other formal occasions. The other lecture halls were built in the other wing. Student quarters were located on the second and third floors, as well as in the attic, and the back buildings in the courtyard were also used as living quarters. The courtyard also contained stables, carriage houses and woodsheds. There was a total of ten rooms fitted out as student quarters. The nine largest were divided by means of wooden partitions such that each had a workroom and a smaller bedroom; here students lived in twos, each having his own bed, a washstand with a basin, a desk and chair, a book shelf, a stool, a fire iron, a metal bowl with a pair of scissors, and light sconces in which to set candles.

Formerly student accommodations were on some of the properties owned by the University in the Vaterland quarter, and the move from there to the Mariboe Building was considered a great advantage. The teachers had to some degree undertaken a kind of *paterfamilias* role and now they could check up more easily on the boys' lives. Since the move to the Mariboe Building, the many complaints that formerly had come from the citizens of Christiania about the students' wild goings-on and noisy nights had almost completely ceased. In order to promote diligence and hard work among the students it was decided, in addition, that whoever had been given a place in the University community must annually submit a report, preferably in Latin, on how their studies had progressed. In 1822 the students petitioned to be freed from this requirement but the Collegium decided that it could not accede to this request. To the contrary, the Collegium held that these mandatory reports would be of particular consideration in relation to the allotment of places and stipends. Unfortunately, these reports have not been preserved in the historical record, and thus Abel's view of his own progress is not to be found, although one can find in the archives that Student Abel signed a powerful protest petition mounted by the students in the autumn, 1822, against the dismissal of Holmboe's brother, Henrik, from a lab assistant position in the chemistry laboratory.

In terms of accommodations, Niels Henrik now probably had better conditions than ever before, at least he was his own master and could more easily meet with friends and fellow students to play cards, chat, or just "loom large" as they said back home at Gjerstad. We know that Abel played cards together with three friends who were all his age, Johan Lyder Brun from Bergen, Arnt Johan Bruun from Kristiansand, and Johan Fredrik Holst from Trondheim. They had all taken *examen artium* together and received accommodations at Regentsen. Two of Abel's classmates from the cathedral school, the professor's son, Jacob Worm Skjelderup, and the pharmacist's son, Carl Gustav Masch-mann, seem to have joined in the card games as well. Of course the latter two lived at home but came every day to the Regentsen building. Every day there were lessons to prepare toward *Andeneksamen* (the Secondary Examination) which they had to face the following summer if they were to go on to public service-oriented studies. In short, Maschmann had decided to take over the Elephant Apotek from his father, while for Niels Henrik, who was certainly clear on the future course of his own studies, there was in existence no course of studies of a public service nature. For the others, theology would have seemed to have the best course for putting bread on the table.

Normally the relationship between professors and students was quite distant. The status of professor had about it a social aura that subdued the students and kept them at a distance. Most people also felt that professors were more friendly toward Denmark and the Danish culture than was common among the general population. The professors repeatedly pointed out that it was Fredrik VI who had founded the University, and even after the separation of the two countries, the King of Denmark had given Norway's university library a large duplicate collection of books from the Royal Library in Copenhagen. During this summer of 1821 many in the capital feared a coup by Karl Johan's forces. For instance, Professor Hersleb of the Theology Faculty stated that, were this to happen, he would return to Copenhagen and, if it should prove necessary, sustain himself by private tutoring.

Nor did the students of the 1820s encourage the professors to attend *their* gatherings. The professors complained frequently about the way the students behaved, and the students complained that their teachers planned lectures more to satisfy their own interests than to serve their students' needs and requirements. However, the situation seems to have been quite different for Student Abel. Professors Rasmussen, Hansteen and Skjelderup not only understood Niels Henrik's mathematical genius, but they also understood his economic situation. Consequently they supported Niels Henrik with monthly contributions during these first years until in the spring of 1824, the Collegium gave him a stipend that he could live on. These three probably also mobilized others to support Abel, and among these was no doubt Pharmacist Masch-mann, who was well-off and an engaged man of science who, for a period, had been responsible for the teaching of chemistry at the University.

27

Examen Philosophicum:
The Course of Studies and the Teachers

The *examen artium* was now behind him. *Examen philosophicum,* the Secondary Examination, that most took after a year at the University, was in reality an expansion of the *artium,* and most got better marks this time than in the Primary Examination, the *artium.* Moreover, it was almost certainly the same professors, for the most part, who assessed the boys in both these examinations. Niels Henrik was one of those who obtained the same mark of 3 as before, but both his friends, Maschmann and Lyder Brun, for example, had improved their results from a main mark of 4 to a new mark of 2.

In this Secondary Exam, marks were awarded in ten different subjects: theoretical and practical philosophy, Latin, Greek, history, mathematics, astronomy, natural history, and physics. In physics, which also included chemistry, marks were given for both the theoretical and practical aspects. Niels Henrik got top marks in the theoretical, but only a *good* in the practical. The teacher in this subject was Professor Jens Jacob Keyser, with whom Niels Henrik had, to be sure, discussed mathematical textbooks. Keyser held his lectures in a chemistry-physics laboratory on rented property beside the garrison hospital on Øvre Slottsgate (Upper Palace Street) near the old city square. Here physics was taught in the mornings and chemistry in the afternoons three or four times a week. These were the only lectures not held at the Mariboe Building. Keyser's lectures were said to have been all right, if lacking a little in originality, although his experiments failed quite often, and when "this Choleric Professor" became angry" this Anger resulted in a flood of insults sweeping over his unfortunate Laboratory Assistant, who could never satisfy him." The dismissal of Holmboe's brother arose from one such situation. And indeed, Keyser served up many quite improbable stories in his lectures and when they became too incredible the students stamped their feet, such that the professor became livid and reprimanded the whole auditorium with Latin expressions for puerile behaviour and boyish pranks. Apart from this, Keyser was one of the few professors who gave private tutorials for a fee, and some felt that with regard to the examination in physics it would certainly be of benefit to participate in these. Niels Henrik was hardly among those who were of this view. When Keyser was staying in Copenhagen in the spring of 1822 in connect with a government task, Hansteen took over the physics teaching for Niels Henrik's class. Of course Hansteen was also responsible for teaching astronomy and here he taught four classes a week in spherical trigonometry and celestial mechanics, closely linked to geodesy and land-surveying. He went through the practical use of the sextant and chronometer for the determination of locations, dealt with topography and meteorology as well, and of course, Hansteen would have spoken about his geo-magnetic studies: more accurate navigation was highly significant in a sea-going nation like Norway. Everyone seemed to hold Hansteen in high regard; it was always interesting to listen to him. Students reported that he had a good and pleasant voice, and that he explained clearly and illustrated his assertions with particular and apt examples. Besides, everyone knew about Hansteen's dangerous journey from Copenhagen in 1814, that had become a sort of ensign for national patriotism: in order to avoid pledging fidelity to the Swedish king, and with important papers for King Christian Fredrik sewn into Mrs. Hansteen's dress, he had taken the sea route, buying the lugger *Mazarina*, and he had taken along Norwegian sailors who had been imprisoned for six years by the English. They had been pursued by a Swedish corsair and were apprehended by a British frigate, but the English captain took pity on both the Norwegian sailors on their way home from six years of imprisonment, and on the Norwegian scientist and his wife, and the *Mazarina* had reached land unscathed at Langesund. There Hansteen had met Professor Christen Smith, who was awaiting an English ship to take him to England on his way to his ill-fated expedition to the Congo. Hansteen was therefore later able to say that he had been the last Norwegian colleague to see and speak with Christen Smith. About the student Abel, whose genius was indeed already clear to Hansteen he was to have expressed disappointment that he could not give Niels Henrik top marks in astronomy, in the *examen philosophicum*.

But in mathematics there was certainly no doubt: Niels Henrik loomed above all the others, and there was little that Professor Rasmussen could teach him. Apart from his main task which in particular was teaching all the students trigonometry and algebra, Rasmussen was particularly interested in arithmetic series. In recent years, for those who were studying the science of mining, he had been teaching the kinds of arithmetic series used in the calculation of stacked cannonballs, and he had lectured on what were called curved lines; that is, the classical conical sections, and about the use of infinitesimal calculus in the determination of tangents and the radiuses of curves on curved surfaces. Higher equations and differential- and integral-calculus had also gradually entered Rasmussen's course of lectures. But B. M. Holmboe had already learned all of this from Rasmussen and had long before passed it on to Niels Henrik. Besides, Rasmussen was more and more becoming the government's financial expert and his work with Norges Bank in no way served to deepen mathematical profundity. But Professor Rasmussen clearly recognized Niels Henrik's abilities, and apart from contributing to the monthly support, he gave Niels Henrik 100 speciedaler two years later, in the summer of 1823, to travel to Copenhagen and meet Professor Degen and the other mathematicians there.

The University in Christiania was the only institution in the country where scientific expert knowledge was to be found, knowledge that the authorities certainly were in need of. Rasmussen became the government's man of finance. Hansteen was active in mapping the new nation of Norway, and was adviser on questions of weights and measures. He also lectured at the military college and was responsible for the almanac. The authorities also knew that Norway's various economic sectors, its business community, agriculture, animal husbandry, forestry, fishery and hunting could develop much more with the introduction of rationally- and scientifically-based methods and procedures in place. *The natural sciences* were considered those university subjects that could lead the advance and development of the country's industry; and for these reasons Christen Smith and Jens Rathke were hired as professors. But Smith was gone and Rathke had difficulty fulfilling expectations. By way of providing an introductory approach to the problem, Rathke was to lecture on plant physiology, botanical terminology, zoology, the zoological encyclopedia, and the geology of the fatherland. The students thought his lectures were boring, and rumours got about that he consciously made natural history uninteresting, otherwise, as he was supposed to have said, "far too many would have thrown themselves into it." Many told about how Rathke came stealing into the auditorium from the natural science cabinet in the room next door, hiding something against his chest, beneath his jacket. When he had mounted the lectern he would suddenly brandish the hidden object, that would be a stuffed creature, a fish, a bird or a bat, and holding it aloft in triumph, would intensely declaim, "Now my Friends, let us investigate this Masterpiece of Nature!" Rathke also had a plethora of definite expressions and formulations that he exployed in his lectures. It was said that he was a great pedant and a confirmed bachelor who was excessively courteous and gallant. On the examination board his benevolence and courteousness were almost excessive. Despite this, Niels Henrik did not receive more than a 3, *good*, from Rathke.

Cornelius Enevold Steenbloch, a professor since 1816, taught history, and he was preferred for the post, after a sharp competition, over N. F. S. Grundtvig. Steenbloch was a popular bachelor, not only for his many spiritual anecdotes in the lecture hall, but he was also a twinkling star in the social life of the city, as well as upon the stage at Grænsehaven. In any case, Niels Henrik had seen him decked out as the Goddess Minerva in the audacious production of Holberg's *Ulysses*, during the spring of 1818, when Venus and Juno were also played by men, and the women in the audience had left their seats when Venus showed no shame in the exhibition of her shapeliness. Steenbloch was also known for playing and composing melodies, and for writing poetry, the price of which was wine, women and song. He even called himself "The Rune Master and Saga Man from the black School of Harald Hardråde's[9] city." Apart from this he was a gifted lecturer and it was said that he devoted himself to teaching, and for this reason, gave up his scholarly work once he had been appointed professor. He took the students through the history of Antiquity, of the Middle Ages, as well as more contemporary history in the course of six semesters, then he would turn over the pile of lecture papers and begin again. But he also had a shorter course for all students on the history of Norway, Denmark and Sweden. The only criticisms made about his lectures were that he never revealed his sources, that he never went beyond entertaining anec-dotes, and he did not try to ask the students questions about their own studies. For Abel the latter criticism would no doubt have been an advantage. Niels Henrik had the desk in his room covered with mathematical books he had borrowed from the University Library. And even if he did have a taste for Steenbloch's histories, he perhaps did not manage to place them in time. In any case he received a mere *good* from Saga Man Steenbloch.

In the language subjects, Latin and Greek, Niels Henrik would no doubt have tried to float through on the basis of what he had learned at the cathedral school. The twenty-three year-old Søren Bruun Bugge taught Latin. The year before he had been the first in the country to take the public service exami-nation in philology, a completely new course of studies that Professor Sver-drup had initiated and taught. The same year S. B. Bugge had been appointed lecturer and taken over the teaching of Latin that until then Sverdrup had been responsible for. Later, when S. B. Bugge was coming to be one of the leading Latinists of the period, he said that his teaching in the first years had been meagre and boring. Niels Henrik probably did not distinguish himself in the Latin lectures either, where he got to go through Roman literary history with Tacitus, Sallust, Cicero, Horace, Virgil, Ovid and Tibullus on the mandatory reading list. Beyond this, the students were drilled in expressing themselves in Latin, through exercises in both writing and speaking.

---

[9] A hard and military man, Hardråde became Harald III in AD. 1047. He is often considered the founder of the city of Oslo. He died fighting at Stamford Bridge, England, in 1066, in the famous battle that was won by the Norman, King William (The Conquerer).

The lectures at the Mariboe Building usually began at eight o'clock in the morning every weekday and followed one another in rapid succession, sometimes into the afternoons as well. The one professor would relieve the other with only a few minutes' interval; longer pauses were not even discussed. The professor that Niels Henrik and his classmates saw the most of was Georg Sverdrup. He had three subjects on the timetable: theoretical philosophy, practical philosophy and Greek. Besides instructing and preparing the students for preparatory tests, Sverdrup also led the philological seminar that was preparatory to the public service examination; in addition he was the University's librarian. It was said that Sverdrup astonished all the new students with his handsome witty lectures. His deep sonorous voice, and the lustre that remained about him after his participation in the political life of the fatherland certainly contributed to the heightening of the impression his lectures made. But for Niels Henrik, Sverdrup was probably associated with memories of his father's bitter struggles in Parliament in 1818. Moreover, critical voices would maintain that Sverdrup now had developed an exaggerated propensity to criticize the writings of everybody else because he himself did not have time to write anything.

More than half the student body would go on to study theology, and because such a large portion were prospective theologians, the study of philosophy, as expounded from Sverdrup's lectern, became a broad, tailored, resonating discourse on higher spiritual life, a question of theoretical speculation in relation to the conception of God. The students were to get an overview of the sciences, and in relation to the history of philosophy, the catchwords in the lectures were logic, aesthetics, ethics and metaphysics. It seems that Niels Henrik enjoyed the reasonings in this general philosophy, and he received a mark of 2, *very good*, from Sverdrup in both theoretical and practical philosophy. As for Greek, which had been Sverdrup's original favourite subject, things did not go as well. Here the prospectus on the course of studies was somewhat overwhelming: by way of preparatory study, Sverdrup took them through the literary history of Classical Greece and the private lives of the Greek people, assorted selections from Plato's dialogues, Theophrastus, Hesiod's poetry, Homer's Iliad and Odyssey, the Homerian lyrical songs, Aristophanes' comedies, the tragedies of Euripedes, Aeschylus, and Sophocles, the odes of Pindar and the Gospel According to St. Luke.

It was during one of these lectures that it was later said that Niels Henrik suddenly got up and stormed out the door, shouting, "I have it! I have it!" Or perhaps he was just attentive enough to use Archimedes' own word: "Eureka! Eureka!"

28.

Student Life and Circle of Friends

When Niels Henrik began his studies at the University there were about a hundred students in Christiania, and the great social gathering point was the Norwegian Student Society, which admitted all those who were studying. They could become members by paying a subscription that was determined quarterly. Abel was probably not among those who spent the most time at the Student Society. His name is not mentioned in the papers that had to do with the Society's activities during these years, although he quite certainly would have followed events, and several of his friends were involved in various tumults and debates.

In the same year that the university had begun offering lectures (in the autumn of 1813), the students had been unanimous about founding a student society. To this end, nineteen people had each paid 15 daler. Every Sunday a gathering was held, with lectures, treatises, and recitations. The first thing they obtained was a tea service with teapot and yellow cups and spoons; later, metal bowls, candlesticks, a cupboard, a tobacco cellar made from pewter and some punch glasses were purchased; in 1815 they established a book collection. The leaders of the Society in these first years had been the students Hans Riddervold, John Aas, Søren Bruun Bugge and Conrad Nicolai Schwach. In addition to the Sunday gatherings, regular festivals were held each quarter; these celebrated the founding of the University on September 2nd, and the Reformation festival of Luther's birth on November 10th, but Karl Johan's birthday on January 26th did not enjoy unanimous support, at least not in 1822.

At the outset, the students were favourably disposed toward Denmark; they had given money to Grundtvig's planned *Saxo* and *Snorre* saga translations, and they had supported Sverdrup, Stenersen and Hersleb in trying to get Grundtvig to Christiania as professor of history. Now, however, from the highest Swedish quarters came an overture to change this and promote a new high priority task: namely, many students were eager to have a student uniform. Abel's friend Keilhau was among the most zealous on this matter, and the mathematics teacher, Holmboe, also supported the idea of a student uniform, which in the eyes of many was linked to the arming of students that had occurred in the patriotic days of the summer, 1814. The student, John Aas, had, in his time, tried to stop the proposal. In an incendiary contribution he had countered the others' arguments: the desire that a uniform would tie "The Hearts together, and lead the Soul to a Unity of Aspiration and Desire, is just as likely as expecting to bind Heart and Mind together with Stocking Garters." But the issue came up again. In February, 1820, Keilhau and his roommates at Regentsen had stated that "Each and everyone living and studying here at the Student Building believe that the Norwegian Academic Citizenry ought to bear distinctive Attire." The majority in the Society supported the proposal. Keilhau and four others worked out a proposal for a possible uniform. At the end of March, 1820, the drawing was sent to the Academic Collegium. It was an embroidered uniform with a three-cornered hat, a black gown of stylish cut, two rows of buttons, a high, upstanding black velvet collar on one of whose lapels would be embroidered an olive branch in green or blue silk, and a cap flourishing a cockade! But the assembled Collegium had strong objections. Professor Hersleb thought the students had very little need for a uniform; the loss side of the ledger could be considerable. Were it to happen that some blameworthy action was committed by *one* student, would then *all* students be guilty, and would it, for example, be *compulsory* to wear the uniform? The tendency toward the disease of ostentation would not be cured with a uniform; new fashions and more finery would only lead the students to greater expense. To the contrary, Hersleb argued, the students ought *not* set great store in such ornamentation.

The matter was taken up in the Student Society again, and there, by a vote of 44 to 33 it was decided to carry on the work of getting a student uniform. Once again Keilhau was active and was the power behind a long, well-formulated application to the Academic Collegium, which could do nothing but refer the matter further to the Department of Church Affairs. The matter was taken up by the nation's Government. The Head of Government, the Viceroy, was also Chancellor of the University. The Government had the same view as had the Academic Collegium and these recommendations were sent for royal sanction in June, 1820. But at the palace in Stockholm the unbelievable occurred: Karl Johan approved the students' application, out of his dislike for both Government and Collegium. The King remarked only that instead of having a cap with a cockade, the students "should bear a *Hat* with Cockade."

Karl Johan's accommodation and friendship vis-à-vis the students had been overwhelming, and from now on his birthday too was also vigorously celebrated. But the student uniform was little used and never came to play any great role. Could it have been that Keilhau had one made when he came home from the Jotunheimen expedition in the summer of 1820? And perhaps Niels Henrik had a finger in the pie as well: in the student environs Niels Henrik was nicknamed Tailor Niels, for one reason or another, perhaps due to his pale face? Otherwise Niels Henrik could have used new clothes; he went around sloppily dressed, more, it was said, out of indifference than poverty.

From among Abel's friends of this period, Boeck, Heiberg and Skjelderup distinguished themselves in the many debates, discussions and disputations on various subjects that found their place in the Student Society. For example, to please the gathering as a "recuperative Pursuit," the sources of Virgil's *Aeneid* was taken up for discussion. Another wanted to prove "that in all Justice the Christian Religion cannot enjoin Love of the Fatherland." A third posed the question, "Whether and to which degree People in a State of Matrimony have the duty, by right of Law to make love?" Boeck spoke one evening on "some aspects of Gynaecology that deal with different theories of fertility," and on another evening Boeck was to prove "That Time and Space not be seen as coordinate, but rather as sub- and superordinate Concepts." Competitions were mounted for prize papers, as a rule for each faculty, that could be prepared either in Norwegian or Latin, and were judged by a student-appointed committee. For example, one topic for the theologists was "What is a Theologian to do when he in the course of his work must swear on the symbolic Books, when these do not quite tally with his Convictions?" Probing interest in the Constitution lay behind most of the topics given to the law students; they dealt with actual constitutional interpretations and led toward evidence that the Norwegian constitution safeguarded justice and the freedom of the citizenry. A topic that strongly gripped Niels Henrik's medical friends, and which was thoroughly answered by Boeck, was: "Can the Doctrine of Vital Magnetism, or a doctrine that is based upon the Hypothesis of this Thesis, where there is a Dynamic relationship between an organic Body's own Organs, between one such organ and terrestrial non-organic Nature, between one such and everything else in the Edifice of the World's Mass, bring beneficial Results to the practical Exercise of Medical Science, or ought one rather adhere to Experience's Knowledge of the Organs' most intimate Physical, Chemical and Organic Effects?" In the field of philosophy it was simply asked, "Is it good that the People be educated?"

When Niels Henrik became a student, the Society had just reacted particularly strongly against his Latin teacher, Søren Bruun Bugge, who had to leave a committee that had been appointed to judge one such scholarly competition. The students maintained that the newly-appointed Lektor Bugge had refused to sit on a committee together with two students to evaluate a Latin treatise on Homer and Virgil.

The students paid strict attention to one another's behaviour and kept each other under surveillance, and the Society had a strong code of etiquette. It was a period in history when the defense of one's honour was extremely important, but also a time when Norwegian students were unable to challenge one another with pistols; instead, they fenced with thundering words inside the Society, and certainly with bare fists outside. Little by little these verbal duels in the Society took on the character of Icelandic family feuds: when someone, in the heat of battle, let fall a taunting remark, the offended stood up immediately and said, I challenge you! Meet me next Saturday at seven o'clock in the evening in the Norwegian Student Society! And here they would meet in long and stormy meetings, where crime, guilt and punishment were argued and debated, with the aid of helpers and helpers' helpers who advanced apologies and accusations, and expended many bitter words in mounting new challenges. After everyone had expressed himself, the matter that had been the starting point, was put to a vote: first, the question of guilt, and if the Society with a two-thirds majority determined that a phrase had been too defamatory, the aggrieved person could impose a punishment if this were called for in a new, simple majority vote. An acceptable punishment might consist of the Society's formal displeasure, or a penalty, or even stronger, dismissal in the event that the matter had involved a leading member. The strongest punishment was banishment, but that required much more than a simple majority vote.

Niels Henrik never became mixed up in any of these feuds and quarrels; he was merely a spectator and observer. Nor does it seem that he participated in the essay prize competitions. Indeed, there was no one in any faculty capable of giving mathematical problems and themes for competitive solution, nor did he give lectures about his own course of studies before any student gathering. But the fact that Niels Henrik's genius was known in these circles at the University, even though he did not seem to participate in any of the evaluation committees, might perhaps indicate that an aura of respect surrounded Student Abel and protected the peace of mind of his work and his concentration. In the course of the 1820s the Student Society developed and expanded itself increasingly from a literary tea salon, to a social club with gatherings around the punch bowl. It became permissible to enjoy hip flasks of spirits, and gradually a regulated sale of wine and spirits was allowed in the meetings, and the former tea-drinking members were ridiculed: "They drank, Oh, what sin, my Daughter, The hot temptations of tepid tea water." During the winter term the Society now had a permanent meeting place in a building on the corner of Rådhusgaten and Dronningens gate, across from Rådstuen (The Council Chambers from the period prior to the establishment of the Norwegian Parliament). During the lighter months they now began to have gatherings at the beer garden at Hammersborg called Sorgenfri [Free of Care].

It was said that Niels Henrik was always cheerful in social company, and could sing a ballad and enjoyed a pipe of tobacco, but that he was very careful in relation to alcohol. After playing cards in one of the student quarters he and his friends often went to Madame Michelsen's public house, Noah's Ark or Pultosten (Stinky Cheese), as the house at the corner of Lille Grensen was often called. Or else they went to the modest public house, Asylet, where there was only one long table with chairs and tallow candles, and the only items of trade were beer and a cornet of pipe tobacco served in a saucer.

Niels Henrik was deeply concerned about his siblings back at home in Gjerstad. He had not spent more than a half-year in student life at Regentsen before he had concrete plans for his brother, Peder. He now wanted Peder to come to Christiania and live with him, and thus Peder, with the help of student friends, could be instructed and prepared to take *examen artium.* Already before Christmas, 1821, Niels Henrik had sent a request to the Collegium that Peder be allowed to live with him at the student residence and with the request he enclosed a statement from his roommate, Jens Schmidt, saying that he had nothing against the arrangement. Niels Henrik must have been quite certain that no others would oppose his proposal to share his bed with his brother because, before he had received a reply from the University authorities, he sent a message home to say that Peder could come to Christiania. But Peder was ill before Christmas and was thus prevented from travelling, a postponement that perhaps suited Niels Henrik quite well, since for the Christmas holidays he was invited home by the brothers Holmboe at Eidsberg Vicarage. Holmboe Senior had been vicar at Eidsberg since 1801 when he had taken over the post from the teaching priest, Jacob Nicholai Wilse, who was well-known across the countryside. Niels Henrik knew four of the Holmboe brothers: above all, of course, Bernt Michael, as well as Christopher Andreas, who that autumn had been substituting for Professor Stenersen, teaching Hebrew to theology students. It was said that Christopher Andreas was already in mind for a post at the University. The third Holmboe sibling was Henrik, who had been working in the chemistry laboratory, and the youngest brother, Hans Peder, who had taken *examen artium* with Niels Henrik and got the same marks as he did.

Each of the first three brothers, Bernt Michael, Christopher Andreas, and Henrik, in turn, had received the highest marks in *examen philosophicum.* The youngest brother, Hans Peder, seems to have taken a study pause, a year off after the *artium,* as he took the preparatory first-year courses the year after Niels Henrik, and he was probably at home at the vicarage when the others arrived in Eidsberg just before Christmas, 1821. Here Niels Henrik met several of the great flock of Holmboe siblings: an older brother, Hans, had finished theology and was curate in the parish under his father, while the very eldest, Otto, had begun his career as a merchant and lived in the northern city of

Bodø. Two older sisters also lived at home at the vicarage, both were later to marry sons of the vicar, Wilse, the one a short time later in 1822; consequently, her intended, Lieutenant Christian Fredrik Wilse, was also perhaps at the vicarage that Christmas.

Vicar Holmboe and his wife had had seventeen children, nine of whom reached maturity. Hans Peder, born in 1804, was the youngest. It was said that the mother, Cathrine, was gentle and loving, but weak, and she spent much of her time in bed at this period. Father Holmboe was a little jolly uncle who was said to potter around on his own and allow his guests to fend for themselves. However, the Holmboe children were certainly active in creating a festive Juletide atmosphere at the vicarage, with the whole staff participating in the traditional festivities. The celebration of Christmas began with a *mølje* around five o'clock in the afternoon on Christmas Eve. This traditional dish was composed of rye flatbread soaked in a hot liquid made from the thick broth of a rich meat soup. In the evening it was usual to eat roast meat with potatoes, cabbage, and soup or gruel. As well there was beer and spirits by the cask, and again, food to excess. Christmas Day was marked by church attendance, and Christmas festivities rolled on until Twelfth Night, when the traditional candles of the Three Kings were lit. In all likelihood Lieutenant Jens Edvard Hjorth was also a Christmas guest at the vicarage. Bachelor Hjorth lived in the community and was responsible for a series of musical gatherings that concluded happily with dance and "unadorned Joy and Cheerfulness." Lieutenant Hjorth was also the man who precisely that year would start a reading society in the parish in order to develop the social life. For Niels Henrik, coming to a Christmas celebration at a vicarage must have awakened memories of earlier Christmases at home at Gjerstad, and on the surface of things he could have noted many similarities: Vicar Holmboe was popular among the peasants; he had joined with the farm labourers and won a dispute about road work; they had just built granaries for rye, barley and oats; and potato-growing was a large enterprise in the parish. Here also, much of the potato crop was utilized for local distilling purposes. From the war period, it was said that Prince Christian August had lived in Eidsberg in 1808, and Lieutenant Hjorth told about the seventeen year-old Olav Rye from the Telemark Infantry Battalion who had made a jump of 15 alen (9.5 metres) on skis, something that no one had since been able to equal, and which was subsequently registered as the first Norwegian record in ski jumping. Also with regard to the war, they related that the Swedish forces had used the main church as an armaments magazine and a stable in 1814, damaging a number of chairs and parts of the floor. The confessional seat in the sacristy had been broken up and the pieces tossed out into the churchyard, and they had taken along with them a copper vessel of holy water and an altar cloth, both used for christenings.

Another event that was no doubt discussed at the vicarage that Christmas was the Hans Nielsen Hauge religious revival that had swept throught the community. People talked about a blind woman called Kari Rømskogen, who through her own testimony had moved many so profoundly. This revival had also been taken note of in Christiania, and when Cabinet Minister Treschow, a little later in 1822, sent a man out to investigate this movement in Eidsberg, the report he received contained the message that neither Vicar Holmboe, nor his curate son, had laid down any barriers against what had been taking place, and neither could it be said that there was anything unfavourable in these witnesses; it was only to be stressed that the *church* was the correct venue for the public proclamation of The Word.

Even after Twelfth Night and their having fulfilled two weeks of Christmas celebrations it seems that still one more week passed before the students journeyed back to Christiania where, perhaps, they were in time for the theatrical production of Kotzebue's popular *Humanity and Remorse* on January 16th. But the first thing they would have heard about when they returned to the city was quite certainly the great scandal: in order to give the people some Christmas pleasure, the Swedish Viceroy had had the idea of opening his royal residence of Paléhaven for food, drink and dancing, and here, anyone who so desired could come in the evening between six and twelve o'clock, either with or without masks. At Christmas time it was usual, despite the prohibition of mumming[10], that the city swarmed with people in masks, while within Paléhaven musicians were playing for the dancers, and the Viceroy's own servants in full livery served at the bar tables where schnapps and punch were being sold. And everybody had been there, from the merchant director to the menial docker, from the housewife to the tart. The dancehalls and dives

---

[10] The Norwegian *julebukk*, had much in common with mummery, a 'profane' custom practiced at the Winter Solstice/Christmas season across much of northern Europe (and which still occurs in Newfoundland and Labrador in Canada). The following excerpts from The Oxford English Dictionary indicate some of its features: "From O.Fr. *momeur*...1648, in the book *Hunting of Fox*: Like mommers in a mask, make a fair shew but speak nothing...1829, J. Hunter *Hallamsh. gloss.* 67. *Mummers*. This is the name of parties of youths who go about at Christmas fantastically dressed, performing a short dramatic piece of which St. George is the hero...1964. L. Diack. *Labrador Nurse* II xiii 70. From Christmas Day to Twelfth Night...was the season for 'Mummering' or 'Janny-ing'. There was much dressing up and disguising, and parties went round from house to house to entertain and have fun. 1969, in Halpert & Story, *Christmas Mumming in Newfoundland*. Disguise is a central element in Christmas mummering in Newfoundland.

of Grønland had sat in complete silence and emptiness as everyone was at Paléhaven. Consequently, all was not drink and dance in the palace gardens. Massive brawling had broken out, one man had even lost an eye, and certainly the medical student, Frellsen had been a zealous participant in this battle. People had trampled flowerbeds, damaged trees and vandalized the furnishings dotted here and there around the gardens. Many and indecent were the embraces exchanged within the summerhouse beside the sea, where the king was accustomed to drink his coffee, and that now sat unlocked and un-illuminated. But worst of all, the bathing chambers were open, and in there one could, for a price, make an offering to *Venus vulgivaga*.

The visit to Eidsberg with its happy sibling atmosphere seems to have inspired in Niels Henrik a desire to take a trip home to Gjerstad before lectures recommenced. At Gjerstad, Pastor John Aas acted as the middleman between Niels Henrik and the family. Aas was certainly familiar with conditions in Christiania and supported Niels Henrik's plan to give Peder the possibility of taking the *examen artium*. On January 18, 1822, Niels Henrik wrote that perhaps in a week's time he would be making a trip home, it having been so long since the last visit, and besides, Peder could journey back with him to Christiania, and that he would send a definite message by the next post. He urged Pastor Aas to "greet my Mother and my Siblings and let them know I am alive and kicking." One week later he wrote to Pastor Aas to say that the lectures were about to begin again, and that since "the Roads are in such an ugly condition it seems I am compelled to give up this Journey. It is therefore better that Peder come as quickly as he can; I am quite ready to receive him. If I might also utilize this Occasion, I would dare appeal to you to see to it that my Brother brings with him one, or if at all possible, two sets of linen for a Bed for two Persons. In a similar vein, although I have a Pension notice of my Mother's but I cannot receive payment without having a Letter of Attestation to the effect that she is alive and in an unmarried Condition, thus do I pray that you might write such a Testimonial, and send it with my Brother. Please receive my greetings, Your most obliged, Niels H. Abel."

These are the oldest letters written in the hand of Niels Henrik that have been preserved, and they are stamped with a seal: "N. H. A. *Deus et virtus*" - God and virtue.

By this point in time Niels Henrik must have received a clear indication from the authorities that it was permissible for Peder to live in his quarters at Regentsen. In any case it was an old custom for the "Regentsians" to have a younger, studious fellow, particularly a relative, living with them as a "pig" (the tradition in the Danish-Norwegian educational system whereby a student had a younger relative living with him almost as an apprentice student). Thus

Peder came to the capital and quickly sent home a letter to Pastor Aas, stamped with Niels Henrik's seal in red wax: "I met my Brother in the highest state of Well-being, and he took me with him to the Main Square where I saw so much. I have a very good impression of Christiania as (it) is an enjoyable Place."

And Peder was to find things so good as to hinder his adjustment to life in the capital city. It was not long before Peder's parents' weakness for alcohol also revealed itself in his own desires and needs. Niels Henrik looked after his brother as well as he could, helped him with his schoolwork and got his fellow students to read with him. Peder became a university student three and a half years later, in August, 1825, just before Niels Henrik began his grand tour abroad. Peder hung around Christiania until he got a domestic position with his future father-in-law, Vicar Fleischer at Våler, and still later, when Niels Henrik had passed away, Peder Abel became both vicar and mayor of Etne in western Norway.

But what of the bedsheets that Niels Henrik had requested? When Peder arrived in Christiania after sailing for two days from Risør he had no bed linen with him. There had not been a single bedsheet to be had in Risør after their maternal grandfather's bankruptcy and death. However, former friends of the family had promised to help since Niels Henrik's good reputation in the capital tended to encourage kindness. Maternal grandfather Simonsen's friend, Shipowner Fürst, who was also the Netherlands' vice-consul, gave Peder 25 speciedaler. Also Merchant Boyesen, who was the vice-consul for England and France in Risør, said that he would supply the bed linen that Niels Henrik had requested, but Boyesen was not home when Peder Abel was to leave. Shipowner Fürst had then promised to send some bed linen to Christiania at the first available opportunity, but this did not happen. Niels Henrik never did receive his bed linen. And the fact that the Abel brothers had only one sheet and had to sleep without it when it was out for washing, was noticed and remarked upon at Regentsen. Later this bed sheet was to become a vivid image that would evoke Niels Henrik's poverty-ridden student days.

29

Spring and Early Summer, 1822

Niels Henrik followed the lectures leading to the *examen philosophicum*, but most of his time he devoted to the mathematics books he borrowed from the University library. These were Euler's volumes on differential calculus and the teaching books of Lacroix. He borrowed those volumes of Laplace's long, major work, *Mécanique céleste*, that contained all the astronomical knowledge available up to that time. He read *Exposition du système du monde*, and he studied Laplace's mathematical works as had been formulated in *Théorie analytique des probabilités*, and where important formulas from integral calculus were to be found, and the ingenious proof for the Eulerian formula:

$$\int_0^\infty e^{-x^2}dx = \frac{\sqrt{\pi}}{2}$$

But it was no doubt still Lagrange's books on function theory and equation theory that he studied most zealously; the problem of the fifth degree equation still had not been solved. As far as they were able to follow him, Abel discussed mathematics with both B. M. Holmboe, who had become head teacher, and Professor Rasmussen, but most of the time he sat in the completely wretched surroundings of his room and studied on his own. Jens Schmidt, his room-mate, sat on the other side of the half-wall that divided the room; in addition, Niels Henrik had to share the table with Peder.

Niels Henrik seems to have found it difficult to say no when friends invited him to play cards or go out for other distractions, and with the dawning of spring and warmth, new temptations arose. Niels Henrik's friends, Niels Berg Nielsen, Jacob Worm Skjelderup and Carl Gustav Maschmann, who all lived at home with their parents, all came from well-off families, and like most prosperous families in Christiania they had their villa or country place outside the city. To have a country place to go to in the summer months had long been a sign of the urban citizen's independence; these dwellings composed a substantial part of Christiania's luxury. Foreigners who came to the capital considered this hunger for the countryside to be a veritable mania, and the demand for such places pushed the prices sky-high, with the result that there was little money left for fixing up one's house in the city. Visitors waxed eloquent about the beauties of nature around Christiania, and its setting, but they considered the city itself to be less attractive. It was said that the city lacked a real harbour, stone fountains in the streets, and an actual Town Council building. But these villas that surrounded the city functioned as farms for their owners, and allowed them to prepare provisions for winter. They grew potatoes and other vegetables, apple- and pear trees began to be common, while plums, cherries and apricots were still rare. Many had a cow or a horse that they took into the city in the winter months, or had a farmworker to care for it. When, in 1812, the University was planned to be built in the Tøyen quarter northeast of the old centre of the city, it was projected that each professor be given a villa with 40 *mål* (10 acres) so that they would be able to keep a couple of cows, something then considered essential for a family of a certain social standing. Things had not turned out that way, and times had changed radically, but to have a place in the country was still considered essential. Thus, as early in the season as possible the wives and children moved out, while the husbands carried on their business in the city and came out on the weekends and perhaps for the occasional evening during the week. These villas were not far out of the city but the journey was considered to be too much to do every day by carriole.

In the summer months there was considerable social life at these villas. One ate and drank, played cards and engaged in lawn bowling. Some told of week-long sprees from villa to villa. Those who did not possess the means for a country place went out to country inns in the region, but the great majority could not even manage this. Most lived in small, cramped firetrap houses along the roads into the city where it was possible to build without prohibitive regulations. There were dock workers, sawmill employees, fishermen and casual labourers who, by virtue of their seasonal work, could survive the off-season as shoemakers, tailors, grocers, hucksters and publicans. And the latter could in addition earn good money by means of unlicensed trade with farmers on their way into the city. Many a peasant did not venture further into the city than to the establishments of those engaged in the peasant trade, where they sold their goods and bought what they needed, where they also got lodgings and a place for their horses, and a plenitude of spirits. Piperviken (around the present day Oslo City Hall) and Vaterland (north of Oslo's current central railway station and bus terminal) were the oldest such suburbs. Later came the quarters of Sagene, Grønland, Hammersborg, and in the 1820s, Enerhaugen and Ruseløkkbakken. Otherwise, it was in the latter neighbour-hood, also referred to as the "Tunis and Algiers of Robbers," where a generation later, the pioneer sociologist, Eilert Sundt, began his studies of working class life.

Drammensveien was a popular promenade route toward the villas of the west side; it led out over the broad Tullinløkka fields where one could still go snipe-shooting or blast away at clay pidgeons, and further, over Bislett Brook that formed a little waterfall before running down Tullinløkka, and after turning left below the hillside toward Bellevue where the Royal Palace would be relocated, the road passed through an area of beautiful country houses of Sommerro [Summer Rest] and Skinderstuen [Cottage in the Light], that Bernt Anker had built, and where the rich merchant, Hans T. Thoresen, now hosted mammoth binges. Arbien's Sophienberg lay across the road, and further down toward the sea was the country inn of Sommerfryd [Summer Joy], as well as the government lawyer Morgenstierne's villa, Munkedammen, and further out, Merchant Heftye's Filipstad. To have an *expansive* view was more sought-after than to have a *beautiful* view, and to live beside the sea was less desirable. Of the august members of the bourgeoisie only Heftye at Filipstad and Commissioner of War Hetting on the island of Malmøen had their villas by the fjord. Otherwise, the great discussion in Christiania was now about where the Royal Palace would be constructed. To be sure, there was a marvellous view from Bellevue, but the fact that Karl Johan would consider having a royal residence in so lonely and isolated a place on the wrong side of Bislett Brook awakened astonishment in Christianian society. People maintained that the city ought rather be built out toward the north in the direction of Hammersborg. And now in 1822, when Parliament granted the first monies for the building, discussion developed about it being perhaps cheaper to outfit Akershus Fortress as the royal residence.

Further out along Drammensveien lay both Solli, which belonged to the wife of Bishop Pavels, and Petersborg of the Nielsen family. There was also a popular inn here. There were many in addition to the student, Niels Berg Nielsen, and his friends who came out to the Petersborg villa. Many came here to admire the uncommonly beautiful masonry of its grey stone wall that Agent Nielsen had gotten Swedish prisoners of war to build in 1808. It was like a prison wall, 4 alen high [two and a half metres] and was an example of top quality stone masonry. It was put up to demarcate the lot's boundary, something unusual at the time. Those who had beautiful views took great care of them, and worked to beautify their surroundings further. Skillebæk [Boundary Brook] lay further out along the road into the countryside; here Professor Skjelderup bought land and built his large villa. Here was also Dragonskogen [Dragon Woods] with its renowned arbours, and Villa Punschebollen [Villa Punchbowl], and to the right, with a staircase up to its tower, was Wilhelmsborg, the villa of Postmaster Wraatz. Merchant Nicholay Andresen had just purchased Villa Nøisomhed [Villa Moderation], and towards Skarpsno where there was also a holiday area, the English consul had settled at Frognerhaug, the villa considered to be the most beautiful in the area. At the crossing point to Ladegård Island,[11] today called Bygdøy (today the home of museums, other institutions and the royal stables) was where the pharmacist Maschmann had his villa which extended from Skøyen Farm to the more westerly of the two rocky points facing toward Ladegård Island. Sophienlund lay next to Maschmann's property and belonged to the high court lawyer, Jonas Anton Hielm; his brother, Hans Abel Hielm, had otherwise in the autumn of 1821 suspended publication of *Det Norske Nationalblad*. The villas of Councillor Ingstad, Surgeon-General Thulstrup and Paul Thrane, the shipping merchant's son who had just been made Justice Counsellor, were located on Ladegård Island.

When one walked back toward the city one could look along the Frogner River at the elegant park that surrounded Frogner Manor with its fifteen horses, its sixty cattle and forty head of smaller livestock. To secure help with haymaking and harvesting, the manor let out small plots, the annual fee being from six to twelve days' farmwork. Other than this there was a confusion of villas, plots and country dwellings running in all directions down toward Incognito. Here there was a terrace with several square reservoirs trimmed with grass sod, shaded by overhanging pine trees, and bright green weed and circling carp. Uranienborg was a good way from the main road with uncommonly many out-buildings and sheds, behind which stretched Uranienborg Forest with its rich hunting grounds. Otherwise Uranienborg was an excellent goal for an evening's stroll. There, from a large stone on the top, one had a magnificent view out across the city to the green wall of Ekeberg Hill. Coming closer to the city, one passed Hegdehaugen, which was almost a little village because the main house was surrounded by other buildings; proceeding further one came to Bolteløkken and Pilestredet that was at the time a little narrow winding lane, impassable during the spring thaw. Professor Sverdrup's place, Frydenlund, was on Ullevålsveien, and beside it the newly-rich grocer, Jacob Meyer had the Steensberg property with its huge asparagus fields all the way down to Pilestredet. On the other side of Akerselven (Aker River) lay Nedre Foss (Lower Falls) or Kongens Ny-Mølle (King's New Mill) with its large main building and its beautiful baroque gardens; then came Mellom-Tøyen where Niels Treschow lived. At Lille-Tøyen lived the city's newly wealthy large

---

[11] During much of the long period of Danish rule, Ladegård Island, later called Bygdøy, functioned as a farm estate for the military garrison, Akershus, which lay across the harbour to the east.

timber merchant, Christian Eger. And Merchant Steensgaard had Vålerengen, and up toward Etterstad's military parade grounds. And all these villas and country houses were surrounded by large farms. Wolves roamed in the surrounding forests, often in packs, and farm dogs were frequently found torn to pieces. And it had not been long since the bear that so interested the natural sciences teacher, Flor, had been shot up at Bogstad. The approach to Bogstad, they said, was framed on both sides with particularly beautiful weeping birches.

The great celebration of summertime, Mid-summer or St. Hans Eve, was greeted with decorations and crowns of leaves, barrels of tar, fireworks and parties everywhere. The official celebration took place at Mærrahaugen, or as it came to be known, St. Hanshaugen. In 1821 the Viceroy and his entourage took part, and later, in 1828, the king himself was there.

For the boys and others who could not gain admission to the bathing rooms at Bjørvika, there was bathing on the open beach in front of Akershus and along Piperviken. Niels Henrik, who was said to have been a zealous swimmer, no doubt exhibited his prowess outside the southern walls of Akershus. But it was probably a great shock to him to find that his confirmation priest, Jens Skanke Garmann drowned while bathing during the summer of 1822. Sailing on the fjord now began to be a regular form of recreation. A younger son of Professor Skjelderup, Jens C. H., was a particularly avid sailor when he was home from naval service down the fjord at Fredriksvern. Maybe he took his brother, Jacob Skjelderup, and Jacob's friends with him out on the fjord. In any case, that summer after the *examen philosophicum* in June, 1822, Niels Henrik and Peder sailed down the fjord to Risør, on their way home aboard a freight schooner.

30

Summer, Autumn, Winter, Spring

Many of the stories that people later told about Niels Henrik Abel must have occurred during the summer of 1822: that he happily and with ease called upon all the farms, that he was good at predicting the weather, that he had these instruments with which he was trying to measure the distance to the sun. Moreover, there was a great deal of sun that summer. The spring had come early; the farmers had begun to plough their fields at the end of April. It looked as though it would be one of the rare dry summers. Maybe that was why people got the impression that Niels Henrik could predict the rain clouds that eventually came. It is likely that it was also this summer that, in order to show his respect for this young man of science, and with hat in hand, Pastor John Aas met with Niels Henrik and they solemnly and politely walked the hills overlooking Gjerstad Lake.

Conditions had not changed at the dowager farm of Lunde where his mother lived. His mother was unstable and intoxicated, and her friend, Jørgen L., was busy with haying and the running of the farm. Niels Henrik's older brother, Hans Mathias was scarcely lighter of mood on seeing his two younger brothers come sweeping home from Christiania. The brother, Thomas Hammond, who was less gifted than the other two, was probably home as well, after having failed at the several job opportunities Dean Krog had tried to arrange for him in Arendal. But what was probably worst for Niels Henrik to see was their twelve year-old sister Elisabeth in these sad surroundings. Perhaps in the depths of his calm mind, Niels Henrik had begun to think of how he could get her as well to the capital. The youngest brother, eight year-old Thor Henrik, seemed to be happiest when he had a flute to play on. It is highly probable that in the course of the summer, Niels Henrik sought out his old teacher, Lars Thorsen Vævestad, who in April had been appointed sacristan and teacher for the village and probably asked Thorsen to see that Elisabeth and Thor Henrik got that portion of regional schooling to which they were entitled. Many things indicate that Niels Henrik now tried to fulfill his father's wish that things be done in order to improve the lives of his nearest and dearest.

When he stood bent over the measuring instruments balanced on the chair he had brought along in one place or another out under the open sky, it would not have been strange that he attracted the curiosity of Gjerstad's youth. Subsequently it was said around the community that in an unobserved moment the height of the chair had been altered surreptitiously by the youths, who had been greatly amused when young Student Abel had cursed without inhibition, swearing that the sun must have moved. But had he not perhaps been happy to be the centre of curiosity? Perhaps he had spoken about how he could calculate the heights of the mountaintops and distances over land and sea, and how far it really was, perhaps all the way to the sun or moon? He had learned to calculate geographical distances from Hansteen, and perhaps he wanted to practice a little in front of his fellow inhabitants? But above all, what he also wanted to determine was the moon's influence on the lines of magnetic attraction, and the instrument he was using was quite certainly Hansteen's specially-constructed swinging apparatus. That he joked about measuring the distance to the sun was perhaps a sign that he was working without the greatest of enthusiasm. After all, he forgot to calculate the moon's effect on the earth, and he therefore calculated the moon's gravitation to be sixty times greater than it actually was.

Niels Henrik's evaluation of his own factual views that summer must otherwise have been as follows: he had now finished with all the obligatory preparatory studies at the university, but in mathematics with which it was so crucial that he continue, there was no one who could teach him what he wanted to learn, and thus a future position and a means of livelihood remained up in the air. On the other hand, his former teacher, B. M. Holmboe, had certainly not taken any public service exam, and C. A. Holmboe and S. B. Bugge had both obtained positions at the university at quite young ages. Bugge was barely twenty-two. Probably similar sorts of views were advanced about his friends Boeck and Keilhau as well. Moreover, it was the actual teaching of mathematics as had been included in the university's *plans,* that still had not been properly fulfilled. Niels Henrik could probably, with quite reasonable security, have foreseen that in time there would also be an opening for him at the new Norwegian university.

Niels Henrik's desire to follow what was transpiring both in contemporary literature and mathematics, was perhaps reflected in what he was actually borrowing from the university library when he returned to Christiania in September, 1822. He borrowed what was the newest of the new in contemporary literature, Maurits Hansen's *Othar of Brittany,* Norway's first novel, published in 1819. The book had the subtitle, *A Crusader Story,* and in many respects dealt with what we today would call "youth and sex." The main character's love for his sister and his relations with the object of his desire, his superior's wife, the empress, took centre stage in the story. The erotic temptations of these characters were the work of Satan, and according to the book, the youth can only recognize and be delivered from perplexity and confusion by means of personal Christian faith. The fulfillment of the last sentence in the novel, "May God give us all His Grace," was an essential requirement for understanding the driving force in human beings, and to attain that bliss that life *can* give. The novel is an allegory and in many ways is an exemplary assemblage of all the Christian virtues and patterns of behaviour that any youngster, through catechism lessons, had stuffed his or her head with. But the novel complicates life; it is never so easy to see the difference between good and evil helpers, and there are so many almost imperceptible ways to be led astray.

Around the same time, and in addition to *Othar of Brittany*, Abel borrowed the works that the French mathematician, Hachette, had assembled under the title *Correspondance sur l'École polytechnique*, I-III. This was a collection of *acta* from l'École Polytechnique where Hachette had taken over the professorate of descriptive geometry in 1799, from the great Gaspard Monge. The books covered the period from 1804 to 1816, with information about the changes in teaching personnel at the school, lists of students and of the students' later professional track records, appointments and publications. But *Correspondance* also contained a number of mathematical papers, most often by teachers and students at the school. Since most were the students of Monge and Hachette, most of the papers dealt with the application of geometric analysis, on curved surfaces, skewed curves, and the surfaces of the second degree. Among other things, Hachette contributed a long article on another legendary French mathematician, Fermat, and his cannonball constructions. The loan records show that Niels Henrik had finished with these volumes before the end of the year.

There were two great events on the theatre front during autumn, 1822. On the 27th of October, Karl Johan visited Grænsehaven, and the moment he stepped into the hall he was greeted with a sea of voices singing Bjerregaard's crowning glory, the national song, "Sons of Norway's immemorial Realm,/ Sing to the ring of the festive Harp!" The hall was decorated, and furnished with golden carpets, and the most glittering of costumes were everywhere. The poet priest, Johan Storm Munch had written a prologue that was recited by the student, Colban. The play that was performed, *Maria of Foix*, had been staged in both 1818 and 1821. And at the end of November The Dramatic Society commemorated the hundredth anniversary of the performance of Holberg's plays[12] with a performance of Holberg's *The Amateur Politician*, in which O. R. Holm played the main character. But first they performed the new play commissioned from Bjerregaard, *Memories of Holberg*. Here, in the name of Philomen, the classical Greek comedian, the writer invited a number of characters from Holberg's comedies to a gathering to celebrate the great writer's birthday. Various characters milled around the stage: the salon-scholar and intellectual snob, Stygrotius, the superstitious Roland, the arrogant penniless Don Ranudo, the doggerel poet, Rosiflengius, the fussy Vielgeschrei, the marriage-sickened Magdalone, as well as the glib, roguish Henrik "guzzling buttery bread" as he fulfils his great appetite for life.

It is uncertain where exactly Niels Henrik and Peder spent the Christmas holidays that year. Perhaps it was in Christiania where the big event this year was the death and funeral of Bishop Bech; perhaps they went home to Gjerstad, but the roads were not particularly good for sleighs before February and March.

At the beginning of February, 1823, the first number of *Magazin for Naturvidenskaberne* came out. This was the country's first journal of the natural sciences. The editors, Gregers F. Lundh, Christopher Hansteen, and H. H. Maschmann, introduced it by arguing the necessity of having this forum as a means to build support for the sciences in Norway. Abel was already well-informed of the plans for the journal and his name stands at the top of the short list of subscriptions. He had probably already been promised space for his own formulations. In *Magazinet*'s second number Niels Henrik Abel's first article was published; his debut work was entitled "A Common Method for finding Functions with one varying Size, when a Property of these Functions is expressed as an Equation between two Variables." Later in the year, divided over two issues of the magazine, he published "The Solution of a Pair of Propositions with the Help of definite Integrals." At the same time, his friends, Boeck and Keilhau also published articles in the magazine. Keilhau's title was "Geognosy in Norway," while Boeck wrote on "Observations of Phenomena that appear to indicate an intrinsic Functioning among certain Aquatic animals." Hansteen himself wrote on a series of practical matters and made public the plan and design for a new system of weights and measures for Norway.

Abel's articles were surely incomprehensible to almost all of *Magazinet*'s readers, but Hansteen defended Abel's contribution the first time by writing, "It might perhaps seem that in a Journal that professes to deal with the Natural

---

[12] Ludvig Holberg (1684–1754), born in Bergen, was a prolific poet, philosopher and historian, as well as being the Molière of Norway-Denmark. His plays burst upon the public with the opening of Den Danske Skueplads in Copenhagen in 1722.

Fig. 18. Professor Michael Skjelderup's summer villa at Nedre Skillebæk, drawn by bookseller, C. Hartmann in 1827. Professor Skjelderup was one of those who felt sorry for the poor student, Niels Henrik Abel, and supported him economically. Moreover, his son, Jacob Worm Skjelderup, was a friend and classmate of Abel's. A younger son, Jens C. H., was among the first recreational sailors on Oslofjord. Drawing from N. Rolfsen, *Norge i det nittende aarhundre*, Vol.I. Kristiania: A. Cammermeyers forlag.

Sciences, a Treatise in pure mathematics is out of Place. But Mathematics is the pure Formulation of Nature, it is to Nature-Inquiry what the Scalpel is to Anatomy, an altogether indispensable Tool without the Help of which man can seldom penetrate the Surface of things. Nature's Builder has set a Canopy to hide the entrance to its Inner Sanctum, bearing the same Inscription that the Greek Philosopher of yore placed over the Entrance to his Lecture Hall: no entrance to those not conversant in geometry. (Hansteen here gave the citation in Greek as well.) The majority who try to enter without this Ticket of Admission can never get beyond the forecourt and must turn back. With this Tool's Help, Galilei, Huygens, and above all, Newton, penetrated further than their Predecessors, and their Descendants can only go, with confidence, deeper into these Labyrinths if they have the help of, and are equipped with, a set of Guiding Principles. In this respect, Contemporary French Physicists, one and all, provide an Example that, following these principles, whatever the task they have in Hand, they advance more quickly and securely toward the

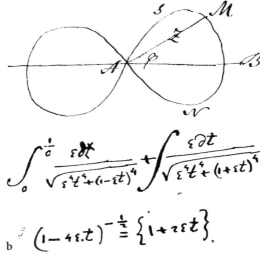

$$\int_0^{\frac{1}{0}} \frac{\varepsilon \partial t}{\sqrt{\varepsilon^2 t^2 + (1-\varepsilon t)^4}} + \int \frac{\varepsilon \partial t}{\sqrt{\varepsilon^2 t^2 + (1+\varepsilon t)^4}}$$

$$b \quad (1 - 4\varepsilon.t)^{-\frac{1}{2}} = \left\{ 1 + 2\varepsilon t \right\}.$$

Fig. 19 a–c. The University's Mariboe Building, facing Kongens gate and Prinsens gate. From a lithograph of the university, by P. F. Wergmann, 1835. Aschehoug forlag archives.**b** From Abel's workbooks. *The Lemniscate.* The geometric locations for the points P if the distance of two permanent points has a given product: a curve that Abel studied and found how its circumference could be divided into n parts of equal size. National Library of Norway, Manuscript Collections, Oslo Division. **c** Elliptical Integrals. Over the integral sign Abel uses the notation 1/o instead of ∞. National Library of Norway, Manuscript Collections, Oslo Division.

Fig. 20: What Abel was engaged with here, about 1820, is difficult to say, and how finely executed the circle is has engendered different opinions. In a heated discussion between Holmboe and Hansteen in 1836 – on the practical application of mathematics – Hansteen maintained that many from the academic schools could not even handle a simple calliper compass, and that

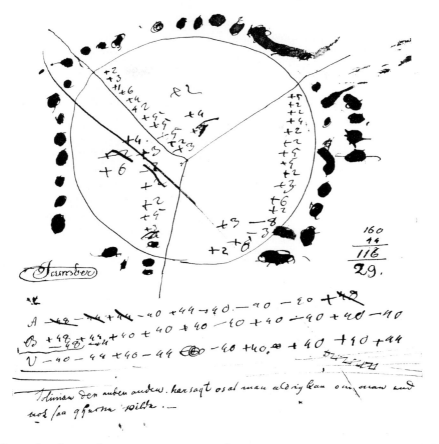

Fig. 20. One frequently encounters unmotivated sentences in Abel's workbooks, and expressions about *Soliman the Second*, as here: "Soliman the Second has told us that one never can [do] as well as one really wants." Soliman, or Sulieman was the name of a Turkish sultan, and Suleiman II, also called Suleiman the Great (1494-1566) invaded Rhodes, beseiged Vienna, and consolidated Ottoman Turkish rule over Hungary, among other things. He was a brilliant organizer, supported science and art and wrote poetry. At other places in Abel's notebooks we find: "Soliman the Second is one of the great scoundrels of all Eternity," and "Soleiman the Second is a helluva fellow. I say this in full Lewdness. Amen." National Library of Norway, Manuscript Collections, Oslo Division.

consequently, their circles looked like potatoes, and that their geometry was therefore not worth a pipeful of tobacco. Holmboe then drew attention to Abel, but in order to reverse the argument: Abel's circles, he pointed out, could resemble something much worse than potatoes and thus would his geometry not be worth "a Pipe's worth of Tobacco"?

Achievement of their anticipated goals than do scientists in other Nations. I therefore think that the Magazine, in addition to Materials, ought to include Tools that can enhance Cooperation, and facilitate the fullest possible handling of the Materials that the learned Public now has Occasion to become familiar with in a Work from the inspired Hand of this talented Writer."

Abel's first paper also elicited a comment in *Morgenbladet's hefty appreciation of Magazinet,* in the following words, "A study that justifies the greatest Expectations for this young Mathematician."

Surely it was enough for the magazine's readers to know that Abel was an exceptionally promising mathematician, and he would not become greater by filling space with material that almost no one could benefit from. Again at the beginning of the second year of *Magazinet* the publishers also had found it necessary to provide a clarification: after first stating that Norwegians formed a small nation with a common culture rather inferior to those of countries like France, England or Germany, where a scholar could expect a considerable recompense from the marketplace for valued projects and scientific services rendered, with obvious allusion to Abel and his mathematical course, the editors went on to explain: "We predict with absolute certainty that many of our Readers will find their Expectations unfulfilled. Many wait in expectation of popular Treatises; abstract Research finds Interest only among the Few. Whereupon we alone would like to answer: whoever works to advance Scientific Development follows an arduous and to some degree untrodden path; from this Point of View, whatever one once gained, one cannot, except by the greatest of Sacrifices on one's Part, manage to dislodge. In doing so, one would lose more than others would gain. But by means of his Future projects, there is always much the researcher can elucidate; to abandon the once staked-out Line that seems to lead toward the Goal is counter-productive. As for those, who for the Sake of their own Advantage write for the entertainment of the Reader, the Matter is quite different; they must align themselves with the Taste of the Reader the same way the Manufacturer does with his Market. Therefore, we have maintained from the beginning of this endeavour what we now reiterate, that we consider our Subscribers' Support as a kind of painful Sacrifice, to be willingly donated in order to elicit serious Studies and intellectual Enterprise in the Fatherland."

All the same it seems that Abel's work was not particularly good material for *Magazinet.* Beside the infamous treatment of the moon's influence on the movement of the pendulum, there was only one later paper of Abel's in *Magazinet,* and that consisted of seven pages during the autumn of 1825, material Abel had already probably submitted at this time during the spring of 1823.

Fig. 21. Charité Borch, Mrs. Hansteen's sister, of whom Abel thought so highly. The picture was painted by F. Vermehren in 1855. Charité was then 54 years old and married to the Danish poet, Frederik Paludan-Müller. Vilhelmine Ullmann, who told about the sensation that Charité evoked with her bold extravagance in Christiania during the 1820s, also wrote in her memoirs (Kristiania 1903) - and probably about this same picture - "When I now see the meagre Portrait of Mrs. Paludan-Müller in a literary work, I cannot reconcile this with earlier memories I have of the fine youthful Creature that I saw at that Time." Strangely enough, no pictures of Mrs. Hansteen, Professor Hansteen's wife and the mother of Aasta Hansteen, who became Norway's pioneer campaigner for women's rights, have survived. The National Historical Museum at Frederiksborg, Hillerød, Denmark.

These first works of Abel's, written in Norwegian, were of course included in the two later publications of Abel's collected works, published in French, but in the mathematical literature they not only represent beginning efforts without particular scholarly significance. Abel's second work in *Magazinet* was the solution to a mechanical problem with the help of what is known as integral equations. This was probably the first time in mathematical history that an integral equation was solved. Moreover, this special integral equation that Abel studied, has proven to be, with a simple conversion, identical with what is known as the Radon Transformer, which is the mathematic foundation that modern radiology is based upon (and for which Cormack and Hounsfield received the Nobel Prize in Medicine in 1979). And we do know that during the spring of 1823 Abel wrote a paper in French that was quite certainly of the greatest significance; some mathematicians even feel that this treatise would have been counted as part of his main work if it had been found. In a word, the paper was lost without a trace, probably disappearing into piles of papers, on its way back and forth between different departments. In the records of the Academic Collegium for March 22, 1823, it is stated that: "Professor Hansteen appeared before the Collegium and presented a Manuscript of Student Abel's, containing a Treatise, whose Goal is to give a common Account of the Possibility of integrating all possible Differential Formulas, and asked if the University would find it appropriate to support the Publication of this Work. The Manuscript was passed into the keeping of Professors Rasmussen and Hansteen to see if they could together form an opinion of the Value of this Work, and at the same time to suggest a Means by which this could be appropriately supported, if subsequently they find it deserving."

That Hansteen personally presented Abel's work in the Collegium, and that he expressed the wish that it be published abroad must in any case mean that Hansteen at least considered the treatise important and "deserving." Professor Rasmussen was certainly of the same view. But the university's constant economic caution was such that the treatise trailed back and forth amid a wake of dossiers and queries about support and stipend funds for Student Abel, until eventually the paper disappeared. At Regentsen, Abel's comrades were astonished that he was capable of writing a paper in French. They knew that he had not distinguished himself particularly in languages, but Abel would quite certainly have turned it into a joke, saying that the work was so full of formulas that there was only the occasional French word inserted here and there.

Professor Rasmussen must have reacted to the Collegium's excessive caution. Once in the course of the month of May, Rasmussen had told Abel that he would, from his own pocket, give Abel 100 speciedaler so that he could travel to Copenhagen and meet Danish mathematicians. Abel wrote right away to the *Collegium academicum*:

> The Undersigned hereby takes the Liberty to report to the high Collegium that he intends to undertake a Journey to Copenhagen during the Summer holidays. My Aim thereby is partly to visit the Family I have living there, partly to develop my mathematical Education as much as Time and Circumstance will allow. The Journey will be of about 2 Months, such that I am thinking to be back again in the Middle of the month of August. Respectfully, N. H. Abel.

31.

Copenhagen – Travel, Summer, 1823

Niels Henrik was thus to travel to Copenhagen, and it was Professor Hansteen, above all others, who helped him with the necessary preparations. One Sunday afternoon they strolled up together to Tøyen to visit Cabinet Minister Treschow, Hansteen's cousin. Hansteen felt it might be useful if Niels Henrik were able to give Treschow's greetings to colleagues and acquaintances in Copenhagen. To be sure, philosopher Treschow could not himself attest to Abel's specialized knowledge. Mathematics was a subject that did not enter the common culture in the same manner as philosophy or politics. Abel's scientific scholarship was not part of the learning and culture that Treschow believed that a student aimed to attain through his reading and self-denial. The 100 speciedaler that Abel had received from Rasmussen was considered adequate for the journey; in any case, Treschow did not contribute any extra support, but Hansteen, surely as a result of prompting from his wife, took Abel to a tailor and had two new suits made.

In two letters to B. M. Holmboe, Abel explained how he prepared for some of his experiences that summer. Three or four days after he had reported his intended journey to the University Collegium, he was on board Captain Emil Trepka's sloop, *Apollo*. Trepka was the one who most regularly sailed the freight route between the two capital cities in these last years before the advent of steamboats. "That first day we only managed to force our way three Miles [34 km]. The second Day we came to Drøbak, where we lay for two Days." Abel wrote further that he utilized the time and opportunity to attend a party at the home of Heinrich Carl Zwilgmeyer, who had participated in the American War of Independence and was a language teacher at both the Military School and the Cathedral School in Christiania. Abel commented to Holmboe that Zwilgmeyer had "three well-ornamented Daughters." One of these girls said much later that Abel had been astonished that her twelve year-old brother, Carl Theodor, spoke such good French. Thirty years later another brother, Peter Gustav, would become the father of Dikken Zwilgmeyer, a Norwegian woman writer of some renown in the latter part of the nineteenth century.

Following its stay in Drøbak, the *Apollo* was blown out of Christianiafjord on a good wind during the third day of the journey, and after a further two days of strong winds they arrived in Copenhagen, on June 13th. Two days later Abel wrote his first letter to Holmboe, saying "I arrived in Copenhagen on Friday and hastened to Mrs. Hansteen's Sister, Mrs. Fredrichsen where I was especially well received. She is an exceedingly pleasant Wife; she is very very attractive, and has four Stepchildren but none of her own; her Husband left for the West Indies a short Time ago. In a Week's Time she is to journey to her Mother's at Sorø, and she had asked me to accompany her. I am thinking to honour the Invitation."

Abel was well-received everywhere. "I have been extraordinarily well treated here in the City; I am lodging at my uncle's, Captain Tuxen's; they have offered me free room and board for as long as I want to stay. His family are very loquacious and interesting, so I hope to get on very pleasantly. He has eight children." Uncle Tuxen, Aunt Elisabeth, and Niels Henrik's eight cousins lived in an elegant house at Dokken in Christianshavn. Peder Mandrup Tuxen was now a man of high repute, responsible for all the instruction in mechanics and hydraulics at the Naval Academy. After having had study tours to Holland, France and England, Tuxen had designed new ropework for the Danish fleet and he had been responsible for buying the machinery that could make the good ropes and hawsers that the ships were so dependent upon. Also, in the course of rebuilding the Danish fleet following the bombardment by the English in 1807, it was Marine Captain P. M. Tuxen who was in charge of testing new cannons. In addition he had the knowledge necessary for storing gunpowder, and it was he who eventually developed an effective on-board pumping system for the navy's vessels. For all his work, and his various improvements to the fleet's artillery, Tuxen received regular pay increases and bonuses. At home at Dokken he had a great collection of copperplate that he had purchased in his travels; he was a member of the city's art association, and he had a large book collection; his favourite writers were Holberg and Wessel. He boasted that even in the dark of night he could find any book he wanted from his bookcase. Tuxen recounted stories from England about the many steamships he had seen, studied, and travelled upon. In particular, the shipping route from Glasgow to Liverpool was exemplary, while in the south of England, steamships were still not in use.

Tuxen also talked about a sport that was peculiar to England, namely, boxing or "fisticuffs", a type of competition in which the public places bets on the outcome. Under normal conditions, padded gloves were used and the sport was not considered to be very dangerous. Boxing without gloves was illegal, but Tuxen had once been witness to such a match. In some very remote place, the boxing ring had been demarcated with posts and ropes, and thousands of spectators had flocked in to see the match between one Oliver, with the nickname "Champion of England," and a younger man called Spring. But just before the match was to begin the authorities arrived, and everything had to be moved to a different jurisdiction, five English miles distant, and there the audience hastened on foot. Naked to the waist, wearing only thin trousers, shoes and socks, the two fighting cocks met in the ring. Each had two handlers or seconds with him. They shook hands amicably and then the match was underway. Whenever one of them was knocked to the ground there followed a little pause during which the seconds gave each a shot of spirits, washed them down, with a sponge dipped in spirits, and in addition, poured spirits all over their backs. Gradually it became evident that the young Spring had the upper hand. He lifted his opponent into the air with one blow, and tossed him right out of the ring, such that one could only conclude that the wretched Oliver had been smashed to bits. The seconds had had to drag him to his feet, but after a while they did the same thing all over again. Nevertheless, in the end, Oliver had to give up, completely exhausted and by now in possession of quite unrecognizable facial features due to the blows and wounds that had almost been the end of him. The match had continued for over three-quarters of an hour. The winning gentleman received five hundred pounds sterling, and a carrier pidgeon was dispatched to London with news of the outcome. Thousands of pounds had been wagered, and Tuxen could well understand why such a barbaric and loathsome sport was forbidden, and as far as he knew it was no longer practiced even in secret.

Indeed, the theatre was a much better place of entertainment. Tuxen and his family subscribed for a theatre box in Copenhagen, and this Niels Henrik later would put to good use, but now in the month of June the theatre was closed.

After two or three days in Copenhagen Abel reported that he had met his former professor, Krumm/Keyser, who was on his way home to Norway. "That was one stiff Fellow," Abel commented. During these first days he had also visited Professors Degen and Thune, and he had already formed an impression of them as people, as well as an opinion of the mathematical milieu in the city. Degen was "the most amusing Man You could imagine. He paid me many Compliments, among other things, saying that he would learn much from me, at which point I became very shy." He had " a lovely mathematical Library." Otherwise, Abel had not been to the city's libraries, "but from what I have heard, they are not particularly well equipped with mathematical Books, which is a bad Thing." Professor Erasmus Georg Fog Thune was "an exceedingly good-natured and kind Man, but to my way of thinking, some-what pedantic. He received me with the greatest of politeness." In addition to

Degen and Thune, Abel mentioned three other Danish mathematicians: August Krejdal, Henrik Gerner von Schmidten, and Georg Fredrik Ursin.

In this milieu Abel would quickly discover that he was beyond everyone else in the field of mathematics, but perhaps the Danish men of science had other qualities that intrigued and inspired him. Professor Degen, who was at the forefront in these circles, was a very versatile man who had studied law, theology and philosophy. He was not only well-versed in mathematical history, but also had mastered Hebrew, Greek and Latin, and in addition to the normal Germanic and Romance languages, could read Russian and Polish. As a student in 1792, Degen had entered the University's competitions for work in both theology and mathematics, and had won prizes in both subjects. He had taught mathematics to the young Prince Christian Frederik while he continued his studies and took his doctorate with a treatise on the philosophy of Kant. Since that time Degen had been head teacher of mathematics and physics at the Odense Cathedral School and rector at Viborg before becoming Professor of Mathematics at the University of Copenhagen in 1814, and he worked continually to gain a greater role for mathematics in both teaching plans and learning schedules. For his part, Professor Thune, in addition to mathematics and physics, had also studied theology. After studies in Germany, under among others, the astronomer, F. W. Bessel, Gauss' good friend, Thune had taken his doctorate and became lecturer in mathematics, and later professor, at the University in Copenhagen. Beginning in 1823, he was in charge of the astronomical observatory in the city. As for Krejdal, Abel wrote, "Krejdal is a Teacher at Odense School and studies mathematics with all his powers; which will no doubt interest you as much as it does me." Krejdal was known as a promising mathematician after having won a gold medal for a mathematics treatise in 1814, at the age of 24. But he gradually became most interested in the philosophical aspects of the discipline, and would explain all the world's phenomena with what he considered the most advanced mathematical theory, namely differential and integral calculus. Von Schmidten was originally from the military and had become lieutenant before he had become acquainted with Degen and Krejdal and thereupon developed an interest in more scientific and scholarly education. He had gone on to teach cadets and students mathematics, while at the same time following Degen's advice to supplement his mathematical skills with knowledge of Latin and Greek. After having won a prize in 1819 for a paper published in French on linear differential equations, von Schmidten received a stipend and was, during that summer of 1823, in Berlin after having visited Gauss at Göttingen and Laplace in Paris. Abel wrote, "v. Schmidten is now in Berlin and is expected here on the first." Whether Abel met von Schmidten that summer is doubtful. But while in Berlin that summer, von Schmidten came to know Crelle and took part in the first gatherings that aimed to found a German mathematical journal. This was the plan that Crelle was to realize after meeting with Abel a little more than two years later, when von Schmidten had encouraged Abel to go to Berlin and look up Crelle.

About the fifth man in the Danish mathematics milieu, Abel wrote, "I have still not talked to Ursin. He should not be particularly stiff. He has recently married." Ursin was at this time the observator at the University Observatory, Rundetårn, and above all was known as an excellent teacher and lecturer who could popularize his material and hold his audience's attention. He had taken the land survey examination, been a student in 1816 and won the University's gold medal for a paper on regular polyhedrons. After taking the Secondary Examination, on his own initiative he had gone to Schumacher in Altona; subsequently he had studied astronomy under Gauss at Göttingen before returning in 1819 to Copenhagen where he took his doctorate, writing a thesis in astronomy; he subsequently obtained the position at Rundetårn.

Professor Degen, who had already seen Abel's brilliant mathematical gifts a couple of years earlier, and predicted he would be a "Magellan-Voyager to...one and the same monstrous analytic Ocean," seems quickly to have understood Abel's mathematical superiority. But for many others it was perhaps not easy to believe that any brilliant light could emanate from Norway, and Abel had to sharpen his argumentation to convince some people that it was possible for anything outstanding to germinate in the soil of the Norwegian fatherland. He wrote, "The Men of Science here think that Norway is pure Barbarity and I do everything in my power to convince them to the contrary."

It was evident as well that Niels Henrik had an eye for people's marital status. He wrote, "Degen is married, which I would not have thought. He has a pretty Wife but no children." Abel also commented that Ursin was newly-married, in fact only some months earlier; yet, about the women of the city he was astonishing general in his description: "The Ladies here in the City are extremely ugly but nonetheless, very nice." About Copenhagen in general, he wrote, "There are many Manufacturers of Wind here in the City. Everything here is inferior to Christiania." The meaning of the expression "Manufacturers of Wind" was closer to what the French call *farceurs*, that is, buffoons, rather than magical warlocks manufacturing wind. As for the sentence that followed, due to a fold in the paper on which the letter was written, it could equally be read as "Everything in Christiania is inferior to here," and this could well be interpreted as a direct quote from the buffoon-like Wind-makers. In any case, the letter ends with warm thoughts and greetings to Holmboe and his brothers in Christiania.

Abel accompanied Mrs. Hansteen's sister to Sorø, where he subsequently met their youngest sister, Charité. He was well received at Sorø and enjoyed so much living near the venerable Sorø Academy that two years later, on his way to Berlin he again visited Sorø and Mrs. Hansteen's mother. Mrs. Hansteen's mother was called Anne Margrethe Rosenstand-Goiske; the daughter of the dean of a diocese. She had in 1778 married a man from the Trondheim area, Caspar Abraham Borch, who nine years earlier had arrived in Sorø and become teacher and professor of Latin and history at the Academy. He had passed away in 1805. They had three sons and six daughters, and led an active social life. Mrs. Borch had decided that upon her death all the paupers in the city should be given free beer at her expense. One of the daughters married Count Holsten-Rathlau, and it was while teaching the Count, that Hansteen met his own wife.

But what did Niels Henrik experience in Sorø? It was summertime and everything was exuberantly green; the city was full of people, and it was no doubt idyllic along the shore of the sparkling lake. A huge fire had destroyed most of the Sorø Academy's buildings in 1813, and the school was not opened for teaching again until 1822. There now followed a period of furious construction and the Academy was once more ready to become a national centre of teaching and culture. In 1822 the writer, B. S. Ingemann, had become lecturer in Danish literature, and under Grundtvig's influence his accounts of Nordic history during the Middle Ages became widely read and discussed. Besides this, the altarpiece in the Sorø church was much discussed, and Ingemann with his sense for utilizing local popular sentiment, had already written a gripping account of this famous furnishing, and entitled it, "The Altarpiece at Sorø." This altarpiece, which stood and was admired at Sorø, had been painted by the Dutch artist, Carl van Mander, whom Christian VI had brought to Denmark in order to raise the artistic level of the realm. And thus it was that van Mander had studied the region's everyday life such as he encountered at the local inn, in the city, or out in the woods, and he used these impressions in the work he was painting for the church.

Ingemann recounted the story of Judas, one of the characters in the altarpiece's painting of the Last Supper, and related this to the withdrawn and moody gamekeeper of the Sorø forests, Franz, who in his youth had been in Italy and travelled through many countries in the service of Count and Countess N, and their son, Otto. But for more than ten years, Franz had been a gamekeeper, and lived with his beautiful southern daughter, Giuliana, within the forest, when one day Count Otto came to Sorø. The young count sought out the former family friend and his daughter, who for some happy years had been his playmate, and the two young people fell deeply in love, much to the intense despair of the girl's father. Franz forbade Count Otto to see Giuliana!

Gamekeeper Franz, who had converted to Catholicism while living in southern Europe, never went to church in Sorø, but he had a taste for great art, and together with Count Otto, went to the church to see the new triptych altarpiece that everyone was talking about. As they entered, the priest was in the process of concluding his diatribe against those who only paid lip service to Christ, while by their deeds, they betrayed him as heinously as did Judas Iscariot! And at this moment, Gamekeeper Franz recognized himself as Judas in the altarpiece painting, and he stormed out and hanged himself from a tree. And when it came time to read Franz' last will and testament, a frightful past was revealed: Count Otto and Giuliana were actually brother and sister! Franz had had a love affair with Countess N and in order to prevent the Count from finding out that she was pregnant, Franz had murdered him by gradually giving him lethal increments of opium so that by the time he got home from a long journey, he was very ill and his end was nigh. Ingemann's story ended with a deeply unhappy Count Otto moving to America and never returning to Europe. But the papers he left behind revealed that death and eternal reunion with his beloved was his only wish. For her part, Giuliana also came to know the real nature of relations and her love remained pure and unblemished, and took the form of a heavenly angel who one day would open the blessed gates of Paradise to her.

Perhaps Abel had a feeling for Ingemann's search for the infinite which had its point of departure in historical events and local legends. But Ingemann's hanging of papers that rustled from the treetops outside his house in Sorø in order to be reminded of the floating presence of the angels, as he was said to have done, was perhaps going a bit too far. We do not know how long Niels Henrik was in the old and stately town of Sorø, nor the amount of history that was brought to mind, nor whether or not he had any inkling of his distant relative, Jens Kraft, who during an earlier heyday in Sorø, in the 1750s and 1760s, had been a prominent teacher at the Academy and, in a rather inspired way, related the development of mental consciousness to the variation of mathematical functions in the treatise called *On the Immortality of the Soul*.

What we know in general about Abel's stay in Denmark that summer are the following facts: after a certain number of days at Sorø he returned to Copenhagen. He had several meetings with Danish men of science, particularly Degen, and he took part with Danish students in the 200th anniversary celebrations of Regentsen. "800 Bottles of Wine were bravely consumed on that occasion," he reported. Twice in the month of July he attended the theatre, and at one point or another, probably at a dinner party at Dokken in Christianshavn, he had met Christine Kemp, who subsequently became his betrothed.

In all, Abel wrote four letters that cast some light on these summer months. Two letters were to Holmboe; the second was dated: "Copenhagen, Year $\sqrt[3]{6.064.321.219}$ (Take account of the decimal fraction thereof)." He wrote one of the letters right after he had returned to Christiania in September, to the student, Fredrik Christian Olsen in Copenhagen; and the fourth letter was written to Degen, dated March 2, 1824, thus eight months later, wherein he apologized for not writing earlier and offering his thanks for his "courtesy and willingness to help," and he also sent his greetings to von Schmidten and Thune, but otherwise spoke mostly about the mathematics he had been working on.

Working out this cubic root $\sqrt[3]{6.064.321.219}$ gives 1823.5908... but what actual date and what the meaning was of this jest, is uncertain and has been the basis for different speculations, not the least because Holmboe dated his letter June 24th, but obviously must have meant July 24th since Abel in his letter referred to events in the month of July. There is much to indicate that Holmboe did not take the time to determine the date when he received the letter, and that probably after Niels Henrik had returned home, Holmboe had written the date on Abel's letter, probably as they sat together over the letter to discuss the mathematical formulas that the letter contained. Thus July 24th could have been Abel's own assessment of when he wrote the letter. Working out this cubic root: 1823.5908...of course gives the year 1823, and the decimal figure, .5908, comes out to a little more than 215 days of a normal year, something that would actually work out as the fourth day in the month of August. But however, were July 24th to be correct, then this would indicate that while he was in Copenhagen Abel oddly made a mistake in his calculation of the playful rendering of the date. He most certainly calculated 210 days to August 1st - seven months at 30 days each - and instead of *subtracting* five days, he had *added* them, and so found himself at the 215th day of the year! But the cubic root of 6,064,300,000 also gives a little more than 215 days, so Abel must also have thought about the hour when he wrote the last decimals and had asked Holmboe to take them into consideration in his calculation. And if *that* is found to be correct, one also has to know whether his starting point was a normal year of 365 days, as 1823 certainly was, or whether he calculated according to a Julian year, a Gregorian year, or a solar year! No ordinary clocks indicated seconds at that time, therefore it is most probable to believe that Abel took as the basis for his calculations, a time expressed in hours and minutes, and on this basis, the calculation is done with the greatest facility if one used a solar year of 365.2422 days. In that case, Abel would have sat himself down at the writing desk at 19:05, on July 24, 1823, even if Abel had made a mistake by calling this the 215th day.

About his theatrical experiences he wrote: "Two Comedies have been performed. I have been to see both of them. The last Time the Play was booed and hissed." The theatre in Copenhagen opened again on July 15th, and the two performances that Niels Henrik reported in this letter must have been on July 15th and 22nd; the next performance was not until July 29th. The plays performed on July 15th were *Peter and Paul* and *The Bourgeois Rendezvous* and on the following week the playbill offered *The Siege of Saragossa* and *The Prudent One*. Niels Henrik had probably seen Kotzebue's comedy, *The Siege of Saragossa, or Tenant Feldkümmel's Wedding Day* in Christiania, in any case it had been staged at Grænsehaven in February, 1819. But it was precisely this play that the Copenhagen audience was now booing. Contemporary reports said it was "audibly Disapproved." Could it not be that someone in the hall, or in one of the boxes, had compared it to a superior performance in the capital city of "barbaric" Norway?

In this letter Abel sets out his observations of the field: "Mathematics is not exactly flourishing here. I have still not managed to ferret out from among the Students any who are very talented, and far fewer, any who mess around with Math. ex professo. The only one here who understands Math. is Degen, but he is also a Devilish Fellow. He has shown me several of his Small Projects and they betray a great Finesse." In 1817 Degen had given out a book on the solution to a certain class of undetermined equations of the second degree of what were known as Pell's equations, and later, in 1824, he published a book with tables for the calculation of probabilities. Apart from this, Degen published a scattering of papers in the journals of the Danish Association of Scientists, the Petersborg Academy, the Danish War Sciences, and in Schumacher's *Astronomische Nachrichten*. Abel also showed some of his own work to Degen, among other things an equation that indicated how many different factors a number has, "but he could not grasp how I had found it." And he continued in his letter to Holmboe: "That little Treatment, that You must remember, dealt with *the inverse* of Functions of Elliptical Transcendents, in which I have arrived at something that must be the wrong outcome - I asked him to read through it; but he was unable to discover any Erroneous Inferences, or understand where it is that my error is rooted, God only knows how I am going to get out of it." This must mean that already before his trip to Copenhagen, Abel had written a paper where he had put down the basic thinking for what later would be so fruitful, exchanging argument and parameter: instead of considering an integral's value as a function of the upper limits, he considered the integral's upper limit as a function of the value of the integral. And this brilliant transposition of the problem would open up a great unexplored aspect of functions, "an immense analytic Ocean."

He also mentioned in the letter two mathematical works that he had recently been studying: *Application de l'analyse à la géométrie* by Monge, and *Essai sur la théorie des nombres* by Legendre. In his book, Monge analyzed differential equations and he developed an integration of partial differential equations. About Legendre's book on numbers theory, which was published in Paris in 1798, Abel felt it was "a great Shame that it is not to be found in Christiania." He had been particularly impressed by Legendre's theorem that quite accurately indicated how many primary numbers there would be in a given quantity of numbers. Abel wrote a formula in the letter and asked Holmboe to ponder over the proof until they met again. He also wrote about another "pretty Theorem" on primary numbers, and went on to say he felt that there were many others in Legendre. But again he complained that the libraries in Copenhagen were not "well-furnished with mathematical Books; but they have a good Deal of the Science Society's writings," and he discovered that "the Englishmen are not as horrible in mathematics as I had thought." He wrote that the mathematicians, Herschel and Young, "are extremely bright," and Ivory, "among the best now alive." Abel's prejudice toward English mathematicians could have been due to the general isolation of mathematics in Britain, that seems to have been the result of a competition between respectively the countrymen of Newton and those of Leibniz, about who deserved the honour of having discovered differential calculus. Among continental European mathematicians, it was Leibniz' version that had been accepted and built upon.

Apart from these books, Abel had read three of von Schmidten's treatises, that turned out to be "not as good as I had imagined," and he had studied several papers by the German mathematician, Gruson, "a horrible Trumpeter" who had had "the Impudence to steal a thesis from Parseval." Gruson, he said, had "shown that e is irrational." Abel also told about his own work, and filled the letter with four theorems of number theory, later known as the *Abelian equations*. As he had earlier worked on the solution to fifth power equations, he now began to consider the other most famous unsolved problem, namely, the French mathematician Fermat's hypothesis of 1637: to prove that the equation $x^n + y^n = z^n$, has no solution in whole numbers x, y, and z, apart from zero, where n is a whole number equal to or greater than 3. When n is equal to 2 there are interminably many solutions; for example, the numbers 3, 4, 5 or 5, 12, 13. But when n is greater than 2? Abel did not manage to arrive at proof for Fermat's great theorem, but he formulated some theorems of differences that postulated that if whole number solutions existed they had to be enormously large.

In his time, the French mathematician, Pierre de Fermat had put forward many exciting connections between the whole numbers 1, 2, 3, 4, 5, etc. but he proved few of his elegant whole number theses. His assertion that n^p - n is divisible by p where n is an arbitrary whole number and p is an arbitrary primary number was proven by Leibniz at the beginning of the 1680s. As well, Fermat's assertion, that any primary number of the form 4n + 1, that is, the numbers 5, 13, 17, 29, etc. is a sum of two squared numbers, was proven by Euler in 1749. Fermat had noted the great theorem of 1637 in the margin of Diophantus' classic work on numbers theory, *Arithmetica*, and there he asserted that he had a magnificent proof, but that the space available did not allow him to note it down. Since that time the problem has occupied mathematicians everywhere, both before and since Abel. In 1908 a German professor offered a one hundred thousand *deutschmark* prize for whoever could prove Fermat's great theorem, and in October, 1994, sure enough, a British professor, Andrew Wiles, finally proved Fermat's great theorem in a work utilizing advanced theory and over two hundred pages of proof.

In reference to his speculations about whole numbers during that summer of 1823, Abel wrote, "I have been hobbled. I cannot get any further with the enclosed Theorems," which showed that the eventual solutions, when they came, were to be of considerably large size. Thus it was that Niels Henrik felt himself hobbled. Was this perhaps also what he felt upon meeting Christine Kemp for the first time?

It is highly probable that Abel met Christine at a party at Uncle Tuxen's house or at a ball for the Naval Department's employees, families and acquaintances. Besides, Christine Kemp's father had always had strong ties to the marine life, and, until his death in 1813, held the honorary title of War Commissioner. Christine's mother, Catherine Christine, born Koch, had consequently been left with nine children, the eldest being seventeen, and the youngest, barely a year old. However, War Commissioner Kemp's widow continued to live in Christianshavn and the Tuxen family, who were known for their compassionate care, quite probably stretched out a helping hand. Two of the Kemp children died young, three became university students, one of whom became a priest, and the fourth son became a skipper in Iceland's Vestmannaeyjar Islands. Christine's eldest sister married the doctor, Niels Koppel and lived in Aalborg, and her second sister lived there as well. Christine was confirmed at Holmen Church in October, 1819, with the following attestation: "Christianity and Conduct: Very Good." It is quite certain that the nineteen year-old Christine was still living at home with her mother in 1823. She was said to have been both healthy and lively and had training in German, French and handicrafts.

One day during this summer she must have groomed herself to attend a party, a party where she knew as well that there would be dancing. And it was here consequently that Niels Henrik at one point or another must have seen her, approached, and asked her to dance. But it so happened that the orchestra now played the newest of the new popular dances, a dance form that neither of them had learned, the waltz, and thus they just stood and stared at each other, and according to descriptions by friends, they backed away and left the dance floor. Did he feel again that his feet were bound and hobbled?

What happened beyond this is unknown. Abel mentioned neither the dance episode, nor Christine, in any of his letters. Certainly this could mean that he met Christine *after* the letter-writing day of July 24th, but it could equally indicate that he did not know how to formulate his point of view on what had happened.

One day at the end of August or beginning of September, Abel left Copenhagen. It appears in the police records that on August 30th a travel permit was granted to the student, N. H. Abel, for a journey to Christiania. His surety was Uncle Tuxen, and the fee was one daler and twenty-four shillings. The letter that Abel rapidly wrote upon his homecoming to Christiania, to Fredrik Christian Olsen, his new friend in Copenhagen, was dated "Christiania, 13th August, 1823." Abel might again have mixed up the dates; it must obviously have been September 13th when he sat down to write to Fredrik Christian Olsen, who was his own age and a prominent young man in the intellectual student life of Copenhagen. Olsen studied philology and had been one of the arrangers of Regentsen's two-hundredth anniversary celebrations, a celebration in which 400 students had participated, and which was said to have been the warmest and most evocative celebration in the whole of Regentsen's history, and where, consequently, there had "been drunk 800 Bottles of Wine."

Abel wrote to F. C. Olsen: "Good Friend! You are probably astonished not to have heard from me about my Journey home and I pray that you might forgive me; thus it was that I came away so suddenly that I scarcely had Time to obtain my Travel Permit. My Journey was hardly the most fortunate. I was on route for ten Days and had Headwinds the whole Way. At the same time we still managed to come through it." Otherwise the reason for this letter was an errand with which Abel wanted Olsen's help. He wanted Olsen to go to the Dean of the Faculty of Philosophy in Copenhagen and request a written transcript of the *examen philosophicum* results for the student, Jens Glatvedt, an examination that Glatvedt had taken in 1813. "It is extremely pressing that he receive this Attest, and (I) must therefore beg You to do this and send it to me via Trepka at the nearest possible Opportunity." Making this request must have been a pure act of friendship on Abel's part. There is nothing to indicate that there was any special contact between him and the ten-year older theology student, Glatvedt, who was toiling away with his studies, which he completed in 1825. Jens Glatvedt, the son of a priest from Norderhov, had belonged to the last year of students who had had to travel to Copenhagen to take *examen artium*. That had been in 1812. The following year Norway would have its own university.

Glatvedt was said to have had a good head on his shoulders, but was also known to be lazy and immoral. Stories were told that while he was in Copenhagen he had won a bet in a highly unusual manner. Student Glatvedt had lived in a garret room overlooking the square, Kongens Nytorv, right across the way from the Charlottenborg rooms where the high stakes lottery was drawn. One day he had gathered a number of his fellow students at his place to watch, like a play in the theatre, the scenes that developed as the winning numbers were drawn. When the students had expectantly gathered there, Gladvedt wagered that he could get all the people who were now participating in the lottery draw to glance up at *his* window! The wager was accepted and Glatvedt had opened his window, pulled down his trousers and sat over the windowsill with his bare backside facing Charlottenborg; the students chimed in with a roar of laughter, the sound aroused attention, and in a few seconds all eyes, amid proclamations of "Such a Pig!" were turned toward the student's window. But why this Glatvedt now needed the 1813 transcript, is not known. Would it be for a new wager? And for Abel, Jens Glatvedt was perhaps a pretense for sending thanks to his friends for the hospitality he had received while in Copenhagen. Abel ended the letter by asking Olsen to greet Henningsen and Krarup "and whoever else you know I am acquainted with." The two so-named, most certainly Henrik G. Henningsen and Otto Christian Krarup, both studied theology, and both had also been together with Abel at the great student celebrations. From Christiania Abel reported, "I am living very well" and "Skjelderup is also living well and likewise sends his greetings. Your friend, N. Abel."

32.

The Quintic Equation and Future Plans

Abel was back in Christiania among old friends and daily routines at Regentsen in the Mariboe Building. During this term, he was assigned Room 3, the student residence's only single room. The stay in Copenhagen had inspired him to go further in his mathematical work and he now concentrated particularly on two or three areas. The work on equation theory was not laid aside; he did additional work on the thesis entitled *Integration of Differential Equations,* which was still awaiting pre-publication peer review; and a third area, closely related to integral calculus, was that of elliptical functions.

He continued to read Legendre. Immediately following his return, on September 12th, he borrowed Legendre's *Exercises du Calcul intégral sur divers ordres des transcendantes et sur les quadratures* from the university library. This was a work in three volumes that had come out in Paris between 1811 and 1819. What he had read in Copenhagen and complained about not being able to obtain in Christiania was Legendre's earlier work from 1798. The concept, *elliptical functions,* was not particularly appropriate for the field of study, but it had the following basis. Known types of integrals that contained square roots and expressions of the third or fourth degree had played an important role in many applications of integral calculus. The easiest problems were linked to calculating the length of an elliptical curve, hence the description, elliptical integrals. Euler had found many exciting properties of such integrals, but now it was Legendre in Paris who had gone most thoroughly into such integrals. Instead of investigating the integrals themselves, Abel now studied the reverse functions, which thus naturally received the description of elliptical functions. This transposition of the problem would shape the direction of much of his work in the coming years. Learning about the elliptical functions first of all became an excellent way to generalize trigonometry. Many properties of elliptical functions were analogous to the recognized laws of sines and cosines, but Abel also discovered completely new properties for which there were no known models. Some of Abel's notebooks from this period have been preserved and it seems that Niels Henrik had already, during this autumn, managed to work out many of the seminal results he later published. Only the time to work out these visions was lacking. It is quite certain that he was already clear about the remarkable relationship that elliptical functions have: two different periods, a double periodicity, and certainly he had set in process the course which would become his great "addition theorem," his Paris Treatise.

But it was the studies around the fifth degree equation that finally, would now bear results. After he himself had found errors in his formulas in the spring of 1821, he had taken his analysis further and gradually had come to feel that an algebraic solution to the general fifth degree equation was perhaps not possible. Perhaps there was no solution to be found. Perhaps there were no roots to the equation? Abel posed the question: is it possible that the *form* the roots must necessarily have in a fifth degree equation *can* satisfy the equation? Once such a question has been posed, an answer had to be found; the answer had to be either yes or no. Here is the way that Abel expressed the task: instead of looking for a correspondence that may or may not exist, one should really ask if such a correspondence really *is* possible. Abel arrives at his proof by finding the commonest expression that can be formed from the coefficients in the fifth degree equation, by carrying out an endless number of times, the operations of addition, subtraction, multiplication, division and the taking of square roots. This expression is thus written in the appropriate form and is used to substitute for x in the equation. By means of a series of acute conclusions where he, among other things, takes advantage of some contributions by Cauchy on the number of different values a rational function of several variables can assume when these variables are substituted by all possible means, Abel succeeded in showing that the formulation of a solution leads to an impossibility when the equation's degree is greater than four. Thus, the answer was no.

At some point before Christmas, 1823, Abel had finished writing up this proof, in French. He certainly knew that the work would attract attention abroad and hoped that the treatise on the fifth degree equation would give him access to an international forum of readers. The only problem was how to get the treatise printed. He knew by this point that it would be impossible to obtain economic support for publishing because, in the months before Christmas as Abel was finishing his work on the fifth degree equation, Hansteen and Rasmussen too were finishing up the task they had been given in March by the Collegium; namely, to assess an eventual publication of Abel's treatise, *Integration of Differential Equations*. These two professors, who had both been Niels Henrik's loyal helpers, had now broadened the question of publishing to a more inclusive plan of securing, for the mathematical genius that they considered Abel to be, a permanent position. In a letter to the Academic Collegium dated December 19, 1823, Hansteen and Rasmussen clearly stated that the treatise was of such absolute merit as to be published, but that the Collegium's appropriation could not be expected to cover the costs "before a long Time had passed by," and the best solution would be to publish this in one or another scientific journal, preferably in Paris. The two professors brought to mind that Abel had been receiving private monthly contributions since he had come to University, but that now he lacked "the great Support necessary to become that Pride of the Land of his Birth, the Man of Science that, from his Aptitude and Potential, one expects him to become. We are of the opinion that a Stay abroad in those Places where the most remarkable Mathematicians are, would most excellently contribute to his scientific and scholarly Education." Then, in four points, Hansteen and Rasmussen set down their recommendations:

a) 20 *Specier* monthly from Jan. 1st, 1824, until such Time as he can undertake a Tour abroad.
b) 150 Sp. for Equipping himself one month before his Departure.
c) 50 silver speciedaler a month during his Stay Abroad, that ought not to be more than 18 months.
d) 30 Sp. monthly for the first 6 Months following his return Home, unless in the meantime he manages to obtain Employment accompanied by a greater Income.

This has to be regarded as a generous recommendation: 20 speciedaler in addition to free room and board at Regentsen corresponded to the lowest adjunct positions at a cathedral school, and the travel abroad was expressly suggested to begin during the coming "Month of May, or with the Beginning of Summer."

The unfolding of events would radically change these plans in many ways: the Collegium forwarded the recommendations to the government on January 11, 1824, with its own endorsement, but without the important fourth point, support following his homecoming. The Collegium concluded by maintaining that, if the government now gave this support to Student Abel, "the Academic Collegium would venture with Certainty the expectation that the following would be gained: a remarkable Man of Science, a Jewel for the Fatherland, and a Citizen, who with extraordinary Proficiency in his Subject one day shall abundantly recompense this Help that now must be granted him."

Niels Treschow in the Department of Church Affairs now accepted the Collegium's recommendation and, in a few words, sought the Finance Department's "agreeable Consideration." Finance Minister, Jonas Collet replied on February 13th that it was quite correct to give Abel support for study abroad on the basis of the "rare Disposition to mathematical Studies" that the Collegium had pointed out. Student Abel ought also to receive support for the period *preceding* such travel. But the Finance Department had reduced the recommendation of 20 daler per month, to 200 daler per annum. At the same time, this was an unusually positive decision for the Finance Department to take. As a rule, one got only a "no" from this quarter, and it is not unlikely that Rasmussen, who was the government's expert on money matters, influenced Collet in this issue. Another influence perhaps was that Abel's friend Boeck, was now engaged to Elisabeth Collet, the Finance Minister's daughter. In addition, the two young people were cousins. But meanwhile the Finance Ministry had added a new clause to the matter: The Department, in its response said that Abel ought to stay home for the period of a *a couple of years* before travelling, and educate himself "at the University in Languages and other Associated Disciplines, that it is probable he in his current youthful Stage of Life has not mastered to the Degree that is considered desirable." The matter was thus sent back to the Department of Church Affairs for ratification, and Minister Treschow undertook, as it were, the correction of the words of the Finance Department when he went back to the Collegium and sought their opinion. The Collegium declared itself in agreement in a letter dated February 23rd, and really did not have any choice in the matter. Nevertheless, there may have been a little protest against Treschow and the government *postponing* the study tour abroad. The Collegium replied: "that for Student Abel, although he has received a good Grounding in the Humanities, still by our Consideration it would not be without Benefit for him to spend another one Year's Time here at the University to further his Scholarly, Scientific Education, particularly, perhaps, in further Study of the learned Languages."

Actually, *a year and a half* would pass before Abel would begin his travels abroad, and only then after he had sent a personal letter to the king requesting permission to travel. In the first round, things went according to the government's wishes. Treschow sent the Collegium's answer to the Finance Department with a note to the effect that could be interpreted as though he too considered Abel would benefit by a couple more years at the university, and by royal resolution on March 29, 1824, Abel was granted an annual support of 200 speciedaler for up to two years "toward the Continuation of his Studies at the Norwegian University, for besides, the study of learned Languages and other subjects for the broadening of the Scholarship of his main Course of Study, Mathematics." This resolution then made its way back via the State Secretariat to the Department of Church Affairs, and from there to the Academic Collegium, which then reported the decision to Abel. Thereupon Abel thanked the Collegium in a letter dated May 1, 1824: "With particular Happiness and Gratitude have I received the Information communicated to me by the Collegium that a stipendium is to be granted me; but as I do not know where or how I shall cash this Stipendium I therefore take this opportunity to ask how to obtain this payment and give thanks to the high Collegium with regard to the Information therein contained, about my approaching Freedom." There is no reply in the archives to this, Abel's communication about a practical method of payment, but Abel eventually received his speciedaler. And there are no records to confirm whether he really did study "learned Languages."

The records from this period that had to do with Niels Henrik, from the autumn, 1823, to the summer, 1824, are the following: apart from working on the fifth degree equation, integral calculus and the elliptical functions, he was also engaged in numbers theory speculations. The record of loans at the University library show he borrowed the most recent and greatest work in this field of mathematics, namely Gauss' work entitled *Disquisitiones Arithmeticae*, that is, in the field of arithmetic investigations. In the letter to Professor Degen of March 2, 1824, we find the following: Abel was definitely clear that the treatise on integral calculus would not be published, and he added, "God knows how I shall get it published, and I wanted it so intensely, as I believe that my own Work would be the best Introduction I could have during my Stay abroad, that I have Grounds to believe will take its Beginning in a Year's Time or thereabouts." Abel wrote to Degen that in addition to integral calculus, he had followed the professor's advice on studying elliptical functions and that in this work he had found general properties in "the Whole of transcendent Functions," and that he had thought of sending a treatment of this to The French Institute. In a mention of these findings, Abel uses the words, "I have come by way of a random Occurrence (*Hændelse*), to where, in addition, I can express a Property of all transcendent Functions of the Form..." Was the word *Hændelse* that is to say, "chance" or "stroke of luck", Abel's way of saying "inspiration"?

In spite of the fact that he knew there would be no travel abroad that year, Abel went once or twice in the course of the fall, 1824, to the book publisher, Grøndahl in Christiania, and paying from his own pocket, ordered the printing of his treatment of the fifth degree equation. Abel must have decided that precisely *this* piece of work "would be the best Introduction I could have during my Stay abroad." The desire to reach an audience weighed more heavily than the possibility of having to go through some completely threadbare days together with his brother Peder, who was still living with him at Regentsen.

But it appears that at this point there may have developed a difference of opinion between Abel and Professor Hansteen about what would best serve Abel's needs, that is, what should constitute his admission to Europe's learned circles. Hansteen, who knew the milieu best, was aware that there was great interest everywhere in new discoveries in the heavens, and consequently he took Abel's paper, "On the Moon's Influence on the Movement of the Pendulum," and had it published in *Magazin for Naturvidenskaberne*, and immediately sent a copy to the astronomer Schumacher in Altona, Hamburg, in the hope that it might be published in German or French in his *Astronomische Nachrichten*. At the same time Abel seems to have held fast to his intention to get the work on the fifth degree equation into print, published at Grøndahl's, where *Magazinet* was also printed. Hansteen and the other editors of *Magazinet* probably considered Abel's work on equation theory to be much too specialized for a general audience. Hansteen may also have not been convinced that Abel had this time provided error-free proof for his reasoning in relation to the fifth degree equation.

In order to make the printing at Grøndahl's as economical as possible, Abel had shortened the proof to only six pages. Only the main points had been dealt with. It is not certain *when* the printing was ready; Abel's treatise was not included in the semi-annual list of Norwegian publications for the spring season of 1824, and when it was announced in *Det Norske Rigstidende*, the word "equations, or *équations*, in the title: *Mémoire sur les équations algébriques ou on démontre l'impossibilité de la résolution de l'équation générale du cinquième degré*, came out as *épurations*; in other words, "purifications." It is likely that very few copies were sold, but Abel sent quite a number of copies to his friends in Copenhagen, and through Hansteen, some were also sent to Schumacher, and Schumacher saw to it that Gauss got to see Abel's work at the end of July, 1824. This happened around the same time that Schumacher received Abel's article "On the Moon's Influence on the Movement of the Pendulum." It was in a letter of August 2, 1824, that Schumacher fired off the crushing verdict on Abel's astronomical effort: he would absolutely not publish Abel's work that calculated the lunar effect to be 60 times greater than it actually was! And it seems that these events were also unpleasant for Hansteen, something that would dog him for a long time. First of all, he himself ought to have discovered Abel's error in the lunar calculations, and a long time later in *Illustreret Nyhedsblad*, in 1862, Hansteen stressed the fact that Gauss too, upon first receiving Abel's work on the fifth degree equations, had been negative, and he had said that he himself could probably prove the possibility of finding a solution to the quintic equation. But Gauss had little by little come to see that Abel was right. And the probable reason for Gauss' initial discouraging response to Abel's compact proof, apart from his constantly receiving prospective solutions to this popular mathematical challenge, was that Abel's title was too short; it lacked the important precision in denoting that with regard to employment of the *radical sign*, a solution to the fifth degree equation was impossible.

Be this as it may, the fact remained that not even his treatment of the quintic equation gained Abel an *entré* into learned Europe. Abel would make a further two efforts to take up the subject, and give a more fleshed-out version of his proof: in *Crelle's Journal* in 1826, and in a manuscript found among his surviving papers, and included in his collected works.

The discussion that no student in Christiania could avoid participating in that spring was the dispute over the celebration of Norway's National Day, May 17th. It was ten years since the Constitution had been adopted at Eidsvoll, and all the students had prepared a commemoration, but such a celebration would be construed as a protest againg the Union with Sweden. Consequently, Chief Administrative Officer Sibbern, Justice Minister Diriks, and Prosecutor-General Christian Magnus Falsen had refused permission to celebrate the day. Since Prince Oscar and Crown Princess Josephine were in the city, Parliament too had abandonned all official commemorations. Professor Sverdrup, who knew the students, was aware that this would not restrain them. He asked them urgently to postpone the planned celebration until May 18th, but this was met with little or no sympathy. In the afternoon of May 17th, about fifty students had gathered at Sorgenfri in the Hammersborg district. They celebrated with a dinner, sang songs, and toasted the Constitution, the King, the Parliament, the Crown Prince, Norway, the University, and Themselves. In the evening, around six o'clock, another fifty-odd students arrived, and according to the protocol recorded by The Student Society: "In this manner the whole Society, composed of about 100 Persons spent the Evening, partly at the Cards and partly in Conversation and finally, late into the Night, parted company with all in good Humour and everything in good Order." Meanwhile, all night long the police had patrolled the area around Sorgenfri, but the only thing untoward that was reported was two broken chairs. Such was the protest the students mounted against Swedish rule. Seven days later it was Parliament's turn: they rejected King Karl Johan's motion on the right to a royal veto.

Nevertheless it was later maintained that the students had been behind disturbances in the main square in relation to the visit by Crown Prince Oscar and Crown Princess Josephine. Whereupon a sharply-worded petition protesting against these allegations was circulated. Abel's was among the signatures on this document.

On one occasion or another during this spring or early summer in 1824, Abel must also have come to know that Christine Kemp had obtained a position as governess at Son, a small coastal settlement on the eastern side of Oslofjord between Drøbak and Moss. During this period, one of Abel's fellow students at Regentsen, Morten Kjerulf, privately tutored the son of Chief Excise Officer

Strøm at Son, and because Kjerulf was not particularly strong in mathematics, he got Abel to teach the boy, Johan Fredrik Strøm in that subject. Perhaps in this way Abel had managed to find out that there was a family in Son looking for an acceptable governess. Christine's date of arrival at Son is not known, but at midsummer her name is listed among those who accompanied a certain Miss Thorne from Son to Christiania. Presumably Christine had obtained a position with the affluent Thorne family of Son.

There was another person whose arrival in Norway that early summer probably attracted Abel's attention, as it did that of the others in his closest circle of friends. This was the great Professor Henrik Steffens, who after being away for more than thirty years, had returned to the land of his birth. This professor of natural philosophy, famous across Europe, had the soul of a romantic; he never tired of calling himself a Norwegian, and expressed his longing for the great, wild, innocent land of his childhood, and for the Norwegian mountains, which he now also had scientific reasons to want to see again. For Abel's friend, Keilhau, Steffens could be an important gateway to the wide world beyond Norway. Keilhau had long been inspired by the observations of the Prussian, Leopold von Buch, that the area around Christiania would probably be revealed to be the most important region for geology in the Nordic countries. That is to say, in the years between 1806 and 1808, in the Christiania area, von Buch had discovered layers of granite between younger types of stone, something that defied the dominant theory of the time, the theory championed by the German geologist, Werner, whose thesis was that the earth's original core was granite. Keilhau had written two papers about this in *Magazinet*, and his idea, that granite was formed epi-genetically by means of the transmutation and metamorphosis of sedimentary rock types, was beginning to take form. Now the great Steffens would give new inspiration to Keilhau and his like-minded associates.

After having been honoured in Sweden with great social gatherings during Pentecost, steamboat tours among the many islands and skerries outside Stockholm, and in the glittering circle at Uppsala around the poet, Atterbom, Steffens arrived in Christiania and was received by his old friends, Professor Sverdrup and Count Wedel-Jarlsberg. Keilhau would be Steffens' guide into the mountain fastness of Norway. They met at Hamar, sailed across Mjøsa, drove westward up through Valdres, and proceeded further by horseback in order to climb Synnfjellet. Here they reckoned they were in the middle of Norway, equidistant from Trondheim to the north, and Lindesnes to the south, the Swedish border to the east, and the fjords to the west. From here they could survey a chaos of naked mountains: Rondane to the east, Dovrefjell to the north, Hurrungane to the west, purple-red in the radiance of the sun, and southward toward Hårteigen. The view was overwhelming: "Suddenly there appeared something so exultant - I had never seen the like of it before," Steffens later wrote. The general concept advanced by the geologists was that the earth's surface had undergone a series of catastrophic changes, that man therefore had to reckon with several prehistoric worlds that in turn had superceded one another. The proof for the existence of these primal worlds and their downfall was to be found, they thought, in the mountains of Norway.

Keilhau also accompanied Steffens on a geological tour in the region around Christiania. There was always great pomp and finery surrounding Steffens; he was invited several times by Crown Prince Oscar to the regal residence of Paléet, and at the beginning of August there was a large farewell party. When Steffens left the capital city by boat, Keilhau was among those who accompanied him down the fjord as far as Drøbak, so as to prolong the leave-taking.

Steffens' visit had stimulated among the broad public an already lively interest in the natural sciences. In the pages of *Magazinet* there were constant reports about comets, the northern lights, "Optical Illusions," "Shooting Stars by Day," and "The Star Cover in Christiania." A great earthquake that had shaken Christiania and its surroundings on November 24, 1823, was much discussed. The quake had last three seconds and had been so strong as to cause a candlestick to fall off a table in Drammen. Now, right after Steffens' visit the following description was to be found in a letter from the writer, Maurits Hansen to his colleague, C. N. Schwach: "You have no doubt heard all about what greatness he (Steffens) has made of Keilhau. He has stated that all those of Rocky Expertise must now lay down the Pen and await the results of Keilhau's Ideas. As a Norwegian, my Heart fibulates when I hear Such a Thing. About Abel, on the other hand, one hears nothing, even though he as well is doing no less."

A short while later, Keilhau was made a state stipendiary, and he received money for a study tour to Germany. Keilhau wrote to his friend Boeck: "I have now got them, really - the Monies - 600 silver speciedaler for a Year's Travel. Hurra! I had only asked for one year and then the Second came much more easily that the First, or so I think. I have Steffens to thanks for everything. I am in Treschow's good books; he himself told me that Steffens had threatened the Gentlemen of the Cabinet, saying that the whole of Europe would blame them if they did not allow me to travel." A couple of years later Keilhau's theories on granite were published as a book in Leipzig.

33.

Betrothal and Preparations

There are no written records indicating how Abel spent the summer of 1824. Perhaps he saw no reason to journey home to Gjerstad, as that would be only to visit the same sad conditions about which he still could do nothing. His older brother, Hans Mathias, had become so listless that he could scarcely be talked to; his sister, Elisabeth, still had not been confirmed, and consequently, she was too young to obtain a position in service in Christiania; and it seems that the nineteen year-old Thomas Hammond was not someone whose labour was in great demand. Nonetheless Niels Henrik may well have gotten his brother Peder to go home during that summer of 1824. The student friends who helped Peder with his lessons would in any case now be away from the capital. The weather had been fine in late spring and early summer, perhaps from the perspective of the farmer, it had approached a drought. But then the remainder of the summer was warm with quite a bit of precipitation, and in Gjerstad nobody complained about the way the crops grew and matured that season. As far as Mother Abel at Lunde was concerned, the weather no longer seemed to play a role in her life. According to Vicar Aas, Mrs. Abel was so dirty and dishevelled that the few times he had invited her to the vicarage, out of decency, he had had to send his servant over to her at Lunde before the visit to see that she was tolerably presentable.

There is much to indicate that Niels Henrik sat peacefully in Room 3 of the Mariboe Building that summer, concentrating on mathematical work. In any case, he did not accompany Keilhau and Steffens on their excursions into the surrounding area, but he might well have mounted his own excursions, for example to Son to visit his Christine. In Christiania he regularly sought out Mrs. Hansteen at her home on Pilestrædet, and now more and more, Mrs. Hansteen seems to have become his most trusted friend.

In September Abel sought a renewal of his place at Regentsen, and from the university library he borrowed issues of *Mémoire de l'Institut de France* from various years, but at the same time he tried to become familiar with the mathematical literature of other countries. There was very little in Norwegian mathematical history that could interest him, but he most certainly knew

Rector Fredrich Christian Holberg Arentz' name and reputation, and also because Arentz, in the vehement debate in Parliament in 1818 on the academic schools, had, in an article in *Nationalbladet,* supported the point of view that Niels Henrik's beleaguered father had defended against the views of the majority. F. C. H. Arentz, Rector of the Bergen Cathedral School, was considered a learned man, well-schooled in several disciplines, and he had published treatises on questions of physics and mathematics in both Norwegian and Danish *Videnskabernes Selskabs Skrifter.* Arentz' use of mathematics to refute the infinity of space and the eternity of time had made quite a sensation. He argued that conceptions of an infinite world would lead to delusion and come in conflict with theological statements on Creation. Thus, using mathematical concepts of size and infinite divisibility, Arentz argued, in a hair-splitting manner, that time was finite, and that the world had not existed through all eternity. The world, he argued, must have had a beginning in time and space, but without having actually said that the world must necessarily come to an end or a cessation. Abel probably had little feeling for this use of mathematics to fuel the old philosophical discussions about eternity. Rather, Abel would have been more at home with the work of a quite different Norwegian mathematician, the land-surveyor, Caspar Wessel, the brother of the poet - that is, if Abel were acquainted with his work. Through his practical work as a surveyor, Wessel actually found a new method of treating directional line segments that was an alternative to normal trigonometry. Wessel found an analytical form of expression that characterized a segment of a straight line by means of *length* and *direction*, and this expression could easily be used in practical calculation. But the fact that this analytical expression simultaneously represented an understanding of complex numbers was not noticed, despite the fact that the work, Wessel's only treatise, was published in the Danish *Videnskabernes Selskabs Skrifter* in 1799. The honours for the solution of complex numbers represented in the plane went to the French mathematician, Jean Robert Argand, who in 1806 published a less well-worked out analysis than Wessel had done. Later Gauss, as well, made it known that he developed the same ideas in an unpublished manuscript of 1799. Almost a hundred years were to pass before Caspar Wessel's contribution was taken note of.

Abel had been instructed to undertake "a further Study of the learned Languages," but whether this was done in ways other than by reading mathematical books published in various lands is difficult to trace. Above all he read French yearbooks and journals: *Mémoire de l'Institut de France, Journal de l'École polytechnique* and the one commonly known as *Gergonne's Annals,* which referred to the journal edited by Joseph-Diez Gergonne since 1810 and called *Annales de Mathématiques pures et appliquées.*

The first theatre production that autumn, at the end of October, was a play that Abel may have already seen in Copenhagen where it was performed while he was in the city at the end of July, 1823. This was a play now called *Still Water Runs Deep*, but in its original version was called *To Rule a Wife and Have a Wife*. The critic in Christiania reviewed the play as containing "raw renderings of Character, hackneyed Situations, and Horrendous Bordello scenes." At the end of November Enevold Falsen was honoured for all that he had meant to the Dramatic Society. H. A. Bjerregaard had written a prologue and, in addition, during the weeks leading up to Christmas, what was staged were performances of Holberg's *Gert Westphaler* and J. C. Brandes' *Ariadne of Naxos*. Certainly Abel was present for one of these double bill performances, both to renew his acquaintance with the garrulous Gert, and to see Mrs. Hansteen play a dryad in *Ariadne*. Mrs. Hansteen must have known the role well from performances five years earlier, as this was barely two weeks after she had given birth to Asta, or Aasta, as she would later call herself as she became known for her work on women's rights.

Niels Henrik spent Christmas of 1824 with Christine in Son, and when he returned, his friends at Regentsen were astounded by the news that he was engaged to be married to Miss Christine Kemp. This surprise would not have been confined to the sudden nature of the engagement, but also would have extended to the fact that they would not have believed Niels Henrik capable of such an initiative. It seems that he was reserved in relation to the rather coarse approach found in student life toward the other sex, both verbally and in relation to those women in the capital who were easily prostituted. By means of his betrothal Abel no doubt managed to eliminate from actual discussion that which he considered too private. By means of an engagement he had secured a degree of safety, and was left in peace whenever discussions arose about women. And by the same fact, it was a foregone conclusion that this engagement would continue for some time: the hovering question was, *when* one's economic position would allow for the possibility of moving into one's own living quarters.

Brother Peder later told that from about this time at Regentsen, he would be awakened by Niels Henrik crawling out of the bed where they both slept, and about the shouts of "Light the candle!" or, "Now I have it! Now I have it!" and that he would sit at the table for hours at a time in his night dress. And on one occasion when Peder sat freezing and gazed unhappily around the sparsely heated little garret room, Niels Henrik was supposed to have said, "Pull on your overcoat, Lad, so you don't freeze to death!" This was a means of keeping warm that he himself often used. Brother Peder also recounted that Niels Henrik always kept a piece of chalk in his pocket, and wherever they happened to be around the city, he might spring into the first or the best entranceway and begin to fill the walls with mathematical symbols.

In January, 1825, Oehlenschläger's *Freya's Altar* was staged at Grænse-haven. This was a work that Abel had borrowed from the cathedral school library as early as the autumn of 1816, and that he was now probably eager to encounter again in this new version. This jolly musical play, *Freya's Altar*, had its source in the desire to portray Nordic mythology, and had contributed to Oechlenschläger's definite status as the greatest poet of the Nordic countries in 1805, and was later to become an element in the bitter struggle between Oechlenschläger and Baggesen, and between their respective supporters.

Abel must certainly have been in the theatre in March to see Mrs. Hansteen play the French Pauline in *The Cunning Exchange of Letters*, a play written by one Fabre d'Eglantine, that had been performed earlier at Grænsehaven. But first and foremost Abel now worked to get ready for his tour abroad. During the spring, 1825, he made a push to have the "couple of Years' Time" that was suggested for his continued studies in Christiania, reduced. Abel wanted to meet the foreign mathematicians, above all, the French, and not least of all, he wanted to show them his own work.

There were also practical reasons for Abel travelling now. Two of his friends, Keilhau and Maschmann, were already in Germany, and three others were on a foot tour. Boeck, who had spent a year as a military doctor, had received support to travel to Germany and France to study veterinary medicine and animal husbandry. His study tour stipend was a direct consequence of Parliament's decision, taken in 1818, to found a veterinary college, and due, not least of all, to the efforts of Søren Georg Abel. Carl Gustav Maschmann had begun to study pharmacology at an institute in Berlin and intended to become a pharmacist and take over his father's "Elephant Apotek." Keilhau wrote home enthusiastically on May 19, 1825, about the rock conference he had attended at Freiberg and about the school's methods of study. He reported that that summer he would be going out with two German geologists to map the stretch between Middelberge and the Saxony border, a region that contained "the Class of Stones that Norway completely lacks." Keilhau would be in Germany for one more year, and for the coming summer had planned a foot tour together with two Norwegian students of geology who were now preparing to leave Norway and travel abroad. They were Nils Otto Tank and Nikolai Benjamin Møller. N. B. Møller had been born in the same year as Abel, grew up in Porsgrunn on the west side of Oslofjord, and entered the University in 1820 after having been privately tutored together with Abel's classmate, Niels Berg Nielsen at the home of the father of the poet, Schwach, in Solum, near Skien, Porsgrunn's neighbouring town. Møller had studied mineralogy and taken the public service examination in 1824; in addition, like Keilhau, he had undergone practical learning at the Kongsberg Silverworks, and like the other "promising youths," had published in *Magazinet for Naturvidenskaberne*. Nils Otto Tank, son of merchant and cabinet minister, Carsten Tank from Fredrikshald, probably knew Abel less well than the others did. The son of a rich man, Tank had been sent to England to obtain an education in trade and commerce, but had come back, developing an interest in physics and mineralogy and was now probably attending lectures in these subjects at the University. Repelled by his father's amoral way of living, young Tank had been attracted to the Hutterite Brethren, a religious sect that was enjoying great popularity under the charismatic leadership of Niels Johannes Holm in Christiania. Like Keilhau and others, Tank had been enthusiastic about Steffens' visit the previous year, and had been in the party that saw Steffens off by boat the whole way to Drøbak. Despite the economic downturn that had affected the trading house of Tank in Fredrikshald, young Tank had ample funds with which to travel.

Abel, on the other hand, could do absolutely nothing until he received his travel stipend which, a year earlier, all had agreed would be given to him. He used the following method of approach to get these funds released: he gathered together letters of reference and requested of the king that the travel grant be released as soon as possible. Professors Hansteen and Rasmussen supported Abel's plan, and both sent along the most glowing recommendations to the king. In his statement, Hansteen mentioned Abel's treatises in *Magazinet for Naturvidenskaberne*, and at greater length, explained as an example of Abel's "uncommon Gifts," that he had worked out "an improved Method of Integral-Calculus." Hansteen added, "Likewise, his Character and Moral Being are deserving of Praise, with which I have had the Opportunity to become familiar through personal experience." He wrote that the Fatherland would have "the most profound Hope that in him it was getting a scholarly Man of Science who would give both Honour and Benefit." Rasmussen followed up, saying, "Few of the young who are Studying can promise as much as Student Abel, who quite reasonably will quickly make himself favourably known in Europe with Inquiries into Theoretical Mathematics. With particular pleasure do I support Student Abel's Request..." that is to say, the request to have the stipend paid out. But Rasmussen, who certainly knew the discussions in the government departments, did not mention the 150 daler for equipping the student for travel. He referred only to Abel's request for the 600 silver speciedaler for the two-year study tour. Abel himself explained his studies and future plans in the letter he wrote from Christiania, July 1st, 1825:

To The King!

Right from my earliest Schooldays have I had a great desire to study the Mathematical Sciences, and I have continued this Study into the first two years of my academic Career. My rather unfortunate Future Prospects provoked the academic Collegium to recommend I be granted Assistance from the State Coffers, that it pleased Your Majesty to so graciously offer, so that I, in a Period of Two Years at the Norwegian University could continue my Studies, and in addition, further improve my knowledge of the learned Languages. I have, according to my Abilities, from that Time on, studied the old and newer Languages in Relation to Mathematics; among the Latter, particularly the French Language. In this manner, here in the Country, with the help of the currently available Facilities, I have made great Efforts to approach the aforementioned Goal as it would be of special Benefit to me to be able to Stay abroad at different Universities, in particular, Paris, where so many remarkable Mathematicians are found, and that I might come to know the newest Productions in the Field of Study, and enjoy the Counsel of Men who in our Times have brought it to such outstanding Heights. Thus, for the above-mentioned reason, and with the enclosed, accompanying Letters of Recommendation, I obediently petition Your Majesty most graciously to have granted to me a Travel Stipendium of 600 silver speciedaler per annum, for a period of Two Years in Paris and Göttingen to deepen further my studies in the Mathematical Sciences.

Most obediently yours,
Niels Henrik Abel

Abel took these papers to the Collegium on July 2nd; the Collegium recommended the request and sent the papers to the University's Pro-Chancellor two days later. Niels Treschow, who had resigned as Cabinet Minister, now sat in this position, and the Pro-Chancellor's endorsement was short and sweet: " This Matter is hereby proposed to the Royal Norwegian Government's Department of Church Affairs and Education for its most gracious Resolution." But the new Department Chief, Cabinet Minister P. C. Holst, who now for three months led both the Church Department and the Naval Department did not let the matter pass lightly through his hands. He sent a letter back to the Collegium asking whether the University could not take over the payment of the travel stipend and whether the amount of the stipend could be reduced somewhat since Abel's programme of studies "did not require him to engage in elaborate Travels, but merely required his Stay in two definite Places." In other words, Göttingen and Paris. The Collegium answered that they by no means had the resources for such a stipend, and that a lesser sum than that given to C. A. Holmboe in 1821 to study Persian and Arabic in Paris, would absolutely not be sufficient for Abel. And the Collegium added that Abel "not only needed Help to live and maintain himself in foreign Parts, but also that he must, in addition, equip himself and purchase several indispensable and costly Works for his Studies." The Church Department subsequently asked the advice of the Finance Department, and Cabinet Minister Jonas Collet answered that there was nothing to hinder Student Abel "from receiving from State Funds a Travel Stipendium of 600 silver speciedaler annually for a period of two years, calculated from 1st July, 1825, until 1st July, 1827, to study the Mathematical Sciences at the Universities of Göttingen and Paris." Cabinet Minister Holst now responded, saying that it would be best to pay out the stipend from the date that Abel commerced his travels; not from July 1st "so that no Time would be wasted before his Stay at the foreign Universities," and he asked Abel to set up a travel plan. This was on August 5th. Two days later Abel sent to "The Honourable Mr. Cabinet Minister P. C. Holst, Commandeur of the North Star, etc." a "Draft of a Plan for my Travel Abroad that you instructed me to prepare. It is highly likely that Acquaintance with particular Circumstances could make certain Changes necessary, but still I think that on the Whole this is what ought to be followed." This draft for Abel's Travels later disappeared from the archives. The Department sent the draft to the Collegium where Professor Rasmussen agreed to comment upon it, but this occurred after the whole formal matter had come to an end. On September 5th, His Majesty made the following command with respect to the proposal: "That it be graciously granted to Studiosus Niels Henrich Abel a Travel Stipendium from the Coffers of the State, for two Years the sum of 600 Silver species annually, for the Study of the Mathematical Sciences at the Universities in Paris and Göttingen." The announcement was sent from the State Secretariat to the Church Department, which sent a writ to the Finance Department and to Abel. On September 6th, Abel gave the Collegium the message "that I hereby relinquish the Place that I have hitherto enjoyed at the University's institution." Abel was ready to travel, and the following day he did so.

There had been a remarkable travel fever in many parts of the country that summer, not only among Abel's friends. About the same time that Abel was sending in his request to have the stipend paid, on July 5, 1825, the sloop, *Restaurationen*, sailed from Stavanger with 52 people on board: the great immigration to the new land of America had begun. Could perhaps Abel's zeal to travel have contained a certain element of flight? That summer as well, Mrs. Hansteen received a visit from her sister, Charité, with whom, everything considered, and according to later stories in the Hansteen family, Abel had fallen in love when they met in Copenhagen and at Sorø two years earlier. Everything points to the fact that at this point in time he was torn between

Christine in Son and Charité at Mrs. Hansteen's on Pilestredet Street. Professor Hansteen's sister's daughter, Vilhelmine Ullmann, writes about the 1820s in her memoirs: "Charité Borch came up here from Copenhagen to visit her sister…in her whole Manner she (Charité) bespoke a higher sphere of Refinement, a richer and deeper Intellectual Life, than we were used to. For all that, it was not these Qualities, which she was unaware of, that introduced a Revolution among us. The first time I saw her, she had on a red and white striped Calico Dress that did not appear to be tight, but both Waist and Skirt were gathered in the front at the Middle. This was called a Smock dress and was extravagant and immeasurably daring, but surely worthy of imitation." This was the new Empire Style of Europe's ruling class women, here being introduced to Christiania. Liberation from the corset followed in the wake of the French Revolution; the style with the highly destructive confinement of the midriff, had shifted to short high waistlines that stopped under the breast, and from here downwards, the material hung free around the body, often accompanied by plunging necklines, something that in France developed into an almost "naked" women's dress before the opposing reaction set in at the end of the 1820s.

Whether it was Charité's boldness or her sphere of refinement that attracted Niels Henrik, is of course unknown, as is the depth of his love. Tradition has it that Mrs. Hansteen now convinced Niels Henrik to stay with Christine, perhaps due to the difference in both levels of refinement and extravagance, but probably most because Abel was engaged and because Christine had come to Norway for his sake. Abel had certainly made a definite choice, but despite that, it seems that in the midst of everything he found it difficult to get Charité out of his thoughts. The later diaries of Mrs. Hansteen indicate that many of her discussions with Niels Henrik were about his relationship to Christine. Vilhelmine Ullmann, who was born in 1816, describes her only meeting with Niels Henrik as: "It was at the Niemann Building in Pilestrædet, second Floor, where the Hansteens lived. I was sent up with a Message to Aunt (Mrs. Hansteen), and when I entered the Sitting Room she was sitting and talking with Abel. He looked at me, smiling, while I delivered my Message, and when I had curtseyed and was halfway out the Door, I heard him ask, Who was that?"

Niels Henrik had scarcely had time to travel home to Gjerstad during that summer of 1825. The request to the king was sent at the beginning of July; at the beginning of August he had, at short notice, to make his travel plans for Cabinet Minister Holst, and then finally, at the beginning of September the papers were in order.

Meanwhile at Gjerstad, his sister Elisabeth was finally confirmed. She stood as confirmant Number One on the floor of the church. Vicar Aas wrote about her: "Remarkably good Knowledge, constantly Hardworking and Well-behaved", that is to say, she was the best of the confirmants. And even though Niels Henrik had not managed a visit home, he had arranged in any case for his sister to find a household position in Christiania. Mrs. Hansteen promised to take Elisabeth in and look after her until a suitable position arose in another place. Later that fall Elisabeth Abel journeyed to the capital and was warmly received by Mrs. Hansteen. Six months later she found a position in the household of Niels Treschow at Tøyen, and stayed there for almost six years.

And Niels Henrik also cared for his brother Peder, with the help of Mrs. Hansteen. At the beginning of August, Peder had taken *examen artium* and passed all the individual tests, only in geography did he have to continue studying, and now he was to study theology. Of course Peder had to move out of Regentsen when Niels Henrik would no longer be living there, and he had found himself a room at a Mrs. Tode's in Voldgaten. Before Niels Henrik left, he deposited 50 daler of his stipend with Mrs. Hansteen so that she could dole it out in suitable amounts as Peder's needs arose.

And as a final preparation before leaving, Niels Henrik borrowed an English book on European history from the university library, on August 29th. It was William Robertson's *The History of the Reign of the Emperor Charles V*, by the Scottish social philosopher, William Robertson, and it bore the subtitle, *With a view of the progress of society in Europe from the subversion of the Roman Empire to the Beginning of the 16th Century.*

PART VI
Europe-Travel

Fig. 22. Gendarmenmarkt, Berlin 1825. Pen and ink drawing by F. A. Calau. Kupferstichkabinett, SMPK. Bildarchiv Preussisher Kulturbesitz, Berlin.

34

Through Copenhagen and Hamburg to Berlin

Accordingly, on September 6th Abel gave up his quarters at Regentsen, and left Christiania the next day. He had wanted to say his farewells to Christine in Son and had journeyed there by land. Some days later, Boeck and Møller came down Christianiafjord by boat, stopping late one night at the quay in Son to take Abel on board. It was most likely Trepka's sloop "Apollo" that on this occasion too was carrying passengers and freight to Copenhagen. They were blessed with a fair breeze at the beginning, but then out off Marstrand they lay becalmed until a stiff wind set in, whereupon the most terrible seasickness assailed the passengers. One of the passengers was so certain his end was nigh that he wrote his last will and testament, twice. Abel and Møller were also sick, but Boeck strolled around like a good-hearted Samaritan and impressed them with his medical skills. By the fifth day, the sea had pacified itself and they could see Kronborg Castle. Abel and Boeck were up at three o'clock in the morning to enjoy the sail in past Helsingør and through the many sailing ships in Øresund. Perhaps Niels Henrik thought about his father who had *lived* there in his youth, in that beautiful landscape.

Now in Copenhagen, Tank, who had come overland, completed the company. They had had a discussion on the means of travel before they had left Christiania. Tank had urged everyone to take the land route and be his guests at Røed Manor, the most splendid estate near Halden, before they journeyed through Fredrikshald (Halden), and carried on south through Sweden, and across Øresund to Copenhagen. However, Boeck had wanted to take the sea route.

Boeck, Tank and Møller now left Copenhagen immediately for Hamburg, while Abel, for at least two reasons, remained in the city a good week. First, he had a scientific errand to complete, a letter from Hansteen to the well-known physicist, Hans Christian Ørsted, and second, there were family and friends with whom he very much wanted to renew acquaintance. On this occasion as well, he stayed with his Aunt and Uncle Tuxen at Christianshavn.

H. C. Ørsted happened to have one of Hansteen's magnetic calculating devices at his house. Ørsted had used it on a trip to England, and now

Hansteen wanted Abel and his friends to take this "oscillation apparatus" with them so that they could measure magnetic intensity at different locations in the course of their southward journey across Europe. During that summer and fall, Hansteen himself was away on a magnetic survey around Torneå on the Bottenvika portion of the Baltic, between Sweden and Finland, and as early as April he had written to Ørsted introducing "Herr Studiosus, N. H. Abel, our mathematical rising Sun." And Abel now visited Ørsted and picked up the "oscillation apparatus".

Abel also visited Professor von Schmidten while he was in Copenhagen, and received good advice for the road, not the least being a recommendation to, and the address of, "Geheimrath"[13] Crelle in Berlin. Abel also came to hear that Professor Degen's great library, which he had admired two years earlier, was to be auctioned off. A catalogue of Degen's books had been drawn up following his death in April, and now, at the beginning of September, Professor Thune had sent copies of this catalogue in two separate packages, to Abel's address in Christiania so that he might distribute them among interested buyers in Norway. Upon learning this, Abel wrote home quickly to Holmboe and asked him to take the catalogues, and at least go to Georg Sverdrup, who in his role as the university librarian, would, hopefully, be interested in a purchase. "Do it as quickly as You can for the Auction is to be held on October the 5th," Abel wrote, continuing, "On the 13th I am journeying to Sorøe to visit Mrs. Hansteen's Mother and Sister, and I shall be there a Couple of Days. On Friday week I am taking the Steamboat to Lübeck, and from there onward to Hamburg." He urged Holmboe to please "not be angry at me for issuing Orders to you to fulfil these Commissions" and he promised next time to write a "more ordinary Letter." He closed by sending greetings to Mrs. Hansteen and her sister, Charité.

That Abel now, during this stay in Copenhagen, took time to go out into the countryside to Sorø to visit Mrs. Hansteen's old mother and sister can scarcely have been inspired by an ordinary feeling of friendship, or a greeting imposed upon him by the sisters in Christiania. Perhaps it was his situation with regard to infatuation and love that required still one more round for the sake of clarity; perhaps at Sorø he might find a series of memories that he had to work through in order to know for sure that he had acted correctly and made the right decision.

In one way or another the stay in Sorø and Copenhagen seemed to have matured Abel such that he could take an independent decision in relation to the future. He no longer felt himself bound to the travel itinerary that he had

---

[13] "Geheimrath", formerly a term for "privy councillor" is used here as an honorary title.

sketched out for the cabinet ministers and professors back home in Christiania. He would not proceed directly to Paris or to Gauss in Göttingen. No, instead he would go to Berlin! All his travel companions were to spend the winter in Berlin, or the vicinity. Abel wanted to be together with them, and besides, he now had the address of the engineer, Geheimrath Crelle.

He travelled from Copenhagen to Lübeck by steamboat, and this was the first time that Abel had been aboard such a modern vessel. This was followed by an exhausting journey by post coach from Lübeck to Hamburg over extremely muddy roads, and on many occasions the coach was almost overturned. Then, safe and sound and reunited with his friends at the inn called "Zum grossen wilden Mann" in Hamburg, it became evident that several of the screws in Hansteen's "oscillation apparatus" had shaken loose in the course of all the rattling and bumping of the journey. The apparatus now had to be repaired at an instrument-maker's establishment in the city.

They stayed in Hamburg for several days. The troops now assembled, as it were, they obeyed Hansteen's command, and paid a visit to Schumacher in Altona, which at that time was a city independent of Hamburg, and which possessed a thriving shipping industry. The well-known astronomer gave them a kindly reception even though he was under the weather the day of their visit. Nothing seems to have been said about Abel's mistaken lunar calculations. Abel's own report on the meeting was very short and to the point, saying only that the professor "received me in a very obliging manner, although he was not at that moment in completely good health." But Schumacher must have formed a strong impression of Abel because he later wrote to Gauss that Abel "was as admirable as a human being as he was remarkable as a mathematician." This, however, was expressed after Abel's death. A complimentary word from the great Gauss in Göttingen in 1825 would have opened many doors for Abel. But what, via Schumacher, Abel came to learn, was the rumour that Gauss had not even bothered to read his treatise on the fifth degree equation, and had merely mumbled something or other about "still another of these eternal attempted solutions", and that one day he, Gauss, would demonstrate a solution to the fifth degree equation, himself.

Abel also met the assistant at the observatory in Altona, a Thomas Clausen, "who," Abel commented in a letter home, "surely had an excellent Foundation in Mathematics, but as far as I could make out, he had not a great deal of Learning."

The trip from Hamburg to Berlin seems to have been uneventful; in any case Abel made no remarks about it. And once in Berlin, on October 11th, they took rooms that Maschmann seems to have arranged beforehand, in Am Kupfergraben 4, and this became their headquarters for that autumn and winter.

Abel wrote to Hansteen for the first time on December 5th, and therein had to defend his travel itinerary:

> You have perhaps wondered about why I first journeyed to Germany; but this I did partly because I thus could live together with Acquaintances and partly to be less susceptible to not utilizing Time in the best Way possible, for I can leave Germany at a moment's notice, whenever, in order to travel to Paris, which ought to be the most important Place for me.

Meanwhile at home in Christiania a discussion had broken out about a new university teaching position in mathematics, a matter that would have great significance for Abel. The following had happened.

A good month after Abel had left Christiania, Professor Rasmussen was appointed Government Paymaster, or "Zahlkasserer," and he now had responsibility for the main exchequer of the government's various offices. A month later the Collegium approached the University's Pro-Chancellor with a request to have a new university teacher appointed in mathematics. Pro-Chancellor Treschow sent the matter to the Church Department with the observation that Hansteen and Rasmussen could no doubt give the best information regarding "the Person's Learning and Gifts, that ought to come into particular Consideration with reference to this Matter." The Church Department wanted to have the Collegium's views on who they considered best qualified for the position, and "whether the said candidate ought to be employed as Lektor, Docent or Professor, or whether, rather, the position ought to be constituted at one of these levels only for a period of Time."

In the proposals submitted in 1812 about the university, the plan was to have twenty-five permanent teachers, professors and lecturers divided between eight faculties. But in the new "Bill concerning the University", passed in 1824, the number of faculties had been shrunk down to four: theology, jurisprudence, medicine and philosophy. The latter was a catch-all faculty ranging from philosophy, classics and contemporary languages and history, to geology and other natural sciences, mathematics and economics.

The Faculty of Philosophy was now asked to make a recommendation in relation to the vacant position in mathematics. This body was composed of Georg Sverdrup, Jacob Keyser, Jens Rathke, Søren Brun Bugge and C. A. Holmboe, who acted as secretary. Their report was completed and signed on December 6, 1825, and contained the following: they considered that there were two men well-qualified for the position, Head Teacher B. M. Holmboe, and Student N. H. Abel. On Holmboe, the report stated that for many years now he had "exhibited remarkable Competence" and showed an "extensive and well-founded Knowledge of Mathematics." On Abel, it acknowledged his "rare Talent for Mathematics and his great Future in this Science," but note was also taken of the fact that at this moment he had begun a lengthy journey abroad that he could not be called back from without harm, and that one felt as well that in this respect "he would not with Ease be able to adapt himself to the younger students' Ability to Conceptualize, and thus, not bear Fruit in terms of lecturing in the Elementary levels of Mathematics, which is the main feature of the above-mentioned Teaching Position, as would a more experienced teacher; notwithstanding, one considers him to be pre-eminently qualified to occupy a Teaching Post in Higher Mathematics, that one perhaps may dare Hope, in Time, to see established at the University."

For these reasons then, Head Teacher Holmboe was recommended, but the Faculty of Philosophy nevertheless considered it a duty to give consideration as to how "important it was, both for Science in its Generality, and for our University in its Particularity, that Student Abel not be lost Sight of."

35

Berlin, October 1825 to March 1826

In 1825, Berlin was a political and cultural centre, replete with military processions and masquerade balls, luxurious houses, palaces and churches. The lodgings that pharmacy-apprentice Maschmann had found in Am Kupfergraben, right beside the Spree, were respectable in every way. For the Norwegian students the quarters came equipped with a man who was able to show them around the city. Professor Georg W. F. Hegel lived on the floor above, and the ground floor appears to have been a beerhall. None of the Norwegians mentioned either the beerhall or the famous philosopher in their letters home, but then Hegel was not well known in Norway at the time. In Berlin however, Professor Hegel gathered around him a great number of adherents, apart from this, he sat at Am Kupfergraben and worked out his new lecture series. In the course of time he would also make his pronouncements on the Norwegian students living downstairs.

One of the first things that Abel did in Berlin was to look up August Leopold Crelle, the engineer, road builder and the builder of one of the first railway lines in Germany, from Berlin to Potsdam. Crelle was a man known for his enthusiastic influence in mathematics, his technical skills, and his great influence among the Prussian authorities. Crelle had many irons in the fire, and when Abel knocked on his door at the Gewerbe Institute, Crelle glanced up and examined the shy young man who was not very successfully articulating his errand, and considered him to be only a student who wanted to take an examination. It took quite some time for Abel to pull himself together and in halting German, state bluntly, "Nicht Examen, nur Mathematikk." Then for the first time Crelle looked up from his papers.

Such was Abel's first meeting with A. L. Crelle, the man who would come to have the greatest significance in Abel's scholarly and scientific career. Crelle had just then translated Lagrange and other French mathematicians whose work Abel knew intimately. Abel had also read one of Crelle's mathematical works, *Analytishche Facultäten,* and had made comments that had made Crelle curious. Crelle became still more intrigued when the conversation, which was conducted in French, came around to higher equations and Abel outlined the problem of fifth degree equations and his proof of their insolubility. Abel gave him a copy of the small, self-published treatise that he had brought with him, but Crelle did not understand Abel's account and doubted the results. This was something that prompted Abel to proceed with a more detailed version of the proof. For Crelle, who was eager to promote and develop mathematics in Germany, and who had plans underway for a mathematical journal, the meeting with Abel was a great inspiration, and a warm friendship developed between the two. Abel got to use Crelle's library as though it were his own, a library that contained the latest in mathematical literature – among other things, Baron Ferrusac's *Bulletin universel des sciences et de l'industrie*, where Abel found information on new books and new mathematical results. Abel was also quickly drawn into Crelle's scientific and social circle in Berlin.

When Abel wrote to Hansteen on December 5th, he recapitulated his experiences in Berlin as follows:

> I have not exactly made a considerable Catch here in Berlin with respect to public Libraries, because from the mathematical Point of View, these are superlatively bad; one finds almost nothing there of the newer works, and what there is [is] sorely incomplete. Our Library, if I may say so, is better furnished. Despite this I am, in another Respect, exceedingly well pleased with my Stay here in Berlin. I have namely been so fortunate as to make the Acquaintance of a pair of admirable Mathematicians, that is to say, Geheimrath Crelle and Professor Dirksen. Von Schmidten had described the former to me as an extremely excellent Man in all Respects, and thus, when I came to Berlin I proceeded to him with as much haste as possible. It took some time before I was able to render understandable to him what the real Reason was for my Visit, and it seemed that it would come to a miserable End, but then I took Courage when he asked me about what I had already read in mathematics. When I had listed a couple of the most eminent Mathematicians' Writings to him, he became particularly obliging and seemed genuinely pleased. He launched into a longwinded Discourse with me on many varied, difficult Issues that still have not been treated decisively, and then, when in the course of this, we came to discuss the higher Equations, I told him that I had proved the Impossibility of solving those of the 5th Degree, he in general would not believe this, and he said, to the contrary, he must object to this. I therefore presented him a Copy, but he said that he could not grasp the Basis for several of my Conclusions. Several others have told me the same thing and therefore I have undertaken a Revision thereof. He also spoke at length about the low Position of Mathematics in Germany and said that most Mathematicians' Knowledge

had been reduced to a Pinch of Geometry and something they called Analysis but is nothing other than the science of Combinations. And yet he went on to say that a more fortunate Period for Mathematics in Germany was now about to begin. When I expressed my Astonishment that no Journal for Mathematics existed here, as it does in France, he responded that he had long had in Mind to take upon himself the Editing of such a journal and wanted as soon as possible to set this into Practice.

This publication became the *Journal für die reine und angewandte Mathematik*, a leading mathematical publication during the rest of the nineteenth century, not only in Germany but also across the whole world, and it is still published today. Indeed, it was for this, *Crelle's Journal*, as it commonly came to be known, that over the course of the coming three months, Abel wrote six brilliant papers that were published in the first issues that came out in 1826, the first appearing in February of that year. It was also widely acknowledged that due to Abel's contributions, the journal rapidly achieved renown. Most of Abel's works were published in *Crelle's Journal*, and if it had not been for this publication, it would not be easy to see how Abel could have gained inspiration for his further work.

Abel wrote in French and Crelle translated. Abel gradually was offered honoraria, "which naturally I had not Counted upon, and that I also declined; and yet I think, however, I could detect that he very much wanted me to accept them." That, by this means, he had some income in Berlin, must have been a great piece of news to write home about, to Hansteen and the others who would have preferred that he had proceeded directly to Paris. In his letter to Hansteen in December he had still more to say by way of defending his itinerary:

With regard to the Form of my Treatises, Crelle thinks that they are clear and well-written, which particularly pleases me, when I have always been anxious about the difficulty I have in expressing my Thoughts in a plain Manner. Still, he counselled me to elaborate a little more, particularly here in Germany.

Further, about the social and scientific life in Berlin, Abel wrote, "I am invited to Crelle's every Monday Evening. There is a kind of Assemblé at his place and the Major Occupation there is Music, something I unfortunately do not understand very much about."

For Abel, singing and playing music, constituted an unknown culture. He was not attracted much to the instrumental music of the day. It was true that the piano had replaced the spinet harpsichord, but Abel could not, for example, evaluate and comment upon the "sonority" of a melody against the accompanying resonance of an oboe. But nonetheless he enjoyed himself. He wrote that both sexes attended such gatherings and that he always met a couple of mathematicians he could talk to, and "indeed, I gain much Practice in German, which I sorely lack, and which is not exactly going well." Earlier, there had also been a weekly gathering at Crelle's of *only* mathematicians, but these had to cease due to the "dreadful Arrogance" of Professor Martin Ohm, brother to the famous physicist. Abel commented, "It is an incontrovertible shame that one Man can in this way lay barriers across the Road to Science."

Abel had remarked on how all the young mathematicians in Berlin "nearly deify Gauss. He is for them the quintessence of all mathematical Excellence." Abel went on to comment, "It may be granted that he is a great Genius, but it is also well-known that he gives rotten Lectures. Crelle says that all Gauss writes is an abomination, since it is obscure to the point of being almost impossible to understand."

This was an evaluation that Abel was not opposed to passing on to others. He concluded his December letter to Hansteen saying that for the sake of the *Journal* he would remain as long as possible in Berlin, that there was no "Place in Germany that would be of greater benefit to me. Göttingen certainly has a good Library, but that is the only thing it has to its credit, for Gauss is the only one there who knows anything and he is absolutely inaccessible. Though I realize I must go to Göttingen."

Enclosed with the letter to Hansteen was one to Mrs. Hansteen: "You will see from my letter to the Professor how things are going with me. But beyond this, I have a Favour I would ask of You. You have always been exceedingly good to me. God bless You, for never forgetting my Brother! I am so worried that things might be going badly with him." He asked Mrs. Hansteen to give Peder money, if necessary, more than the 50 daler that he had already deposited with her; and that he, Abel, one way or another would find the money for this, and he bid her greet Peder and urge him to write: "He can send the Letter to my Fiancée, and she will send it on, and the best thing is for him to send it to her unfranked, and she will look after it." We do not know how much Abel knew about his brother's conduct in Christiania, as we do not know which letters Abel received from home. However, from the police reports in Christiania that autumn it appears that Peder Abel and another student he shared quarters with, were called in "in Relation to the fact that they often held Parties wherein there was Drunkenness that gave rise to loud Brawling and Pandemonium to a Degree that disturbs the public Peace and Quiet of the Nighttime." And the records report Peder's response: "The apprehended affirmed that the alleged activities did take Place, and have made most solemn assurances that such shall no longer happen."

Apart from this, Abel wrote to Mrs. Hansteen from Berlin on December 8th, saying:

> Otherwise I am living exceedingly peacefully and am quite into my work; but on occasion I have a terrible Longing for Home, which is supplemented by the fact that I so very seldom hear anything from that quarter. My dear Sister is doing well. She sends me marvellous greetings. And I hope with all my Heart that the good, sweet-tempered Charité is also keeping well. Keep well, dear Mrs. Hansteen. I cannot write more, I am really quite melancholy.
>
> Adieu, and do not be angry with me; I must strike you as quite odd.

Perhaps Abel felt himself to be in an "odd" mood rather frequently at that period of time. It was also at this time that he, according to the later descriptions by Boeck, had the habit of getting up every night, lighting the candle, and writing down the ideas that had abruptly awakened him.

In the rainy weeks leading up to Christmas, 1825, Boeck, Abel and Møller shared rooms in Am Kupfergraben. Keilhau was in Freiberg and Tank in Breslau in order to "fly a bit with Steffens," as Boeck put it. The plan was that they should all gather in Berlin to celebrate Christmas together. Tank arrived on December 20th, and Keilhau, who had first given the message that he did not have the money to travel, but who then received an unexpected advance honorarium for a book of articles that would be published in Leipzig the following year, was now in a condition to join his friends for the Christmas celebrations. The spirits were high, the atmosphere jubilant, and they celebrated Christmas in the "Old Norse Way" with *julegrøt*, the sweet milky Christmas pudding and no doubt a plenitude of drink: Boeck, Møller, Abel, Maschmann, Tank and Keilhau. Rudolf Rothe, a Danish student, had also joined the Norwegian colony. Rothe had a small stipend to support his studies to become a park gardener, and during that autumn of 1825, he had worked in the gardens of Sans-Souci at Potsdam. He later became one of the leading men in the care of parks and gardens in Denmark. But, as already mentioned, Hegel lived directly above the merry party at Am Kupfergraben No. 4, and it was most probably now that the great philosopher sent his serving girl to investigate what he considered to be a drunken brawl in full swing downstairs. The servant girl returned and reported that it was Danish students and Hegel was said to have growled, "Nicht Dänen, sie sind russische Bären." Hegel thus defined the lively Norwegian party as a gathering of Russian bears, but it was certainly not the first time that he had been wrong. Earlier, he had seen the hand of the Four Horsemen of the Apocalypse when Napoleon had ridden into Jena; he had also attacked the astronomers for daring to search for an eighth planet, after the seventh, Uranus, had been discovered in 1781. Anyone who possessed the smallest particle of understanding about philosophy would immediately realize, according to Hegel, that there were only seven planets to be found, neither one more nor one less; the research of astronomers who did not heed his pronouncement was merely a meaningless waste of time.

The Norwegian students had quite certainly heard of Hegel and quite likely had also seen him in the building. It is factually very probable that Abel had met Hegel in social gatherings at what were called literary salons in the city. Abel does not mention Hegel in his letters, but neither does he mention his weekly participation in Madame Levy's salon on Cantianstrasse. From Boeck's report to Hansteen we learn that Abel at least at the beginning of December, in addition to Crelle's musical Monday evenings, also visited a rich lady, a "Madame Levi or some such," every Saturday. This was quite certainly Sarah Levy, daughter of the prosperous banker, Daniel Itzig; she in turn had married a banker. She upheld the custom of arranging weekly gatherings of prominent men and women who assembled and engaged in social exchanges and conversation. Such salons had played a considerable role in the German Romantic Movement, particularly around the turn of the century and into the economic downturn that followed the Napoleonic Wars. Women were very important in these salons. One of the most prominent and famous was Henriette Herz, famous for both her beauty and her intellect. In addition to the classical languages, she had taught herself Sanskrit, and her understanding of physics and mathematics was likely also considerable. Her husband was a highly esteemed medical doctor, and the gatherings, which Henriette later took over, began with her husband holding seminars at home. And there had been many such salons, where age, sex, social status, and the religion one professed, were not issues of foremost importance. Indeed there was, at these gatherings, a generous atmosphere and a romantic fellowship of free spirits, nourished and cultivated by discussions about all the affairs of the world. Two of Sarah Levy's sisters, Fanny and Cecily, were married to Jewish bankers in Vienna, where they too conducted literary salons. Theodor Körner, the poet and champion of freedom, had been captivated by Fanny's charming, melancholy daughter, and during the Vienna Congress of 1814–15, not only "Chancellor" Metternich and the French foreign minister, but Bishop Tallyrand also frequented Fanny's salon.

In Berlin, in her day, Sarah Levy had Mirabeau living in her house, the count who had become the leader of the Third Estate during the first phase of the French Revolution; in addition, during her period of exile, Madame de Staël had been a frequent guest of Madame Levy. Many other French refugees sought shelter here as well. It was also reputed that Madame Levy's salon had been the venue at which the city's most well-known couple, Bettina and Achim von Arnim, had become engaged to marry. During this particular autumn (1825), the sixty year-old Henriette Herz attended the Levys' salon, and perhaps it had been rumours about the phenomenal mathematical gifts of the young student that caused Abel to be invited into what had gradually become a somewhat exclusive milieu; none of the other Norwegians were invited. Hegel and the poet, Heinrich Heine, and Bettina von Arnim, all frequented Rahel Varnhagen's salon at this point in Berlin's social life, but a bit of interchange between the various social circles seems not to have been uncommon.

Most of the people Abel met and mentioned at the outset were rather unknown, but he was well-acquainted with the history of the French Revolution, certainly also Mirabeau with his popularity among common folk, and de Staël had been discussed in several Norwegian journals. Before Karl Johan had become king, he had been among the friends of de Staël, and in 1814 she had written to Karl Johan to urge him to ratify the Norwegian Constitution. Madame de Staël and her circle now hoped that the democratic ideas that Napoleon had done away with in France might find fertile soil in Norway. Those in the liberal circles who criticized the old Europe also held up the United States of America as a land of freedom, but then, the fact that a small, little-known country like Norway had been able to give itself a constitution which, to a great degree, could stand as a paradigm for the rest of Europe, had engendered an astonishment which the Norwegian students, at many places in the course of their travels, were to remark upon. (Be this as it may, this lauded Norwegian Constitution excluded Jews from the kingdom.) There was probably as well, among Abel's friends, an attitude of such a character, toward Jews that he could not easily mention Madame Levy without finding himself in an even more difficult defensive position. Abel had more than enough about which he had to defend himself from the authorities back home in relation to his sojourn in Berlin.

It is evident, however, that Abel held the great and multi-facetted Berlin milieu in high esteem, and that he happily crossed traditional religious and social barriers. His friends had never seen Abel so even-tempered as he was at this time. Many of his dreams seem to have found form in Berlin. During the last months of his life he spoke fondly and with the anticipation of becoming a resident there.

His friends complained about the unending rain in Berlin, and they longed for home with its winter and its snow, while Abel's mood seems to have fluctuated in great swings, depending upon the nature of the thoughts that were assailing him. But the celebration of Christmas had been great in all ways: first was the celebration by these friends at Am Kupfergraben, and then there was an excellent ball at the Crelles'. Abel wrote about this to Mrs. Hansteen two or three weeks later:

> During Christmas I was at a Ball at the home of Geheimrath Crelle but I did not manage to dance, although I had made myself presentable to a degree that I have never been before. Think of me new, from Top to Toe, with a waistcoat and stiff collar, plus Spectacles. You see, I have begun to conform to Your Sister's Injunctions, and I hope to have achieved perfection by the time I reach Paris.

It is almost as though Abel remembered how badly things went when he danced at the ball in Copenhagen; in any case, he recalled Charité and her well-known relationship to clothes. Apart from this, inside the Crelle family, it was said that Abel always got off on the right foot with the ladies of the household, and that they spruced him up before gatherings when one or another of them considered him to be in a particularly scruffy and disheveled condition.

In addition to using Crelle's library freely, Abel went out on Fridays around dinnertime for walks with Crelle, and they were most often accompanied by another young man, a promising mathematician whom Crelle had taken under his wing, namely, the Swiss Jacob Steiner, who was also to become a prominent contributor to *Crelle's Journal*, and who in 1834 was appointed professor in Berlin. When the three of them were out on their weekly promenades, it was said that people would look up and say, "Look! Here comes Adam with his two sons, Cain and Abel!" – that is, Crelle's middle name was Adam.

Their discussions ranged over many topics, but a prevailing theme must have been the new demand for logical stringency in mathematical proofs, spearheaded without doubt by Gauss in Göttingen. Discussing mathematics seems to have sharpened Abel's critical faculties; he became bolder and seemed more self-assured, or perhaps it was just that thoughts and doubts that had been brewing up now came to the surface. As well, the possibility of at long last getting his works published seemed to have infused Abel with an undreamed of inspiration.

During the month of January, 1826, Abel made ready six articles for *Crelle's Journal*, perhaps more. On January 16th he wrote a long letter to Holmboe, in which he tried to defend what he had been busy with. Four months had now passed since Abel's last letter to Holmboe, and that had been only a short one from Copenhagen about the book lists and book purchases in relation to Professor Degen's library. Also Abel excused himself for allowing so much time to go by, but the reason was, he did not want merely to write about how the trip was progressing in practical terms, but also:

> how in the overall sense, my Foreign Tour has been going. Besides, I also wanted to share with You one and another recent Investigations I have been engaged in, into several interesting Materials that have been occupying my time. I do not want to entertain you with descriptions of my Journey's events, of which there have been precious few in the Pot (= experiences) and about which You anyway must have heard enough from Prof. Hansteen.

But before Abel got into the mathematical materials, he once again described Crelle and the fantastic help and support Crelle provided. Abel wrote: "I am as close to him as I am to You and the others who are the best of my Acquaintances." And then he wrote down a litany of Crelle's translations from French; these were books Abel had now obtained. And these books, together with Crelle's textbooks in mathematics, and his published articles, Abel would soon send home, and he urged Holmboe to hold them in safekeeping for the duration. Abel again mentioned the Monday evenings at Crelle's home and the promenades every Friday:

> It is a time for dealing with mathematical Matters, as you can imagine, and it goes by as rapidly as my unGerman Tongue will allow. Though in so doing, I stick my neck out. He cannot get it into his Head that I can understand everything he says, even though I cannot myself speak correctly. This Berlin Language is, otherwise, not exactly the best; in some Respects, rather hard, and in other Respects unusually soft and flat. Besides this, they place a "j" at the beginning of words which normally begin with "g", and in this way they sound damnably odd. F. ex. O! Jot! which one hears every Second. The following expression gives an idea of how Berliners sound in this Respect: "Eine jute jebratene Jans ist eine jute Jabe Jottes" [A yood-cooked yoose is a yood yift from Yod]. Another thing that also makes for a strange Effect is that they mix "mir" and "mich"; "dir" and "dich" with one another; similarly here one incessantly says "sind" instead of "seyn". My Servant says: Wollen Sie so jut sind mich Jeld zu jeben; ich werde jleich hier sind. [In a thick Berlin dialect he asks for his pay, for which he has waited a long time.]

Following these linguistic observations, he approached mathematics by expressing his great joy in the mathematical journal that was now about to come out in Berlin, and in which he himself would have a couple of articles in each of the first issues: "I have already finished working up 6." In the first issue, to appear in February, Abel's expanded treatise on the impossibility of solving fifth degree equations by means of square roots, would be highly conspicuous; the proof now extended to well over twenty pages. Abel also had a lesser article in the first issue, on symmetrical functions with two variables. In the second issue, which was now waiting to go to press, Abel had three articles, and in the course of the first year's publication (the four first issues) Abel published altogether seven papers in *Journalen*. The first major one was of course, his treatise on the fifth degree equation, and he commented to Holmboe in the January letter:

> Crelle said about this treatise that it was credible, but he could still not really comprehend it. I find it so difficult to express myself in a quite understandable manner with this Material, which still, as far as I am concerned, has received so little work. Since I arrived here in Berlin I have also tried to solve the following common Problem "of finding all the Equations that can be solved algebraically." I am still not quite finished but as far as I can tell it is turning out well. As long as the Equation's degree is a Prime Number it does not involve so very great a Difficulty, but when it is a Composite Number it is the Devil to solve. I have made Applications of the Fifth Power Equations and have successfully solved the Problem for this Occurrence. I have discovered a great Number of Equations, apart from the already well-known ones, that I have managed to solve. When I get the Paper finished to the degree I want, I flatter myself that it will be good. It is rather ordinary though, and it seems to me that the Method is the most important part.

This problem of determining which classes of equations *are* capable of algebraic solution was a problem that Abel kept returning to all his life: what constitutes a solution? What does a solution look like? However, he would only have time to make public one great work on equation theory, the one published in *Journalen*, in the second issue of 1829, that came out only a few days before he died. This is the treatise entitled *Mémoire sur une classe particulière d'équations résolubles algébriquement*, that Abel had dated: Christiania, 29th February, 1828, and then quickly dispatched to Crelle. Abel had found a way of expressing the rational relationship that must necessarily exist between the roots of equations which made algebraic solutions possible. It is this class of algebraic equations capable of solution by formula that later came to be known as *Abelian equations*. In papers left behind at the time of his death there were computations and treatments for still another great work in the field of equation theory: *Sur la résolution algébrique des équations*. Here

Fig. 23. Oil painting by Johan Christian Dahl: Moonrise seen from Bruhl Terrace, Dresden, 1825. Dahl made many paintings on the theme of Dresden by moonlight, as viewed from both right and left banks of the River Elbe – a favourite motif of this romantic painter. The Bruhl Terrace, an old fortress wall, was famous for its beautiful outlook over the Elbe and Neustadt. Abel visited both Dresden and J. C. Dahl in March–April, 1826. Here he also met the Danish poet, Jens Baggesen. Oil painting on paper, laminated to a card backing. [National Gallery, Oslo. Photo: J. Lathion]

Abel set out several theorems on the form that an algebraic solution to an irreducible polynomial to the power of n must have, and he indicated evidence for these. Abel operated according to what he called "an arbitrary field of rationality," and later mathematicians have felt that in this they detected the first germs of the concept, numbers field, and other important related mathematical tools. Be that as it may, in the course of his European tour, Abel had brought into use the auxiliary component that later would be called Galois'

Fig. 24a–c.
a. August Leopold Crelle, Abel's good friend and benefactor in Berlin. Painted in 1815. From *Beilage zum Journal für die reine und angewandte mathematik*, V.203:3/4, Berlin, 1960.
b. Below: Nils Otto Tank. The picture is dated Dresden, 1826, and was probably painted by J. C. Dahl. Oil painting. The State Historical Society of Wisconsin, Madison. (X28)1020.
c. Below: Jacob Steiner, Abel's close friend in the circle around Crelle in Berlin. Steiner was later a professor and the founder of what came to be known as synthetic geometry. [Contemporary lithograph. Bildarchiv Preussischer Kulturbesitz. Berlin]

b   c

resolvent, named after Évariste Galois, who studied and elaborated upon Abel's equation theory work.

But the mathematical problem Abel dedicated most attention to in January, 1826, concerned theories that had to do with the development of series; and within the field of infinite series theory, Abel would also come to exhibit far-reaching insights. In mathematics, a line of numbers with sizes that follow one another in a definite order are simply *a series*, and the individual sizes are

Fig. 25a–d. Clockwise from top left: The Danish mathematician, C. F. Degen (**a**), who spoke about the "Magellan voyages" (Copper engraving by F. John after a painting by Agricola, The National Historical Museum, Frederiksborg Palace, Hillerød, Denmark); the astronomer, J. J. von Littrow (**b**), Abel's close friend in Vienna (Lithograph from the Picture Archive and Portrait Collections of the Austrian National Library, Vienna); the famous C. F. Gauss (**c**), whom Abel never did manage to visit in Göttingen (Archives of Gyldendal forlag); and the astronomer, H. C. Schumacher (**d**), who published *Astronomische Nachrichten* in Altona/Hamburg (Archives of Gyldendal forlag).

Fig. 26. "Neue Aula" in Vienna, with the astronomical observatory. It was here that Abel visited J. J. von Littrow in the spring of 1826. "In the afternoon I *walk the cadaver* with him (wie man so zu sagen pflegt [as they say here])" wrote Abel in a letter to Holmboe on April 16th, and later he was often at von Littrow's, as a rule, from seven o'clock in the morning. Anonymous copper engraving, 1756. [Archives of the University of Vienna]

called the series' *terms*. A numerical series in which the difference between one term and the previous one is constant, is called an arithmetic series; a series in which the quotient between one term and the previous one is constant, is called a geometric series. When the number of terms in an arithmetic series increases beyond all limits, the sum of the series will also increase beyond all limits. However, the situation with an infinite geometric series is not quite so simple: the sum of the *n* first terms *can* approach a finite limit when *n* increases beyond all limits, and when the series is said to be convergent, it is, on the contrary, divergent. To decide whether a series is convergent or not, is often difficult, and to set up secure convergence criteria can be an extremely complicated mathematical task. In relation to this question, there was much cheating in Abel's day. The simplest convergence criterion is: if a number value of a quotient in a series is less than 1, the series will always converge; but in the use of geometric series, and particularly in power series, as are used to work out known functions, here the core question arises in those instances where the quotient approaches the finite limits of 1. In Abel's time many calculations of such series were taken as given, without any demonstration of whether or not it was even meaningful to speak about convergence. For example, the calculations of the trigonometric series *sinus* and *cosinus*, etc., were based on power series. Before Abel embarked on his European tour he too had a rather old-fashioned view of infinite series.

He had read the old master, Euler, who was not particularly careful in the treatment of series. Abel, in his notebooks from 1822–23, repeated Euler's equations; for example,

$1 - 1 + 1 - 1 +\ldots = 1/2$, and the series: $1^2 - 2^2 + 3^2 - 4^2 +\ldots = 0$.

Perhaps Abel had had his doubts about the validity, but it was only now in Berlin that, for the first time, he would seriously concentrate upon unveiling the weaknesses. He had found inspiration and impetus for this task in A. L. Cauchy's book, *Cours d'Analyse*, that had come out in 1821, and in other works by Cauchy that had been published in recent years. In a letter to Holmboe now in January, 1826, Abel said:

> On the Whole, Divergent Series are the Work of the Devil and it's a Shame that one dares base any Demonstration upon them. You can get whatever result you want when you use them, and they have given rise to so many Disasters and so many Paradoxes. Can anything more horrible be conceived than to have the following oozing out at you:
>
> $0 = 1 - 2^n + 3^n - 4^n + $ etc.
>
> where *n* is a whole Number? Risum teneatis amici. It has been an Eye-opener in an extremely amazing Manner; hence, when one accepts the very simplest Instances, for Ex: Geometric Series, where – out of the whole of Mathematics – there exists scarcely a single Series wherein the Sum is determined in a strict Manner: in other words, the most important features of Mathematics stand without Substantiation. Most of them are correct; that is a fact, and it is exceedingly surprising. I am striving to find a Basis for these. An extremely interesting task.

Abel used Horace's words *risum teneatis amici* (hold back the laughter, my friends) in characterizing the old condition of the theory of series. He himself would go on to produce works distinguished for their method, precision and universal validity, and he succeeded to the degree that one of the next generation's great mathematicians, Charles Hermite, said that Abel had given mathematicians enough to keep them busy for several hundred years.

In his work on series, Abel quickly got down to work on one of the most well-known of all: the binomial series; this is the series $(1 + x)^m$, that Newton had discovered, and about which now, in order to advance series theory, Abel would make fundamental discoveries. Euler had treated this series with a real variable x and any exponent m; Cauchy had worked out detailed investigations of complex variables; and Abel now took the process to its full conclusion by allowing both the variable and the exponent to be randomly complex numbers, and he solved the convergence questions completely. This work was published in *Journalen's fourth issue for 1826, and from this paper stem the concepts Abelian partial summation* and *Abelian theorem of limitation values*, also known as the Abelian theorem of continuity, and he was a forerunner of the extremely important mathematical concept, *uniform convergence*.

That editor Crelle was nowhere near Abel's level is seen in the following. Crelle had personally translated Abel's paper on the binomial series into German, but still, after Abel's death, he published another paper on the same theme, wherein he seems ignorant of the fact that the whole problem had found its definitive solution with Abel. Crelle was somewhat impeded in his thinking by old, pre-Abelian conceptualizations.

In the first issue of *Crelle's Journal* for the year 1828, Abel revealed to the public a little paper on infinite series, that was a reply to an earlier paper in the journal by the mathematician L. Olivier. Abel disproved so concisely and simply the convergence criteria that Olivier had set down, that this little paper of barely three or four pages was pointed to by several mathematicians as paradigmatic of Abel's clarity of insight and fabulous ability for original thinking.

In the January letter to Holmboe, Abel had just described the work with the binomial formula and was deep into the reasoning about how poor many series were in mathematics. He wrote:

> To illustrate with a common Example (sit venia verbo[excuse me for saying this]), no matter how poorly one reasons, and how careful one necessarily must be, I want to choose the following Example: – I had got this far when Maschmann came through the Door, and because I have not had a Letter from home for such a long time, I stopped in order to ask if he had anything for me (He was, you see, our faithful Letter-Carrier) but there was nothing.

However he himself had received Letters, and among other News, he related that You, my Friend, have been selected to become Lektor in Rasmussen's Place. You must accept my sincerest Congratulations, and be secure in the knowledge that none of your other Friends could be as enormously happy for you as I am. You must understand that I have often dreamed of a Change of Position for you, for to be a Teacher at a School must still be something rather frightful for One, like You who are interested so very deeply in your Science. Now as well you must see to it that you find yourself a Fiancée too. I hear that your Brother, the Priest, has got One. I do not deny that this surprised me greatly. Please give him my warm greetings and congratulate him profusely. – Now, back to my Example again. Let $a_0 + a_1 + a_2 + a_3 + a_4 +$ etc. be any infinite Series, and thus You know that a very useful Manner of adding up this Series is to seek the sum of the following: $a_0 + a_1x + a_2x^2 + a_3x^3 +...$ and then later, put $x = 1$ in the Results. This is correct; but it seems to me that one cannot accept it without Proof, thus...

and here he went into a lecture on convergence criteria again, and concluded by saying that "It is going rather Well and interesting me exceedingly." And so, after having said that he would probably remain in Berlin until the end of February or March, and that he thereafter planned to travel via Leipzig and Halle to Göttingen, not for the sake of seeing Gauss, but rather to visit the library there. He ended the letter:

I wish I were home, which I miss terribly. Write me now at last a long Letter, telling me Every kind of Thing. Sit down and do this as soon as You get my Letter. Tomorrow I am going to the Comedy theatre to see Die schöne Müllerinn. Goodbye, and greet my Friends and Acquaintances.

Your friend, *N. H. Abel*

But to receive this news that Holmboe had been awarded the position at the university at home in Norway, was not easy; it must have felt like a slap in the face. Perhaps at the beginning he had had his eye on it, at least since the professorate had been scaled down to a lectureship, a hope that there could be two positions in mathematics at the university, but the bitter facts were soon made clear to him: visions of a permanent job at home in Norway now, at best, lay in the distant and uncertain future; the hope for a development of the mathematical field was only a vain wish hovering out in the blue. The letter to Holmboe is dated January 16th, but once again Abel is imprecise with dates. The letter was written on Saturday, January 14th. On Sunday the 15th he went to see *Die Schöne Müllerinn*, and on Monday, the 16th, he wrote a letter to Mrs.

Hansteen: "Yesterday I saw your Husband's favourite Madame Seidler in 'Die Schöne Müllerinn' and she was really very sweet." As a postscript at the end of the letter he wrote:

> In my letter to the Professor I have perhaps, and I swear by The Lord it was unintentional, come to use Turns of Phrase that he would not like. Regarding such, kindly be my Talisman and excuse me and set things to right.
>
> May You Keep well, and greetings to Charité.

Not only the contents, but also the work of the signet on the sealing wax, and the nature of the fold in the page indicate that this letter to Mrs. Hansteen was enclosed within a letter to Professor Hansteen, and as such, these two letters, together with the letter to Holmboe, were sent from Berlin on January 16th. On the letter to Holmboe the following is also written: "I apologize that this Letter has become so thick (and therefore so costly)." But the letter to Hansteen has not survived, and the expressions that the professor "would not like" are not to be found. In all probability, Hansteen destroyed this letter now or at some later point, as it is likely he would have felt that it did damage to Abel's reputation. The desire to edit also seems evident in Hansteen's later publication of Abel's letters in *Illustreret Nyhedsblad*, 1862. The underlying reason for this seems to have been that it *was* a great disappointment for Abel not to get the permanent appointment in Christiania, and that he had used strong words to this effect in his letter to Hansteen. As Boeck wrote in the supplement to Holmboe's Abel obituary in the autumn of 1829: "Although the Tour Abroad was for him a great Inspiration, and although in Berlin he found himself particularly satisfied with things as pleasant as they were useful, which the Exchanges at Crelle's certainly were, yet his Spirits so often sank into the uttermost Darkness. Contemplation over his Prospects clearly depressed him, and he often told me about his probably unfortunate Future."

What had happened in Norway was this. After the Faculty of Philosophy had submitted its recommendations on the 6th of September, 1825, the Collegium, without further comment, sent the proposal on to Pro-Chancellor Treschow, who said that he "must absolutely concur with the Collegium's Proposal regarding the mathematical Teaching Post Appointment." On January 6, 1826, the Church Department's recommendation was submitted. This document accepted the expression by the Faculty and the Collegium, that Student Abel was an "especially competent Mathematician," but that his tour abroad ought not be interrupted, and "that he would not with Ease have the necessary Ability to adapt to the younger Students' level of Conceptualization." All were in agreement then, that Bernt M. Holmboe be awarded the post of new lektor at the university. Here for the first time the post is officially referred to as a lectureship. The formal appointment was issued from His Majesty's office a month later, and this made no mention of Abel or a possible expansion of the study of mathematics at the university. In reality, the matter had already been settled on January 6th, and a week thereafter Abel also got the news in Berlin. During this whole appointment process, Professor Hansteen had scrupulously kept himself out of the matter. His only involvement had been when Pro-Chancellor Treschow had named him and Professor Rasmussen as persons competent to give advice on who would be qualified for the position. There are no papers in existence that document the fact that Hansteen was actually asked to give advice on this matter, but informally, in the corridors, it would have been difficult not to encounter Hansteen's views, if he had any. It was also discussed among the Norwegians in Berlin: at the end of October, Boeck had tried to put in a good word for Abel. At that point, by way of his cousin, Johan Collet, who was the Finance Minister's son, Boeck had heard that Rasmussen had been made "Paymaster," and everyone knew that there would be a vacant position at the university. After having given his report to Hansteen on the magnetic observations made in the course of their journey, Boeck had written: "Can there be any Hope for Abel that upon his Return home, he might come into that Post, or has perhaps Holmboe stolen that particular March? This seems reasonable, as in some Respects it could be, but still, it is quite right that Abel stands Head and Shoulders above Holmboe."

Disappointment over not having received the permanent position in Christiania seems not, however, to have dampened Abel's lust for work, his lust for travel, or his longing for home. Abel was known for his great mood swings, and through everything he would hope that things turned out for the best; he believed his dreams would be fulfilled. After he had written to Hansteen on January 16, 1826, and had probably complained in writing for the first and last time, he wrote in the letter to Mrs. Hansteen:

I care so much for You, dear Mrs. Hansteen, that I must send You at least a Couple of short Lines. But am I still in good standing with You? I am so very much afraid that You will not be thinking very highly of me, for I have still heard nothing from You, something that would please me inordinately. Yet I certainly have Hope, for my Intended wrote to me that You had in Mind to honour Me with a little Letter. I also have another Reason for Hope, for I dreamt last Night that I had received a Letter from You, and can believe that nothing other than my Dream can come to Fulfilment, and that explains why I have been so particularly happy.

Then, in this letter to his "second mother," he recounts the previous day's events, the actress, Seidler in *Die Schöne Müllerinn* (an opera written in 1821 by Duveyrier and Scribe, with music composed by Garcia and Schubert), and the Christmas ball at the Crelles', and then, after he had mentioned the coming journey to Paris, he returned to what seems to have been the constant background to his emotional life:

> Rather I had already been there (in Paris) and were home again. It is still so odd to be among Foreigners. God only knows how I would survive if I were to be separated from my Countrymen. And that will come to be with the first intimations of Spring.

Abel's plans were now to travel with Keilhau, who was still in Berlin, to the mining centre, Freiberg in Saxony, and then journey back to Berlin for:

> a day's Tour with Crelle to Göttingen and the Rhine Regions. I would spend only a short time in Göttingen as there is nothing to be had there. Gauss is inaccessible and the Library cannot be better than in Paris. It is likely that Crelle will also journey to Paris, which for me would be exceedingly lovely.

Abel wrote this to Hansteen on January 30th, in a short letter enclosed with Keilhau's letter to Hansteen. The use of the expression, "there is nothing to be had", in the sense of nothing to come to grips with, amounts to an extremely simple means of expressing his attitude to Göttingen and Gauss. It was very important for Abel to show his industry and work progress to Hansteen, and the treatise on integral calculus that Hansteen and Rasmussen had earlier not managed to find a publisher for, functioned almost as a trump card in that Abel now wanted to show Hansteen that it had been a stroke of luck that he had planned his itinerary via Berlin. Abel wrote:

> Journalen is going very well. You will receive the first Issue in April. You shall see that I have been working with all my powers. Every issue will include 3–4 of my Papers. If I can get a Publisher, I would like to let my Research into Integral Calculus be printed as well, but that would presumably be extremely difficult, as such Things, especially here in Germany, are not very much in demand. Nonetheless, I am going to see what I can do with Crelle's Help. He has given me Hope that Things will begin to move once I have written one or two things in The Journal. Therefore I am beginning with my best Treatises. It does not appear that I will lack for Subject Matter for the immediate Future. I wish I could get Everything printed, but as is said, this is [a] difficult Matter. If Journalen had not come

into production, the situation would have been even worse. I pray very much that you will give my greetings to Your Wife and Charité, and with the wish for amiable conclusions.

Yours faithfully, *N. H. Abel*

From my Fiancée's Letter I see that she had been to Christiania during the Market Days. It would make me very happy were she to have found Favour in your Eyes.

It seems as though Abel and his friends took part rather avidly in the carnival time that now gripped Berlin with balls and masquerade parties. Boeck wrote home to his father and reported that there were masked balls almost every week, where royalty appeared with their whole entourage, and Boeck indicated that they expended a little money and time on this, "but still one must also see Something when one is Abroad." Keilhau also wrote home to his mother on February 17th, about the great opera house that accommodated two thousand people and where the five Norwegians were "simply masked with a Silk Domino outfit over our black clothes. There was much dancing, and most of the Ladies were of dubious propriety. This had earlier aroused my great expectations. On Tuesday I am leaving, accompanied by Abel, for Freiberg via Leipzig."

And thus, Abel left with Keilhau, probably in order to find peace and quiet in which to work, as well as to meet the mathematician, August Naumann, professor at the metallurgical institute in Freiberg.

36.

Via Leipzig to Freiberg

Keilhau and Abel left Berlin early one Wednesday morning, as passengers on a crowded post coach in the company of loquacious merchants and *Geschäftsleute* [business people]. They arrived in Leipzig a little after seven in the evening and took accommodations at the inn "Stadt Berlin" located right across from the post house. They were tired, and Abel, after having consumed a glass of punch, was unfortunate enough to break the chamber pot.

The next day they made their way to the geologist, Carl Friedrich Naumann, brother of the mathematician in Freiberg. C. F. Naumann had undertaken bold and astonishingly long foot tours in the Norwegian mountains during the summers of 1821 and 1822. Without guides, he had almost made it to the peaks of Skagastøltindene, and found his way over from Stordalen to Lom, all in the high mountain country of Jotunheimen; in addition, he had explored Numedal, from Kongsberg through Telemark; he had been in Ullensvang, described an ascent of the glacier, Folgefonna, and regions around Bergen, Voss, Aurland, the islands of outer Sunnfjord and Romsdal. He had roamed over Fillefjell, been up through Hurrungane and on the glaciers near Jostedalen and Lodalskåpa. He had roamed around Dovrefjell, been at Snøhetta and Røros, and throughout he had kept basic records that he published in two volumes under the title, *Beyträge zur Kenntnisz Norwegens*. Here, side by side with scientific descriptions, are accounts of the environment and people he met on his wanderings.

Naumann gradually mastered Norwegian well enough to conduct chats on political and social relations. Naumann stressed the Norwegian's self-esteem, which, he felt, was expressed already in young boys eight or ten years old when they came bouncing toward him, asking forthrightly, who are you? where do you come from? what are you doing? Naumann praised the Norwegian mountain peasant as a true master of a thousand arts who, with his own hands, made everything he needed, and he seems to have enjoyed the many varied costumes he saw. For the most part, he considered Norwegians to be cheerful and lighthearted by nature, and trouble and hardship could never impair the light atmosphere in which the Norwegian began and ended every activity, or so he maintained. In the same way, Naumann wrote, the Norwegian's whole identity was dominated by a certain dignity, that, like expressions of his cheerfulness, could lead to great exuberance when this Norwegian happened to be on a drinking spree. This disposition toward drink, that Naumann understood to have come into full play when it became permissible for anyone in the country to distil spirits for his own use, was a stain on the otherwise fine, clean portrait. When the Norwegian was drunk, one had to be circumspect and guard against discussing with him, for in those conditions, his self-esteem melted into insolence, his willpower into wild rage, his openness into meaningless impudence, and his friendliness into repulsive importunity. But, Naumann asked, which other country did not reveal itself in the same light?

In May, 1822, Boeck had introduced C. F. Naumann to the Christiania Students' Association. He was the first German accorded this honour. Abel too had probably met Naumann on that occasion, and it also seems that Keilhau regarded the same-aged Naumann as a close friend. Now, in Leipzig, in 1826, Naumann had translated Keilhau's geological articles into German under the title, *Darstellung der Uebergangs-Formation in Norwegen*, and he had found Keilhau a publisher in Leipzig. The book was furnished with seven copper engravings and would soon be printed. Keilhau hoped to receive an honorarium for this right away. This was the money with which he had planned to finance his further travels. On the whole it now seemed that it was Keilhau who determined Abel's and the others' travel plans. During a good three weeks in Berlin, Keilhau had not only managed to convince Abel to accompany him back to Freiberg, but also to convince the others to accompany him on studies in what he called some of Europe's most beautiful mountain scenery: in Bohemia, the Tyrol and Switzerland, and to Predazzo and the Fassa Valley in northern Italy, where it was important for Keilhau to study geological formations according to his ideas on the origins of granite. This itinerary included stops in beautiful cities like Prague, Vienna, Trieste, Venice, Basel and Lyon, before their August arrival in Paris. It was particularly important for Keilhau to have Boeck with him, his earlier travel companion during exhausting excursions in the homeland. They had agreed to gather in Dresden around Easter time, that is, at the end of March, and then go on together. This must have sounded seductive to Abel, but he still held fast to his plan to return to Berlin, and then, hopefully, together with Crelle, travel to Göttingen and Paris.

Keilhau and Abel stayed in Leipzig the whole of Thursday. It transpired that the printing of Keilhau's book was late, and an arrangement was made whereby Boeck would collect the honorarium and the free copies when he came through, taking the same route, in three or four weeks' time.

On Friday, after the midday meal, the post coach carried on to Freiberg, and Keilhau and Abel were passengers, together with a journeyman shoemaker. In the middle of the night a passenger whom Keilhau characterized as "a very Obscure Person" boarded the coach, and towards dawn they were also joined by a blind man. Very tired, the foreigners reached the mining city of Freiberg about half-past nine Saturday morning, February 26th, and according to Keilhau were "that same Afternoon together with Jürgensen and drank shamelessly."

The next day Keilhau wrote to Boeck in Berlin both about this and the journey, and the task Boeck had in store for him in Leipzig, and he reported: "Today Abel had his first Conference with Naumann, whom he finds to be a really gifted Fellow." In a postscript to Keilhau's letter Abel wrote: "The suitcase did not arrive but I guess it will be coming. Don't forget my Brooch, and my questionable girl's Head together with the Clothing."

Abel was to stay almost a whole month in Freiberg, and the reason seems primarily to have been good working conditions. He worked continually and corresponded eagerly with Crelle in Berlin. He was proud that he could now write in German. It was probably here in Freiberg that he got ready his great treatise on binomial series, and maybe he even began the great Paris treatise. The problems that were also preoccupying him at this time were about which equations were capable of solution by means of square root. And certainly he imagined that one might find lines of connection between the different complexities of problems he had taken up.

The plan to travel together with Crelle to Göttingen and Paris came to nothing. The extremely busy Crelle sent the message that he had been prevented from accompanying Abel. Abel now faced the choice between travelling west alone or to leave together with his friends for Prague, and then to points south into Italy. Duties toward the granting authorities at home in Norway pulled in one direction, and his desire in another. When, on March 17th, Keilhau wrote to Boeck, Abel had made up his mind. Keilhau addressed this letter to their friend, Naumann in Leipzig, where Boeck and Tank were expected in a few days, and wrote that "we three" in Freiberg wanted to leave for Dresden on Wednesday, March 22nd and stay at "Stadt Berlin" or, if it was full, at "Im Deutschen Hause" on Schleffelsgasse, and he hoped Naumann managed to squeeze money out of the publisher, and that they, Boeck and Tank, could as soon thereafter as possible come to Dresden with the money and the free copies of the book. After the shortest possible stay in Dresden they would, in a united body, travel onward to Prague. "We three" in Freiberg were besides Keilhau and Abel, Comrade Møller, who had just arrived in

Freiberg via the porcelain city of Meissen, where again he had forgotten a *Regenschirm* [umbrella]at the post office.

37.

Through Dresden to Prague

The five of them met in Dresden as arranged. Boeck seems to have managed to bring along the money from the publisher in Leipzig, and great pleasure was found as well in the fact that he had brought a letter with him from Hansteen, the first that any of them had received on the whole trip. At home Hansteen had informed the readers of *Magazinet* of their journey, for the most part with citations drawn from the letters they had sent, and with an overview of the measurements they had made with the magnetic oscillation device. "Our travelling young Scholars" is what they were called in Norway. In the letter Hansteen apologized for not having written before, but the reason was all the meetings, lectures and matters that had sprung up after he had returned home in the autumn from his expedition to Bottenvika, and all the extra work that had fallen to him after the resignation of Professor Rasmussen. Now in addition he had been appointed Dean of the Faculty of Philosophy. But his busy-ness was also due to the fact that he had been eagerly working up the travel data, as well as the fact that he had been in ill health, and that there had been difficulties with *Magazinet*. But in the letter he now commended the "young Scholars" and the travel itinerary about which Keilhau had informed him: Prague, Vienna, Trieste, Venice, Tyrol, Switzerland, Lyon, Paris. The itinerary "is utterly excellent in its magnetic Respect," Hansteen stated. The letter was addressed to Boeck in Am Kupfergraben in Berlin and also included greetings to Abel: "Greet the good Abel in the most amiable way on my behalf and tell him that I will write to him by the next post. He can feel secure about his Future Destiny. I have published One Part of his Letter in Miscellany (*Magazinet*'s short notes and "mixes"). I have urged Miss Kemp to keep on giving him reassurances." Hansteen did not indicate he knew that Abel intended to accompany this "excellent" tour for a further period, but he seems to have decided already that Abel had spent too much time following his friends. In the letter to Boeck, Hansteen wrote another reference to Abel in addition to the friendly greetings: "But – Thor in the Talus! – what a swing through Leipzig and the Rhine Lands is good for, I have no idea. It is only undertaken for the Sake of Feelings, and he wobbles along in his own irregular

Manner like another Butterfly that is more Wing than Body." Quite possibly Hansteen had intended this only in jest, and if Boeck thought Abel "was becoming too sensitive to my jest" he ought not to show these lines to Abel.

The expression "Thor in the Talus" certainly sprang from a legend that had just been published in *Magazinet*, about the god, Thor, who while angry, tossed down the stones from mountains at Urebø near Lake Totak in Telemark, lost his hammer in the rock scree, and as the story was popularly recounted, in order to find it again, he tossed stones to the side, to right and left, such that there remained a sort of road across the talus slope.

The stay of the five "young Scholars" in Dresden ended with anger and division in their ranks, and in this misunderstanding it seems as though Abel stood between the two camps, between Keilhau and Boeck on one hand, and Tank and Møller on the other. The outcome of what had begun as their friendly reunion in Dresden was that one week after Abel, Keilhau and Møller had arrived "in this City of Taste", the five friends parted company, Tank and Møller staying behind, and Keilhau, Boeck and Abel continuing on toward Prague.

Everything had begun very well in that beautiful city on the banks of the Elbe. There was gaiety and celebration, the reunion was heartfelt. They toured the city sights together: the gallery of art with Raphael's Sistine Madonna, and the great collection of old masters, the natural science cabinet, the collection of sculptures of Antiquity, as well as the great collection of sculptural castings. They attended midnight mass before Easter Sunday at the Catholic church, Hofkirche, and had seen the king and entourage and been in agreement that the whole affair was somewhat pompous. They had been to the theatre and seen a German play, but their stay in Berlin had made them critical and blasé. On another evening they heard an Italian opera, and they had looked up Denmark's *chargé d'affaires*, the Norwegian, Herr Irgens-Bergh who gave them tickets to an aristocratic casino in Dresden. They had also been to a party with their countryman, Professor of Painting J. C. Dahl, and the Danish poet, Jens Baggesen.

In many ways it seems as though Abel did not want to acknowledge the conflict between his friends. On March 29th, two days before the team separated in Dresden, Abel wrote a long letter to Hansteen:

Hochzuehrender Herr Professor!
Many thanks, Hrr. Professor for your cordial Greetings to me in Boeck's Letter. I was truly worried that I had seemed to express myself somewhat oddly in my last Letter to you, and perhaps I have indeed done so. On the Whole, I urge You to take me with a pinch of salt in many Respects, especially in relation to Formalities. You have completely set me at ease in

Relation to my Future, and in this way You have done me a great Deed for I was somewhat worried, even perhaps to an excessive degree. I would be happy beyond belief to come Home again and have Occasion to work in Peace. I hope that things will turn out well; I have enough Materials to keep me busy for several years; I suppose I shall obtain more in the course of the Tour, because precisely at this Time there are many Ideas going around in my Head. Pure mathematics in its purest Sense ought most certainly become my Future Course of Studies. I will engage all my Powers to bring more light into the enormous Darkness that is now unquestionably to be found in *Analysen*. It lacks almost all forms of Plan and System such that it is really a great surprise that it can be studied by so many, and now the situation exists wherein it is handled without the least degree of stringency. At the outside there are few Propositions in higher Analysis that are proven with convincing Rigour. Above all, one finds the unfortunate Method of deducing the General from the Special Occurrence, and it is strange to the extreme that by following this sort of Approach, so few of the so-called Paradoxes are to be found. It is really extremely interesting to look into the Foundations of this phenomenon. According to my Thinking, the reason for this state of affairs lies in the fact that Functions, that until now have occupied the full attention of Analysis, have, for the most part, been expressed by *Powers*. As soon as other functions come into Consideration – this however not often happens – things usually do not go very well, and this in turn gives rise to false Conclusions wherein a number of inter-linked, yet incorrect Propositions arise.

Abel had already corrected several of these "Propositions" in works that would later come to be published in *Journalen*. And then he turned back again to Crelle. He could not praise highly enough "his lucky stars" for having brought him to Berlin:

I am however a basically very lucky Person. Indeed, only a few people have really shown interest in me, but these Few are for me truly precious because they have shown me such particularly great Kindness. I do hope that I shall live up to the Expectations they have with Respect to me; for it must be hard indeed to see all one's well-intentioned Efforts go down the Drain.

Abel now communicated to Hansteen the fact that Crelle desired that he live permanently in Berlin. Crelle had tried to convince Abel of all the advantages of this, and he included the proposal that Abel take over editorial responsibility for the *Journal*, which presumably would come to have a sound economy, but naturally I refused this. Although I felt I had to put this in the Form that I might do this, if I were unable to obtain something to live from back home, which also I would like to procure. In the End he said that he would repeat his Offer at any Moment of my choosing. I cannot deny that this did flatter me greatly, but was this not also rather decent?

But Abel had promised Crelle one thing, to come back to Berlin before he finished his travels. Abel was now in a position to say that Crelle would quite certainly find him a publisher for the larger treatises:

and to think, for once with an ample honorarium…is this not marvellous? And do I not have reason to be pleased with my trip to Berlin? I cannot say I have learned from others on this Tour, but then I did not look upon that as the main Purpose of my Travels. Building acquaintances for the Future ought to be the main Purpose. Are you not of the same opinion?

Abel briefly mentioned his stay in Freiberg, and the mathematician Naumann "a very nice Man and we got along well." He then got down to what was most difficult, namely explaining his further itinerary. Abel had presumably heard of Hansteen's remarks about "Thor in the Talus" and about "another Butterfly with more wing than body." And *this* had been Hansteen's reaction when Abel had only journeyed to Freiberg and proposed a tour to the Rhine regions together with Crelle on the way to Paris. When the journey planned with Crelle came to nothing, Abel wanted to accompany the others to Vienna, and he wrote: "From Vienna to Paris I will most likely travel in Company with Møller, and the coming Winter there I will spend in the company of Keilhau. We will throw ourselves rather fiercely into Work. I think it will go well."

Abel had calculated that the whole journey would bring him to Paris two months later than originally planned, but he thought that with hard work, he could make up the time. Also from Professor Dirksen in Berlin he could obtain a letter of introduction to Alexander von Humboldt and others in Paris. As for his personal reasons for remaining a little longer among his friends, Abel wrote: "After all, I am so constituted that I am either completely unable, or at least not without the greatest difficulty, to stay alone. Alone, I become considerably Melancholic and then I am not precisely in the best Mood to get anything done."

These were the feelings that Abel perhaps hoped Hansteen would recognize from his spouse, and thereupon perhaps accept them as a valid explanation. Moreover, Crelle had given him extremely flattering letters of introduction to Vienna's two great mathematicians, Joseph Johann von Littrow and Adam von Burg, about whom Abel reported to Hansteen:

They are really brilliant mathematicians, and apart from this, I shall travel abroad but once in my Life. Can one then blame me for wanting to see a little of the Way of Life and Activities of the South? I am able to work extremely well in the course of my travels. If I am in Vienna, and travelling from there to Paris, the most direct way is through Switzerland. Why should I not see a bit of it? God knows, I do not lack all sense of the Beauties of Nature!

About social life in Dresden he wrote:

We have all been in Attendance at Hr. Irgens-Bergh's. Last evening he was so gallant as to pass on tickets to us to the aristocratic Casino, *ou l'on ne danse qu'en escarpins* [where one dances only in courtly attire], and thus we became Nobles for the Night, Monsieur de Keilhau, Monsieur d'Abel, etc. There we saw the whole of Dresden's *Incroyable avec Elegance*. At mid-day today we were invited to dine at Bergh's, where we met Baggesen. He is extremely weak. They say that the Reason for this is the Bottle. We have also become acquainted with the painter Dahl from Bergen. He is about to journey to Norway, which he will not forsake until 1827.

Abel ended his letter with the hope that Hansteen would write back quickly and had arranged things such that Maschmann in Berlin would send any letters on to Vienna.

I do so look forward to hearing something from You and my second Mother. From time to time I will take the Liberty to write to You, not because I dare flatter myself that You will be have any Interest therein, but because I myself find great Satisfaction in it. From me, You cannot expect any interesting Travel Observations accompanied by Aesthetic Descriptions. That I must leave to my more gifted Travel Companions, *in specie*, Keilhau. I bid you greet Holmboe warmly but also tell him that it is not a pretty state of affairs that he has not written to me. Perhaps I do him wrong. Perhaps his letter is on route. Farewell, Hr. Professor, and may You live as well as You do in my wishes.

Yours faithfully, *N. H. Abel*

Would You be so kind as to give the enclosed letter to my Sister? My sincere and respectful greetings to Professor Rasmussen, and above all, to Mrs. Hansteen and Charité.

The bad feelings and discontent within the group of friends were an established fact when Abel wrote this letter, but he hoped that the personal disagreements could be cleared away and forgotten as quickly as they had arisen. In any case, that was Møller's hope. In a reconciliatory letter to Boeck some weeks later, Møller, who remained in Dresden, wrote that he hoped everything would be forgotten, and that after all, on their last day before travelling, they had "drunk together rather efficiently." But that had not happened. Recollections of what really had happened continued to be bitter and unpleasant, and Keilhau would never be able to become any kind of a friend again. Møller wrote, "Wir sind geschieden auf immer geworden" [We are forever separated]. What had happened?

The days in Dresden had been filled with new experiences, large parties, and passionate discussions. For those interested in culture, Dresden had long been a centre of new modes of thinking. It was here that the talented classicist, Johan J. Winckelmann, in the 1750s, had written his great work on imitations of the ancient Greek works in contemporary painting and sculpture. These attitudes were later advanced by Friedrich Schlegel, the early flag-bearer of romanticism, who had described his meeting with Dresden as one of the most fortunate and important things to have happened to him. The poet Novalis had also been part of the city's cultural life. Heinrich von Kleist had had his richest work period in Dresden, and it was here together with his learned friend, Adam Müller, that von Kleist made his greatest contribution to literary romanticism, the journal *Phoebus*. As well, E. T. A. Hoffmann had lived in Dresden, and Schopenhauer had written one of his most important works while he lived in the city.

The poet, Ludwig Tieck still lived in Dresden. Tieck's hospitable house was a gathering place for all the culturally interested who were impelled toward the greater sensitivities. It was reputed that, for the curious who came to see him, he was in the habit of reading them selections from his own dramas, and then reading from a Holberg piece. The story was also told that Tieck had been challenged to a duel with pistols because he had argued that an affair had occurred between Hamlet and Ophelia.

J. C. Dahl had arrived in Dresden in 1818, inspired by the artistic milieu where the central personality was Caspar David Friedrich, and where Friedrich's innovative style of landscape painting was highly esteemed. Dahl lived in the same house as Friedrich, beginning in 1823, and they expressed their common love of nature by saying it was not *nature* itself that they painted, or would paint, but rather their own *feelings*. Now, at the end of March, 1826, Dahl was getting ready for a lengthy, much longed for journey home; he longed for the Norwegian mountains and the Norwegian landscape. What he knew of Norway was only Bergen, its immediate surroundings, and the im-

Fig. 27. L'École Polytechnique in Paris, a definitive centre of mathematical research and scholarship. Lagrange, Monge, Fourier and Laplace were there from the beginning, in 1794, and the host of outstanding mathematicians and men of science who gradually emerged from the school tended, later on, to develop a feeling of inspired fellowship. They called themselves the former pupils of l'École polytechnique, and through the journal, *Correspondance sur l'École polytechnique*, their careers and works could be followed; but what was more interesting, a number of mathematical treatises were published here. Abel had borrowed and studied the journal at home in Christiania but remarkably enough it seems he did not seek out the school while he was in Paris, as, for example he might have, to hear Cauchy's lectures. (Lithograph. National Library of France, Paris).

pressions he had formed on the boat journey south along the coast on his way to Germany. Many of the pictures of his homeland that he painted in Dresden were actually inspired by C. F. Naumann's sketches from Norway. Dahl left Dresden on April 10th, and following a lengthy stay in Copenhagen, he arrived in Norway in October, 1826. This, his first journey back to Norway would be a meaningful occasion: Dahl sketched and expressed Norway as no one had before; he put Norway on canvas, and his sense of nature, the old art of building wooden houses, and memories of times past, made lasting impressions in his homeland.

Consequently, the five "travelling young scholars" met Dahl in the days before his first departure from Dresden, and Dahl made a strong impression on them. Boeck felt that he had never met another person as open-minded as

a   b

Fig. 28a–c.
a. (top left): Johan Gørbitz (self-portrait, 1838, oil on paper. Bergen Art Gallery, Bergen, Norway). Gørbitz had lived more or less continuously in Paris for 17 years and he now helped Abel find passable lodgings at St. Germain-des-Près, in the house of Monsieur Cotte. Gørbitz made a portrait of Abel, the only one made during his lifetime.
b. (top right):Alexis Bouvard, director of the astronomical observatory in Paris, whom Abel, armed with a letter of introduction from von Littrow in Vienna, sought out shortly after arriving in Paris. Bouvard showed him around this well- equipped observatory. He also introduced Abel to the foremost mathematicians in the city. Roger-Viollet, Paris.

c

c. (Above): J. N. P. Hachette, one of "the great men" Abel met in Paris. Hachette, the son of a bookseller, had been active in the founding of l'École Polytechnique, as assistant to Monge, as professor, and as editor of *Correspondences*. Collections of the Academy of Sciences, Paris. Photo: Jean Loup Charmet.

Fig. 29 a–d. Other persons important to Abel's stay in Paris, from top left, **a.** A. L. Cauchy, who managed to lose Abel's great treatise (Roger-Viollet, Paris); **b.** F-V. Raspail, who later became the very popularly admired biologist and politician, and after whom, since 1913, one of the most famous boulevards in Paris has been named (National Library of France, Paris); **c.** G. P. L. Dirichlet, the Prussian who called Abel his countryman and had found a benefactor in the person of Alexander von Humboldt (Drawing by Wilhelm Hensel, Bildarchiv Preussischer Kulturbesitz, Berlin); and **d.** É. Galois, who in a brilliant manner elaborated further Abel's equation theory work, and who died in a duel at the age of 21 (drawing by his brother, Alfred Galois, 1848, Roger-Viollet, Paris).

Fig. 30. The University of Berlin, 1820 (ten years after it had been established). The principles and organization that were set down as the foundation for this university, not least by Wilhelm von Humboldt, chief executive of Prussia's education authority, were uncommonly clear: there should be no division between teaching and academic research; the university's teaching *should be* research, and professors and students should work together as re- searchers in training, using the form of instruction that worked best, namely, the *seminar*. Von Humboldt wanted to leave higher levels of instruction to specialized colleges outside the university. (Etching by A. F. Schmidt, after a drawing by J. H. Forst, 1820. Bildarchiv Preussischer Kulturbesitz, Berlin.)

Dahl, and "when he speaks on Art he is an Enthusiast of the highest Degree." About this occasion, Abel confined himself to reporting, "He is about to travel to Norway, which he will not leave again until 1827." And it was together with Dahl that they met the old poet, Jens Baggesen, who had maintained that his position in his homeland had been lost to his poet colleague, Ohlenschläger, and he had chosen to go into exile. Although he was weak and ill, he had a sharp tongue, and few had crisscrossed Europe as much as Baggesen had. They all quite certainly knew his much-printed book, *The Labyrinth, or Travels through Germany, Switzerland and France*, written more than thirty years earlier. Baggesen was a captivating man about town. He could rhapsodize about mountains, perhaps quoting his own descriptions or Rousseau's praise for the Swiss Alps, or Ossian poetry from the Scottish Highlands; Montesquieu had long maintained that one's disposition towards Freedom and the Fatherland always matured best in mountain country! And Baggesen said that he would love to die in the mountains, preferably among the wild peaks of Norway.

Also together with Dahl and Baggesen was the helpful Irgens-Bergh, the priest's son from Tønsberg and Ringsaker in Norway, who had become a Danish nobleman, and had the much-coveted position of *chargé d'affairs* at the court in Dresden. Irgens-Bergh praised the Norwegian freedom, that according to his views, was certainly appropriate for Norwegians, due to their political innocence.

The distinctive stamp of things Norwegian, the Norwegian mountains and relations to nature in general seem in all ways to have run together in the social gatherings that the Norwegian students encountered in Dresden. In addition, Henrich Steffens' conceptions of nature were presumably a flashpoint for the sharp contradictions that arose among Abel's friends. Tank had sat before the lectern of that natural philosopher in Breslau; Keilhau had been out with Steffens making painstaking observations in the Norwegian landscape: and who had understood the master best? In all likelihood they had strong disagreements in the reasoning for what was the core issue in a debate that arose in many a tempestuous discussion at the time: which has priority: the outer, observable world of the senses, or the infinite inner reality of the self? Keilhau must have used drastic words to criticize Tank's customary propensity for airy speculations which were not supported by any empirical reality that could be used as the positive basis for actual investigation. Boeck stood on Keilhau's side but was not as adamant, and much less blunt of expression. Keilhau must have gone ahead and openly made a fool of Møller, who sympathized with Tank. In the conciliatory letter Møller sent to Boeck some weeks later, he wrote: "Had Keilhau realized how much I thought of him he would not have treated me as he did." And regarding Tank, Møller wrote that he "speaks no more about Steffens now that he is finally fed up with talking to deaf Ears, otherwise he is sincerely honourable and kind as always."

When Abel wrote to Holmboe two or three weeks later, he mentioned nothing about this disagreement, only that they had been to a social gathering at Bergh's together with Dahl and Baggesen, that they had eaten a great deal and been "purely Danish and Norwegian."

On March 31st Abel, Keilhau and Boeck left Dresden. Tank and Møller remained there. According to the above-mentioned letter they, together with J. C. Dahl and new acquaintances, had had a marvellous time in this city of culture. Their friend Rothe came from Berlin bearing greetings from Maschmann, who was bored beyond belief and longed to return home now that all his friends had left.

When Boeck reported to Hansteen about what had happened, he maintained that "Tank's practical-painterly, metaphysically-aesthetical – religious-natural philosophical ideas had come to such a tough and vicious Point of Impact" that further travel *together* was out of the question, and "Møller did not take kindly to Keilhau's Way of Speaking." Therefore, Boeck stressed to Hansteen, it was very important "for our Magnetier" that they had Abel with them, for in this way they were in a position more often to "observe with both Instruments." Moreover, Boeck continued, "Abel could not travel very well alone; he would have become bored and would have fallen into a Bad Mood."

Abel described no other stretch in such detail as the journey out from Dresden, past monasteries and large fruit orchards, with ruins of fortresses on the hilltops:

As soon as one crosses the Bohemian Frontier everything changes, the countryside, the people, etc. etc. When we were on Erzgebirge it snowed, and when we came down into the Valley below, it was the loveliest Road in the World, and an especially beautiful, extraordinarily fertile Region. As soon as one enters Bohemia, one begins to see images of saints everywhere along the Roadsides. We saw a great Many, and among them Nepomuk [the local Bohemian saint] was much in evidence. But beside these Pillars we also saw a great Number of Beggars, particularly blind ones. They wait along the highway all Day long. The first day we reached Töplitz, which is famous for its thermal Baths. In the summer months a tremendous Number of wealthy people, both ill and healthy, come here. When one journeys on from Töplitz, one comes over Middelgebirge, from which one has an immensely open View out over Bohemia which is almost one endless Plain all the way to Prague, where we had decided to stay a couple of days, but we remained there a whole 8, as Boeck found there no small number of natural history Things that interested him. In the meanwhile I walked around in the City, visited the Theatre, one of the better ones in Germany, etc. There I saw an actor from Munich, Eslair, that one can tell is about to become the most excellent in Germany. I saw him in Schiller's *Tell*, as William Tell, and You should have seen the Acting! I visited a Professor Astronomiæ, David. He was an old Fogey who made noises about being afraid for the Future. From this I concluded that his Knowledge must be extremely thin. Another mathematician, Gerstner, also lives in Prague, who I thought would be a worthy man, but when I heard that he was called a *Veteran*, I grew worried, for this Name is generally applied to whose who once did something, but who now no longer do so. It was also good that I did not go there because, as I later learned, he was almost unable to see or hear. Prague is no ugly City and sits rather handsomely. One Part that is situated rather high up, is called Hradschin and from a Tower standing there one obtains a tremendous View. From there one can see not only Middelgebirge and Erzbebirge, but also Riesengebirge, when it is clear. I was up there but could not see these Things as the Weather was not favourable. Behind Hradschin one finds the Observatorium that Tycho Brahe made use of, the Building now used as a military Installation. Tyche Brahe's tomb is also on view in one of the City's innumerable Churches. Otherwise, it seems that Manners in Prague are awfully raw. Hats on in the Theatre etc. And things are not very pretty in the Eating Quarters. There one finds one bad Habit after another; Women with great tankards of Beer in front of them etc. Beer-drinking is particularly strong in the Austrian States that we have been in until now. The first question one gets at a public Place is: Schaaffens Bier Gnaaden; but we always keep to the Wines that, according to my Tastes, are very good, and not very costly. Two bottles of good Wine cost about one and a half Norwegian Marks. But one can also find Wine that costs 4 Ducats a Bottle.

It might seem strange that in this overview of persons worthy of visiting in Prague, Abel did not mention Bernhard Bolzano, the mathematician much talked about in Berlin. When later Abel arrived in Paris, he expressed his astonishment at Bolzano's well-defined concepts and stringent methods of work.[14] As early as 1816–1817, Bolzano had published from Prague important works of mathematical analysis, and his lectures on continuous functions approached what later would come to be called uniform continuity. But as head of the Faculty of Philosophy at Prague, Bolzano had been accused of heretical views on religion and orthodoxy; he had been forbidden to publish some of his writings and was under constant police surveillance until the matter came to an end in 1825. In recent years during the summer, and also from time to time at other seasons, he stayed at a friend's house in the city of Techobuz, in southern Bohemia. Perhaps Bolzano was there when Abel was in Prague, between April 2nd and April 8th, 1826. This "veteran", Frans Joseph von Gerstner, had been Bolzano's teacher of mathematics; and "the old fogey", Aloys David, was the Professor of Astronomy, and Director of the Observatory in the city.

---

[14] While in Paris, Abel scribbled in his notebook, "Bolzano is a clever Fellow, from what I have studied." See Figure 44 below.

38.

To Vienna

On the way to Vienna, Abel reported:

From Prague we travelled onwards by means of *ein Lohnkutscher* [a hired carriage]that was to take us to Vienna for about 24 Speciedaler which is not so expensive as the distance is almost 40 Norwegian Miles [ca. 450 km.[15]]. The drive was particularly comfortable in our *Glaswagen* [closed carriage]. When we were some Miles from Prague we came right along beside the Elbe, and could at the same Time see Riesengebirge which was bedecked with Snow. We had Warmth of almost 20 Degrees [on the Réaumur temperature scale; or 25° C.], that particularly bothered us with our magnetic Observations which, for the most part, we undertook twice a Day, at Noon and in the Evening. One sees an enormous Number of Cities and Towns on the way from Prague to Vienna, the like of which would, at home [in Norway] be considered not so insignificant, but here they are considered Unremarkable. At the inns we came to we were treated in the Common fashion: well and cheaply, though one found things far from the state of Cleanliness of North Germany. The country South of Prague is not as flat as the northern Portion, but extremely fertile. On the other hand, when one comes into Mähren, one has a terribly sterile view that ressembles many Regions in Norway. But all at Once it changes when one crosses into Austria. This is the most fertile country I have (seen), and so well-cultivated. There is not a Scrap of land without either Grain Fields or Vineyards. We often had Occasion to look around about us, and as far as the Eye could see there was nothing other than fields of grain. One encounters pasture land extremely rarely. After a Drive of 4 Days, we came to Vienna a little before Sundown. Even from afar we could see the Spire of St. Stephen's Tower, which is tremendously high. For some Time thereafter we saw the Capital pass in Review, and [it] was not long before we passed an Arm of the Danube. After having undergone a light "Visitation" [by the

---

[15] At that time, 1 Norwegian mile = 11,295m.

authorities] we drove through Leopoldstadt, over Ferdinand's Bridge and into the Capital. We dismounted at one of the most expensive Hotels in the whole City, called *Zum wilden Mann*.

It was on the evening of April 14, 1826, that Abel, Boeck and Keilhau saw Vienna's proud landmark: the vast spire of St. Stephen's Cathedral, glistening in the sunlight across the fertile landscape. For the next six weeks they would stay in the city, which was much bigger and more stately than any other they had yet visited. Møller and Tank came from Dresden at the end of April and joined the others again, and, on May 25th, the five of them were ready to continue the journey southward.

Abel wrote *one* letter while he was in Vienna, to Holmboe, about the journey (quoted above) and his first impressions of the city, and this was written two days after their arrival. (He added a bit more to the letter on April 20th.) Abel was eager to reply to the letter he had just received from Holmboe via Maschmann in Berlin, who had forwarded it to *poste restante* in Vienna. Information about the stay of Abel and his friends in Vienna is also found in letters by Boeck and Keilhau to Hansteen, and in a letter from Boeck to his fiancée, Elisabeth Collet.

Even more than the stately buildings, the multitudes of people, and the steep prices, what struck them first about Vienna was the ingrown bureaucracy. In order to get their papers in order for a residency permit, Keilhau went to the Foreign Bureau on their first morning in the city, and he found over one hundred people already gathered there. Many were labourers waiting for passports and stamps allowing permission to work, and they were held in check by a pair of gendarmes. Keilhau pressed himself into the crowd but was stopped by an officious young man with the message that he and his friends had to prove that they possessed the necessary means for such a stay in Vienna before they could be awarded a residency permit. This attestation had to be taken to the banker, where diverse forms had to be filled out, and that procedure was followed by an exhaustive examination at the central police station: name, fatherland, birthplace, religion, latest address, travel permit from where, etc. etc. Some days later, Boeck received a sharp order to appear again at the police station because his papers were not in order, and that was the first time he resorted to flourishing the certificate of his academic degree, in Latin with large red seals, in order to be accorded the status of an upstanding person whom Vienna could officially tolerate within its city limits. Keilhau was outraged; he thought the bureaucracy had been something relegated to the past, but: "Such a useless, illiberal Organization, long since obsolete in less oblique and obscurantist Governments, is here very much part of the Agenda of the Day." A Viennese assured them that if it was difficult for them to come *in*, there were still vastly more obstacles facing any Austrian who wanted to try to get *out* of the country. In his letter to Holmboe on their second day in Vienna, Abel wrote:

> People here are very attentive to Foreigners, and one is thus examined to such a degree that it feels strange to us. Keilhau was asked who his Father was, and had to tell his whole Life's Story. In order to get Permission to stay in Vienna one must prove that one has enough to live on...One uses a dreadful Heap of money here. One must live in Hotels and that is a frightful drain on us, and besides, here in Vienna there is every Opportunity in Hell to live lavishly. The Viennese is especially sensual and makes particularly much of Food and Drink. The Man who was visited upon us ate, and as everybody is always eating, very shortly everything had been consumed. The other day I made Note of the fact that a Man began his Meal by first unbuttoning his Trousers. He packed a frightening amount into himself. Vienna is a grand Place, and Berlin is quite eclipsed when it comes to being what they call *städtisch*. An endless Swarm of People on the Streets, that in Part are narrow with high Houses (5, 6 and 7 Storeys), and one with endless Shops, Churches, etc. St. Stephens, the highest Spire is the tallest I have ever seen. I am living in the Vicinity. Inside it is most magnificent, and it oozes with Catholicism from every Pore. The Holy Mass really is highly ceremonial, and it is not at all surprising that Many are swept up in it.

After traipsing around the city for a couple of days and reading the advertisements attached to the gateposts of houses, about *Wohnung, Zimmere, Stuben, Betten* and so on, and getting to know numerous caretakers, housewives and masters of the house, the three "young Scholars" eventually found two passable rooms in a building in which more than two hundred families lived. The landlord was probably some sort of teacher at court, but even three weeks after they had moved in, they still had not met him, but on the other hand, they were more and more frequently in contact with the wife and daughter. Keilhau and Boeck felt that the wife and daughter fussed over them to a degree that was in excess of what was necessary in relation to washing, morning and evening ablutions and other household matters. According to Boeck and Keilhau, the daughter of the house had soon revealed her love for Abel, without Abel seeming to have done anything about it. But Keilhau, who rigorously scrutinized all the bills, had come to the conclusion that the wife was taking advantage of them, and he had very strained relations with her. In the letter to Holmboe Abel wrote:

You may well imagine that it is wrong for me to while away so much Time with travelling; but I do not think it can be called whiling the Time away. One learns many strange Things on such a Tour, things of which I can find more Use than if I were purely studying Mathematics. Besides, I always need to have what you know as Deaf Periods, in order to pull myself together again with renewed Energy.

The first person that the three students looked up in Vienna was Møller's paternal uncle, Jacob Nicolai Møller. They were very warmly received at his house, and later during their stay in Vienna, they had other contact with him and his twenty year-old son, Johannes. Through the person of Jacob Nicolai Møller, a great and well-known gallery of personalities was woven together: J. N. Møller had been born in Gjerpen near Skien, not a great distance from Gjerstad where Abel had grown up. He had been born on the same farm where their travel companion Møller had also been born. Møller the uncle had later been educated by his father, the district doctor, confirmed, and at the age of fourteen, sent to Copenhagen. That occurred in 1791, at precisely the time that Abel's father, Søren Georg, had been at the University in Copenhagen. J. N. Møller became a distinguished student in 1793, and two years later took the juridical public service examination, but, because he had always exhibited such interest in philosophy and the natural sciences, he had been awarded a large and generous foreign travel grant to study geology and mineralogy. Thus he travelled to Berlin with his childhood and university friend, Jacob Aall. Here he met another comrade from his student days, Henrich Steffens. Møller assisted Steffens economically and they travelled together in 1799 to the geological seminar at Freiberg, where the distinguished geologist, A. G. Werner, the great spokesman for "Neptunism", had assembled top scholars from across Europe. Møller and Steffens stayed together for two years in Freiberg. They attended lectures together; they went on joint outings to the five-to-six hundred year-old mines in the district; they went to Dresden, to Jena and Bohemia, and they both became followers of the Schellingist natural philosophy that had gripped Germany, particularly after Schelling's book, *The System of Transcendental Idealism* came out in 1800. Both Møller and Steffens wrote for Schelling's *Journal of Speculative Physics*. J. N. Møller also became acquainted with a number of the leaders of the Romantic Movement: Novalis, the Schlegel brothers, and Tieck. Møller became suddenly gravely ill during a visit to Hamburg, and while bedridden, he was cared for by a priest's daughter, Charlotte Alberti, and a sweeping transformation happened *in* and *to* him. Earlier, in discussions with Tieck and others, he had been a fanatical defender of Protestantism, but now, evidently everything that he knew about the Catholic Church revealed itself in a completely new light. And when he was well again, he married Charlotte Alberti, the sister of Tieck's wife, and converted to Catholicism. That was in 1804. Then he lived for six years in Westphalia, before, at the initiative of Novalis' brother, becoming a teacher at the Nürnberg gymnasium in 1812. During the years that Møller was there, this gymnasium was headed by Hegel, and Møller still counted Hegel a friend, even though he had become more and more critical of Hegel's philosophy.

After a number of years in Nürnberg, at a salary too meagre for a man with a family to support, he accepted a position as steward for a rich Bohemian, Count Kinsky. Møller was assigned to teach the count's young son and heir, and in this connection he spent a few years in Prague, where his own son, Johannes, underwent the same instruction as the young count. In addition, while he was living in Prague, Møller became acquainted with Bernhard Bolzano, and had attended his lectures, where, in a common-sensical way, Bolzano had among other things wanted to demonstrate that Catholicism was the best of all religions. Møller was also deeply engaged in the question of determining the limits of the capacity of common sense, and finding out where evident truths took over. Møller took one of his pet sayings from Bolzano: "the finite is relative; in the final analysis, all that is relative must have its limit in an absolute. God is the absolute, consequently everything finite has a relationship to God." Perhaps it was J. N. Møller's discussion of the religious philosopher, Bolzano, that brought Abel almost within sight of Bolzano's mathematical writings? If Abel was not much concerned with the various philosophical systems, the many interpretations of reality and human life and history, was it not in any case inspiring to associate with so many philosophers who constantly expressed their ability to solve the universe's enigmas and expose the clear significance of human life? There was a great diversity in his own work that also could lead to a vision of unity.

Moreover, there were many coincidences between Møller's life and the fresh experiences of the Norwegian students: when Count Kinsky was ready for university in Prague, Møller's work came to an end and he moved to Dresden. Here he moved in the same circles at his brother-in-law, Tieck, and he quite probably met his countryman, J. C. Dahl there. Friedrich Schlegel had given him hope of better prospects in Vienna and he and his son moved there in 1822. He had lost his wife during their time in Prague. And here in Vienna, which the Catholic world regarded as a German-speaking Rome, Møller had first become a teacher at a Catholic institute, and then he managed a boarding establishment for young noblemen. He thrived both in the work and in the social life of the city, but now bureaucratic ordinances had made the situation such that he was planning to leave the city. His son Johannes had become a student in Vienna in 1825; he wanted to study history but as a foreigner he had no hope of finding a position in Austria. For the Catholic Møller who had been living abroad for so many years that he could only express himself in his native tongue with difficulty, it was not especially realistic to consider returning to his homeland. In the course of 1826 both Møllers, father and son, moved from Vienna to Bonn.

Apart from this, they went to the races and saw Napoleon's son who, they felt, ressembled his father. They reported that he was simply clad and very cheerful of disposition. They climbed the St. Stephen's Cathedral spire, one of the highest in the world. Otherwise, it was so cramped up there that Keilhau was almost hit on the back by the clapper of the large bell. But the great entertainment here, as in the other cities, was the theatre. When Abel wrote to Holmboe after having been in Vienna for two days, he described it as follows:

Vienna has 5 theatres and of course I must visit them all, 2 in *der stadt* and 3 in the suburbs. One of these, in Leopoldstadt stands out, and is where one has the Opportunity to study the Viennese, for here almost the only Plays they perform have to do with Vienna, particularly about its immigrants hailing from the lower Classes. This theatre is heavily attended. It is called Beym Casperl [chez Casperl] because the perpetually comic Role there was a rather stupid Sign Painter of the name Casperl. But nowadays one more often sees Umbrella-Maker Staberl, the personification of the state of craft work in Vienna. An infinitely comical Person. I have been there once and enjoyed myself immensely. The audience is turbulent to an awe-inspiring degree, clapping and shouting all over the place. Apart from this, most of the Plays that are performed there are an endless general weaving of the most preposterous Things and exaggerated Caricatures. But the players are excellent. I have been to another Theatre as well, namely, K. K. Hoftheater, which is huge. A very Good Play was performed there, such that one afterwards felt oneself to have been deeply moved by the experience. Such good theatre therefore yields rather remarkable Enjoyment. It is something that perhaps we in Norway lack and quite probably will never have. It is also good to go there for the Sake of the Language. There one hears the purest and the best. I can also say that what German I have, I learned at the Theatres in Berlin, otherwise I have had extremely little Opportunity to hear it. Now it is going rather well and I can deal with almost anything without coming to Grief. I am more afraid about my French, though it will probably go well once I arrive there where one has to use it.

During one of their first days in Vienna they also visited the official who represented their native land in this metropolis, "the Swedish Envoy", C. J. D.

U. Croneberg, whereupon they were invited to dine at his house some days later. Keilhau found Croneberg to be an "extremely good-natured and perhaps a more candid Man than the Swedes usually are." Abel merely referred to him as *Baron* Croneberg, but he was not a baron, although his wife was a very rich baronized banker's daughter, and at the dinner it was Keilhau who escorted her to the table. On April 20th, Abel made the following comments:

> Yesterday we ate Dinner at the home of the Swedish Ambassador, Baron Croneberg. There were not many guests, only the three of us and Three Ladies, the Baron and his Wife. It was a very Viennese Meal. We ate monstrous amounts, particularly the Baron's mother-in-law, a native-born Viennese. I did not neglect my plate either, although I do not exactly have the Desire to repeat the experience.

Neither did Abel neglect his stipend obligations; he was eager to meet the city's available mathematicians. He had letters of introduction from Crelle to von Burg and von Littrow, but beyond this, he was no longer reluctant to look people up: "I am no longer afraid to go and see People. At the beginning it felt a little odd. But when one is travelling, one manages to summon up the necessary degree of Impertinence."

In any case, Abel made good contact with von Littrow, who seemed to remind him a great deal of Hansteen, but von Littrow could also be prickly and "When he is opposed to something he erupts immediately into Fire and Flames," Abel commented. Von Littrow spent the greater part of the day at the university's traditional observatory, and often also met Abel up the "Stern-warte" clocktower at seven in the morning and they chatted about "a considerable number of Things." Abel also visited von Littrow's home. Von Littrow was a highly respected man in Vienna; he was among the most learned of his day. Before becoming astronomical observatory director in Vienna, he had been observatory director in Crakow, Kazan and Ofens [the old German name for the city of Buda in Budapest]. While at the University of Kazan, von Littrow had been among the first to recognize the outstanding mathematical abilities of the young N. I. Lobatjevski. Lobatjevski was twenty-three years old when he became professor in 1816, and he would come to revolutionize the ideas of many mathematicians by demonstrating that traditional Euclidean geometry was in no way an absolute and necessary truth. Von Littrow was a member of several great scientific associations in St. Petersburg, Prague, London, Crakow, Kazan, and he was certainly a central figure in Vienna's scientific milieu. Under the leadership of von Littrow, Vienna's observatory, the Stern-warte, entered a new era, not only in terms of science, but also in the work of making astronomy accessible to the interested man in the street. This was, as

Abel expressed it, both "strict" and "popular" astronomy, and he admired both aspects of von Littrow's works. Von Littrow also had responsibility for publishing *Annalen der Sternwarte* in Vienna, and he asked Abel to write a piece in it,

> and I would naturally enough avail myself of this Opportunity to produce a few Notes. Littrow has a very lovely Wife with whom, although she is only 34 Years, he has 12 Children. She is Polish and takes a lot of snuff; in her younger Days she also smoked like a Turk (as her Husband put it). In retribution she told many pretty stories about him.

Abel wrote this to Hansteen three days after leaving Vienna and he added that von Littrow had also given him a letter of introduction to the observatory director in Paris, the great Alexis Bouvard.

Which other mathematicians Abel met in Vienna is unclear. In April-May of that year the first issue of a new journal, *Zeitscrift für Physik und Mathematik*, came out. It was edited by two professors at the university, Baumgartner and von Ettinghausen; and it is quite certain that Abel met them. In the second issue of this journal, published shortly thereafter, there was a short anonymous article that, according to the writer, was provoked by Abel's penetrating treatise in *Crelle's Journal*. This anonymous article goes through an attempt that was older than Abel's, to deal with the quintic equations: that of the Italian doctor and mathematician, Paolo Ruffini, whose works dated from 1799 and 1813. Ruffini's proof was extremely unclear, and most mathematicians had reacted negatively to the work, except for Cauchy in Paris and subsequently these two professors in Vienna, who felt that such an important theme ought to be illuminated from many sides. Taking everything into consideration, Abel was not aware of Ruffini's work earlier, which despite its insightful considerations suffered from the weakness that it accepted, without proof, that the general n^{th}-degree equation can be treated in such a form that root expressions that occur are polynomials in the roots of the equation. Abel wrote a little article about this some months later while he was in Paris, in *Bulletin de Ferrusac*.

But this mathematician, von Burg, to whom Abel had a letter of introduction from Crelle, and about whom von Littrow also spoke well, he must have met at some point in Vienna. With the passing of years, Adam von Burg became a local great man, and in public opinion, a prominent mathematician. When the mathematician, Leo Königsberger, famous for having taught the young mathematical genius, Sonja Kowalewski, arrived in Vienna as professor in 1877, he looked up the old von Burg, who then told him that once in the middle of the 20s, he had been visited by a young man, Abel, who wanted his cooperation with a journal of which he, Abel, had offered to be the main editor. "As a Person, Abel seemed to me to be obviously intelligent," von Burg was to remember, "but how could I have entrusted *my* work to such a neophyte?" the old man asked, and was astonished by what had happened to the young man, and wondered what had happened to the journal. Von Burg, the local great man, seems to have been unaware that much of mathematical science of the next fifty years had been built upon Abel's discoveries, and that *Journal für die reine und angewandte Mathematik* during this period had been the organ most central to mathematical production of the period. Von Burg's own works had in the meanwhile had long ago slipped into oblivion.

Nonetheless, Abel's six weeks in Vienna in April and May, 1826, had gone well. The friendship with von Littrow and family was heartfelt. That Tank and Møller had eventually returned to the fellowship of friends was gratifying, and a good sign for the coming period, and what Abel experienced besides this, in his hostel, or at the imperial and royal K. K. Hoftheater, or in the city, the sources do not tell us. The first weeks in Vienna had been cold, it was unusual to be so cold at that time of year. On the 20th of April, Abel wrote:

> my Fingers are so cold that it is all I can do to write. I pray that you therefore excuse the fact that my Handwriting is somewhat illegible. I would not like to live permanently in Vienna. The wind blows all Day long, and it is quite terrible. There is a Dust such that one can scarely bear it. Consequently a frightening Number here die from Infections of the Lungs.

But over a month later when he was leaving Vienna late in the night, he described his feelings thus: "It is a strange Feeling to be leaving such a great and diverse city forever, especially a Place where one has enjoyed oneself so much. I was in a bad Humour and spent a fatal night, almost sleepless (as You can well imagine)."

Abel's dwelling on the departure, the reluctance at having to leave people and places he had become familiar with, an unwillingness to end things, to place the final period at the end of the sentence, is almost an echo of the finale of a symphony by the city's great composer, Beethoven. It is not possible that Abel could have avoided experiencing a little of Vienna's overwhelming musical life. Hayden was a well-known name from the musical life back home; perhaps he had heard stories about the virtuoso, Mozart; and perhaps he caught a glimpse of the old, deaf Beethoven, if not at the theatre, perhaps on a stroll through the city streets or the city's beautiful surroundings, in Heiligenstadt for example. Otherwise, that spring of 1826 was the last in Beethoven's life. He died at the end of March the following year, during a furious Viennese snowstorm.

39

From Vienna to Graz and Trieste

The plan had been to go on foot from Vienna to Graz, but as there was still snow in the passes, it was only the experienced mountain men, Keilhau and Boeck, who held fast to this plan. On May 18, 1826, Abel, Møller and old J. N. Møller drove together with Keilhau and Boeck in what was called a *Gesellschaftwagen* [private company coach] to Baden, a spa outside Vienna. There they waved goodbye to the two rugged mountain wanderers and made their way back to the city. Then a week later Abel also took his final leave of Vienna, about ten o'clock in the evening with what was called the *Eilpost* [the postal express], accompanied by Møller and Tank, and with plans to catch up again with Keilhau and Boeck in Graz.

When once again we got a glimpse of the Day the first thing I did was observe my Travel Companions, and after a period of Study I found that in addition to the three of us, the Wagon contained 2 Germans and 3 Italians, all of them foul Fellows, particularly a "Kaufmann von Venedig" [Merchant of Venice] who kicked up a frightful Row. About halfway between Vienna and Grätz one crosses a pass through the Alps called Semmering, which is the Frontier between the Austrian provinces of Österreich and Steiermark. Here the Surroundings begin to be very pretty; I thought I was in Norway, so much does Steiermark have a Resemblance to it. The road follows a tremendously narrow Valley, through which the River Mürz flows, that contributes greatly to one's experience of the Scene, where every Moment a new lovely Situation appears; but while the Land is lovely, the People are not. Everywhere, one encounters People covered with Sores. It is horribly repulsive. They say that it comes from the Water. This Illness is found more rarely south of Grätz. We reached Grätz, rather tired, about 8 in the evening, and following the intake of a Meal, we went to Bed.

The following day they wandered around the area, and among other activities, they climbed a nearby mountain in order to admire the city and above all, the beauty of the landscape; then, when they had returned to the inn and were sitting down to dinner, Keilhau and Boeck suddenly strode into the room. They had "by accident," according to Abel, found the same inn in the city and were eager to tell all about their experiences from hiking overland. It had been extremely strenuous due to all the snow, but nonetheless they had "enjoyed themselves to the hilt." One of the high points had been when by chance they had come to the pilgrims' church at Maria Zell precisely on the day that the great annual Mass was taking place. Numbers of pilgrims with ornate sculptures of saints under canopies came toward the church in processions complete with songs and prayers. Not all had been of a solemn frame of mind. Among other things, they had observed a choirboy who made small balls out of the pilgrims' burnt-out votive candles, and amused himself by throwing these balls at old women who were promenading up and down after having fulfilled their modicum of prayers.

The five "young scholars" stayed a couple of days in Graz and they wrote letters home. Abel wrote briefly to Hansteen, telling about his warm meetings with the von Littrows, and he concluded:

> Møller and I will reach Paris in about 6 Weeks. By then perhaps I will flatter myself with a Couple of Words from You or Your Wife, although unfortunately until now I have not been so fortunate. It would give me considerable Encouragement, and I cannot say how much I would cherish it. When I reach Paris I will write You a Couple of Lines regarding what I am engaged in there, together with another Matter that I do not yet dare burden You with. *Mes complimen*[t]*s les respectueux à votre épouse et à Charité.*
>
> Yours, Abel

> I pray that you will greet B. Holmboe, and my sister when You see her.

The five Norwegians were eager to continue on to Trieste. They longed to see the ocean and the sea again, but before they left they were also eager to visit the theatre in Graz, since they now "were about to take Leave of the German Theatre," as Abel expressed it. Something that distinguished the German stage was a clear and crisp enunciation of the correct language. As for Abel himself, he felt that he had learned to speak German by attending the theatre. It was not simply the stage illusion that was vital; the theatre should not give an unguarded slice of life. The players should not forget that they were in the theatre, but rather they should be part of a theatrical art in which the spoken expressions between characters on stage were directed toward the audience. In preference to naturalism, the theatrical arts should rather be illustrative, beautiful and harmonious, and lead to something indicating a truth or ideal of humanity. There were critical voices that, conversely, felt the declamatory impact emanating from most of the German stages easily became excessive and bombastic, and a species of musical lecture lacking in scope for individual expression and insight. But for Abel it was both instructive and entertaining, and as a foreigner, he was more in the mood to be an observer than a participant, even in Vienna's most popular theatre.

On May 29th they left Graz in a hired *Lohnkutscher* that was to get them to Trieste in four and a half days and cost 44 florins, or approximately 21 speciedaler. When, a couple of weeks later, Abel described the journey from Graz to Trieste in a letter to Holmboe, he wrote:

> We had a very pleasant journey. The Region was extremely beautiful. Bountiful Fields, great Rivers (Mur, Sau, Drau), and high Peaks made a good Effect. On the other hand, the nights we spent on route were not very pleasant, for the Inns were bad. Everything was so filthy, although not at all Costly. The most remarkable Thing we saw on the way was the famous underground Passageway at Adelsberg, some miles from Trieste. This Cave extends several Miles into the mountain and one needs 24 hours to reach the point to which at present one can go. It extends even further but one is hindered by the depth and breadth of the Cave. We were able only to venture a little way in. A river runs through the same Mountain, running 3 Miles [34 km.] out of sight of Human Eyes. We saw both where it flowed in and where it came out. On the 5th day we entered Italy, and we had Dinner in the first Italian town, Sessana. The people were German but the Food was Italian Macaroni, etc. We had to make do with the Fasting Food as the day was Friday. Red wine here is called Black Wine, and it is quite deserving of its Name. It looked tremendous but was not so good. We were now not very far from the Ocean, and soon we reached a Place where suddenly we got a view of it. We climbed out of the Carriage in order to Better Enjoy the Scene. Suddenly before us lay the Adriatic. Far below us, Trieste. In the harbour, many Ships, to one Side we could see the Coast of Istria, and over on the other, the Venetian shoreline. It is undeniable that the view was particularly lovely, though far from being comparable to that which one has from Ekeberg[16] [in Christiania]. But upon us, who had been so long away from the Sea, it naturally made a pleasant Impression, which arose especially from the fact that it was the Adriatic that lay before us.

They rolled majestically down the hills and were soon in the lively city of Trieste, where they drove to the Plaza Contrado del Corsa, and installed

---

[16] See Figure 36 below.

themselves in the "Albergo all'Aquila nero." They eagerly put things in order so as to celebrate the event by bathing in the sea. But everything took longer than they thought. The language problems were of consequence. They themselves understood very little Italian, and very few Triestians understood French or German. After a considerable amount of fuss they were eventually settled in, but they encountered new difficulties out in the city when they tried to borrow a small boat in order to get out on the sea. In the end, when they had met an English mariner, it was Møller who had explained their errand. "Møller speaks English," Abel commented and seems to have thereby removed himself from the task of trying to make himself understood in that language. Tank, who had even lived in England, could not have been with them on this bathing expedition.

They stayed five days in Trieste, marvelling in the beauty of the city, and were surprised by the lively commercial life, and the many nationalities intermingled with the almost forty thousand inhabitants. Abel confined himself to stating that all the European nationalities were represented, as well as Turks, Greeks, Arabs and Egyptians. Keilhau had a go at determining the characteristics of the people who lived in this city which Napoleon had taken, Austria had gotten back, and now the Italians were fighting about. Keilhau felt that they were surrounded by the hard-working and thoughtful Germans, the lively, dramatic Italians, the serious Turks, the clever Greeks, the brown and guttural Arabs, the daring, devil-may-care Dalmatians, and the broad-cheeked, witty Croats. Four Norwegian ships also lay in the harbour, carrying fish; two from Bergen and two from Trondheim. Abel and his friends went on board three of them, and they invited a man from Bergen, and a skipper, Lars Jacobsen Larssøn, from Arendal, to dinner "during which we lived high, thanks to the Classic Vine."

According to Abel, the Arendalian Larssøn had been in service to the Danish Consulate in Genoa, but had now moved permanently to Italy, probably as well due to his homeland's union with Sweden. Since then, he had been a broker and Danish Vice-Consul in Venice. Through the Bergen skipper, Abel managed to send a number of books to a Lektor Christian Fredrik Gottfred Bohr in Bergen. Enclosed with the books was a note in which Abel requested that the books be forwarded to B. Holmboe in Christiania, and in subsequent letters to Holmboe, Abel asked that he receive and hold on to the books until he himself came home. He had probably been given these books by von Littrow in Vienna, and in this way he could get them home at the minimum outlay. C. F. G. Bohr had been a disciple together with Abel's father at the Helsingør Latin School in the late 1780s. He had arrived in Bergen in 1797 as organist and music teacher, and since then, on his own he had studied mathematics, physics and astronomy and had become a teacher in these subjects at the Bergen Secondary Comprehensive School that he founded, together with Lyder Sagen, in 1806. Bohr was a distinguished natural science researcher and also zealous about hiking and climbing in the mountains, and was also one of the first to have actually described a journey over Jostedal Glacier. That had happened *after* Boeck and Keilhau's account of their findings about Jotenheimen and their attempted ascent of Lodalskåpa. Lektor Bohr had consequently written in *Morgenbladet* in September of the same year, 1820, claiming that he, together with a Lieutenant Daa and two guides from Jostedalen, had climbed one of the peaks of the Lodal Glacier and had calculated its height to be 6,548 Norwegian feet, or 2, 055 French metres above sea level. This had led to a public dispute between Keilhau and Bohr on which peak had really been climbed. The conclusion was that both Lieutenant Daa and the local experts declared that the real Lodalskåpa had not been climbed by Bohr. Abel did not know Bohr personally, and had quite certainly never met him. But the knowledge that Bohr was in Bergen was probably enough for Abel, with a healthy dash of courage, to consider that a lektor in his own country would receive the books he sent. And Bohr, who had now been made senior teacher at the Bergen Cathedral School, quite certainly knew of his former classmate's famous son.

The five of them could not leave Trieste without having been to the theatre, which made greater use of equipment than they were used to: "In Trieste, I saw my first Italian *Comoedie*, 'Il dottore e la morte'. Several of the most amazing painted Scenes and the Title in metre-high Letters. On the 7th of June at 12 o'clock at Night we left Triest[e] by Steamship for the journey to Venice."

40.

Onward to Venice;
via Bologna to Bolzano

8 o'clock, we glimpsed the Towers, and not long after we lay at Anchor in this strange City. I seemed not to have wanted to believe I was in Venice. We lay in the vicinity of the famous St. Mark's Square. We were rapidly encircled by an endless Number of Gondolas, all of the Gondoliers wanted to earn a fare. These Gondolas are long and narrow and have something like a little House in the Middle where one sits and is driven with an Oar. We took one of these only after having negotiated, for if one does not do so, one is cheated: everyone in Venice operates by Extortion. There are to be found so infinitely Many Drifters, Beggars and Scoundrels that one must be on Guard at all times. We stayed at *Hotel Europa*, which was recommended to us as one of the best, but proved poor enough and was rather dear.

The five young Norwegians hired a "penny guide" later to show them around the city: by gondola through the canals, and on foot through the narrow streets. Eighty thousand inhabitants lived in "this strange City", despite the fact that half of the city lay in ruins, as it was said. Abel was not ignorant of the history of Venice and was familiar with quite few descriptions of the city by others. He summed up his impressions in the following way in a letter-report to Holmboe:

There is a melancholy Spectacle pervading Venice. Everywhere one sees Signs of former Glory and contemporary Wretchedness. Magnificent Palaces quite destroyed, and many almost fallen to pieces. Awful, ugly Houses where perhaps one or two Rooms are inhabited. Ruins of decaying and collapsed Buildings, that once were beautiful. Everything gives testimony of decay.

Abel felt that the most remarkable place was St. Mark's Square, the square that Napoleon had called Europe's prettiest salon.

This is an inordinately beautiful Square ringed by the most beautiful Buildings with endless Colonnades. This Square is especially lively from Late Afternoon until long into the Night. There people frequent the numerous coffeehouses that are located below the Colonnades. On one side I counted 25, of which several were extremely large.

Abel and his friends climbed to the top of St. Mark's Campanile and had a monumental view of the city. All around them they could see water, water and land stretching away into the distance. They went into St. Mark's Cathedral and were impressed by all the marble, all the mosaics, and the ornamentation everywhere. They probably also entered the Doge's Palace. The only thing that Abel commented upon from the Doge's Palace was the prison. The leaded chambers in Venice had been the most infamous in Christendom, as the leadwork trapped the horrible heat generated by the broiling hot sun in summer and the cold, bone-chilling dampness in winter. Abel knew about this prison from the biography of Giacomo Casanova. Casanova's memoirs had begun to be published in Germany in the 1820s, and the escape from the leaded rooms during the night of November 1, 1755, the same day that a violent earthquake had taken the lives of 30,000 in Lisbon, is one of the high points in this exciting and eventful story. It is quite likely that Abel had read some of Casanova's memoirs; possibly he knew of the story of Casanova's attempt to become a mathematician. He had thrown himself into the doubling of cubes, obviously without any persuasive results, although he had managed, so to speak, to emerge from the leaded chambres of Venice, all in one piece, but, Abel declared, "they have now been destroyed by the French" and thereupon he mentioned Napoleon's sacking of this city built upon a hundred islands in the Adriatic Sea.

Ever since the tumult that had surrounded the questions of the solutions to tertiary and quadratic equations, this city with the proud lion in its arsenal had been in decline and decay. But nonetheless it had been here in Venice that the mathematician, Niccolo Tartaglia, had run around with his morose and angry visage, disfigured by French soldiers three hundred years before Napoleon. Later it would be said that Tartaglia in his middle age was a megalomaniac to demand the right to keep these discoveries to himself. On the other side of the controversy, Cardano had perhaps been too much involved to believe that what he wrote and thought deserved to be made public. Cardano, the most famous man of science of the day, could pick and choose between the best offers in Europe; wherever he journeyed he was met with honour and esteem. He published 131 books, indeed was reputed to have burned 170 manuscripts, but nevertheless left behind 111 book manuscripts that were published in Lyon in 1663 in ten fat volumes. Cardano had treated allergies;

he was the first both to give a clinical description of typhus and a treatment for syphilis; he had described the processes of infection and been strongly opposed to continual blood-letting. His popular science books on medical practice, on the art of healing, on different cures, on poison, on water and air, on daily fare, diet and personal hygiene, on preventive medicine, on plague, urine, teeth, music and on gambling with cards and dice (a pastime that the great Cardano took up almost daily) were sold in great numbers once the art of printing had become common. His books were also illegally copied and translated from Latin to different national languages, and they sold very profitably.

But Cardano himself was to become a man who suffered great misfortune: his wife, Lucia, died in 1546, only thirty-one years old, and he was left to raise two sons and a daughter on his own. The children ranged in age from three to twelve at the time. Cardano had then immediately written a book on child-rearing, but his own children did not turn out well. The eldest son trained to be a doctor, and married of girl of low reputation. When they had had three children, she publicly stated that her husband had not been father to all her children. Young Cardano poisoned his wife with arsenic that had been administered in a piece of cake, was arrested, tried and conducted to the scaffolding amid derision and torture, despite the fact that his father had engaged the best of lawyers on his behalf. Cardano seems never to have recovered from the loss of this, his favourite son. Certainly he had accepted a professorship in Bologna and had been made honorary citizen of the city, but all he wrote were elegies to the dead. The daughter was said to have made a common marriage and brought her father neither joy nor sorrow. The youngest son lived at home for a long time, but by means of friends and constant escapades, he wasted and gambled away large sums of money and gradually became much more criminal of mind and practice. He ravaged and stole, and was in and out of prison in various cities, and even broke into his father's house.

At the age of seventy, Cardano himself was put in prison for the crime of heresy. In many eyes at this time, the art of medicine was only a hair's breadth away from witchcraft and magic, and Cardano had been convicted of dealing in hocus-pocus. Given the prevailing social climate in those post-Reformation times, Pope Pius V was eager to set examples by way of warning to those whose world views strayed from those of the Church, and it was not difficult to find doubtful passages in Cardano's many writings. Among other things, Cardano had been thinking of working out the horoscope of Jesus, and he had praised Emperor Nero who had been such a despot against the Christians. But Cardano had never criticized the Church; he regarded himself as an ordinary, traditional believer, and young students testified on his behalf. Following the proceedings, Cardano was given a mild sentence, only a few months in prison, without torture, and then was allowed to serve this sentence at home. He was under house arrest and he had been forbidden to publish and to teach. His last happiness was to be honoured by the medical circles in Rome a short time before his death at the age of seventy-six. As for Ferrari, Cardano's pupil, who had even solved the quadratic equation, he had died at the age of only forty-three. It was rumoured that he had been poisoned by his sister and her husband in the hope that they might gain a little something by his death.

"Everything gives testimony of decay," Abel now wrote in June, 1826. He saw around him the remains of an empire. Perhaps after having seen the gondolas rocking with their violin necks, and after having counted the colonnades of the venerable St. Mark's Square in the night streets, he got a whiff of the secretive *Carbonari*, the national liberals who had been active in uprisings in Naples and now planned actions in Bologna.

On June 10th , Abel, Boeck, Keilhau, Møller and Tank left Venice. Clambering out of two gondolas, they came ashore at Fussina, where they had ordered what was called a *veturin*, a large, roomy carriage that took them onwards to Padua.

> We drove along beside the River Brenta and through the most fertile and most cultivated Land that one could imagine. The whole Country was as flat as a Lake, and perhaps a Sea. Every Scrap of Land was covered with Vineyards and Orchards. After a Journey of about 6 Hours we were in Padua, which is a terribly ugly City, the ugliest I have yet seen.

"Fair Padua, nursery of Arts," as Shakespeare described it, seemed to have seen better days, in the long and distant past. Abel and his friends visited some churches, and the house that the Roman historian Livy had occupied, and "which is still maintained to this day," Abel wrote in a footnote, thinking of his old course of studies in Latin. Apart from this, they spent the night at a guesthouse that seems to have been both bad and expensive. The history of Padua's ancient university, where both Dante and Galileo Galilei had studied, seemed not to have raised comment among them. The next day they passed through "quite enchanting Country," arriving at Vincenza about dinner time. They had their mid-day meal and reached Verona by evening. Abel found himself impressed by the Roman monuments: the massive town gate and the bridge "built by Vitruvius over the River Estch [Adige]," and towering over everything, the immense "Ampitheatre from Ancient Times where the voices of 2,300 People roar."

On June 12th they left Verona, followed the Adige, and continued into the Tyrol. On June 14th they were in Bolzano. The plan was now to spend a couple of days touring around in the Val di Fiemme and Val di Fassa, and the mountains in the vicinity. It was here that Keilhau intended to find support for his geological theories. For Abel it was high time that he took the straight road to Paris.

41

In the Dolomites, over to Innsbruck, and on to Mount Rigi

Abel followed along through the Dolomites. And some of the "young scholars" became so boisterous as a result of the magnificent scenery that at the inn "Goldenes Schiff" in Predazzo, they registered themselves as what, one day, they hoped to become. They wrote: "Keilhau, professore della mineralogia. Boeck, professore dell'arte veterinaria. Abel, professore della geometria." The landlord at "Goldenes Schiff" merely noted five "studenti da Norvegia." In geological circles, the town of Predazzo was well-known, and Abel and company stayed there three days. It was important for Keilhau to be able to study the Agardo Mine in addition to the mineral formations.

The five young men travelled back to Bolzano and it still seemed very difficult to have to part company. For Abel the road led inexorably to Paris, and Tank would throw in with Abel and go to Paris as well, while Møller now decided to stay with Keilhau and Boeck, who in the coming weeks would continue their alpine studies and go on foot a goodly distance north. Schaffhausen, in northernmost Switzerland, was a town that they all, in one way or another, would pass through. For Abel, Schaffhausen lay on the direct route to Paris, and in Schaffhausen, Keilhau and Boeck planned to part company. Keilhau would then go on to Paris, while Boeck would carry on to Munich and his own field of study, veterinary medicine. So as to lighten the packs of the sturdy mountain men, Abel therefore took with him part of his friends' luggage in a bag that he would leave for them in Schaffhausen. On June 27th he waved farewell to Boeck, Keilhau and Møller, and together with Tank continued his journey to Innsbruck and westward to Bodensee/Lake Constance, "but when we came to Bodensack [Breganz] it seemed to us more interesting to go via Zurich to get to see a little more of Switzerland…", wrote Abel in a short letter he sent from Zurich on July 5th, addressed to Keilhau, and addressed, "Poste restante, Schaffhausen." Abel explained that now, instead of Schaffhausen, their luggage would be deposited in Zurich. In haste before they left Zurich, and in indistinct handwriting, he described their journey so far: "We had a terrible journey coming over Innsbruck and the Spinning Wheel has had its Way. We found the Tyrolian Mädchen [maidens] very cheerful, having had one in the Carriage for a Stretch. The route from Bodensack to Zurich is also very nice but filled with a Mass of Muck-Cookers." It is most probable that the "Spinning Wheel" was, in the five friends' slang terms, Hansteen's magnetic oscillating instrument, that they still felt compelled to use; and "Muck-Cookers" was a popular expression for heaps of farmyard dung.

The route that Abel and Tank had now planned to follow went through the city of Zug, along Zugersee and toward Mount Rigi, where there was said to be a lookout over a gorgeously beautiful view of the alpine world, and over Vierwaldstädtersee/Lake of Lucerne, and then they would continue on to Lucerne. Rigi had become one of the most famous alpine regions of Switzerland, and more and more visitors came to visit Mount Rigi, "the Queen of Mountains," to be impressed by the breathtaking landscape. Interest in Rigi had begun with pilgrims to health-restoring natural springs in the region. It later became evident that the region was also of interest geologically and botanically. However, what attracted Abel and Tank most was the glorious view in every direction under the sun. Already, by 1816, a hotel had been built at Kulm, the summit of Rigi, and later, in 1871, Europe's first funicular cable-car system would be built here in order to convey people more easily to the top.

Abel and Tank were *tourists* on Mount Rigi. Otherwise, "tourism" was a description and an activity that Abel in his lifetime encountered only minimally. The word "tourist" probably first entered a Norwegian dictionary about 1840, and the first wave of tourists, predominantly English, came to Norway in the 1850s. The aim of Abel's foreign travel was after all a study tour, following traditional prototypes, but in Italy and Switzerland he was part of the tourism, the identity of which increasingly began to coalesce into a more permanent form. Following on from the old Christian pilgrimages and the European ruling class cultural tours, the prosperous bourgeoisie of the late 1700s began to follow their precursors. Now, as earlier, the aim was to visit Italy and the Mediterranean region, the cradle of civilization, the cultural landscape of the Biblical and Classical worlds. The sense of the might of nature, fundamental to Romanticism and embodied in the Swiss and French Alps, had led to an almost passionate religious fervour for wild and untouched landscapes. Nature's own temples, cathedrals and altars now, in their own right, became goals of travel as legitimate as Rome or Jerusalem.

When four or five weeks later Abel wrote a letter to Hansteen, he recapitulated the tour, writing, "I was at Innsbruck, beside Bodensee, and I caught a Glimpse of Switzerland: do You blame me for this? It cost me two Days and a modest few Shillings more than a straight route through would have…Indeed I do not regret this little Detour."

Abel and Tank had stood upon the roof of Europe, on Mount Rigi, and their respective jubilations at the landscape and nature were in harmony, happy and unsuspecting that their respective courses in life would come to seem like night and day. A good twenty years later Tank would, as an enormously rich man in America, build what he thought would become a new Jerusalem. He called the place "Ephraim, the very bountiful" in America which, in the eyes of those who would critique the old Europe, stood for "the land of freedom" and "the Promised Land" across the seas. Just a few days after their experience on Mount Rigi, Abel and Tank parted company, and probably never met again.

From Mount Rigi the journey took them to Lucerne and quickly on to Basel. There Tank received the news that a catastrophe had occurred in his hometown: large parts of Fredrikshald had gone up in flames. Young Tank was afraid that much of Tank & Co.'s holdings had also been lost, and he set out immediately for the journey home. Two weeks later he reached his family at Fredrikshald and reported that most of the family's holdings and valuables had miraculously been saved, but that most of the city's inhabitants had become homeless, and there were now about a thousand people living at the family estate at Røed.

After Tank's departure from Basel, Abel now went on alone, and in three days and four nights "continuously so to Paris."

42.

Paris, July 10th to December 29th, 1826

"I have finally arrived at the Focus of all my mathematical Desires, in Paris. I have already been here since July 10th." Abel wrote this in his first letter from Paris, dated August 12th, to Professor Hansteen. Abel was still afraid that Hansteen would think it had been stupid of him to have taken the long way via Trieste, Venice and Switzerland, and he wrote: "It makes me feel exceedingly wicked to have done something which has not received your Approval, and now that it is done, I have to seek refuge in your Goodness." He did not supply any excuses with his apology, apart from saying his strong desire to see a little of the world had been decisive. "And does one travel only in order to study the strictly Scientific? After this Tour I am working with ever more Vigour."

After a month in Paris Abel had found out that it was vitally important to master the French language in order to become known and accepted in Paris. And in order to learn to speak better French, Abel had taken lodgings with a family, Monsieur et Madame Cotte. For 120 frances a month he also received clean clothes and two meals a day. Abel had suddenly recalled that Johan Gørbitz lived in Paris, and subsequently it had been Gørbitz who had helped him find this room in rue Ste. Marguerite, No. 41, in Faubourg St. Germain, right across from St. Germain-des-Prés. Abel himself characterized his room as "extremely plain," and he was far from satisfied by the two meals a day. Perhaps the reason that he ended up at the Cottes' was because Monsieur Cotte wanted to give the impression of being interested in mathematics. In any case, Abel referred to him as "a half-schooled Pirate in Mathematics," and it was Monsieur Cotte who first took Abel to Legendre. After having lived with the Cotte family for a couple of months, Abel described the relations as:

The Man is a bit of a Mathematician, but extremely stupid, and the Wife a Madcap of 35 years or more. At the table One always speaks with Equivocation, about *les secrets du menage* etc. The other Day it went so far that a Lady said that the Goose that had been [on] the Table in the Morning had turned into an Étronc [turd]. Discussions about Chamberpots, etc. are among the most respectable of topics. I always drink Coffee from *mon petit pot du nuit*. Otherwise I eat well, but only Twice a Day. Before noon *un dejeuner à la fourchette* and in the afternoon, at 5 1/2 hours, a long *diner*. 1 to 1 1/2 Bottles of Wine every Day.

It was the summer holiday period when Abel arrived in Paris in July. People were away and the libraries were closed. Nevertheless, Abel and Gørbitz met several times both in the city and at home *chez* Gørbitz in rue de l'Université. Gørbitz had lived more or less continuously in Paris since 1809, and he worked in the atelier of the reputable Jean-Antoine Gros. J.-A. Gros had won honour and renown by painting a series of portraits of Napoleon, the most famous being "Napoleon in Battle at Arcole." After Napoleon's time, Gros continued to be accorded high respect in Paris, but a certain critique had caused him to abandon historical paintings and to move more toward the demands of pure classical form. Gørbitz created both interiors, landscapes and portraits in oil and pastel, and his specialties were finely-wrought miniatures that were exhibited in le Salon de Paris. In the course of the autumn of 1826, Gørbitz painted a portrait of his countryman, the only one of Abel to be found. Gørbitz longed for home and its Norwegian motifs, just as J. C. Dahl did in Dresden, and he spoke about taking a trip back to Norway the following summer.

Some weeks after Abel's arrival his travel companion, Møller, reached Paris. He was tired of travelling and wanted to go home: "and I cannot say other than that I too am beginning to long passionately for home, particularly as it becomes evident that Paris is not the most pleasant Place to Stay, when it is so difficult to make proper Acquaintance with People. It is certainly not like Germany..." Abel wrote to Hansteen after having been in the city a month. But still, after three months of travel without many opportunities to work, Abel must now have been eager to write down the thoughts that had come to him en route, and to work out his mathematical ideas: "Write all Day long and in the middle make only a little Tour au Jardin du Luxembourg or au Palais royal. You have to believe that I have still not been to la Comedie. Talma[17] has been near Death, but now is out of Danger."

In the course of his first month in Paris, despite summer, the holidays and his own work, Abel had visited the observatory director, Alexis Bouvard, the elderly mathematician, Legendre, and the journal publisher, Baron Ferrusac.

Abel went to Bouvard with a letter of introduction from von Littrow in Vienna. Bouvard received him graciously, showed him around the well-ap-

---

[17] François-Joseph Talma (1763–1826) was the leading tragedian at La Comédie Française. A friend of Napoleon, he had come to France's foremost theatre in 1787 and was known for playing Shakespeare and the French classics with a new fidelity to historical detail.

pointed observatory, and said that he would introduce Abel to the foremost mathematicians in the city when Abel came around to the Institute. The Institute was l'Institut de France, formed when l'Académie des Sciences and three other academies were amalgamated in 1793. Director Bouvard was certainly an important figure in the natural sciences in Paris. Later he was to be particularly well-known for having been the first to suggest that the irregularities in the orbit of Uranus could be due to the fact that there was another planet to be found, still further out. Neptune was first discovered in 1864.

But Abel hesitated about meeting Bouvard at the Institute. He would very much like, as he put it, to be able "to speak a few Crumbs of French first." The most important thing perhaps was to make ready his treatise on transcendent functions. He hoped and believed that this would be so good that the Academy in Paris would publish it, and if they would not undertake to do so, he himself would try to get it printed and send it to Gergonne's journal, *Annales de mathematiques pures et appliquées* in Montpellier.

Together with his host, Monsieur Cotte, Abel then made his first visit to the now venerable, 74 year-old Legendre. But when they reached Adrien-Marie Legendre's front door, Legendre was then about to go out, so they exchanged only a few words in the doorway. Abel learned that there was a *soirée* at Legendre's once a week and that he was welcome to attend.

Nor was Baron Ferrusac at home when Abel called on him, but Abel must have met others in the circles around Ferrusac's journal, and he got to know that here as well there was a weekly soirée, and that whenever he liked he could come to Ferrusac's library to read the journals and the newly published books. Abel intended to buy the mathematical part of Ferrusac's journal, *Bulletin universel des sciences et de l'industrie*. He also bought a number of mathematical books, especially treatises that one could not obtain "without being on the Spot," as he put it. This was literature that he felt ought to be available in Christiania, and he calculated that Holmboe would be in agreement and would pay.

Otherwise, Abel had observed the mathematician Poisson on the street, and he commented, "he appeared to me to be much taken with himself. But then he probably was not." Abel, who had now been in Paris a month, was looking forward to making a number of new acquaintances: "Now that I have got my French Tongue a bit in Motion. The Frenchmen seem to me very difficult to understand." In the letter to Hansteen Abel reported that he had almost finished writing several papers and that he planned to get some published in *Annales de Gergonne*, some in *Crelle's Journal*, some in von Littrow's *Annalen der Wiener-Sternwarte*, and the complex and time-consuming work on the great integral question he had reserved for the foremost of scientific publications, the Paris Institute's *Mémoires des savants étrangers.* By means of this dazzling list Abel hoped to impress Hansteen that he had worked as intensely as he had promised. But the treatises never did come to be published according to this plan.

Keilhau arrived in Paris on August 20th. He too now boarded at the Cotte household at St. Germain-des-Prés. Keilhau had with him a letter from Hansteen, that he had received at Schaffhausen. The letter contained the good news that Keilhau's appointment to the university in Christiania was well in process, including him being sent, in Paris, a 100 speciedaler advance. But for Abel there was no uplifting news, only: "Now, in Conclusion, My Good Abel, I must thank You for Your cordial Letters. I am precariously overloaded with many Businesses and rarely able to work. You can impute from this the Reason why I have let You wait so long without Reply. Rest assured that I and my Family completely acknowledge all that is good and amiable in You, despite the fact that we may see some small Frailties."

The friends, Abel and Keilhau, had a happy time together in Paris. The intense friendship that, to the end, make Abel bequeath his fiancée to Keilhau, was probably initiated during those seven weeks they lived together in Paris. Keilhau journeyed home on October 16th after a successful and eventful stay. First he got a definite notice that he had been appointed lektor in "the mining sciences…with the Duty to undertake scientific Journeys in the Fatherland's less understood Regions as long as this is considered useful and necessary," as the royal appointment of August 11th expressed it. A short time later news came that Keilhau had won two hundred silver riksdaler, being the prize that the Danish Academy of Sciences had offered, and which had been awarded for a geological paper Keilhau had earlier submitted upon announcement of the Danish Academy's competition. The two friends shared their happiness at the fact that Keilhau was now the first of the "young scholars" to obtain a permanent position. And with bright future prospects it looked as though Keilhau could with good conscience combine his work and his studies. Keilhau and Abel probably went to the Jardin Luxembourg to take magnetic readings, but more frequently they went to the theatre, and they were in agreement that what they saw at Le Théatre Français surpassed what they had seen in Germany. Otherwise the two friends played billiards, *daily,* according to Keilhau, who also reported that "Madame Cotte grumbles about the Amount of her Wine that we drink." This was written in a letter that Keilhau wrote on behalf of both of them, to Boeck in Munich, and Keilhau admitted here that he had never loafed around as much as he now did in Paris.

Playing billiards was an activity they both knew well, particularly from gatherings at the Student Society in Christiania. Some of the first inventory that the students had managed to obtain there had been billiards and accessories: billiard rules, six cues, five small and five large balls, a billiard table, and a billiard cloth made of sackcloth From the Student Society inventory list it also appears that they had three chess boards painted red, with drawers and all the attendant pieces. One would think that for Abel, who loved to calculate plays in card games, it would be tempting to play chess, the game of utmost calculation. There was a milieu in Paris in which people managed to support themselves as professional billiard-, whist-, and chess-players, and in chess the French had long been the absolute leaders. In the years before the revolution, a man called Philidor had revolutionized chess-playing by elevating the apparently weak *pawns* to the very "soul" of chess, and with his new theories and analyses, he had harvested great triumphs and had defeated everyone with his practical playing all over the continent and in England. In his book *L'Analyse des Échecs*, Philidor had shown that an attack by "officer" pieces could not succeed unless it was backed up by the pawns. Philidor himself fled to England during the revolution, and died in London in 1795. Napoleon, who also had been a zealous chess-player in the cafés of Paris had elaborated Philidor's pawn strategy when, on the battlefield in order to arouse the soldiers, he had reminded them that each and every soldier carried the marshall's swagger stick in his rucksack! Since then, players like Deschapelles and Labourdonnais kept French chess at the highest international level, and it was said that whenever one might visit Café de la Régence in Paris, one would always find players hunched over small marble tables.

But that autumn it was a *Norwegian* who won great sums of money for those who dared to wager. This was the fantastic runner who called himself Mensen Ernst, who was being hotly discussed and made famous in the Norwegian and foreign newspapers. He had won great sums of money in road- and cross-country foot races in many European countries. He had covered the stretch from London to Portsmouth in ten hours, London to Liverpool in thirty-two hours; in South Jylland, Denmark, he had raced a man on horseback, and earlier that year in Copenhagen the king had rewarded him with sixteen hundred riksdaler for performing in Frederiksberg Allé, and on the eight-sided palace square where he ran thirteen circuits (nine hundred paces each) of the square in twenty-one minutes. Mensen Ernst had run in Germany, Italy and Spain, and now, with great takings in his pockets from running in Marseille and Toulon, he came to Paris, on September 24, 1826. Five days later he had obtained permission to reveal his skills before a huge public at Champ de Mars, and in the course of that one day the great Norwegian runner made fifteen hundred francs, or almost 300 silver speciedaler. Mensen Ernst lived like a lord in Paris on rue de Montmartre, for two or three weeks, before he ran off in his characteristic feather-decorated cap, toward new and further wagers. It was quite fitting that one of his mottos was never to travel the same road twice. If he arrived in a city from the south, he left in one of the other directions.

The man who helped Mensen Ernst obtain his permit to run, and assisted with the payment of fees in Paris, was the Swedish-Norwegian ambassador, Count Gustav Carl Fredrik Löwenhielm. Löwenhielm was a prominent figure in Paris; he interested himself in the arts and sciences, loved *festivitas* and a good wager. Six years later, in 1832, it was Löwenhielm who backed Mensen Ernst's most famous run. With a firm sense of publicity, he advised Mensen Ernst to run the distance from Paris to Moscow, and after closely scrutinizing maps and making calculations, he wagered that Mensen Ernest would cover the distance between Paris and Moscow in fifteen days! The whole of Paris was astir with excitement. Paris to Moscow! It was equivalent to running in Napoleon's footsteps, and doing what he and thousands of Frenchmen had been unable to do in 1812: capture Moscow! In Paris alone, one hundred thousand francs were bet for and against Mensen Ernst. The runner himself would get four thousand, which he considered was the bare minimum. And after great exertions and a danger-filled run, where at times he had not only robbers, but also wolves and wardens at his heels – fourteen days, five hours and fifty minutes after Mensen Ernst had left Place Vendôme in Paris – he was received by the Kremlin's Commandant.

How much Count Löwenhielm wagered and won on the running sensation, Mensen Ernest, during that summer of 1832, has not been revealed. In any case it is certain that the Swedish-Norwegian ambassador also worked on *Abel's* fame and posthumous reputation during these years. In that same period, Löwenhielm had done his utmost to track down the Abel treatise which had disappeared in one office or another in Paris, and which the French men of science so ardently wanted, and which was to be included in Abel's collected works.

But six years before this, in September-October, 1826, Abel too, had been a guest of Count Löwenhielm in Paris. In a letter to Holmboe on October 24th, he wrote:

> The other Day I partook of a diplomatic Dinner (*un Diner*) at His Ex. Count Löwenhielm, where I, in Company with Keilhau, became a little bit Whipped, but a fine wound indeed. He is married to a French Lady. He said that every Year on December 24th he drinks all his Countrymen under the Table.

To become "a little bit Whipped" was, in other words, to become a little drunk. If Abel had any further contact with Count Löwenhielm that autumn, nothing was reported about it, but Abel *was* in Paris on December 24th that year.

Perhaps the Swedish-Norwegian ambassador in Paris also talked about his planned visit to Norway the coming summer, when he would consequently pass through Abel's home region when he visited Risør and Arendal. But neither social life, spectacles nor billiards were at the top of Abel's list of entertainment activities in Paris. Now, as before, *that* honour went to the theatre. And on the theatre front there was an epoch-making event unfolding in Paris that fall. Already, shortly after his arrival in Paris, Abel had made note of the newspaper accounts about the illness of the famous actor, François Talma, but rumour had it that he was getting better. Gradually it became clear that Talma was dying. It was said that eleven doctors were treating him, and that few of them expected him to recover. The least movement occasioned violent retching. But the doctors seem to have kept him free of pain. Many words were expended in discussions about both his lawful wife, who was now back under his roof, and his many known lovers, particularly a Madame Bazire, with whom he had lived for fourteen years and who had borne him three children. For over thirty years, among those interested in the theatre, Talma had been the most illustrious name all over Europe. In 1822 the Norwegian magazine *Hermoder*, under the byline "On Talma's Earlier Artistic Life", wrote that Talma's superb acting talent had occasioned a theatrical battle and a political uproar in Paris in September, 1790. The traditional players at le Théatre Français, for the most part imperious aristocrats who had lost their privileges and influence in the revolution, would not perform together with the young, handsome Talma, who harvested public exhalation for his natural way of playing and his striking costumes. Through his whole life in the theatre, Talma was an exponent of stage reform in accord with the ideals of the revolution. He wanted to get rid of unnatural and affected elegance. Talma lived his roles, whether they were of Titus or Hamlet, in a way that no one had done before. He electrified audiences, plunging them into madness, sorrow, happiness and the lust for power. Talma and Napoleon had known one another since the time of their youth, and Talma had always prominent among the players that Napoleon brought from Paris to perform wherever he, for the moment, was located, and where the performances were mounted for officers and well-disposed princes in the area. Napoleon's interest in theatre, which he saw as advertisement for the nation, created a new level of prosperity and lustre in the French theatrical world, but perhaps with more stringent order and regularity than Talma had wanted. Talma died late on October 19th, 1826, and two days later an enormous throng of people assembled to follow the great actor's last journey, from his princely house in rue de la Tour-des-Dames to the Père-Lachaise Cemetery. Talma's heart was removed during an autopsy, and indeed, has been kept from that day to this, in a mahogany reliquary at the Comédie Française. In a letter to Holmboe on October 24th, Abel wrote:

> Talma, the well-known, great Tragedian died some Days ago. On this Occasion the Théatre français was closed for two Evenings, and all the remaining Theatres closed with it. An enormous Mass of People followed his Corpse. The Corpse was brought to the Cemetery, without first being taken to the Church, which otherwise is the Custom, but as an Actor he was denied the communion of the faithful. Ludicrous, but to no Avail. He had allowed his Children, all of whom are illegitimate, to be raised in the Protestant Religion. He had great Weaknesses in living his Life. That is to say, to a high degree he was prone to Gambling, Womenfolk, and a Mania for Building. The actors at the Théatre français allowed him to set up a Monument for the sum of 12,000 fr.

Thus it was that Abel did not get to see the great Talma in Paris, but Talma's female counterpart, the idolized Mlle. Mars, he did manage to see on stage. Together with Talma, she had always been the main star whenever Napoleon invited people to the theatre. But while Talma had interpreted tragic heroes, Mlle. Mars' heart was much closer to comedy. It was said that her power lay in the spiritual and in "charming piquancy". In any case she was set upon a pedestal as an expression of the ideal woman, and on stage she was beautiful and enchanting. Abel and Keilhau seem to have been in accord with this ideal of womanhood. Keilhau characterized Mlle. Mars as "more than human." Abel wrote to Holmboe: "I do not know any greater Pleasure than to see a Piece by Molière in which Mlle. Mars is playing. I am really quite enraptured. She is 40 years old but still plays extremely young Roles."

Keilhau had left Paris for Le Havre in a great rush on October 16th. He had found out that he could obtain passage on a ship bound for Arendal. This departure was sad for Abel, but Keilhau yearned for home. He wanted at once to start planning the scientific expeditions that he now could officially conduct in Norway. The first would be a geological expedition to Finnmark in the north of the country. However, the money he had been promised had not come, and Abel had to lend Keilhau the money he needed for the journey home. The agreement was that Keilhau, the moment he got home to Christiania, would pay back what he owed, to Holmboe, and Holmboe would send it on to Abel. Keilhau also took with him a large red suitcase for Holmboe. Abel had filled this case with most of the books and papers he had purchased and collected. Among the books was also a gift for Hansteen, the newly published fifth and last volume of *Mécanique celeste*. Abel knew that Hansteen had the first four volumes of this great work by Laplace, France's Newton, and in the letter to

Holmboe, Abel commented upon Laplace's work: "Thus, Mécanique is finished. He who has written such a Book can look back with Pleasure at his scientific Life." This work, which Laplace toiled over for twenty-six years, and which among other things, he wanted to show that the solar system was a gigantic *perpetuum mobile*, a stable system with planets eternally revolving in their complicated orbits – this work stands as one of the most powerful scientific monuments of the time.

Abel also sent another gift with Keilhau, to his sister Elisabeth, who continued to be in service in the household of Cabinet Minister Treschow in the Tøyen quarter of Christiania. This was a pair of bracelets, a belt buckle and a ring. As well, Abel wrote a letter, the only surviving written testament between the two siblings:

I think of You so very often, my Dear Sister, and I always wish You Happiness. You are living quite well, are you not, among the excellent People of the House? But how are things with my Mother and Brothers? I have heard nothing about them. It is a long time since I have written to Mother. I know the Letter reached her, but I have heard nothing from her. What is Peder doing, and how? I am very worried about him. When I departed things did not look good with him. God only knows how many Times I have been distressed on his Behalf. I have never been able to give him quite enough, and this makes me feel very bad, even though I have never Willfully done Anything to harm him. Listen, Elisabeth, write to me and tell me all about him and my Mother and the other children. Here in Paris I am living quite pleasantly. Am working diligently and once in a while I take in the Marvels of the City, and take Part in those Enjoyments that please me, but otherwise, I long to come home, and would willingly Travel today if this were possible, but I must still remain here a terribly long Time. I am coming home in the Spring. I should really be staying abroad until next August, but I am aware that there is no particular Utility to be had from being longer abroad. So I shall be coming home by sea, or perhaps by Land, via Berlin, where I very much want to go before I come home, but I do not know [if] the Money will stretch that far. I have not heard much from my Fiancée for a terribly long Time. She is now in Aalborg at her sister's. I am already beginning to worry but I still have hope that she is well. She has indeed written, but the Letter must have been lost. How is Mrs. Hansteen? I am sure she is doing well. Finally, do not forget to give her my most affable greetings. And the same to Professor Hansteen. I wrote to him some Time ago. You probably go there sometimes. Give my sincerest respects to the Cabinet Minister and his Wife. Keilhau has been so good as to be the bearer of a little Gift for You. I wish that it might have been more, but I do not have the Means therefor. It consists of a Pair of Bracelets and a Buckle that can be set upon a belt for around the Waist; together with a little Ring. Please do not sneer at them, and give the occasional thought to your

devoted Brother, *N. H. Abel.*

Abel gave his address and wrote that to send a letter to Paris would not cost more than two shillings, and finally, "Keep well, my dear Sister, and write back the very moment You get this Letter." Whether she replied, we do not know.

Abel felt homeless in the big city, and he waited passionately for letters from Norway "to console and cheer me in my Loneliness", as he wrote to Holmboe, a good week after Keilhau had left. "Although I am in the most boisterous and lively Place on the Continent, I feel as though I were in a Desert. I know almost nobody."

In reality, Gørbitz did not live far away. Besides, Abel knew another countryman in that "boisterous and lively Place," a man called Hans Hågensen Skramstad, who was in Paris for further studies in music. About this Skramstad, who at home in Christiania, at least before his stay in Paris, had been looked upon as a promising virtuoso pianist, Abel wrote: "He lives together with 3 Swedes in a Suburb of the City. He goes around dressed as a peasant from Hedemark in Hose of blue wool and a striped Vest. I have not seen him but have had him described to me. He speaks Swedish." There was also a third Norwegian mentioned in Abel's letter, Brede Müller Grønvold from Holmestrand, a town on the west side of Oslofjord. He was in Paris for an education in trade and commerce. It seems as though Abel had a certain degree of contact with this Grønvold, at least to the extent that he gave Grønvold's name as someone who might be of assistance to Boeck when later he was to go to Paris.

But Abel was disappointed that he himself had now been in Paris three months and met so few of the French men of science. On October 24th he wrote:

Up to the present Moment I have only made the Acquaintance of Legendre, Cauchy and Hachette, together with a couple of lesser Mathematicians, and the quite bright Monsieur Saigey, Editor of the Bulletin des sciences, etc., and Herrn Le-jeune Dirichlet, a Prussian, who the other Day came to me and said he regarded me his Countryman. He is a very sharp Mathematician. Together with Legendre, he has shown the impossibility of solving in whole Numbers the equation $x^5 + y^5 = z^5$ and other neat Things.

Fig. 31. "Blick vom Kreuzberg auf Berlin." Oil painting by Heinrich Hintze, 1829 (Original in Schloss Charlottenburg, Berlin. *Bildarchiv Preussischer Kulturbesitz*.) Abel stayed in Berlin for eight of the twenty months he was abroad between 1825–27, and here he thrived: He was inspired in his mathematical work and found happiness in the social life. "It is however, most strange to be among Foreigners," he wrote. During the last year of his life, Abel wanted very much, together with his fiancée, to establish himself in Berlin, and go around as "Hr. Professor mit seiner Gemahlin" (Herr Professor and his Lady). Definite news of such a position was sent to him two days after he had died.

This equation was part of the same great work, Fermat's great opus, that Abel himself had tried to solve during the summer of 1823 in Copenhagen. Earlier the equation had been demonstrated impossible to solve for exponents 3 and 4.

But Abel's contact with Legendre did not lead to the scholarly and scientific connections that *could have* been fruitful for them both. In the last year of his life Abel would for the first time come to have a brief but inspiring exchange with the man whom he now in Paris considered to be "an inordinately courteous Man…but unfortunately as old as stone." Abel had most certainly joined the gatherings at Legendre's home, but Legendre, who two years earlier

Fig. 32. Rigi-Känzeli/Kaltbad, 1830 (Department of Prints and Drawings, Zentralbibliothek, Zurich). Abel felt very guilty about the fact that he – together with his friend, Nils Otto Tank – took a detour here to Rigi, "the Queen of Mountains", at the beginning of July, 1826. He wrote home apologetically: "My Desire to see a little with my own eyes was great, and besides, does one travel only in order to study that which is strictly scientific?"

had been deprived of his pension because he had not voted for the government's candidate for a position in the Institute, was perhaps not in the best of humours. His great work on elliptical integrals, *Traité des fonctions elliptiques et des intégrales eulériennes*, had just then been published by the Sciences Academy in two volumes, but unfortunately it did not reach the bookshops before Abel left Paris.

Cauchy was the great mathematician of Paris. He lectured on mathematical analysis at l'École Polytechnique; he was a professor at the Sorbonne, and was tremendously productive in many mathematical fields. He made the stringent presentation of evidence the main feature of his work, and it was this feature that made Cauchy, Gauss and Abel representatives of the definitive new history of mathematics. Abel had also been inspired by Cauchy's writings on ways of integration in the plane of complex numbers that are fundamental to a theory of elliptical functions, and Abel especially held Cauchy's book, *Cours d'Analyse,* in high esteem. Cauchy was born in the revolutionary year, 1789,

Fig. 33. Map of Europe, 1829. Abel's travel route between September, 1825, and May, 1827, included: *1* Christiania, *2* Copenhagen, *3* Lübeck, *4* Hamburg/Altona, *5* Berlin, *6* Leipzig, *7* Freiberg, *8* Dresden, *9* Prague, *10* Vienna, *11* Graz, *12* Trieste, *13* Venice, *14* Verona, *15* Bolzano, *16* Innsbruck, *17* Zurich, *18* Mount Rigi, *19* Lucerne, *20* Basel, *21* Paris, *22* Brussels, *23* Liège, *24* Cologne, *25* Kassel, Berlin, Copenhagen and home again to Christiania. [Map Collections, 9, National Library of Norway, Oslo Division]

and due to the Reign of Terror, his strictly religious parents moved out of Paris to Arcueil, and here Cauchy grew up on a meagre diet and a narrow religious education. Throughout his later life he had to be careful of his health, and his strongly religious conduct was remarked upon by many. But Arcueil was also home to both the great Laplace and the chemist, Claude Louis Berthollets, who had kept his head during the Reign of Terror because he knew everything there was to know about gunpowder. These two men of science in Arcueil had kept an eye on this weak boy's unusual mathematical gifts. Later,

Fig. 34. A view of Paris, painted by Seyfert, 1818. (Musée de la Ville de Paris. Musée Carnavalet. Giraudon/NPS.) Abel stayed in Paris from July, 1826, until the end of the year. He met many scholars and scientists, but his great Paris Treatise was laid aside by his colleagues, and he was not content there. He commented in a letter, "The Frenchman is uncommonly reserved with Respect to Foreigners. It is impossible to come into a *more intimate* Acquaintance with him. And I dare not Count on such occurring. Everyone works by himself here, without bothering others. Everyone wants to teach and no one wants to learn."

Lagrange had also been involved in the education of the young Cauchy. At the age of sixteen, he became a pupil at l'École Polytechnique. Here he was teased for his religiosity, and began to study civil engineering, bridges and roads. He was chosen by Napoleon and sent to Cherbourg to secure the harbours and build up a fleet strong enough to invade England. But after Napoleon's fiasco in Moscow, and at Leipzig the following year, the plans for the fleet were put aside. Cauchy returned to Paris and gradually became the city's leading mathematician, and elsewhere in Europe Cauchy was better-known than was Gauss in Göttingen. But Cauchy was known for being self-absorbed, self-righteous and sanctimonious. Abel, who had met Cauchy in the September-October period, probably at one or another soirée, or in the Institute, wrote:

Cauchy is *fou* [mad] and there is Nothing to come out of him, although he is the mathematician for the present Time who knows how Mathematics ought to be treated. His Issues are excellent, but he writes very obscurely. At first I understood almost not one Bit of his Works, but it is going better now. He has now had a Series of Treatises printed under the title, *Exercises des Mathématiques*. I buy them and read them diligently. 9 Numbers have come out from the Beginning of this Year. Cauchy is tremendously Catholic and a bigot. A sorely strange Thing for a Mathematician. Nevertheless, he is the only one now working in pure Mathematics.

These mathematical issues that Abel mentioned were a journal that Cauchy himself had founded that year in order to have a place to publish his many reviews and works in the fields of pure and applied mathematics. *Exercises des Mathématiques* came out until 1830, whereupon it was followed by a new series.

Director Bouvard had subsequently promised to introduce Abel to members of the Institute, and it seems that later Professor Hachette had also undertaken this role. Otherwise Abel knew well geometrician Hachette's contribution as publisher of the series, *Correspondences sur l'École polytechnique*, which he had borrowed from the university library in Christiania. Abel was now introduced to a series of scholars and scientists but regretted the fact that most of them, apart from Cauchy, "occupied themselves exclusively with Magnetism and other physical Matters." Abel had met the physicists Ampère and Poisson, whom he earlier had seen on the street, and he had been introduced to the mathematicians Fourier and Lacroix. All of these men were around sixty years of age, and with the highest French distinctions for their scientific activity. In addition to these encounters, Abel had seen the venerable Laplace several times at the Institute:

He appears quick and small, but He has the same Shortcoming as Halte-fanden [The Demon] accuses Zambullo of, that is to say, la mauvaise habitude de couper la langue de gens. Poisson is a short Man with a Pretty little Stomach. He carries his Body with Dignity. So too, Fourier. Lacroix is awfully scaldic and remarkably Old.

That Abel, in making mention of Laplace, invoked one of the time's most famous novels, *Haltefanden* or *Le diable boiteux* by René Lesage, might indicate that he did not confine his reading to mathematics. Apart from this, Laplace's expression about the "superior intelligence" that must be in existence to account for the order of the world at any point in time, assuming that all contemporary manifestations and movements in all bodies were predetermined, it was this "superior intelligence" that later came to be known as *Laplace's demon*. The Frenchman with whom Abel came to have the most contact in Paris was Jacques Frédéric Saigey. Saigey was only five years older than Abel; he was a man of many interests and editor of the mathematics and physics section of Baron de Ferrusac's journal. It was Saigey who now got Abel to write for *Ferrusac's Bulletin*: references and *précis* of articles from other scientific journals. It was not something that Abel found particularly to his liking, but among other things, he seems to have felt that he had to make *Crelle's Journal* known in France, while simultaneously making use of Ferrusac's well-appointed library. The first thing Abel did for Saigey was to give an account of his own proof of the impossibility of solving the quintic equation, and in this presentation he had included a note to the effect that Ruffini had some years earlier given a similar but less comprehensive proof for the same proposition. And it was at this time, in *Ferrusac's Bulletin,* that Abel mentioned and corrected his infamous article from *Magazin for Naturvidenskab* on the moon's gravitational effect on the pendulum, and in this manner he introduced the Norwegian journal to Paris. Abel seemed to have most enjoyed being in the youthful circle around Saigey and Ferrusac's library. This was the milieu frequented as well by the Prussian Dirichlet, who considered Abel his countryman, and the self-taught man of science, François-Vincent Raspail. Raspail was eight years older than Abel, Dirichlet was three years younger. Raspail had lectured in philosophy and theology in Avignon, but he had been forced to leave due to his heretical ideas. In Paris after the fall of Napoleon, he had defended with great eloquence the ideas of republicanism. He had taken part in secret organizations, and on his own he had studied botany, biology, and medicine, and all the while supporting his wife and children by private tutoring. In 1824 he had attracted notice with a paper on different types of grasses, and the classification of them. He carried out chemical analysis under the microscope, and later, after 1830, he made several studies in organic chemistry, and he was involved in laying the basis for the knowledge that both plants and human beings were composed of *cells*. Raspail could see that diseases issued from these cells, and he had a refined understanding of parasites, both in the body and in society. His annual handbooks on health were very popular, and throughout his life he was a radical politician who fought against injustices in society, and exposed corruption and incompetence in high places. For this, he was thrown into prison and sent into exile several times, but Raspail returned and became even more of a hero and defender of ordinary people. Raspail became world-famous with his combination of work in science and politics; he became a leading cultural personality in France, and it was as a famous representative of the French Chamber of Deputies that Raspail later used the Norwegian, Abel, the friend of his youth, as an example of how the Academy favoured old men of science who earned a great deal of money, while keeping out the young: the young had to suffer for the self-righteousness of the old. This indeed was, in a fashion, Abel's fate there: in his mathematical work Abel found no encouragement in Paris, quite the reverse.

Right from the day Abel arrived in Paris he had worked on what would become his great Paris Treatise. He wrote to Holmboe on October 24th:

> I have worked out a big Paper on a certain Class of transcendent functions in order to present it to the Institute. I shall do that on Monday. I showed it to Cauchy, but he scarcely gave it a Glance. And I dare say with all Modesty, it is good. I am curious to hear the Institute's verdict. And that I shall communicate to You in due Course.

On Monday, October 30th, Abel appeared at the Institute's meeting. Fourier, the academic secretary, read the introduction to the work that was to be submitted for publication, *Mémoire sur une propriété générale d'une classe tres étendue des fonctions transcendantes.* The paper was signed "par N. H. Abel, Norvégien."

For Abel, the rest of his stay in Paris became a wait. He hoped that the work he had delivered would provoke reactions and give him encouragement for the future. What would it not have meant to come home to Norway with his *mémoire* having been acclaimed in Paris? But Abel received no reaction to his work, not as long as he lived.

The French Academy of Sciences, which had become part of the French Institute, appointed the two leading mathematicians, Cauchy and Legendre to a committee that would evaluate Abel's work, and Cauchy had the task of writing and forwarding the committee's recommendation.

Abel's treatise was larger than any he had previously written (in his collected works it fills 67 large pages). The ideas were new and general and appear to have certainly been completely foreign to Cauchy and Legendre. Abel's handwriting was not always easy to decipher; in other words, Legendre did not read the treatise, not at this time, even though Abel, in the introduction, had courteously paid tribute to Legendre's work on elliptical integrals. Abel's manuscript lay around at Cauchy's, and Cauchy, to be sure, having both his head and his desk crammed with his own epoch-making projects, put Abel's to one side. Cauchy scarcely had any order to his own papers, and despite his sovereign position in mathematics, he had strong opponents among the other members of the Institute, and therefore kept his guard up. Some felt that

Cauchy was over-zealous in constantly seeking to secure his position, while others considered he suffered from mathematical diarrhoea.

Meanwhile, Abel sat and waited; he calculated that the Institute's recommendation would be positive, and he was optimistic that the findings would be ready in a couple of weeks. At the bottom of his manuscript he had written his address: rue Ste. Marguerite, no. 41, Faubourg St. Germain.

Hardly any mathematical paper since has garnered so many words of praise and so many superlatives, as has Abel's Paris Treatise, and the addition theorem that he demonstrated here, and that was later called "Abelian", in many ways marked the high point of his mathematical discoveries. Today one finds few mathematicians who will undertake to explain the insights and consequences of this Abelian theorem, but something can be said about it: just as the lost manuscript in Christiania dealt with the integration of differential equations, the Paris Treatise too, dealt with integral calculation. Very early on in the history of integral calculations, it was clear that there were even many *simple* continuous functions that could not be integrated; that is to say, for which it seemed impossible to find anti-derivatives with the help of ordinary elementary functions. The elliptical integrals were the most important example of such a situation. Legendre had shown that with adequate exchange of variables these integrals could be steered into a standard or normal form. Leonard Euler had found that the sum of two elliptical integrals of the same form could be expressed with only one integral, where the limits of the integration could be expressed by the limits of integration of the two first integrals. Abel generalized Euler's addition theorem. Instead of taking as his point of departure an elliptical integral, that is, a polynomial of the third or fourth degree, Abel let the point of departure be so immensely general that the results he worked out threw a colossal light upon the functions and limits of integration, and made deep correlations and opened new perspectives in the world of mathematics.

But the correlations and connections that could perhaps have secured him a better *life* – those he was unable to see. After he had submitted this work on integral calculation, it became more difficult for him to concentrate on major works. And it would become worse. Simultaneously with the integration problems, Abel had also been working on equations. It was not only the insoluble fifth degree equation that he was concerned with. In the letter to Holmboe of October 24th, he explained:

I am now working on Equation Theory, my favourite Subject and it has finally come to the point where I see a Way to solve the following general Problem: Determiner la forme de toutes les équation[s] algébriques qui peuvent être resolues algebriquement [to determine the form of all algebraic equations that are capable of algebraic solution]. I have found an infinite Number of 5th, 6th, 7th etc. Degrees that no one has sniffed out until now. With these I have found the most direct solution to Equations of the 4 first Degrees with clear Insight into precisely how, and in no other way, they allow themselves to be solved.

Abel had posed himself the general question: What constitutes a solution? And in the search for an answer that was equally general, Abel almost reached the results with which the young French mathematician, Évariste Galois would revolutionize equation theory some years later. Abel continued to Holmboe:

> Moreover, I have concerned myself with the imaginary magnitudes with which there is so very much to do: Integral calculation, and in particular the Theory of Infinite Series that stands on such extremely weak Feet. I do not really expect to get out of this before I have ended my foreign travel and come home to Peace and Quiet, if such happens. I regret that I sought a Stipendium for 2 Years. One and a half would have been ample.

Even before he had submitted his Paris treatise on October 30th, Abel was thus in many ways tired of his exile. He considered that he had make himself "acquainted with everything of importance and unimportance that exists in pure Mathematics." He longed for home and he longed for the peace and quiet which would enable him to develop his ideas further. But as soon as he began to think so far into the future, he became anxious with the thought that he had no permanent position to come home to. If only he were "in Keilhau's Shoes with Respect to the Lectureship," he wrote. From his experiences, Abel seems to have felt that all the possibilities in Paris had been exhausted, that there was nothing more for him to do there. But why, for example, did he not go to Cauchy's lectures in order to awaken interest in himself and his work? Why did he not have the "necessary Degree of Impertinence to look people up", as he had employed in Germany and Austria? Abel wrote to Holmboe from Paris:

> Otherwise I do not feel as good about the Frenchman as I do about the German. The Frenchman is uncommonly reserved with Respect to Foreigners. It is impossible to come into a *more intimate* Acquaintance with him. And I dare not Count on such occurring. Everyone works by himself here, without bothering others. Everyone wants to teach and no one wants to learn. The most absolute Egoism holds sway everywhere. The only things that the Frenchman seeks from Foreigners are the Practical. There can be no thinking without the Frenchman. He is the only one who can come up

with anything theoretical. Such is his way of thinking, and hence you can well grasp that [it] is difficult to be recognized, particularly for a Beginner.

Abel also thought that mathematics was regressing in France, and the reason could be found in the covert obstinacy between church and state, and he felt the Jesuits were particularly culpable; he alluded to them as "a sort of Pact with Satan", and in that respect, his views were in accordance with those of the Norwegian Constitution.

But Abel had met the Prussian Dirichlet who included him among his countrymen, and Dirichlet quite certainly told Abel about his own patron and helper, the German natural sciences researcher, Alexander von Humboldt, and his voyages of discovery. Von Humboldt had firmly promised Dirichlet a position at home in Prussia, where his brother, Wilhelm von Humboldt had important positions in Berlin, both in the university and the department. Alexander von Humboldt was on his way to becoming the central figure in the initiation and promotion of scientific cooperation of that period. One word from von Humboldt could also have changed Abel's fate. But Abel never met von Humboldt in Paris, and the reason lay both in Abel's state of perplexity and other, external unfortunate circumstances. While Abel had been in Berlin there had also been talk about von Humboldt who lived in Paris. Abel had even been promised a letter of introduction to von Humboldt by Professor Dirksen. But as chance would have it, Professor Dirksen had not been in the city when Abel left Berlin and he did not at any point receive a letter of introduction, and without an introduction, Abel did not dare approach von Humboldt when he had arrived in Paris. And when Abel finally could have met von Humboldt, through Dirichlet, von Humboldt was away from Paris. He left Paris in September, and on the way to Berlin he stopped to visit Gauss at Göttingen. A nod from Gauss would also have elevated Abel to the eyes of the world. In Berlin, Crelle worked constantly to try to make it economically possible for Abel to take over as editor of *Journalen*; Crelle too was clear about how important it would be for Abel to meet von Humboldt. Crelle wrote to Abel on November 24th that he had now met von Humboldt in Berlin and had warmly mentioned Abel's name. He urged Abel to seek out von Humboldt as soon as he returned to Paris, before Christmas got under way, and if a letter of recommendation was necessary, then he, Crelle, would send one along. The reason that Crelle did not include a letter of introduction with that very letter was probably because he was not closely associated with von Humboldt who had lived the past eighteen years in Paris, where he wrote up the findings of his voyages of discovery on the American continent. It was not until a year later, in 1827, that Crelle and von Humboldt became friends and worked together to build up a college of higher learning in Berlin, a college to which they agreed they should get Abel as mathematics professor.

Abel never did ask for the letter of introduction, and when von Humboldt returned to Paris on December 21, 1826, Abel seems to have lost all faith that anything good might happen to him in "this most boisterous and lively Place on the Continent."

"The Frenchman is uncommonly reserved with Respect to Foreigners," Abel had written, and he himself must have been imbued with this same attitude. One of the young French mathematicians who visited Abel in Paris that autumn, said many years later that one of his life's mistakes was meeting Abel without getting to know him. This was Joseph Liouville. Liouville was only seventeen in 1826, but ten years later he would found the *Journal des mathématiques*, the only learned journal in mathematics that in terms of professional substance could compete with *Crelle's Journal* in Berlin, and one of his great demonstrations of proof was for the *impossibility* of expressing those elliptical functions that Legendre had put into a normal form in the course of his search for the anti-derivatives of elementary functions.

But the acquaintanceship about which there was the most to regret not having taken place that autumn was between Abel and the young Évariste Galois. Galois was fifteen years old in the autumn of 1826; he suffered intensely from a strong religious school regime. They say he read the mathematics books of Lagrange and Legendre the way his classmates read novels. He seriously discovered algebraic analysis, studied equations theory on his own, and thought, exactly as Abel had done earlier, that he had solved the general fifth degree equation. What could Abel not have shown him had they met? Galois would have only six more years to live. He could not care less about school, such that even his mother felt that his character had changed from being serious, open and loving to taciturn and peculiar. The school authorities accused him of putting on affected airs, being eccentric and ambitious. It was said he was bizarre, that there was something furtive in his character, that he teased his fellow pupils, and was in every way a fish out of water. But unfortunately, young Galois had nowhere to turn, it was either the school or the street. He had still not made contact with the city's scholarly and scientific soirées, and Galois and Abel did not meet, either in their parallel miserable lives, nor in their forays into the brilliant vistas of mathematics. Galois' fate was in many ways even more tragic than Abel's. When Galois first tried to enter l'École Polytechnique, the great home of mathematics, he was refused entry due to the fact that he lacked the general subjects. It was said that the entrance examination ended the following year with Galois hitting the examiner in the face with a blackboard cleaner (a sponge), after a discussion in which probably Galois was right. Tradition also has it that Galois refused to answer the elementary questions on logarithms because he found them to be absurdly elementary, and in addition, it was said that he had difficulty expressing himself verbally and at the blackboard because he worked almost exclusively in his head.

Meanwhile Galois had made public his first paper in *Annales de Gergonne*, and he had submitted a paper on algebraic equations to Cauchy, who promised to forward it to the Institute. But Cauchy put the paper to one side, and when Galois came to get the manuscript back, it had disappeared. Galois felt he was misunderstood and persecuted, and things became worse when his father, the mayor of Bourg-la-Reine, right outside Paris, lost his position as a result of a mendacious and malicious flood of gossip and rumour spread by the local priest. In the grip of a persecution complex, Galois the father ended his life by imbibing carbon monoxide, and at the cemetery, as his father was lowered into the earth, the young Galois witnessed the wildest uproar as the population of Bourg-la-Reine took the side of the deceased mayor and threw stones at the priest.

Young Galois wrote a new equations theory paper and gave this to Fourier, who in his role as the secretary of the Academy, promised to read it. This was in January, 1830; in May of the same year, Fourier died, and once again Galois' manuscript disappeared. Then the July Revolution broke out, and Galois, who was a zealous republican, wanted very much to go to the barricades, but did not get to do anything before Louis-Philip was made king. Poisson now encouraged Galois to rewrite the paper that had disappeared, and Galois did so. Four months later Galois received the report that the whole work was almost incomprehensible. A large party was held in May, 1831, on the occasion of the freeing of nineteen republican officers. The previous year's revolution was jeered, but stormy applause greeted the toasts to the revolutions of 1789 and 1793, for Robespierre and the other revolutionary heroes. And then, from the bottom of a table Galois rose to his feet, raised his glass, and toasted: "To Louis-Philip!" At first there was whistling, but then more and more celebrants discovered that Galois, along with his elegant glass had raised a small knife in his hand, whereupon the jubilation and toasting rose still higher, while some, for example, the writer, Alexandre Dumas, left by way of the window so as not to compromise themselves, because Galois' toast was reported and taken as an expression of intent to make an attempt upon the life of the king. Galois was arrested, but set free due to his youth, but following new demonstrations against the king, he was imprisoned at St. Pélague, where among others he met Raspail, who had been sentenced to fifteen months for being unwilling to serve in the national guard, and for contempt of court.

This occurred at the year end between 1831 and 1832; the scandal about the Institute's treatment of Abel had broken out, and the hunt for Abel's Paris

Treatise was underway. Galois expressed his own criticism from prison. He wrote: "But I have to explain how the manuscripts so often go astray in the drawers of the members of the Institute, although in truth it is incomprehensible that such a thing should still go on among the very men who already have Abel's death on their conscience. It is not my aim to compare myself with this famous mathematician, so it must suffice that I merely tell that my paper on algebraic equations, in its final form, was submitted to the Academy of Sciences in February, 1830, while a summary had been sent in in 1829, that one report has been presented, and that it has been impossible for me to get the manuscript back again."

Galois was sentenced to six months in prison, but in the middle of March, 1832, out of fear of a cholera epidemic in the prison, the authorities sent him to a convalescent home, and here he fell in love, certainly for the very first time, with a lady whom, a month later, he characterized as "a miserable tart", but about whom he had nonetheless allowed himself to be challenged to a duel by one of the officers with whom he had been the previous year at the liberation celebration. Galois knew that he would lose the duel. In a letter that he left behind, "To All Republicans" he regretted that he had to die "in a manner other than for my fatherland." He wrote: "Why should I die for so little, something so contemptible!" Galois *was* mortally wounded in the duel, and he died the following day, May 31, 1832. However, on the day, and in the night before the duel he had written his scientific testament: The first was the equation theory that Fourier had not understood; the second was on elliptical functions, where he had elaborated Abel's work, and the third was on those integrals that later would be called *Abelian*. The whole of Galois' production filled fewer pages than Abel's Paris Treatise. Galois had solved the problem that for centuries had plagued mathematicians: Under what conditions can an equation be solved? Abel had made his contribution with his proof concerning the fifth degree equations, and other equations theory works studied by Galois had also been contributory. Galois gave a complete solution to the problem of developing a group theory that would be of fundamental significance for the development of mathematics.

In the obituary that Saigey published in his and Raspail's newly established journal, after the report of Abel's death reached Paris in May, 1829, Saigey had, among other things, the following to say about Abel: once when someone revealed the wounds he had received from some assailants and had urged Abel to be careful with himself and his belongings, Abel was said to have replied with a smile, 'I have nothing to fear from robbers!'

Abel probably spent much time walking the streets of Paris alone during November and December, 1826, and *he too* was tempted by women, coquettes or otherwise, in this big city. Right before the unhappy waiting period had seriously begun, he wrote to Holmboe:

> I also sometimes frequent the Palais-Royal, which Parisians call "un lieu de perdition." One encounters a tremendous Number of "les femmes de bonne volonté. They are absolutely not importunate. The only thing one hears is "Voulez vous monter avec moi mon petit ami, petit méchant." Naturally, as one who is engaged, etc., I do not listen to them, and leave Palais Royal without the slightest temptation. Many of them are extremely pretty.

Whether the temptation increased as the cold of autumn set in, is unknown. (As one finds in his Paris notebooks [see Fig. 43] this may well have been the period when he wrote "Our Father, Who art in Heaven, Give me Bread and Beer. Listen for once.") Be that as it may, his desire to sit quietly at his worktable was not overwhelming. Abel still attended some soirées, and he must have been happy about the papers that were gradually being published at Crelle's in Berlin. The fourth issue of *Crelle's Journal* contained, among other things, his "stringent Proof for the Binominal Formula in all possible Occurrences", and was to be printed in December. But what he was most preoccupied with at this time was undoubtedly his loneliness. The melancholy that he so frequently sank into when he was alone, that he had earlier alluded to in a letter to Mrs. Hansteen, must now have overwhelmed him completely during his last months in Paris.

And he knew he was not in good form. He was coughing a great deal. Perhaps he sought out medical knowledge at the science soirées he attended, and it seems that in one or another place he had been diagnosed as tubercular. But why should he have believed such diagnosis? Probably there was talk of blood-letting as a practical treatment. Blood-letting had received a sort of new lease on life during the Napoleonic period and continued to have its adherents. Raspail and other young men in his circle of acquaintances must have had recourse to other forms of treatment, but they were probably not consulted, because Abel would not have wanted to know that he was seriously ill. He was coughing and had developed a cold. That was all. But the fact that he was poor must have been a definite factor: the little money he had left would not be used for anything in Paris. He wanted to go to Germany, and be back among friends in Berlin. There were a few remnants to be had from Boeck in Munich, and Holmboe had written that he was willing to give a small loan in addition to that which he would be sending on behalf of Keilhau. Abel thought that these small sums would perhaps sustain him until spring, but never until the end of the stipend period, September, 1827. At the beginning of December Abel wrote to Holmboe that indeed he would like the use of a loan when he reached Berlin:

> In a short Time I am thus leaving Paris, where I have no more fish to catch, and am going to Göttingen to see what is obstructing Gauss, and whether or not he is seriously beset by Arrogance. And I now want very much to be in Germany to learn more German, which will be of the greatest Importance to me in the Future. I shall survive the French language at least to the extent that I am able to write *une Mémoire*, something I would very much prefer to do in German.

Despite his doubt that anything would happen there, the reason that Abel still kept to the old plan of visiting Gauss, was probably that the mathematical study he was now engaged in had something to do with a discovery in Gauss' legendary *Disquisitiones Arithmeticae* from 1801. It was said that Dirichlet, Abel's friend in Paris, carried around a copy of this great work, and that he kept it under his pillow in the event that he might awaken in the night with a new understanding of an obscure or difficult point. With this work, Gauss had given numbers theory the status of a discipline in mathematics, side by side with algebra, analysis, and geometry, but so much was expressed in such a short and concise form that it was not always easy to grasp, and it would be precisely Dirichlet who later made *Disquisitiones Arithmeticae* more accessible. Dirichlet would also become Gauss' friend, and ended his days as professor at Göttingen. Gauss had solved the age-old problem of dividing the line of a circle into equally great parts; he had, that is, given a reliable answer to which regular polygons it was possible to construe with the help of compasses and rulers. In relation to this, Gauss had also produced an obscure allusion to it being possible to demonstrate something of the sort for the classical curve called the lemniscate. The lemniscate is a fourth degree curve in the form of a figure-eight, and is much used in analytic geometry. And the letter to Holmboe in December was full of enthusiasm for what Abel could write about, and what he had discovered "about the Division of the Lemnicatic-Arcs" and about Gauss' "Mystery":

> You shall see how pretty this is. I have found that one can divide, with the help of Rule and Compass, the lemniscate into $2^n + 1$ Parts when this Number is a Primary Number. The Division depends upon an Equation whose Degree is $(2^n + 1)^2 - 1$. But I have found its full Solution by means of Square Roots. In the same Vein I have approached the Mystery that has reposed over Gauss' Theory of the Division of the Circle. I see as clearly as Day how he has got it. What I have told you here about the lemniscate is

one of the Fruits I have harvested from my Endeavours with Equations Theory. You will never believe how many lovely Propositions I have found there...

Abel consoled himself with "lovely Propositions". There was no more said about the girls at Palais-Royal, or about beer and bread, and he did not look up von Humboldt after December 21st when the latter returned to Paris, even though Crelle had written: "Er (von Humboldt) kann Ihnen nützlicher sein als vielleicht irgend ein Anderer."[He (von Humboldt) may be more useful to you than any other person]. *Perhaps* Abel went to Count Löwenhielm's on Christmas Eve, *for sure* he awaited a reaction from Cauchy and Legendre about the treatise he had submitted to the Institute. "But that would never be finished by these slow Men.[...] And then the Berlin Journey has snuck up on me like the wolf on the lamb," he added.

On December 28th Abel received a long letter from Keilhau who wrote that things had not gone well for him in Christiania; rather, he longed to be abroad once more. "Curiously enough," Abel commented, who did not completely believe Keilhau's reassurances that he, Abel, would find his situation better when he got home. On December 29, 1826, Abel waved goodbye to the city that ought to have been the "Focus of all my mathematical Desires."

43.

The Homeward Journey via Berlin and Copenhagen

> I travelled from Paris on the Diligence to Brussels via Valenciennes. I was alone for the whole Journey with a Danceuse, not from the great opera, but from one of the lesser Theatres. A dangerous Neighbourliness in the Night. She slept in my Arms, of course, but that was all. Otherwise I had extremely edifying Discussions with her, about ephemeral Things of the World.

Abel wrote this to his friend Boeck in Munich, and he continued:

> I stayed a night and a day in Brussels, which is a very nice Place, and I ran around in the City the whole Time. Thereupon I continued on as before with the Diligence to Achen via Lüttich. For company I had a neat and tidy Fellow from Frankfurt am Main. Right up to Lüttich the whole World spoke French.

He arrived in Aachen and felt himself immediately more at home among the German-speakers, and when he arrived in Cologne, he went to the theatre but did not consider that it was a particularly good performance. Cologne was otherwise:

> Köln am Rhein: an immensely old and ugly City with many Whores. Stayed one Day and two Nights, and travelled then with the Post to Cassel via Elberfeld and Arnsberg. This Region should have been extremely nice, but Night and Winter hindered my Observations of it. We had an accident between Elberfeld and Arnsberg wherein the Carriage drove over a boy of 7 or 8 Years of Age. He lay dead on the Spot. The Carriage had driven over his stomach. We continued on our Way without stopping.

Why allow oneself to stop, why complain about accidents that irrevocably have happened? This is like an echo of his own sad experiences over the recent period, hope and expectations dashed, while everything nonetheless must go on.

Abel continued on with the carriage. He arrived in Kassel, but that is as close as he ever got to Göttingen. It seems he no longer considered any visit with Gauss. He later said he had lacked sufficient money to stay in Göttingen, that he had to take the direct route, and that he only had 14 *Taler* left (1 Prussian taler = 1 Berlin taler = 2/3 speciedaler). In his letter to Boeck he said only:

In Kassel, which is a very nice City, I spent the Night and went to the Comedy. The Theatre was extremely nice and the acting was good. In Köln I was also at the Theatre, but it was bad. From Kassel [I] travelled on by the Extrapost to Magdeburg in the Company of a Kaufman [merchant] who was on the way to Berlin and Königsberg. We came by way of Hartzen. It must be very beautiful there in the Summer. In from Qvedlinburg to Magdeburg there is the most excruciating Road I have ever travelled. We were two in the Carriage and even though we had a team of 4 Horses, we were in dire straits and Scarcely made it through. In Magdeburg I again spent the Night and from there I took the Lohnkutscher to Berlin. The Road was terrible, but the Company was even more disreputable: a Shoemaker, A Glovemaker and a demobilized Soldier. They drank Spirits continually. I was bored and nobody was happier than I after being in Motion for two Days to come into Potsdammer Thor in Berlin. [...] A Quarter of a Hour after I had arrived in the City I sat at the Köningstadter and became happy to see familiar Faces and hear familiar Voices.

This was written on January 15, 1827, five days after his apparently radiant arrival in Berlin. He lived at Französische Strasse no. 39, "right beside Gens d'armen Markt" [Gendarmenmarkt], and Maschmann, who was continuing his studies in pharmacy in the city, and who had become a wizard at speaking German, introduced Abel to new Norwegian friends in the Prussian capital. First there was a merchant, Hans Backer, from Holmestrand, who from his surplus lent Abel 50 *Taler*. And then there was the pharmacist from Bergen, Georg Herman Monrad, who was in Berlin with his wife and his mother, who hoped to find a cure in the city for her disease of the eyes. Monrad was nine years older than Abel, was interested in the natural sciences, and most certainly knew of Abel's renown as a mathematician. Abel and Maschmann were now together a great deal with the Monrad family. They played cards together, probably a couple of nights a week. Abel was the most frequent winner and would pocket the small change, commenting, "I cheated them, but I am in need, and it was easy."

Beyond that, in the course of his first days in Berlin, Abel went to Schauspielhaus, and certainly had visited Crelle. As usual there were gatherings every

Monday at Crelle's home. And Crelle continued to do his utmost to come up with a better economic footing for *Journalen,* among other things, he was trying to get the Education and War Department to subscribe copies to the colleges and libraries; and at the same time Crelle was trying, in one way or another, to make it possible for Abel to live in Berlin. In this letter to Boeck, Abel also commented upon the letter from Keilhau he had received on his last day in Paris:

> I still believe that Foreign Countries are the best. When we come home, what happened to Keilhau will happen to us. He predicts that You will be sorely Taxed to mix in when You come home. He says my Position is the best, publicly perhaps, but between us, it must be said that I privately see many Horrors yet in store. I worry dreadfully about the Future. I almost have the Desire to stay here in Germany forever, which I could do without Difficulty. Crelle has bombarded me mercilessly to get me to remain here. He is a Touch exasperated with me because I say no. He does not understand what I shall do in Norway, which he seems to think is another Siberia.

Keilhau's predictions would come to be revealed as somewhat faulty. The following spring, Boeck was appointed lektor at the projected veterinary college in Christiania, while Abel continued to be without a permanent appointment. Crelle's efforts to obtain a permanent position for Abel in Berlin continued on in their own slow pace.

Abel ended his letter to Boeck, January 15, 1827, with the hope of a rapid reply and a forwarding of all the *money* of which Abel was sorely in need. Apart from that, it was as though Abel tried to take care of all his old contacts during those first weeks in Berlin. He would write to his fiancée in Aalborg, to Hansteen, to Keilhau, to Møller and to Holmboe, and perhaps to Skjelderup and the family in Gjerstad. In his letter to Holmboe on January 20th, he urged him earnestly for the money with which Holmboe had promised to help him: "As I am in a Hellish Pinch, I am naturally in want of as much as You can manage, and as quickly as possible....Do not be angry that I am plaguing You so excessively, but what else should I, poor Teufel that I am, do?"

But the first money to arrived came from Christiania on February 25th: 293 marks sent in the form of Hamburger-Banco (Hamburger-Banco were silver-value speciedaler of the day). The following day, Abel again wrote to Boeck and urged him to "send the little Money I can have from You." Abel could also now tell his friend in Munich that he had just received a long letter, six full quarto pages, from Mrs. Hansteen and the professor. The letter had been sent from Christiania on January 25th, to Abel's address in Paris, "but there

was no great News in it, inasmuch as most of it was between the Mrs. and me."
Abel went on to say anyway that Mrs. Hansteen had had a son, Viggo, that
Hansteen had become a member of two learned societies, in Copenhagen and
Edinburgh, and that in *Magazinet* a hefty debate had broken out between
Sommerfeldt and Rathke, and that Hansteen hoped "Rathke comes out of it
with his Ears clipped." Beyond this, Hansteen in Christiania was in full stream
with the planning of his Siberian expedition, and was upset that the Norwegian
Cabinet had not rapidly earmarked the money for the trip. Consequently he
had had to write to the King's State Secretary, and to Berzelius of the Academy
of Sciences in Stockholm to expedite a royal resolution about the appropriation.

Otherwise, Abel could report from Berlin that it was terribly cold with
much snow, that he had been sick and bedridden for some days, but was well
again. "Am once more my old self," Berlin, 26th February, 1827." And: "so I
am leaving here in May, out of Need, and without Reluctance. Hansteen thinks
that I shall be appointed to the University, once I have come home, but there
has also been talk of putting me on the rack for the Period of One Year in a
School. The latter would certainly set me cleanly up on my Hindlegs."

But what of Madame Levy and her hospitable house in Berlin? She most
certainly continued to keep her salon in Cantianstrasse, and Abel most likely
had sought it out, even though he gave no hint of this. It would also have been
amiss if the rumour had gone out that he had sought refuge at the home of
prosperous Jews in Berlin. Abel was wan and pale and penniless. When he
thanked Holmboe for the money on March 4th, he wrote:

> Many many thanks for Your Kindness. I have derived an amazing degree
> of Wellbeing from it, as I was poorer than a Church Mouse. Now I shall
> live on it here as long as I can, before leaving and making my way north. I
> shall stay for a Bit in Copenhagen where my fiancée is going, and so home,
> where I shall arrive so empty that I will have to set the Collection Plate in
> front of the Church door. I am not allowing myself to be disconcerted, for
> I am so terribly used to Misery and Wretchedness. It will turn out well.

He hoped Holmboe had seen his papers in *Journalen*, and he formulated his
clear goals in the work: "I have sought to be so stringent that no basic
Objections can be made."

During his time in Berlin, Abel carried out small tasks for Saigey and his
division of *Ferrusac's Bulletin* in Paris; he also mentioned to Holmboe his
studies on the lemniscate, and "I have continued with Equations Theory and
have solved the following Problem, which contains all others within it: to find
all possible Equations of a given Degree, that allow themselves to be solved

algebraically: I have encountered many lovely Propositions in this connection."

And above all else, he had completed a portion of what would, in the overall scale of things, be his greatest work, *Recherches sur les fonctions elliptiques*, of more than 120 quarto pages, that once again initiated "Magellan voyages" into a great ocean of functions. Among other things it was this work that he had hoped to find a publisher for, but for now he prepared it as long articles. These were published in two issues of *Crelle's Journal*, in September, 1927, and in May, 1828. In March, 1827, Abel wrote from Berlin:

> But the greatest beauty [I] have is in [la] Theorie des fonctions transcendantes en général et celle des fonctions elliptiques en particulier. But I must forget about this until after I get home and acquaint you with it. On the whole I have made an outrageous Number of Discoveries. If only I had them in order and written out, but most of them have not progressed further than my Head. They are not something I can think about until I come into some honourable Order at home. Then I shall slave like a Workhorse, but of course with Pleasure.

About his everyday life and what was in store for him, he had this to say: "I lead a terribly boring Life, for there are no Variations to it. Study, eat and sleep, and otherwise nothing of any magnitude."

> I long to come home now that there is no particular Need for me to continue on here. When one is home one condemns oneself to Conceptions of Foreign Countries other than those one ought. They are not up to much. On the Whole, the World is flat, but terribly straightforward and honest. There are no places from which it is easier to emerge, out into the world, than Germany or France; among us it is ten Times worse.

To be "not up to much" was an expression of the day for not being of the best quality, and the word "flat" was used in the sense of "tepid" or "dull".

At the beginning of March Abel also wrote a letter to Mrs. Hansteen. The first part of the letter has been lost, but the contents probably had to do with thanking her, etc. etc. for the long-awaited letter she had sent via Paris. Abel openly admitted his joy at soon being able to see her again, and

> I have the feeling I shall be coming to visit you often. Truly that will be one of my greatest Enjoyments. Goodness gracious, how many Times have I not wanted to go to you but have not had the courage? I have been at your Door many Times, and turned away again out of Fear of burdening You

with my Troubles, for that would be the Worst Thing that could happen to me, that You would become sick and tired of me. It is very good that I have managed to avoid this happening. I am now here in Berlin, and in this I am happy, for the Frenchmen do not please me. They are some of the coldest, most prosaic People. They deal with every possible Thing in the same Manner. With equal Importance or Unimportance they speak about the most serious Matter as though it were the most inconsequential. No intimacy among them. A Frenchman has almost the same acquaintance with all people. Monstrous egotists. When they hear that Foreign Countries possess something they have or do not have, they marvel at it, and say Diable! and in this way they come to marvel at everything. And as for the dear Fair Sex. They are so nice, so ingratiating, and dress so beautifully, but Voilà tout. The Modesty and Shyness that Men want so much to see in Womenfolk, to be sure, they lack very much. The Frenchman says as much himself. They say, Les étrangères sont plus modestes que les française. In this respect the Germans are certainly to be preferred.

Abel was thinking of Mrs. Hansteen's two sisters, Mrs. Frederichsen and Charité, and he wrote:

You must know that I regard them both so cordially. I am intensely happy to have the Pleasure of seeing them again when I come to Copenhagen, which shall not be in such a long time. My fiancée who is now in Aalborg will also come there. I have always lived the most pleasant Life in Copenhagen.

He bade Mrs. Hansteen greet his sister, Elisabeth, "in the most loving way", and after the sentence, "I always remember her" some words have been written, then crossed out and made illegible. And Abel ended as though this were to be his last letter from abroad: "Now, adieu, my very most dear, motherly Exhorter, and find a little tiny Place in Your Heart for Your Abel."

Abel left Berlin at the end of April, or the beginning of May, and travelled as quickly and cheaply as possible to Copenhagen. Whether or not his stay in Copenhagen attained the level of "the most pleasant Life", there is no certain indication. One of the stories in the repertoire of Hansteen family stories was that Abel had obviously been happier to see Charité than his dear Christine, and that the engagement had almost come to be broken off. Be that as it may, we do know that Abel left the portrait that Gørbitz had made of him in the safekeeping of Charité and Mrs. Frederichsen. This may, however, have occurred because the portrait was the only thing that Abel had to give as a sign of gratitude for their evident hospitality. In any case, it seems that after some weeks in Copenhagen the bonds between Abel and Christine were once again strong, even though the aim of getting married continued to hover out in the blue ether.

As previously, Abel stayed with his aunt and uncle, the Tuxens, at Dokken in Christianshavn. Christine probably stayed with her mother in Prinsengade, Christianshavn. And it so happened that Uncle Peder Mandrup Tuxen, through his brother, Ole Tuxen of the Løveapotek (Lion Pharmacy) in Arendal, knew the owner of the Froland Ironworks, Sivert Smith, who it was said was looking for a governess. The plans for Christine Kemp to be employed as governess at Froland Ironworks were probably made during these weeks in Copenhagen. And thus it was that Abel could return home with a certain equinimity of temperament.

Otherwise, the Tuxen family was soon to go to Norway too. Uncle Tuxen had received a leave of six weeks beginning on May 20th, and perhaps they talked about travelling together with Abel. Now, in this year, it had also become much easier to travel to Norway. The steamer *Prinds Carl* had begun regular traffic between Copenhagen and Christiania on April 1st, and at the Norwegian naval base of Fredriksvern at Stavern, on the west side of Oslofjord, the *Prinds Carl* corresponded with the steamer *Constitutionen,* which plied the route all along the coast to Kristiansand. It was these ships, still not officially blessed, that Finance Minister Jonas Collet, Boeck's uncle and prospective father-in-law, had contracted two years earlier, without Parliament being consulted. Now Parliament was preparing to impeach Collet while the public celebrated and sang about the amazing new means of transport. Abel probably also felt a bit easier and happier when he left Copenhagen on the 18th of May, and beyond that, he was no doubt delighted by progress everywhere when, on board the *Prinds Carl*, he passed Goteborg and Marstrand, where twenty months earlier he was preparing himself to be shipwrecked while aboard Trepka's little sailing vessel.

Abel returned to Christiania on May 20, 1827, perhaps better dressed than when he left, and eager to tell his story, eager to find peaceful working conditions.

PART VII

The Last Years in Norway

Fig. 35. Arendal. Lithograph by C. Müller from a book of drawings brought out by Chr. Tønsberg, 1848: *Norge fremstillet i tegninger.* Christiania.

44.

Back Home

On May 20th, Mrs. Hansteen was at the quay along with her husband and eldest daughter when the steamship docked. We also learn from Mrs. Hansteen's diary that Abel, together with Holmboe, came to the Hansteen house on Pilestrædet later that day. Abel was bearing gifts from Charité in Copenhagen: a hat for Mrs. Hansteen, and toys for the children. Between these pieces of information, Mrs. Hansteen wrote: "Hansteen's Disgraceful Behaviour to be deplored." But her husband's ill-humour was hardly directed at Abel; quite the opposite: it seemed to Hansteen that Abel had matured and become more in control of himself than before his travels abroad. Travel companion Møller wrote to Boeck in Munich that he had recently met Professor Hansteen who had talked "exceedingly kindly" and spoken "with extremely Good Will about You all. He was very pleased that Abel had acquired a degree of civility. Whereupon I gave the Honour to You." At that moment, N. B. Møller was the only travel companion of all the "young Scholars" who happened to be in Christiania when Abel arrived home. Møller had found a position as manager for Snarum's Cobalt Works and had entered into a company in order to take over Blaafarveværket, the cobalt works at Modum. Keilhau was off and away on his Finnmark expedition, and Tank, in accord with his religious predilections, had gone to the headquarters of the Hutterite Brethren at Herrnhut, a day's journey from Dresden.

Abel visited the university library on May 28th and borrowed one of von Littrow's books on astronomy: almost as a little tribute to the helpful and obliging professor in Vienna, or perhaps to savour his good memories from abroad. In any case, Abel took the book with him, and von Littrow's book was not returned to the library until after Abel's death.

The best assessment is that Abel lived at Holmboe's for the first period after returning home. His brother Peder, who had had a room at Mrs. Tode's on Voldgaten when Niels Henrik went abroad, was no longer there. His sister Elisabeth continued to work at the Treschow's at Tøyen, and no doubt she had a lot to tell about Peder's disorderly conduct. He had quite certainly abandonned Christiania and his unpaid debts.

To pay a visit to the respectable Treschow household was otherwise some-thing that Abel, out of courtesy, was bound to have done. Many considered Treschow a superannuated old fuddy-duddy, ready for the grave, but he was still able to awaken interest with his writings on Christianity and cognition. And perhaps Abel was happy to pay visits to Treschow, who was still pro-chancellor of the university. In any case, it was said that on the occasion of a whist party at the Treschows, when, to the disgust of Abel and his calculations, the host was still sitting with a jack, Abel was said to have shouted out, "Isn't some swine sitting here with the Jack of Clubs?!" And then in a flash, Abel was said to have added, "I *am* sorry, excuse me. I thought I was sitting at Regentsen [the student residence]!"

Abel was greeted with goodwill and kind words wherever he showed himself. He appeared to be a kind, good-natured, somewhat unconventional fellow known to be a genius, according to professors who did not know what his genius could be used for. No concrete plans had been laid for Abel to obtain work and have something to live from. Therefore Abel was probably advised to report his homecoming to the Academic Collegium, so that the matter of his future could in this way be taken up once more. In a text sealed with a signet mark holding the initials B. M. (quite probably for Bernt Michael Holmboe) Abel informed the Collegium on June 2nd :

> that I have now returned after having struggled to pursue the best degree of Refinement possible, which was the Goal for the Tour abroad. Hence I do offer my most sincere Gratitude for the Collegium's Involvement in the Attainment of this Goal. I continue to commend myself to the Collegium's favourable Consideration.

In the course of the twenty months that Abel had been abroad, he had published several epoch-making treatises, but he had no written *attestations* from foreign authorities to present, and the fact remained that his travel plan lay at variance with the goal of seeking out Gauss in Göttingen. He seemed to want to forget his Parisian defeat. And what he *had* published was, after all, for the most part, only to be found in this new mathematics journal in Berlin, *Crelle's Journal,* which still had little prestige in university circles in Chris-tiania. Those who sat in the Collegium well knew of Abel's difficult economic situation, and only three days after he had reported his homecoming, the Collegium took up the matter. They made it clear that the University did not have the economic resources to offer Abel anything, but they stressed that it was important "to retain him for his Fatherland, in specie, for the University," and the Collegium sent an urgent request to Pro-Chancellor Treschow with the proposal that Abel had to get "interim Support from the Official Side", with reference to the support Abel had received *before* his travel abroad. Treschow went directly to the government with the recommendation of support for Abel until he received an acceptable position, such that "the Fruits of both his extraordinary Talent for higher Mathematics and of the already disbursed Costs to the Fatherland not be lost." Treschow sent the matter on to the Department of Church Affairs, that in a formal manner could urge the Finance Department's "agreeable Consideration." The head of the Church Department was now Cabinet Minister Diriks. The Finance Department's response was a brief four lines on June 20th: "That nothing could be given from the State Exchequer for the aforementioned Purpose." It was Minister Collet who here showed no mercy, at a time when he himself was facing a trial for having used almost one hundred thousand speciedaler on the two steamships.

At the same time that the Collegium had taken up Abel's case, a discussion of principle was also occurring on whether it was to be *scientific studies* or *giving lectures* that should be the primary duty of the university's teaching staff. What caused this discussion was Professor Fougner Lundh's dissatisfaction with the fact that he had so few students in his field ("Political Economy with the Teaching of Farm Economy and Technology") that the Collegium had to decide whether they should recommend that Lundh leave the university and take a position as an army quartermaster. Professor Hansteen made an interesting contribution to this discussion. He felt it was important to support Lundh's scientific *research and studies*, and not least of all now when the university had expressed its desire that it would like very much at some point to have *Abel* employed at the university "although for all that it would not in the least be expected that he [Abel] would ever be in a Condition to give Lectures, at least the kind that Someone could understand." Others also expressed doubt about Abel's ability to give understandable lectures, and so saying, stressed the university's purely scientific tasks. It is, furthermore, natural to wonder if Hansteen was also thinking about the forthcoming Siberian expedition that certainly was to the highest degree scientific *research*.

There is much to indicate that Abel remained in Christiania until this first round concerning his economic support had come to an end. From Mrs. Hansteen's diary one learns that Abel had come to visit her six times in his first month back home, and that the whole time she had worried about him. She writes that she spoke to her husband about Abel's future. On June 18th she informs with her characteristically cursory entry style: "Likewise I sad and faint and Headache. Abel here this Evening. Said farewell. Left for Sandeherregaard. Late Evening talked with Hansteen which comforted me exceedingly so I went to sleep with a light Heart."

Abel's customary ardour and expectancy after getting down to mathematical work seems not to have been particularly assuaged in these first weeks after his homecoming. Not having his own lodgings, and without a secure income, it seems that after about a month in Christiania, Abel took to the land: through Drammen to Sande, and further south along the Vestland main road, to Gjerstad. This means of travel also proved to be cheaper than the new steam-boat route.

Abel arrived in Gjerstad, probably around Midsummer Night, and it is most likely that on this occasion too, Pastor Aas was very friendly toward him. Despite the fact that he had now become a famous man of science, he remained in the eyes of the local people, simply Niels Henrik, the priest's son and a student without a calling. According to tradition, Niels Henrik wandered around among the farms, chatted with people and predicted tomorrow's weather. At his mother's home at Lunde there was not much he could con-tribute this time either. Economically, he was in the same dire straits as his mother. The unpaid portion of the half-cask of grain that Abel senior had pledged to the university still hung ominously over the Lunde farm. There was a mortgage clause involved such that the authorities now, whenever they pleased, could legally invoke a mortgage auction in order to recoup what was owed them. The outstanding grain contribution went back, in fact, to 1813, and now amounted to almost 26 speciedaler. In 1821, the year after her husband's death, and with Pastor Aas' help, Mother Abel had requested the university to free her from this obligation. But she had received a letter in response from the revenue office that stated the payment could be annulled only if the family undertook a planned sale of the book collection Pastor Abel had left behind. And there the matter still stood, except that Mrs. Abel had already sold off both books and most of the domestic furnishings from her abode. Much of this had been immediately purchased by Pastor Aas for his own use in the vicarage: chairs, beds, duvets, curtains and bread baskets, a tiled stove, a table centrepiece fashioned from stone, a clock, fire tongs and ash spade, and a roasting spit. From papers in the diocese archives one sees as well that Pastor Aas also bought the Luther Bible for 7 speciedaler, and all in all, he acquired things to the value of almost 70 speciedaler. For her part, Mrs. Abel had received grain and potatoes from Aas, and more than 25 speciedaler in the summer of 1824, that had been used to equip her son Thomas with a frock coat, hat, boots and travel money. This was the time that Thomas, with Tuxen's help, went off to try his luck in Copenhagen, but after a short time he had come home dejected and in failure. Thomas had not managed to control himself during either working hours or leisure time. Now, in the summer of 1827, it looked as though Thomas was home at his mother's, at Lunde, together with the oldest brother, the deficient Hans Mathias, and the youngest, Thor

Fig. 36. Christiania seen from Ekeberg, and facing toward Akershus Fort, painted by P. F. Wergmann, 1835. The view from Ekeberg was admired, painted and described by many. When in June, 1826, Abel and his friends came to the heights above Trieste and caught sight of the Adriatic Sea, Abel described the experience in a letter to Holmboe: "It is undeniable that the view was particularly lovely, though far from being comparable to that one has from Ekeberg. But upon us, who had been so long away from the Sea, it naturally made a pleasant Impression, that arose especially from the fact that it was the Adriatic that lay before us." (Museum of the City of Oslo.)

Henrik. Quite probably brother Peder also made visits home to Lunde that summer while he was looking for a post as tutor. One can imagine that these were not cheerful conditions to come back to. Many years later, when his mother and all his brothers were long dead, the youngest brother, Thor Henrik, wrote a poem about his birthplace:

Ah, You, my place of birth,
So rich in bitter Testaments!
In search of blessed mirth,
Away from You one time I raced;
Came back, and O! with what Grief I sense
The anguished Youth you long embraced.

Fig. 37. Christine Kemp, Abel's fiancée, painted by Johan Gørbitz in 1835. (Archives of Gyldendal forlag.)

Niels Henrik was certainly relieved to travel on further, to Froland, with the errand of trying to secure work and residency for Christine in Norway. He might even have travelled together with his aunt and uncle: on Sunday, July 8th, the Tuxen family had arrived in Christiania by steamship from Copenhagen, and shortly thereafter they travelled south along the coast. But Mrs. Tuxen no longer had any desire to see the sister who had fallen into drink; it is likely that they merely stopped in Risør, and travelled on to Arendal where Tuxen's pharmacist brother lived; from there they would have journeyed inland to visit Smith, the owner of the enterprise at Froland.

In any case, Niels Henrik did arrive at Froland and either with or without the Tuxens' assistance he managed to arrange things such that he was quickly able to inform Christine in Copenhagen that Smith's children at Froland were waiting for her: at the first opportunity she could take the *Prinds Carl* to Christiania. That this was made clear seems to indicate that Abel would quickly return to Christiania, both to meet Christine and to get down to his mathematical work. The return journey to the capital was taken on the *Constitution* from Arendal, the steamer that now had its proud name printed on the paddlewheel housing. But before he left Smith and Froland and Tuxen and Arendal, he no doubt heard the new scandalous story about Count Löwenhielm, the ambassador to Paris who had boasted that every Christmas

Fig. 38. The portrait of Abel painted by Gørbitz in 1826. All subsequent depictions of Abel stemmed from this portrait, the only one made in his lifetime. The story was told in the Hansteen family that Abel had this portrait with him in May, 1827, when he began his journey home, and that he had deposited the picture with Mrs. Hansteen's sister in Copenhagen. Nevertheless, one way or another, it came to Christiania and was among Abel's household effects. In any case, Christine Kemp took the picture away with her when in June, 1829, after Abel's death, she was in Christiania to fetch his belongings. Upon the death of Christine Kemp Keilhau, the portrait was said to have gone to Thekla Lange, the daughter of Niels Henrik's sister, Elisabeth, who in turn passed it on to her daughter, Elisabeth Lange, who took it with her when she moved to the USA and married a Norwegian engineer in 1890. Her daughter, in turn, brought the portrait back when she married in Norway in 1912. When she was widowed in 1966, she sold the portrait to Norske Liv, where Fredrik Lange Nilsen was director. He presented the picture to the University of Oslo and since 1992 it has been on display in the mathematics library, Abels Hus, Blindern Campus. (Foto: O. Væring.)

he drank all the Swedish and Norwegian guests under the table. In the course of his visit to Norway that summer, Count Löwenhielm had gone to Arendal, and at a reception at Sophienlund, the summer residence of the merchant baron, Anders Dedekam, on the island of Tromöy, and suffered the same fate that he had inflicted upon so many of his countrymen in Paris.

There was always a great deal of drinking at the summer parties at Sophienlund, and one of the reasons was a silver cup which had a windmill soldered to it that was set in motion when one blew through a thin pipe fastened to the outside of the silver cup. The trick was to blow in the pipe to set the windmill

Fig. 39. The Froland Works, painted by Lars Berg in 1831. (Original in the Aust-Agder Museum, Arendal.) The home of Sivert Smith was one of the most prosperous of the time. Froland had received permission to mine iron ore in 1763, and had been owned by the Smith family since 1786 – first under Magistrate Hans Smith, who in 1791, put up the imposing main building. Beginning at that time, Hans Smith's wife, Magdalene Marie, had had Danish and Swedish landscape gardeners lay out a magnificent set of gardens in the Baroque style, with a gazebo at the high point on the grounds. The son, Sivert Smith, took over in 1820. Old methods of production and a lack of technical knowhow and knowledge of ore-refining led to extra difficulties in the business and financial upheavals that afflicted the country after 1814. The golden age did not return. Froland tried to survive the pressures of the market by shifting from cast iron production to the production of iron bars, and by diversifying into forest products, but a continuously more difficult economy forced creditors in Copenhagen to take over the enterprise in 1835.

in motion, and finish drinking the contents before the wheel stopped turning. But usually, before a guest learned this manoeuvre, the silver cup had to be filled several times. At the time of Löwenhielm's visit there, a North American was living in Arendal. It so happened that Anders Dedekam served as consul both for the United States of America and the Netherlands. At table there had been a discussion about the displeasure Norwegians felt about Karl Johan and the King's attempts to put an end to the Norwegian Constitution. The American, who had not been able to follow the whole discussion and its implications, had nonetheless understood that they were talking heatedly about the Constitution, and he had raised his glass and proposed a toast, with the words: "May that hand wither and that heart die, that will alter a single letter in the Constitution of Norway!" Everyone had emptied his glass with joy, except for Löwenhielm, who interpreted the toast as a "Death to the King!" Löwenhielm had shoved the glass away from him and declared in a loud voice, "No, God damn it, that is a toast I will not drink!" The American, who was an unusually big, powerful man, took this as an insult, and in a flash, indicated that he intended to toss the skinny Count out the window. Meanwhile the other guests had thrown themselves between them, and little by little it was revealed that the American had not intended to say anything unpleasant about the *King*, and had offered to settle his difference with Löwenhielm amiably with drink. And thanks to the special silver cup, this was done with a vengeance. The result was that Löwenhielm had to be carried down to the boat when, late into the night, the party came to its conclusion and they were to return to Arendal.

The *Constitution*, with Abel on board, slid past Sophienlund and Tromøy on Friday afternoon. Normally the boat spent the night at Risør, and early the next day, crossed to Fredriksvern. On Sunday mornings, the *Prinds Carl* arrived from Copenhagen, and by Sunday evening it came alongside the wharf in front of Christiania's Akershus Fort.

45

Christiania, Autumn, 1827

On Monday, July 16th, Abel paid a short visit to Mrs. Hansteen's in Pilestrædet, quite probably to convey the news about Christine's arrival. At the beginning of the month Professor Hansteen had left for Copenhagen to make preparations for his Siberian expedition.

Abel had now begun to take out small loans from among his friends. He found himself a little hovel, and decided to make a new thrust at getting public help and support. There was a candid and mildly desperate tone to the letter he sent to the Collegium on July 23rd, not only on his own behalf, but also worried that he would also lose his mathematical ideas:

> For quite a long period of Time it has been my Thinking, almost to the point of persoanl sacrifice of my mathematical Studies, in order to make myself worthy of employment for a period of time, as an academic Teacher. I perhaps dare flatter myself that now, after having completed my Foreign Travels, I have acquired Knowledge that thereupon could be deemed sufficient, and that therefore when Conditions allow it, I would like to obtain Employment at the University. Still, until that Time, if indeed I am to share such Employment, I remain quite without the Means to obtain for myself the barest of Necessities, and have been in this condition since my Homecoming. In order to live I now face the necessity of setting aside my Studies, which would precisely cause me exceedingly much pain right now, when hopefully I am able to complete several great and small mathematical Works that I now have underway. This would grieve me so much more than the fact that I have also had to break off, already while Abroad, a literary Trajectory that I had embarked upon, namely, my collaborative work on the journal coming out in Berlin, "Journal der reinen und angewandten Mathematik von Crelle", of which I take the liberty of enclosing the Issues that have already been published. I am therefore taking the risk of soliciting the Collegium's support on whatever terms the Collegium deems suitable.

Never had Abel formulated a more urgent plea for help, and never expressed himself more clearly on his own qualifications. And the letter did make an impact. The Collegium took the unusual step of renewing the application for financial support to Abel, despite the Finance Department's categorical "no" one month earlier. In its application to the Church Department on July 31st, the Collegium pointed out "that Herr Abel has already long since received such strong Encouragement from the official Side that he ought to continue along this Course, that he by so doing seems to have acquired a claim on its continuing Care, just as he surely considers himself committed to Reciprocation, by superbly offering his Fatherland his Industry and his Talent." The Collegium also pointed out that support from the state coffers would certainly be of short duration "as already now there is the Situation of Professor Hansteen's forthcoming Journey to Siberia that presents the perspective that Hrr. Abel's Service to the University will be sought", and the Collegium proposed, in all modesty, that Abel should receive the same amount that he had been given before his trip abroad – 200 speciedaler per annum – retroactive to the date of his return home.

There now came a message from Christine in Copenhagen that she would be on the boat that arrived in Christiania on Sunday evening, July 29th. But the boat was late and it did not arrive until late into the night, and Christine was not on it. She had gone ashore to stay with friends down the fjord at Son where she had worked earlier. Abel then went to Mrs. Hansteen the next morning. She noted in her diary: "No Miss Kæmp. An Idiotic Blow. Abel took the last of my Money to journey to Son to fetch her." Christine's alighting at Son seems to have been a blow to Abel's strained calculations, and was perhaps a forewarning of the coming difficulties in the relationship. Abel went to Son, and together with Christine, returned to the capital on August 5th. The same evening they were at Mrs. Hansteens. It was a quiet visit.

Christine now stayed in Christiania a week before she took the boat southward to Arendal, and continued on from there by horse and carriage to Froland. All we know about that week when the young engaged couple were together in Christiania was that on August 10th they visited Professor Skjelderup; on the 11th, Mrs. Hansteen noted in her diary, "In the garden with Sewing, together with Miss K. and Abel. Learned from Miss K. how to sew woollen flowers." And three days later, on a Tuesday, Christine left the capital. Abel conducted her on board the ship. Despite Abel's dire straits concerning money and accommodations, it seems that the ties between the two of them had strengthened in the course of the week. In any case, Abel waited for a letter to come from Christine on the first boat, that coming Sunday, and he seemed especially eager to read this letter immediately. When the steamboat was expected back in Christiania on Sunday evening with post and passengers, Abel, together with other friends, had been invited to a dinner party at Holmboe's, and hence he himself could not be on the quay to fetch the letter. However, he knew that Mrs. Hansteen was also expecting a letter that day, from Hansteen, who to her disappointment had journeyed on to Altona instead of coming home. Abel therefore wrote a little note to Mrs. Hansteen wherein he asked that her serving girl also fetch *his* letter. Abel's request to Mrs. Hansteen was written in French, certainly so that no unwelcome eyes might understand the contents.

He excused himself for needlessly causing Mrs. Hansteen to tire at the sight of his insignificant person, but he sensed that she understood that he was eager to get news of her whom he loved above all. The way he put it was: "Avec toute le force de mon ame d'avoir des nouvelles de ce[lle] que j'aime le plus." On one hand, then, it did seem important for Abel to keep this task secret, but on the other hand, he urged that if the letter arrived, she have the serving girl deliver it immediately to him at the party at Holmboe's, on the second floor, across the street from the theatre.

Whether or not a letter arrived from his sweetheart that evening remains unknown.

The Collegium's renewed request that Abel receive a stipend of the same amount as he received before his travels abroad, 200 speciedaler per annum, was apprehended in the Church Department. Minister Diriks knew from the earlier response that there was no budget that allowed such a payment from state coffers, but the minister suggested a way out: the University *itself* could pay Abel the money as an *advance* against the salary that Abel could expect to receive as temporary replacement during Hansteen's planned expedition to Siberia.

The Collegium seized this proposal with alacrity, and consulted the Faculty of Philosophy to see what their thinking was. The same day that this was done, on August 27th, a document was drawn up by the university paymaster, and attesting to the fact that Abel had been to see him and had given a verbal promise that he, when he received the expected employment by the University, would pay back the almost 26 speciedaler owing on his father's grain pledge. In this way, Abel would salvage what was left of the family at home in Gjerstad, and presumably in the course of these economic deliberations he would have made a positive impression.

The Faculty of Philosophy gave the following opinion on the matter, stating "that whereas Mr. Abel, not only in the Fatherland, but also Abroad, is recognized as an outstanding mathematical Genius, about whom there is the greatest Reason to hope that were he to be placed in an appropriate Post this would contribute enormously to the Future of the Sciences, and moreover he is known to most of the Faculty's Members as a young Man, in whom Want combines with all his other Qualities, to make him worthy of the Support that the Collegium may be in a position to grant him." This was signed on August 30th. Five days later, 200 speciedaler were appropriated from the University Scholarship Fund, without conditions that the sum be repaid in the event of eventual permanent employment. But the stipend would not be retroactive to Abel's date of homecoming, but rather from July 1st. He was permitted to take out 100 speciedaler in cash as an advance; the remainder would be paid out at the monthly rate of 8 speciedaler and 40 shillings.

Thus it was that now in September, Abel had 116 daler and 80 shillings in his hands, but this was scarcely enough to pay back all the small loans and credits he had with the shoemaker, tailor, and the couple who supplied his meals. On September 21st he went to Mrs. Hansteen with the money he owed her.

At the same time, in September, the Church Department informed the Collegium that 4,500 speciedaler had been budgetted for Professor Hansteen's Siberian expedition. Hansteen himself had returned from Altona and Copenhagen before lecturing began on September 9th, and a month later was able to inform the Collegium and the Church Department that his Siberian journey was scheduled to begin in March of the coming year and would extend for a year and a half. Consequently it would be necessary to appoint a temporary associate professor in his absence, from the beginning of 1828, and Hansteen announced that Student Abel was willing to take over the lectures in theoretical astronomy for the philosophical examination, and in the event that mineralogy students signed up, he would also teach the most important propositions in mechanics. Both in order to obtain the honorarium, and to lighten the workload for the temporary replacement, Hansteen had already prepared the almanacs for the coming three years. An earlier document from the Department had considered that Hansteen himself should pay for his temporary replacement, but Hansteen felt that the honorarium to cover the employment of this associate professor should be paid by the state, and he got support for this position from both the Collegium and his colleagues.

In the further developments of this matter between the Department, the Collegium and the Philosophy Faculty there was never talk of anyone but Student Abel filling the forthcoming temporary position, but the job description and pay were certainly discussed. The Collegium's final statement on the matter, on December 10th, declared that Student N. H. Abel was quite qualified for the task, made "so very much lighter" by Hansteen already having prepared things with the almanac, but responsibility for the astronomical instruments was assigned to Lektor Holmboe. And since Hansteen's Siberian expedition was deemed an affair of the state, the Collegium maintained that no disbursements in relation to the expedition were to be assigned to the University, and not even the salary for the temporary associate professor. And "an Honorarium of 400 Spdlr yearly as adequate Recompense for Mr. Abel" was proposed. The normal pay for an associate professor was 600 speciedaler. This proposal was swiftly commended to Pro-Chancellor Treschow.

In the middle of December, 1827, Abel was thus in a position to settle down calmly to work while, in the meanwhile, the solving of his economic problems were underway. He had tried to make a little extra money by giving private tutoring. On October 11th and 12th, the following announcement appeared in *Morgenbladet* under "Advertisements":

Here in the afternoon hours it is possible to obtain Information in Arithmetic, Geometry, Stereometry, plan- and sphere Trigonometry, Algebra, as well as the basic foundations of Astronomy and Mechanics. Contact Morgenbladet's General Office. The amount to be paid will depend upon the Number who sign up to participate.

It is quite likely that Abel hoped to use this to reach those who were facing the *examen artium* for school graduation, and those who wanted to sit the University's secondary examination, the *Andeneksamen*. There is a notation at one place in Abel's workbooks that totes up the hours and the receivables: Wilhelm Koren Borchsenius and Niels Wolf Christie are mentioned: both took the *artium* as private pupils in the spring of 1828. But there were no doubt several of them. The student Hans Christian Hammer, who wrote the memorial poem "On Docent Abel's Death", was probably one of them, and according to tradition, several of them were of the opinion that he was a good teacher, his explanations were both clear and easy to understand. An announcement in the *Norske Intelligenz-Seddler* on October 20th was probably also Abel's: "Information in French and German, together with the general School Sciences, is herewith offered by a Student, at very reasonable Rates. Those who hereby consider this offer will kindly leave a Note at the Newspaper Office, with an A printed upon it." But there cannot have been many who signed up, and virtually no extra income was realized this way. During the month of October, Abel was forced to seek out a respectable loan, and he went to his former good samaritan, Søren Rasmussen, who was now head of the Bank of Norway. Rasmussen willingly signed a loan allocation of 200 speciedaler. Surety for this loan was signed by Professor Hansteen, Lektor Holmboe and Professor C. A. Holmboe.

As for Abel's mathematical work during this autumn, the following occurred. A paper Abel had submitted for publication in *Magazinet* prior to going abroad in 1826, had been sent by Hansteen to the Association of Sciences in Trondheim, probably to relieve some of *Magazinet*'s space from heavy mathematical material. Now, in September of 1827, this paper was published in *Det Kongelige Norske Videnskabers Selskabs Skrifter* [Proceedings of the Royal Norwegian Association of the Sciences], and Abel became a member of the country's only scientific association, together with, among others, Lektor Keilhau and Professor Steenbuch. Apparently Abel did not record his response to the honour of receiving this membership; in any case, he seems not to have sent a letter of thanks, as was the normal thing to do upon the receipt of notice of admittance to such an exclusive association. The paper that was published, "A Small Contribution to Knowledge on a goodly number of transcendant Functions," was not professionally important in comparison to what he was now concerned with.

A new issue of *Crelle's Journal* came out on September 20th, in which were printed the first 80 pages of his "Recherches". And now Abel would have been eager to get the second part ready for printing. He probably knew nothing about the stir that this first part of "Recherches" aroused across Europe, and neither, presumably, did he have any idea of the expectations that many mathematicians now held for the second part. Abel had definitively placed himself among the world's leading mathematicians. Part Two of "Recherches" would contain the studies of the division of the lemniscate, and that part of the knowledge of elliptical functions that have come to be called Transformation Theory. It was here that Abel seriously worked out his idea on the great "inversion"; Abel turned the problem from the elliptical integrals that Legendre had studied, to the elliptical functions, and with this new consideration he opened people's eyes to a new understanding of function theory, an understanding that in its turn had great significance for mathematical developments both in algebra, numbers theory, group theory, and analytic geometry. By means of the lemniscatic problems, Abel would also plot out the class of equations that later would go under the prefix "Abelian". As well, Abel would take up the treatment of what is known as complex multiplication in this foundational treatise, "Recherches sur les fonctions elliptiques".

At the same time he was studying infinite series and convergence criteria. In *Crelle's Journal*, the first issue of 1828, Abel's elegant work on series was published: his response to the mathematician L. Olivier's work on convergence criteria. In three pages, Abel demonstrated that Olivier had been wrong in his assertion that there existed a general criterion for separating convergent and divergent series.

Otherwise, the correspondence with Crelle seems to have been sparse that autumn. But Abel saw clearly that Crelle continued to work on the plan to establish a polytechnical institute in Berlin, and that Crelle very much wanted him there. Certainly there were very few in Christiania who had been let in on these plans, and half a year later, when there was serious discussion about a teaching position for Abel in Berlin, this fact precipitated a considerable amount of trouble for him.

It was his dire economic condition that was talked about in Christiania. Right before Christmas Hansteen also informed the military college that from the coming February he would be unable to hold his usual lectures in applied mathematics, and after discussion with the college director, Major-General Aubert, this was divided between Abel and Engineering Captain Theodor Broch. The latter would take up the use of instruments, while Abel would be responsible for theoretical astronomy and mechanics, two sessions a week. This would give Abel an extra income of 11 daler and 13 shillings every month.

That autumn saw a vicious battle underway on the theatrical front. The Dramatic Society's performances at Grænsehaven were no longer the only ones in the city. At the beginning of 1827, Christiania had been graced by a public theatre. Johan Peter Strömberg, from Sweden, had presented his first attraction – H. A. Bjerregaard's *Mountain Story* – on January 30, 1827, in the newly-built premises on the lot owned by the assessor, Motzfeldt, the institution which later would be known as Teatergaten No.1. Despite a certain amount of ill-will occasioned by the fact that a Swede had built a Norwegian national theatre, the first half year was marked by enthusiastic responses and a number of different plays. Strömberg had his stable of Norwegian actors, consisting of seven men and three women, and in addition, a ballet troup of his own pupils, together with a little orchestra led by Waldemar Thrane. In the competition that ensued, many players from The Dramatic Society went over to Strömberg. Thrane, who earlier had been the Society's conductor, was now replaced by the seventeen year-old Ole Bull.

Many had felt that the theatre life in the capital had been given far too poor coverage in the public media. It was said that only *Morgenbladet* paid respectable attention to the theatre, and in order to remedy this situation, Christiania's *Aftenblad*et was now founded; the editors were the poet and high court lawyer, H. A. Bjerregaard, and the translator and department head, H. L. Bernhoft. The first issue came out on October 3rd, with the expressed wish that: "Those who most revere and are most charmed by the Arts, will find in our Pages a refined Perception of the Essence of Art." It is highly likely that Abel read *Aftenblad's* theatre columns, and quite certainly he met the editor, Bernhoft, in the capital's theatrical circles, and perhaps, indeed, they discovered that they had both been born in the same place, in the Parish of Finnøy-Bernhoft, nine years before Abel. Besides, it was common knowledge that Bernhoft had published collections of poems, translated Goethe, Madame de Staël and Novalis, and was now in the process of translating Steffens' stories. The debate about theatre, both in public and in *Aftenbladet's* columns, was focussing upon the language used on stage. On one hand, The Dramatic Society kept its diction as closely as possible to the Danish ("bæve" and "svæve" for example, instead of the more Norwegian "beve" and "sveve"), while on the other hand, it was said that Strömberg was looking for actors who had not "stuck their Noses out the Door," where one could hear the most "vulgar" and "unrefined" of expressions. "Peasant talk from the mountain valleys", for example, was what one heard in *Jeppe paa Berget*, when an actor would say "the ba-ron" instead of "the baron."

The first great defeat for Strömberg came when he staged his own play *The Peace Celebration* or *The Union*, on November 4th, the very anniversary of the union of Norway with Sweden. The play provoked a violent protest of whistling, and in addition to the press criticism, this was put down to the fact that Strömberg had hired two Danish actors: Jens Lang Bøcher and Miss Christiane Hansen. (Miss Hansen shortly thereafter became Madame Bøcher.) After a year and a half, for various reasons, Strömberg had to resign, and the leadership was taken over by Bøcher, whereupon Danish diction persisted in the Norwegian theatre.

Most of the performances of The Dramatic Society that autumn were otherwise plays that The Society had performed before. The primary performances were of a play by F. von Holbein called *Rose of the Alps*, and a play by J. W. Lembert called *Journey to the Wedding*. The titles could have awakened both memories and expectations in Abel.

For what Abel had hoped, in respect to his fiancée, had not turned out the way he would have wished. He did not know what to do. Abel had visited Mrs. Hansteen quite frequently throughout that autumn, and an underlying theme of discussion between them seems to have been the conditions needed for the existence of love and solidarity. In September, Mrs. Hansteen had summed up a magnificent day spent together with friends in the garden and "an interesting Chat with Abel", thus: "A delightful Day. A Compendium of what I wish my Life was composed of, namely a Society replete with Spiritual Dimensions, and the whole framed by a delightful Nature and by illumination from the proper Sunlight of the Spirit, and always accelerating toward an immanent High Point." On November 17th Mrs. Hansteen noted in her diary that "Abel came by towards Evening. Spoke to me with regard to Miss K., whereupon I became very down at the mouth, through the association of Ideas to my own Labyrinths." In the following day's entry one finds only some words jotted down: "Still very melancholy. Finally got on my feet again toward Evening with the help of my dear Husband. – ended the Day trusting and happy." And on November 19th: "Wrote a little at Abel's request on Saturday to Miss K. This quite pleasing."

Christmas, 1827, was approaching. Abel wanted to celebrate the holidays together with Christine, and moreover, an invitation to Froland seems to have been in place. In brief, it is quite likely that Abel took the main road south once more, and celebrated Christmas at Froland for the first time.

But he was back in Christiania on December 31st and signed for the receipt of his monthly stipend payment.

46

From January to July, 1828

When classes began at the military college in Christiania on January 12th, it appears that Abel was not present; his absence was likely due to illness. Shortly thereafter however, he was back, and working full stride.

Even though Abel did not receive the formal message from the Collegium that he had been appointed associate professor with a salary of 400 daler for the period of Hansteen's absence, he in fact began his teaching duties, and began receiving salary payments from February 16th. With these two teaching positions Abel now had an annual income of 533 daler.

The Physiographical Association was founded in Christiania at the New Year, 1828. In many ways this was an outgrowth of the informal gatherings, some eight or nine years earlier, that had formed around Keilhau and Boeck and their mapping of Jotunheimen, and the friends' "small scientific Compilations." Boeck and Keilhau were central to this formal foundation, although Hansteen, Fougner Lundh, Pharmacist Maschmann and a couple of others were also involved. Abel was made a member on February 4th, and he faithfully showed up, on the first Monday of every month at five o'clock for the membership meetings. The meetings, with a lecture on a natural science topic, followed by discussion and an evening meal, were hosted on rotation by the members. For Abel these gatherings seem to have been most appealing, even though it does not seem that he himself contributed with anything from his own field of study, and nor did he host any sessions; rather, he treated these meetings almost as though they were soirées. The association changed its name in the 1850s to the Christiania Society of the Sciences, and later still, to the Norwegian Academy of the Sciences.

The Association's most important task at the beginning was to publish *Magazinet*, and in Hansteen's absence it was necessary to choose a new editor. Boeck, newly home from abroad, was chosen in a sort of competition with Professor Keyser, who wanted to make *Magazinet* into a more popular monthly, to the cost of the pure science content.

Teaching certainly demanded a great deal of time, but scarcely much preparation on Abel's part. None of his lecture notes have survived, his concentration and lust for work was determinedly focussed around his own projects. He had finished the second part of "Recherches" on February 12th. This would be published in *Crelle's Journal* on May 26th. In many ways, "Recherches" contained ideas with which Abel had been preoccupied for a long time. Already in the summer of 1823, in a letter from Copenhagen, he had mentioned the "revolutionary reversal" of the problem: instead of considering the elliptical integral's value as a function of the upper limit, Abel treated the function's upper limit as a function of the value of the integral. Since then, Abel had given himself much time to work out these ideas in the conviction that nobody else would approach integrals and functions in this manner. But now, in an agonizing manner, events were to reveal that this conviction would not hold water. The elliptical functions would occasion a veiled competition between Abel and the German mathematician, Jacob Jacobi, about who discovered what. This competition is a dramatic and oft recounted episode in the history of mathematics. Later, for the most part, it was considered that Abel was the first, and did the most fundamental work. Many considered it important that this be demonstrated, not in the least because right after Abel's death, Jacobi published a large textbook with the title, *Fundamenta nova theoriae functionum ellipticarum*, that became much used in Europe's universities, and where Abel was not given much credit.

Abel would have spent much of his time working on elliptical functions, and would come to write a further seven large and small papers on the theme. Six were published after his death, and the unfinished seventh, "Précis d'une théorie des fonctions elliptiques," covered 100 large-sized pages.

Carl Gustav Jacob Jacobi was two years younger than Abel, born and raised in a prosperous Jewish family in Potsdam. In 1826, shortly after Abel had left Berlin, Jacobi had become a zealous contributor to *Crelle's Journal*, and had received a permanent position at the University of Königsberg. Jacobi certainly taught a number of mathematical subjects brilliantly, and he shared his discoveries in the field of numbers theory with Gauss. And Gauss, who considered numbers theory his pet field, lauded the young Jacobi. Jacobi had also studied Legendre's work on elliptical integrals, and during the summer of 1827, he sent some small contributions, dated June 13th and August 2nd, to Schumacher at Altona and his *Astronomische Nachrichten*. Here, without proof, Jacobi had submitted special and general transformations of the elliptical integrals. As it was without proof, Schumacher was in doubt as to whether Jacobi's contributions could be printed, and he sent them to Gauss who, he knew, had studied elliptical functions. Gauss replied that Jacobi's results were correct, and that they easily followed from his own work, but that they were only fragments of a greater work that he himself hoped one day to have the time to publish. Whereupon Schumacher published Jacobi's formulas in *Astronomische Nachrichten* in the September issue, 1827, about the same time that the first part of Abel's "Recherches" came out in *Crelle's Journal*. Many considered that these two works revealed that something new was occurring in the field of functions. Legendre studied Jacobi's transformational equation and was extremely impressed, and referred to it enthusiastically at the meetings of the Academy of Sciences in Paris in November, 1827. Jacobi's equations were quoted and reproduced both in journals and a number of German newspapers and became the basis for his renown. But Legendre doubted whether Jacobi had the proof for his equations. He initiated a correspondence with him and bid him urgently to make public his methods. Schumacher was of the same opinion; he also wrote to him requesting the proof. In order to defend his scientific honour, Jacobi was now forced to come up with a decent deduction. It was at this point that many, in later times, thought that Jacobi found his saving ideas from the first part of Abel's "Recherches". It can be verified that *Crelle's Journal* had reached Königsberg by the beginning of October, 1827, and this issue, most of which was taken up with Abel's paper, also included one of Jacobi's works.

Jacobi worked intensely to find proof for the equations he had put forward, and at the end of November he had cobbled together a proof, precisely by using Abel's idea of "inversion". A month later, Jacobi's proof was printed, and he himself was appointed professor at Königsberg, at barely the age of twenty-three.

By New Year's, 1828, Legendre in Paris, through Poisson, had become aware of Abel's work, and both Abel and Jacobi were now proposed for membership in the mathematical section of the Institute in Paris. Neither of them were voted in; Jacobi got three votes, and Abel, none.

At the end of January, 1828, Abel got to see Jacobi's contributions in the September issue of *Astronomische Nachrichten*. In this connection he would certainly have noticed that Jacobi's transformational equation could easily have been deduced from his own equations. But when, at one point in March, Hansteen showed him the new issue of *Astronomische Nachrichten* with Jacobi's proof for the equation, Abel realized he had a competitor who was apparently making use of his ideas without a single word of acknowledgement to him. Abel was shocked. According to Hansteen, he became "quite pale and upset, and said that he had to run to the café and take a bitter schnapps in order to recover from the Alteration".

Abel was certainly disgusted, and indeed, felt as though he was in danger of losing yet another of his major works. He was afraid that the Paris Treatise was lost forever. Given this situation, Abel now worked harder than ever before. He thought he had the solutions to considerably more general transformational problems than that which Jacobi had set out. He did not want to attack Jacobi, but rather to establish an *approach to the problem* that was so comprehensive that Jacobi's treatment would be left in its shadow.

Simultaneously, Abel was busy putting the finishing touches to a large work on equation theory. On March 29th he sent off his treatise, "Mémoire sur une classe particulière d'equations résolubles algébriquement", but this sat at Crelle's for over a year and was published just before Abel's death.

On May 27th Abel had finished the work that was to be a response to Jacobi's treatment; he ominously called the piece "The Death-ification of Jacobi". The actual published title was "Solution d'un problème général concernant la transformation des fonctions elliptiques", and naturally enough this was sent to Schumacher and *Astronomische Nachrichten* which had published Jacobi's work on the subject. Enclosed with this submission was a letter of recommendation from Hansteen to Schumacher. Hansteen had just left on his Siberian expedition, and writing this letter was the last thing he was to do for Abel. Hansteen urged that the paper be printed as quickly as possible. He maintained that for many years, transformation theory in its general form had been Abel's turf, and he told about Abel's "alteration" and the bitter schnapps at the café when he saw Jacobi's work. For his part, Schumacher sent Abel's paper to Gauss, on June 6th, with the following caustic comment: "Once You have made known Your Assessment, this will surely cost him more in schnapps."

Both Schumacher and Crelle, as well as others, had eagerly tried to get Gauss to publish *his* works on the subject. But after having read the first part of Abel's "Recherches", the great Gauss had replied that Abel had preëmpted him, that Abel had reached the same results he had in 1798. And Gauss had added: "Seeing as he [Abel] has, with respect to his account, brought so much delicacy, depth and elegance into the light of day, I therefore see myself, for this reason, exempt from publishing my own findings." Abel now probably received an account of this recognition from Gauss, via Crelle, at the time he was finishing his "Solution".

We know really very little about what this time was like for Abel, between January and summer, with its teaching, students, friends and feverish work. At the end of February he helped the young mathematician, Otto Aubert, get an article published in *Crelle's Journal*. This was a ten-page paper that answered some geometric works published in an earlier issue of the journal. Otto Aubert was nineteen, and for the past couple of years, after taking the University's secondary examination, he had been teaching history and geography at the Christiania Cathedral School. Most likely Abel came to know him through his brother, Johan Aubert, who had been teaching at the Cathedral School since 1817, and through other circles around the Aubert brothers at their home in Akershus. Abel wrote to Crelle in Berlin to say that the publication of the article would be a great encouragement for Otto Aubert, who to be sure was quite young, but demonstrably full of talent. Indeed, Aubert's article was published, but not until the year after Abel's death, and when, some years later, Aubert arrived in Berlin he was very well received by Crelle and taken along on his weekly strolls together with Steiner, Dirichlet and Möbius, and on that occasion Aubert was urged to obtain six portraits of Abel.

In recent years in Christiania and a number of other places in the country there had been large-scale celebrations of Norway's Constitution Day, May 17th. In 1826, Bjerregaard's "Sons of Norway" was being sung everywhere in Christiania, and in the evening, the words: EIDSVOLL 17. MAI[18], were illuminated in flaming letters above a ship in the harbour. But a year later, when Parliament had celebrated the day with an enormous festive dinner, this had aroused King Karl Johan's anger to such a degree that he now, at the beginning of May, 1828, had sent out a proclamation against festive celebrations of the day. This was disseminated by couriers sent out in all directions, and proclaimed in cities, towns and villages across the land. But no one knew if the people would acquiesce to the will of the king. The king himself was in Norway's capital at the time, in full uniform from early morning. It was said that when he buckled on his sword that day he had muttered, "God save me from having to draw this out of the scabbard today." Naturally enough, Parliament had postponed all public commemoration of the day, and for the first time since the day began publicly to be observed, in 1824, not even the students gathered to celebrate. The few who gathered privately, and vociferously sang patriotic songs were reported and made note of in police reports. Some disciples of the Cathedral School who had, through the window, greeted some Swedish royal retainers with "Vivat the 17th of May!" were summoned to an interrogation before a assembly composed of the Minister of Church Affairs, the Bishop, the Chief Administrative Officer of the Christiania Diocese, and the Commander of Akershus Fortress. Otherwise, Constitution Day in Christiania passed peacefully. It was only a year later that it came to dramatic riots, in what came to be known as "the Marketplace Battle."

Hansteen had finally journeyed to Sweden on May 19th to commence his expedition to Siberia. He had long wanted to have Keilhau and Boeck with

---

[18] Eidsvoll is the town in eastern Norway where, on May 17, 1814, patriotic representatives for all regions of the country signed into existence Norway's Constitution.

him, but Boeck was about to get married, and the faculty thought that Keilhau could not be away from the university for such a lengthy period of time; consequently they refused the application for leave of absence. The newspaper *Rigstidende* reported on Hansteen: "He was escorted away from home by First Lieutenant Due of the Navy and in St. Petersburg is to collide with Dr. Ermann from Berlin, who will also make the Journey with him, both as a Natural Sciences specialist and an Astronomer." The Norwegian public was kept informed about how things were going during the course of the whole expedition, by means of letters and commentaries, and when Hansteen returned two years later, he was greeted as a popular hero.

This year, 1828, on May 25th, in the midst of Abel's final spurt of work on the "Solution", his brother Thomas came to Christiania, to try once more to find a way to earn a living. After a short time he found it expedient to depart from the capital, leaving behind him unpaid lodging bills that he expected Niels Henrik to take care of.

Abel was expecting a visit from Christine at the beginning of June. She was to come by steamship together with her employer, Sivert Smith, owner of the ironworks, who was on his way to Copenhagen. Abel had explained to her that this time she could stay at Mrs. Hansteen's. Mrs. Hansteen noted in her diary on June 1st: "Busy with sewing and getting things in order for the arrival of Miss Kemp." Abel picked up his sister Elisabeth, and together they went out to Drøbak, a town on the eastern side of Oslofjord, to meet Christine there. They got back to Christiania on June 8th and Christine, who was also known as "Crelly", stayed for three weeks. Many things now seemed to be pointing in the direction of happier times for the young couple: Abel could take the time for quiet discussions at Mrs. Hansteen's, and go for long promenades with his fiancée through the streets and past the outlying properties with the safe assurance that "Recherches" was published and that "Solution" was quite certain of publication in the next issue of *Astronomische Nachrichten*. And a report had come from Crelle in Berlin about Gauss' commendatory words. Whether Abel heard of the laudatory things being said about him at this time in Berlin is less certain. But Crelle, who worked tirelessly to secure *Journalen*'s future, had pointed out, in a letter to the Prussian government on May 14th, the journal's importance to mathematicians, and in particularly he promoted his young collaborator, Abel, and Abel's work on the fifth degree equation and on elliptical functions. Crelle had written that Mr. Abel "is a young Norwegian, and perhaps one of the foremost talents in higher mathematics at the current point in history, but unfortunately he remains all but completely overlooked in his native country and he is struggling against unfavourable conditions." This made such a strong impression on the government in Berlin that Crelle, had already received a reply by June 7th, to the effect that "the ministry would like very much to try to get the young scholar supplied with service to the state here" but in as much as there were for the time being no definite positions that could be offered, Crelle was asked to "find out whether he [Abel] would be disposed towards stepping into service as a private senior lecturer at the local university with a passable annual remuneration." Crelle quickly wrote to Abel and urged him to answer, but without telling him precisely where the enquiry had come from. Abel received this joyful news exactly at the time he was enjoying the presence of his fiancée, daily at Mrs. Hansteen's house and garden on Pilestrædet. Discussions about the bright future and the coming "Herr Professor und Frau Gemahlin" must have delighted them all. Nevertheless, the thoughts of having to leave his homeland and Norway's university seemed to disturb Abel. It was as though he was looking for a clear sign that it was right to abandon Norway: he tried to provoke a reaction that he could relate to. On June 21st, he wrote to the Collegium:

> As, at this Moment, a View of Employment Abroad is opening for me, namely at the University of Berlin; thus with respect to this I take the Liberty of consulting the high Collegium in order to ascertain whether I might obtain permanent Employment here. Of course it is my innermost Desire to devote my Life to my Fatherland in a Manner in which I can be of service; but without such an assurance I believe that I ought not refuse a Means of securing my Future, that here appears to be extremely precarious. Should a permanent Position not now become my Lot I would indeed dare presume that my Constitution at the University ought not be any Hindrance to me seeking to obtain Employment in Berlin. Should at some later point in Time, there open up a secure Livelihood for me here, then there would certainly not be anything on my Part against my coming back, if I even dare cherish this Hope. As I am most urgently importuned to submit my Answer immediately, I would perhaps presume to ask that the high Collegium take up this Matter as soon as possible. It is for me a Matter of Welfare.

On the same day, June 21st, in a letter to the Pro-Chancellor, the Collegium regretted "that at the present Time it is not possible to propose any permanent Employment for this talented young Man, by means of which he could be retained for the University and the Fatherland." Nevertheless, the Collegium felt itself dutybound to find out if perhaps from the public sector there could be any opportunity to offer Abel a position "corresponding to his in the learned World where his Merits have been recognized, leading to Prospects through an Offer at a foreign University, where most certainly, after a short

Employment as a private senior lecturer, he would obtain a professorship." In the end, the Collegium noted that Abel's present position as a temporary senior lecturer would be no hindrance to him accepting employment in Berlin, for Lektor Holmboe had pronounced himself willing to take over all the teaching duties in Christiania.

Treschow had resigned from the post of Pro-Chancellor at the University at the end of March, and the new man in the post was the seasoned politician, Count Wedel Jarlsberg. And now Count Wedel Jarlsberg could only join in the wish that it had been possible to find a position for "this talented young Man", but "as I up to this point do not see for a Moment the Possibility, thus with respect to this matter I do not dare make any Recommendation, and must therefore comfort myself with the Hope that in the Future the Opportunity might arise to get Herr Abel back." This significant signal was written by Count Wedel Jarlsberg when he forwarded this matter to Minister Diriks of the Church Department on June 24th. Abel must have received clarification about where the matter stood, and on June 30th, he wrote briefly to the Church Department, and urged them that the whole matter "should be shelved until further notice." On the same date there is the notation in the Church Department's journal: "Granted and shelved."

But Abel had now made a decision. He would write to Crelle and say that he was now, at any time, willing to take up service as a private senior lecturer in Berlin. In reality the future looked bright, or did it? Had his uneasiness over having to leave his homeland finally left him? Did his bond to the homeland arise from his sense of duty, and was it a release to know that it was persons other than himself who forced his departure? And what about "the circumstances of love"? Did Abel have concrete plans in the direction of marrying Christine, or did he perhaps hope in his innermost soul that external forces would also step in and make the decision on his behalf? We know almost nothing about this for sure. Mrs. Hansteen's diary notations are incomplete and give no hint about discussions and future plans. But had the desire to marry been strong, then perhaps would this not have come to be expressed in ways other than the charming chat about "Herr Professor und Frau Gemahlin"?

Abel was in any case invited back to Froland, along with Crelly, that summer. They would spend a month and a half together there. Thus, on Tuesday, July 1st, they left Christiania on the steamship *Constitutionen*. One of Abel's friends from Regentsen, Morten Kjerulf, who was also on board, later recounted that Abel had had himself fitted for glasses, and now had a more respectable appearance.

47.

The Final Summer

The approach of the steamship *Constitutionen* enraptured people everywhere; people streamed to the quays; songs and poetry were written in its honour. One such poem was penned by C. N. Schwach:

Swim safely through the North Sea's roll
You, most wondrous Work of Art!
Who in your innermost Soul
Conceal an Elf of stoutest Heart
Lodged in friendship, brave and peerless
In the Ocean's very Breast and Bowels,
Who so calmly smiles, fearless,
At Njørd[19] and Ægir's[20] ready Growls.

Niels Henrik and Christine landed at Risør and presumably journeyed through Gjerstad on their way to Froland. Niels Henrik would of course have wanted to show his fiancée the village of his childhood and present her to the family. In any case, he wanted to tell his mother and siblings the good news: Father Abel's debt had been settled once and for all; no debtors' auction could force them from the farm or the land. The remains of the stipend from the university had gone to cancel the outstanding grain debts.

Some days later Abel and Christine reached Froland. After a year's employment, Christine seemed to have been much loved as governess, and the good prospects for Candidate Abel in Berlin must have increased his respectability. On one of the first days they were at Froland, Abel became ill and was confined to bed, but he was soon on his feet again. He had his own room in the Smith house, and here he was able to continue with his mathematical work, particu-

---

[19] Njørd, Njord or Njarðar the Crafty, was the Nordic god of wind, sea, well-being and prosperity.
[20] In Norse mythology Ægir personified the sea. His wife, Ran, kept all drowned souls enmeshed in her net beneath the waves.

larly that which had to do with elliptical functions. In between times it seems that Candidate Abel also functioned as a play-uncle to the children. He had stories to tell from the great outside world, and when this was not enough, he could demonstrate his swimming prowess, as the water was temptingly near. The children held Abel in high regard and among other things gave him socks and pocket handkerchiefs. In their free time, Abel and Christine frequently strolled around the lovely park beside the ironworks, and it was said that they were often be be found sitting in the white garden gazebo, high up in the garden with a view out over the water; people noticing them there assumed that they were engaged in discussions on marriage and the future.

It had been Sivert Smith's mother who, around the turn of the century, had laid out a large and magnificent garden in the baroque style, with the gazebo up at the top. At this time 162 people were employed at the Froland works; in addition, there were other casual workers whose employment depended upon the delivery of charcoal, iron ore, and timber. There were cast iron ovens one, two and three storeys high; here they made iron cooking pots, spatulated baking sheets for bread, and ballast iron, nails and ship's spikes, and above all, every year the works had produced more than thirty thousand cannonballs of various sizes. This was in the golden era before the war years of 1807–14. But there was still great activity now. A similar number of people were at work, a double iron-rod hammer and six spike drivers were in operation. In addition, three large gate-saws made it the largest sawmill in the district.

The weather that summer was very changeable in the region: alternatively rain and sun and enough warmth to ensure a good crop; when all was said and done, it had been called a good summer. But in another area, dark clouds began to gather; something was brewing that would now plunge Abel back into a state of depression and uncertainty.

It began in Christiania. Abel's short message to the Church Department concerning his query about the possibility of permanent position in his homeland had been dismissed, but it had reminded the Minister of another matter. Minister Diriks now found out that in reality Abel had been treated far too generously. Diriks proposal the previous year that Abel could be paid from the University coffers had been expressly premised upon the money being drawn from Abel's future earnings in a position at the University. Now the Department made it clear that since February Abel had been receiving the salary for an associate professor position without having paid back any of the previous year's money. The Department discovered that the Collegium had given Abel the money as a stipend and not as an advance, and Diriks therefore wrote to the Collegium and pointed out that they had acted at cross purposes to the Department's recommendation. Diriks requested that the Collegium "take Care that the aforementioned Advance, in adequate installments be

deduced from Mr. Abel's salary, none of which has yet been recompensed."
The Collegium thought that this was a little too much, and answered on July
14th, that it did not consider it fitting to demand any repayment "with Respect
to the longterm Prospects that Hrr. Abel had for the achieving of some
permanent Employment", and the Collegium pointed out that nevertheless
Abel's present position could be considered only provisional, and that without
doubt it would prove "utterly importunate" for him to have to repay the
money.

This matter now became the basis of a two-month-long feud between
Government and University, a feud in which the Collegium felt the University's whole independence of internal affairs was at stake. Diriks of the Church
Department did not want to admit anything further to the effect that the
Collegium had the authority to disburse the University's funds without the
decision of a higher body, and he wanted, through the person of the University's Pro-Chancellor, a decision about whether or not a support such as the
Abel stipend represented, had to be, or ought to be handled from university
funds. The Collegium answered with Chapter Six of "The University Funding
Protocols" which showed, according to their thinking, they had full rights to
dispense monies for positions up to and including the sum of twelve hundred
speciedaler per annum. The Collegium felt that it had acted within the bounds
of its mandate on this matter, and it stood firmly and unshakeably behind its
decision. When the matter was sent on for the Pro-Chancellor's deliberations,
Count Wedel Jarlsberg was away; in his absence the matter was handled by
Professor Rathke, who signed an expression of full support for the Collegium.
In the Church Department the matter was now calmly shelved. The Department merely stated that Abel's stipend would not be handled by revising the
university's accounting process.

Beyond this, Minister Diriks had a terrible reputation among both students
and teachers. It was said that he was both lazy and autocratic, and very skilled
at getting others to do his work for him. Since he only had one eye, he was
often called "Polyphemus", after Homer's cruel cyclops, and in intellectual
circles it was maintained that the only good service carried out by the cabinet
minister was that he was careful to see that all the free copies of the books
published in Norway, which by law were to be sent to his department, were
assiduously passed over to, and cared for by the main library, Deichmann's
Bibliotek.

On July 7th a new newspaper came out in the capital, *Nyeste Skilderi af
Christiania og Stockholm*, with the bookseller, Jørgen Schiwe as editor and
publisher. In its first issue, *Nyestes Skilderi* discussed rumours about Abel's
prospects in Berlin: "Our young Scholar, Abel, teacher of Mathematics at the
University, who a little over half a Year ago came home from his scientific

Travels Abroad, has received such Esteem in Berlin that he in recent Days received from that Place's scientific Seat of learning, an Offer of Employment which does him and the Fatherland great Honour. One regrets that for the present Time there is no more esteemed Place for him than that which to date has been held open for him, yet we must be happy over the fact that he now in consequence is prompted to consider yonder honourable Offer, and we cherish the Hope that one day, with a famous Name, he will hasten back to the beloved Land of his Birth."

Abel was outraged at this entry. Crelle, from Berlin, had been very circumspect with regard to news about Abel's future position, and it did not serve the matter well that now Abel was being portrayed in public as having entered a safe economic haven.

In Berlin, around the same time that the newspaper item came out in Christiania, there was another item of news that would brutally pour icy water over Abel's expectant hopes for an improvement in his circumstances. On July 20th Abel received word from Crelle that he had a competitor: a second applicant had appeared, like a bolt out of heaven, or "vom Himmel gefallen" for the expected position in Berlin. Suddenly everything seemed to be darker than ever. The next day, Abel in Froland wrote to Mrs. Hansteen in Christiania:

The letter I received yesterday from Crelle and dated the 11th of July begins with "Leider" [Sorrows] – an ominous Word, and with Leider, I must confess, the letter has dispirited me enormously. Es wird nichts daraus [There is nothing to come of it]. Someone else has appeared, as though dropped from Heaven, who has made his Ansprüche geltend [staked his claim], and who inevitably must be dealt with before they think of me. Who this other is, Crelle does not write, and I know nobody of that Calibre. He says that he will not stand up for me for a Second as this would do me more harm than gain – God knows for what Reason. Moreover, the Minister of Education is away and will not be back for 8 Weeks from the Date of Crelle's letter. He says, therefore, that he cannot give me any definite Answer before the coming Month of October. But his Letter is written in such a disheartening Manner than I have given up all Hope. That will say that I am consequently back at the same Point I was before, though somewhat worse, for I have prostituted myself here, and the same thing can happen abroad (see the edifying Piece in the Newspaper given out by Bookseller Schiwe: "Nyeste Skilderie af Christiania og Stockholm", No.1. Page 6). I will say nothing against it in order not to reopen a filthy Matter. Now it will simply be put down to one of the Lies of the Press, et enfin le temps tue tout [in the end, time kills everything]. This will go its own way. It is with difficulty that I look for something else in Christiania. As before, I shall slave through all the commitments that I already have. But I have learned to shut up; that is one good Thing.

Mrs. Hansteen was urged not to say anything about this matter, only that he "had never received an Offer". And if his brother, Peder, should show up at her house, he should not be left with the impression that he ought to abstain from finding a post as private tutor. And Abel went on to say, "What pains me most is my betrothèd's situation. She is much too good to suffer this."

He bade Mrs. Hansteen withdraw his salary for July from the paymaster's office and from the money send him a ten-speciedaler note folded up inside thick paper and an "envelope fixed with sealing wax", and he concluded:

So that the Boy who comes here from Arendal with the Letter will not steal it, it is best to write nothing on the outside about Money. But do not be vexed on my account because I am troubling You so. Indeed, I quite annoy myself as well. [...] Crely sends you her special greetings. She wrote you by the last post. I am trying to comport myself here in a tolerably manly fashion.

Most lovely Mrs. Hansteen, I remain, Your most poverty-stricken Creature,
N. H. Abel

One week later, the following Monday, a post day at Froland, Abel sent a letter to B. M. Holmboe. The subject was the same.

...the Berlin venture. It has gone to Hell, and I, almost with it. Crelle wrote me on Sunday, a Week ago, that vom Himmel her gefallen [as though dropped from heaven], somebody else had advanced his Application, and that this had to be investigated. God only knows who it is, but it is of no consequence, for he has tossed me out, the Swine. Other than this, he writes that while this looks irregular, I must not thereby give up all Hope, as things are possible later. I shall receive a definite Answer in October. But you are not to say anything about this. Only that I never was, nor shall be, going to Berlin, which is very close to the Truth. It seems that Crelle has been unhappy that I have mentioned the matter.

It was important to Abel that nobody was "under the Impression" that he was going to Berlin. There had been talk in Christiania that perhaps there would be a need for a teacher of astronomy at the War College, and Abel was now eager to obtain that position. He urged Holmboe to communicate this: "I must be available in all Quarters," he wrote. Apart from this, Abel wrote that he had received a letter from Schumacher that "My Death-ification of Jacobi" was printed, and he urged Holmboe to congratulate Johan Aubert on the occasion of his appointment as vice-rector of the Cathedral School, and he wrote that he would be leaving Froland two weeks from Friday. In addition, in order to secure a roof over his head, he concluded with the hope: "You will not take it amiss if I spend a couple of Days at your place upon my Arrival, until I find myself some Lodgings, and that I do not pay for the Letter that I have here enclosed for you to take care of?" The enclosure was a short letter to Mrs. Hansteen, asking if she could send her maid down to the merchant, Lars Møller Ibsen, and say that as his brother Thomas had left the city without paying his accounts, that Abel would pay when he returned to the capital. There is also another little note written by Abel while at Froland, probably to Mrs. Hansteen:

I am as poor as a Churchmouse, in as much as I do not have more than 1 Sp. 60 sk., which I must give up as Drinking Money. Yours destroyed,

It appears that Abel left Froland on the morning of Friday, August 15th; in any case, he climbed aboard the steamship from the Arendal quay later the same afternoon. He was back in Christiania on Sunday evening.

48.

The Final Round, Autumn, 1828

Abel was back in Christiania well before the lecturing was to begin. He stayed at Holmboe's and was looking for a place to live. Mrs. Hansteen was in the process of preparing to journey to Copenhagen with her six children. Among other reasons, she intended to save money while Hansteen was away in Siberia by living at her family home in Copenhagen. Her diary gives little information about the period prior to her departure, and it says nothing about whether Abel had visited her. On August 28th it recorded: "To dinner at the Treschows'. Ørsteds there." And on one early September day: "One of those mornings when the loveliest Rainbow stood so close and was so vivid that it had one Foot on the Brochmanns' Roof."

Abel found himself new lodgings, and he continued with his mathematical work. But it was not long before he grew ill again, and was confined to bed. A doctor came to see him, perhaps Skjelderup, who expressly told him that strong exertions would be very harmful. When on the evening of September 15th, Mrs. Hansteen boarded the steamship bound for Copenhagen, Abel was so ill that he could not say goodbye. And with Mrs. Hansteen's departure, Abel lost his helper, confidant and wailing wall. A week later, on September 22nd, he wrote to her in Copenhagen:

You were certainly somewhat surprised (if I should not use a stronger expression), my most wonderful Mrs. Hansteen, not to see me before your departure, but in any case your surprise was greatly exceeded by my disappointment at not being able to go out. I was sick and bedridden from the evening you were at the Treschows' until a Couple of Days ago. But now, praise God, I am very well. It is so strange, I cannot get it into my [head] that You have gone, and many times I find myself about to go to visit You. I am however almost completely alone. I assure You that in the most profound Sense I am not in association with a single Human Being. Nevertheless, this Lack of friends is not foremost in my mind because I have so horribly much to do for *Journalen*. [...] As recently as Yesterday, I received a letter from Crelle, in which he says that there is still Hope that I can come to Berlin, and that he will soon be in a position to determine whether or not there can be anything there. I am to greet You warmly from little Crelly. She herewith sends You a little cap knitted by her own Hands, and which she urged You to accept in its present Condition. That it is not quite finished is due to the fact that she had to send it here a Week ago, in order to get it to You before your Departure. Unfortunately I did not receive it until Tuesday, and that was too late. I have enough small things to write about, but the Lectures make things such that I must end this here. My dearest greetings to sweet Mrs. Frederichsen, and to all the Angels. With constant Wishes for your Wellbeing,

N. H. Abel

Crelly regards You nearly as highly as I do.

Abel still had considerable hopes of a position in Berlin in the course of the autumn, and in that event, he would, naturally enough, also have anticipated seeing Mrs. Hansteen and "all the Angels" again, on his way through Copenhagen.

From Berlin, Crelle did everything he could to keep up Abel's courage and expectations. In his last letter he had listed a series of positive responses to Abel's work from Legendre, Jacobi and other mathematicians he was in contact with. What had happened in Berlin, and who the "vom Himmel gefallen" competitor was, is uncertain, but there is much that points to the fact that Abel's friend from Paris, the Prussian, Dirichlet, had, unawares, been the man who dashed Abel's hopes and expectations of a post. Dirichlet had been at the University of Breslau since April, 1827, and now had the position of professor. A week before Crelle had written his letter to Abel on July 11, 1828, about the "sorrows" of someone else who had an application that had to be dealt with first, the Minister of Culture, von Altenstein, with whom Crelle was in correspondence and for whom he provided mathematical advice, had received a letter from the director of the Military College in Berlin, a Major-General von Lilienstern, with the request that Dirichlet be given leave from his position at Breslau in order to come along and help build up a new military school in Erfurt, something that would soon prove to be impossible to carry through. But some weeks later, on July 27th, the Department in Berlin granted Dirichlet an uncommonly long leave, something that occurred after the intervention of no less than Alexander von Humboldt, who indeed had long assisted Dirichlet and been his benefactor. In this period of leave, from October, 1828, until the end of July, 1829, Dirichlet had said he would be willing to teach some classes both at the Military College and at the University in Berlin.

When Abel wrote that "in the most profound Sense" he had no communication with a single soul in Christiania at that time, he must have meant in his private life, outside of teaching. The meetings of the Physiographic Society had still not properly come into being; the theatre season had not begun; Keilhau had been on a research trip to Nordland in northern Norway over the summer and was now extremely busy with his teaching; Boeck was about to get married: the wedding of Boeck to Minister Collet's daughter, was held on October 6th, and was celebrated in grand style, but whether Abel attended we do not know. Nor are there any reports of Abel's lectures in astronomy.

Abel had "horribly much to do for *Journalen*". On September 25th he set the date on a paper that, for the most part, he had written at Froland, "Addition au mémoire précédant", and now sent to Schumacher and his *Astronomische Nachrichten*, since it was a followup to "la Solution", or "the Death-ification" that indeed was published there. Abel concluded this work in the following words, translated from the French original: "There are still many things yet to say about the transformation of elliptical functions. One will find further developments on this theme, and on elliptical function theory in general, in a paper that is coming out in Mr. Crelle's journal." And now, in an apparent condition of euphoria, Abel produced papers one after the other and sent them off, urging quick publication. And Crelle did what he could to gratify his young friend, who had now become one of the most talked about mathematicians on the continent.

At the same time as Abel's fame was growing, it also became known that the young genius lived with little money and great insecurity in far-off Christiania. In Paris, Legendre looked at Abel's work with growing amazement, and more and more had his eyes opened by this Norwegian student he did not seem to remember having met during Abel's months in Paris. For some time Legendre had tried to regard Abel's inversion as an unnecessary detour; he was not captivated by elliptical functions and he held firmly to his own starting point, elliptical integrals. Legendre had been quite disappointed when he also saw that Jacobi used the inversion (it was at the sight of the same paper whose existence had forced Abel to take refuge in bitter schnapps). In addition, Legendre would not acknowledge Jacobi's proof before it was completely demonstrable by means of his own theories, and this he succeeded in doing, at the end of his days when he was almost eighty years of age. It had been Legendre who first regarded the works of Abel and Jacobi as locked in a duel. He used the expression "the young Athletes". Jacobi had confessed in his correspondence with Legendre that Abel's "Solution" cast its shadow over his own work, and that transformation theory was attributable to Abel. Gradually Legendre began to prefer Abel's breadth, his result-oriented presentations, where underlying methods were clarified, to Jacobi's rather sketchy descriptions of his discoveries. But while Jacobi was working on his great work, *Fundamenta nova theoriae functionum ellipticarum*, Abel had no means by which to have his own assembled works printed. Indeed, Crelle paid for the publication of Abel's articles in *Journalen* from his own pocket at the rate of one ducat per sheet of paper.

A communication was sent from the august Institut de France on September 15th, to Karl Johan, the King of Sweden and Norway, who happened to be of French ancestry. The king was described as a Magnanimous Prince who was well-known for his desire to illuminate scholarship and uncover "those of modest earnings and help those whom fortune has treated badly", and thereupon the sovereign was reminded that he had, in his own realm, "a young mathematician, Monsieur Abel, whose works show that he has mental powers of the highest rank, and who nonetheless grows ill there in Christiania in a position of too little value for one of his so rare and early-developed talent." This communiqué was signed by Legendre, Poisson, Lacroix and Maurice, and concluded with a plea to the king to improve "the fate of such an outstanding man" and at least bring Abel into the well-renowned Academy in Stockholm. The letter from the French Institute was given to Count Löwenhielm in Paris and sent on by him. Nothing indicates that Abel ever knew about this request. Count Löwenhielm had to repeat to members of the French Institute innumerable times that he had received no reply to their letter. The reason seems to have been that Karl Johan passed the matter over to Crown Prince Oscar, who had requested information several times about Abel from Viceroy Count von Platen, who for his part was more concerned to discuss plans for building a canal from Christiania to Eidsvoll. Von Platen had made his name with the building of the Göta Canal in Sweden.

In a letter to Crelle on September 25th Abel thanked him for all the kind words and praises that he had sent on, and he thanked Crelle for his efforts to get him a secure position, and earnestly implored Crelle to report immediately on what had happened "whether it be glad tidings or not". If things did not go the way they wanted, he had to "get down to improving his situation" in Christiania. Abel urged Crelle to provoke Jacobi to produce his methodology, "for it is clear he is in possession of excellent materials."

In Abel's own mathematical work the situation was thus: after having begun a continuation of "Recherches II", he had gradually come to a new discovery, of the whole theory of elliptical functions, and this would be the basis of a new, greater work: "Précis d'une théorie des fonctions elliptiques." To find a publisher for such a book was unthinkable: the work had to come out in sections, in *Journalen*. But Crelle had the treatise on Abelian equations lying unpublished; he was waiting for the continuation, and wanted very much for Abel first to complete the solution of equations by means of square root. Abel confessed directly to Crelle in this letter of September 25th, that for the time being, the equation theory work over-taxed his physical strength. He recounted the doctor's advice, and indeed, by this point Abel may have understood that it may well be a question of how long his health would hold out. He wrote (in German): "That You have accepted my 'Précis' makes me extremely happy. I shall exert myself in order to make the manuscript as clear and good as it is possible for me to do, and I hope I succeed. But do You not think it best to begin with this paper instead of the one on the equations? This is an urgent question to You. That is to say, I believe, first, that the elliptical functions will be of greater interest, and second, that my health will hardly allow me to occupy myself with equations for some time." Work on the equations would cost him disproportionately much more work than the elliptical functions, but he emphasized expressly that if Crelle absolutely demanded the paper on the equations first, then indeed he would get it! "Moreover, the equations will quickly follow, and if You are not opposed, I am thinking to make the papers so short that in each issue [of *Journalen*] there can be something on elliptical functions and something on equations." But for the forthcoming issue, Abel would accordingly send only the piece on elliptical functions.

Crelle did not object to this, and Abel used the greater part of the autumn to edit his "Précis", but he never managed to get it completely ready. In his collected works, the "Précis" filled one hundred large pages, in *Crelle's Journal* it was published in the latter half of the year 1829, in Issue 3 and Issue 4, after Abel's death.

On October 3rd, encouraged by the laudatory reports in Crelle's letters, Abel took courage and wrote to the famous Legendre, the man whom, in a letter written from Paris, he had characterized as "an inordinately courteous Man...but unfortunately as old as stone." The letter has disappeared, but from Legendre's enthusiastic reply it is clear that Abel must have explained his most recent studies on elliptical functions, and in the end, also told about his paper "Remarques sur quelque propriétés générales d'une certaine sorte de fonctions transcendantes", which he had sent to Crelle, and in which Abel developed mathematical knowledge about elliptical integrals that would later come to be known as the hyperellipticals. Here he used an idea from the Paris Treatise in a rather special case, something that Abel also pointed to in a footnote, a footnote that would be given great attention, and that would set in process the hunt for the great work that he had submitted by hand to the Institute in Paris. Legendre received Abel's letter with great delight, and in his reply, written on October 25th, congratulated Abel on "the already beautiful mathematical treatments" published in Crelle's and Schumacher's journals, and he continued his appreciation by writing: "the new Details that You have been so kind as to impart to me about Your more recent studies, have, if such is possible, increased still more highly the claim of high esteem that You have acquired from scholars and scientists, and in particular, the claim upon my own." Legendre also expressed pride in being, in a way, associated with Abel's work, seeing as it had been to a considerable extent through his own work that Abel and Jacobi had found the opportunity to develop their unique talents. Legendre was now eager to send Abel his *Traité des fonctions ellip-tiques*, that came out in two volumes shortly after Abel had left Paris. The difficulty, on one hand, was to send them in a safe manner, and on the other, consisted of his reservations, for: "You will not learn anything from this work, quite the reverse, it is I who must trust in You [Abel and Jacobi] to increase to a high degree valuable discoveries that I never would have been able to attain by my own efforts, for I have reached the age where work is difficult, if not impossible." And Legendre continued: " The end of Your letter bewilders me with the generality that you have comprehended in your treatment of elliptical functions, and including even more complicated functions. I am extremely anxious to see the methods that have led You to such beautiful results. I do not know if I will be able to comprehend them, but it is at least certain that I have no idea about the means that You have been employing to move forward and overcome difficulties like these. What a head this young Norwegian must have!"

Abel never mentioned his Paris treatise in his correspondence with Legen-dre, and Legendre had also forgotten that at some point he himself had been appointed member of a committee that was to assess Abel's work. In Paris, Abel's manuscript continued to snuggle down ever deeper into the feather bed of papers on Cauchy's desk. Abel was no doubt afraid the manuscript was lost and it can now be assumed that he now intended to make it superfluous by publishing his treatments of equations through Crelle. The elliptical functions were only one part of the extremely daring analytical theories that the Paris Treatise contained.

Abel wrote to Crelle again on October 18th. It was an answer to a letter he had just received, and he gave thanks for Crelle's discrete involvement and con-cern with Abel's health, but, he wrote, he was well again and found himself to be so "unusually well" that without harm he could now work as much as he wanted, and this, indeed, he was doing! He was worried that no decision had yet come from Berlin, but he reassured Crelle that he would be patient, that he hoped for the quickest possible reunion in Berlin, sent his greetings to Crelle's wife, to Dirksen, to Steiner and Dirichlet. With regard to his actual situation in Christiania, he wrote: "Hier bin ich so weit von Allen entfernt und habe deshalb keinen mathematischen Umgang [Here I am so far away from everybody and I have no mathematical circle]." Had he been in Berlin, perhaps he would have found a publisher, and thus this could have perhaps been solved in a serious way, he wrote, and explained that he was in "ein Periode der Erfindung [Period of Research]", and that the work gave him considerable delight: "es geht mir so ziemlich von der Hand". He was pleased that "Remarques" would be published in the fourth issue of *Journalen*, and reminded Crelle of the two other papers he had sent to Berlin: "Note sur quelques formules elliptiques", and added "und einen anderen, ich weiss nicht mehr wie ich den genannt habe [and the second one, I do not remember the name we gave it]."

He noted that this paper, that he was thus unable to remember the title of, was the one that he wanted very much to be printed first. And that had already come to pass: the paper, "Sur le nombre des transformations différentes qu'on peut faire subir à une fonction elliptique par la substitution d'une fonction rationelle dont le degré est un nombre premier donné" was published in the third issue of *Journalen*, but it had not yet reached Christiania. Abel said that he was also eager to see the other paper published quickly, but he stressed that it was Crelle who best understood what there was room for. "Note sur quelques formules elliptiques" came out in the first issue of 1829. Abel had sent another paper, "Théorèmes sur les fonctions elliptiques", that was also a commentary on Jacobi, to Crelle at the end of August, and now in October he commented that it was important that this work accompany the "Précis" which he hoped would be printed in the first issue of the coming year's volume.

"Théorèmes" was published in the second issue of 1829. "Précis" subsequently came out in the two following issues.

Abel mentioned his equation work to Crelle once again, saying it would be sent as quickly as possible. He was worried about the postal service between Christiania and Berlin, and he wanted to write something more detailed on infinite series, and suggested in a perfunctory way that perhaps some of the pieces could be sent on to Gergonne's journal. But Crelle in Berlin would not hear of it. Crelle commented flatly that Gergonne in Montpelier would not do well by these pieces.

"I am however almost completely alone. I assure You that in the most profound sense I am not in association with a single Human Being. Nevertheless, this lack of Friends is not foremost in my mind because I have so horribly much to do for *Journalen*." Such were the terms in which Abel portrayed his everyday life to Mrs. Hansteen in September. When he wrote again in November, answering a letter that in the meantime he had received from her (and which is no longer extant), it seems that his external relations in Christiania were much the same. He attended to his lectures and worked intensely to perfect his mathematical theories. He recounted his foreign correspondence, and cited long passages from Crelle's and Legendre's expressions of praise. Right before the citations he wrote: "I have, I dare say, in the recent Period, basked in considerable pride occasioned by a Couple of Letters that I have received from Abroad. I want to indicate this to You by citing a Couple of Passages, though I really do not know how to do it without appearing conceited. And after having cited the most laudatory passages, he commented:

> For to tell the plain Truth, I suppose that in citing the above-mentioned passages I show myself to be a bit pompous, but I do so, partly, Dear Mrs. Hansteen, because I believe it will gladen your Heart to see the Future I have before me, since You share so much of my Weal and Woe. In consequence You must not consider this bragging.

Abel also wanted very much to tell her about his relationship to his fiancée. Almost precisely a year earlier he had asked Mrs. Hansteen to help clear up the incongruities between himself and Christine. Now, in November, he wrote this to his "spiritual advisor" in Copenhagen:

> I have perhaps not behaved toward her as I [ought] but now we are very much in agreement, and have come to adjust to one another openly. I have straightened up significantly and would hope that eventually we shall be able to live happily together. But when that happy Point in Time is to come, I do not know. Just that it should not be too far off. I feel so badly that my Crelly is compelled to slave so.

And Abel himself more or less tried to be a consoler for his "second mother" in Copenhagen:

> It pains me that You are not in the best of Humours down there. I well understand that many things must be streaming into your Feelings, and what makes You particularly anxious is of course Hansteen's Absence. This is completely natural, but think of how very happy You shall be in a not so very long Time. And thus one imagines what one desires, and You, Dear Mrs. Hansteen, have certainly the highest Probability on Your Side.

Apart from this, he told about his economic situation, about his salary and his debts:

> I am still at the level of 400, and am in Debt up to the Ears, but I have nonetheless worked off a little of this. In the meantime, my former Hostess,

The Queen, had not received a shilling and to her I owe 82 Spd. The Bank I have worked down to 160, and the clothing merchant from 45 to 20. Apart from this I owe the Shoemaker and the Tailor and the Man who provides Meals, but otherwise I am not borrowing. But now You are not to pity me for any of this. I shall come out of it somehow.

Of events and gossip in Christiania – what he called "Raüber Geschichter" – Abel had nothing to say:

On the other hand, turning to the Realm of Truth, it was a son of Councillor Saxild who now sits arrested for Thievery, and what is worse, for a Break-in at the home of Prof: Bugge. He took his Silver and smashed a Windowpane in order to get in. In the realm of the Ridiculous, one hears that there is a Priest who has had his portrait painted in full ornamentation, with his Dearest Love sitting in his Lap.

An underlying point here in this connection was probably the fact that Councillor Saxild and Professor Bugge were both active in the congregation of the city's Hutterite Brethren.

Great changes came to the theatre that autumn. Christiania's Public Theatre, as it was now called, under the leadership of the Danish Jens Lang Bøcher, had established an executive composed partly of publicly appointed representatives. The theatre building was restored and beautified, both inside and out, and both the repertoire and ensemble were alluring. The executive had invited the public to take out a subscription for thirty performances, held on Tuesdays and Fridays, and now, in addition there were Sunday performances. But Abel would not have taken the time to sit in the pit of the theatre. Earlier, he had written eagerly about his theatre experiences, but now he did not once report that Madame Bøcher's playing was being compared with that of Madame Mars in Paris! It was Christiania's *Aftenblad* reporter "R", well-known as the actor, Ole Rein Holm, who had also recently been in Paris and there seen the famous Madame Mars in the same role that Madame Bøcher now played in Christiania: Betty, in *The Youth of Henry Vth* by Alexandre Duval, a play that had been mounted many times at Grænsehaven. In addition, Madame Bøcher would be playing Thora in the strong, emotion-saturated final act of Oehlenschläger's *Hakon Jarl*.

At the end of November, on the 25th, Abel wrote to Legendre again. This was a long letter in which, page after page, he laid out his work on elliptical functions, and discussed his general results within integration. He regretted that an already-written major work had not been published: "Unfortunately it is absolutely impossible for me to be able to publish this work, for here I cannot find a publisher who would do this at his own expense, as I myself do not have the means. Quite the reverse, it is difficult for me even to find the bare necessities to keep body and soul together." Abel must have felt the last sentence was very strong as it is only to be found in a rough draft of the letter. But his need for expression and his happiness at having made contact with Legendre were great. Abel wrote that he reckoned it was "one of the happiest moments" of his life when he discovered that Legendre, "one of the greatest mathematicians of our century," paid attention to his work, and he continued: "This has to the highest degree increased my ardent desire for my studies. I shall continue persistently, but should I be so lucky as to make any discoveries, they cannot be ascribed to me, for undoubtedly I would have accomplished nothing without having been led by Your Light."

Abel thanked Legendre for the books he would be sending, and wrote that the package could be addressed to the bookdealers, "Messel og Keyser & Co." in Christiania. It was not uncommon for private post to be received there, but this may likewise indicate that Abel continued to have an uncertain lodging situation, that in order to save money, he planned to give up his current lodgings when, in two or three weeks' time, he planned to journey to Froland for the Christmas holidays. Abel ended this long letter to Legendre with the hope that he had not awakened displeasure by explaining his findings, and said that later he would very much like to communicate to Legendre "a great number of other results, both on elliptical functions and also on more common functions respecting the theory of algebraic equations. When Legendre replied a month and a half later, Abel was too ill to continue the correspondence.

During the course of the autumn in Berlin, there had been little movement in the plans to find a position for Abel. Crelle had tried to keep up Abel's expectations but so far had had nothing concrete to show. The hypothesis that Dirichlet and his unusually long leave of absence was the reason for the dashing of Abel's prospects for security, back in July, are strengthened by a statement that Alexander von Humboldt made in a letter to the astronomer and mathematician, F. W. Bessel, in December, 1828. Here Abel and Dirichlet were mentioned in the same breath, with an assertion that Dirichlet already was teaching at the military college in Berlin, and that hopefully Abel as well would soon be called there.

The institute that Crelle worked so assiduously to have Abel employed by, was still not established in Berlin. A tug-of-war as to who should be selected as the institute's director contributed to putting the plans behind schedule. The exceptional Gauss was indisputably and without doubt the mathematician who would have given the institute prestige everywhere in the world of mathematics. But many, like Crelle, felt that it was, rather, young active teachers who were needed, and who could attract and excite crowds of students. This was not exactly Gauss' strongest quality, and moreover, he had not shown the least interest in administrative tasks, something that to the greatest degree would be demanded of a director of a new polytechnical institute.

The young mathematicians that Crelle had in mind were Dirichlet, Abel, Jacobi and Steiner. With the exception of Abel, they were all in service with the Prussian state. Von Humboldt in Paris had asked Legendre to evaluate the work and achievements of Abel and Jacobi, ostensibly with a view to employment. In his November reply, Legendre had commended both of the two young mathematicians who, by their strokes of genius, had advanced his own, forty-year long life's work. Legendre reeled off Abel's and Jacobi's ingenious works in Schumacher's and Crelle's journals, and felt they both deserved von Humboldt's whole-hearted support. Jacobi may have been the first into print, but Abel appeared to exhibit deeper insights and a greater generality to his enquiries. And Legendre could not refrain from stressing that it had been Abel's remarkable acuity and intelligence that had persuaded Jacobi to admit that Abel's work had cast its shadow over, and influenced his own.

Crelle's first follow-up to his support for Abel came at the end of December. On December 28, 1828, Crelle wrote to the Secretary of State for Education. He called attention to their correspondence of the May-June period, concerning Abel's post, and the young Norwegian's on-going mathematical contributions were again outlined and accompanied by other mathematicians' statements about his work and achievements. Crelle wrote that he felt that it was now necessary to act quickly before others also discovered Abel's remarkable talent and would offer him a position; particularly since there was already talk of this in Copenhagen. Abel had also been very modest in his requests, Crelle commented, and informed them that in confidence Abel had communicated to him that since he had no income, and had run up debts in the course of his studies, and must in part support his siblings, he had hoped at the beginning to receive an annual income of five to six hundred *taler* and a sum of travel money, "perhaps as much as three months' salary." In parentheses, Crelle added that Abel was "the son of an evangelical pastor without means". (The taler was worth 2/3 of a silver specie.)

Before Abel completed his teaching obligations that fall, he submitted a request for a pay increase. He had to try to "improve his condition" before he left to celebrate Christmas with his fiancée and friends at Froland Ironworks. On December 6th, 1828, he wrote a carefully-worded letter directly to the king:

By the gracious Resolution of the 16th of February, this year, I was temporarily appointed, during Professor Hansteen's Absence on a scientific Expedition to Siberia, Associate Professor at the University, and with an annual Honorarium of 400 Spd., to carry out the University Duties appurtenant to this Professor who is on leave. Notwithstanding that this Remuneration was less than that which has been allowed to others at the University who are employed as Associate Professors, I must thus, due to my indifferent economic Condition, consider that, in order to have the Good Fortune of achieving anything whatsoever with my Studies in relation to this Position, which had promised me sufficient Recompense, such that I, in any Case, could have found minimally acceptable before I had been examined in my dona docendi [my actual record as associate professor /docent] and could petition for some Increase to the graciously bestowed Honorarium. Since that Time I have now lectured in Astronomy at the University, and in this way have taken over from Professor Hansteen his related Duties as University Teacher, and now, in part, the passage of Time has occasioned the Opportunity to judge whether or not in my Period of Employment can be considered to have been sufficiently compensated, and partly, the University has continued to have had the Opportunity to experience whether I, in my continuing Assignment, have matured. I therefore humbly risk hoping that my Plea will not be considered untimely or immodest when I hereby submissively solicit, from January 1st of next Year, to be placed at the same salary as the University's other Associate Professors, and as such, that I graciously be allowed an annual Honorarium of 600 Spdlr.

Humbly, *Niels Henrik Abel*

Despite the fact that Abel had written directly to the king, his request had to follow the usual service route: the Collegium approved Abel's performance as a university teacher and recommended a salary increase, and on December 16th, the request passed to the Pro-Chancellor's office, where Bishop Christian Sørenssen, father Abel's old colleague and bishop, signed in the absence of Count Wedel Jarlsberg. Since the increase would have to be paid out of state coffers, Diriks in the Church Department very much wanted to have the Finance Department's view on the matter. And Cabinet Minister Jonas Collet, on January 31, 1829, was unable to raise any objections "so long as he, during Professor Hansteen's Absence, attends to those duties at the University."

Before Abel left Christiania in mid-December, he apparently packed up all his things in a trunk and stored them away, possibly with his sister at the Treschow household at Tøyen, where on one of those last days in the city, he came for a visit. Mrs. Treschow said that Abel was not well. She said he appeared faint and had pains in his feet. But neither she nor Abel's sister, Elisabeth, managed to convince him to drop the cold and exhausting journey down the main Vestland road. The desire to be together with his fiancée at Froland Ironworks was greater than any distress arising from the possible outbreak of a cold or the influenza, or whatever else his body harboured.

On December 19th he glided into the yard at Froland on a horse-drawn sleigh, enveloped in a voluminous black overcoat, a frock coat with wide lapels. Lacking mittens, he had stuck his hands inside a pair of socks.

PART VIII

Fig. 40. Detail from the wallpaper in the "Buck Room" in which Abel spent his last twelve weeks at Froland.

49

The Deathbed

Abel was well-received at Froland, both by his fiancée and the Smith family, and even the household cat seemed to recognize him. By day, Candidate Abel romped in the snow with the children, chatted with the workers at the plant, and joked with the ladies of the house. In the evening he amused himself at the card table, and between times he sat down to his own work. To come to Froland, in comparison with the life he had been living in Christiania that autumn, must have felt like a continuous celebration; it must have made him feel, indeed, that he was situated upon an increasing curve, and for once perhaps he dared believe that everything could really turn out well.

One of Abel's expressions when he was happily in the company of others, was that he felt himself to be "among all the Angels". This was so at Mrs. Hansteen's, and now too, at Froland. Hanna Smith, one of the daughters of the family, was then twenty years of age, remembered many features of that period. Her memories, written in a letter more than fifty years later, are the source of much of the information that we have about Abel's final visit to Froland. Despite the evidence that reveals her desire to view everything in a positive light, it must still have been Abel's actual good humour that, in one way or another, found expression here. When Abel did not hold himself in check, whenever he yielded to fancy and knew that what he did and said would be taken in a spirit of goodwill and sympathy, then he could be both lively and playful. Those who did not know him could even take him to be a person of almost superficial character, or so Boeck maintained in his obituary. In the day room of the Smith house at Froland, Abel would often sit among the ladies of the house and work and write. It was said that he wrote on the thinnest available paper in order to save on postage. And when he got up from his work, there was nothing that amused him more than surreptitiously to steal a lady's handkerchief, or rummage around in someone's work box. His charming use of the second person singular to everyone who did not stand at too great a social distance was something that no one could resist. They said it was impossible to resist or get angry with him, or with whatever he got up to. They felt he was like a child in his rapid shifts from gaiety to melancholy, and would grin from ear to ear when he pulled one practical joke or another, and he would become completely crestfallen at the smallest failure, as for example, when he had opened the pincushion which had been stuffed with iron filings, and in an effort to separate out the largest filings, had come to spill them all. Or when at the dining table, to everyone's great dismay, he snatched the serving terrine from the lady of the house, and fished out the raisins that remained at the bottom. No one could deny him anything and no one took offence: his charm and good-heartedness, as it was called, compensated excessively for whatever he customarily lacked. Around him, gravity and consternation came to be transformed into laughter. Was this perhaps an exaggerated and hectic phase before the illness took him over?

And when, at that period toward evening, he waited eagerly for the candles to be lit so that they could all gather around the card table, because he greatly enjoyed playing cards - certain as he was to win some trifles. And the others could not but be astonished at what they called his constant *good luck*.

People later said that it was following the Christmas balls with their dancing and jubilation that Abel's fragile health became apparent to all. Hot and sweaty from dancing, he had gone outside to cool off, and out in the cold he had begun to cough horribly. He had had to go to bed the following day. It was pneumonia and the beginning of twelve weeks of invalidism.

In between times he had some good days, felt better and believed that he would soon be well again, but objectively, his being bedridden became chronic, deepened, and proceeded on to the final outcome.

January 6, 1829, was the date that the ladies of the house remembered that Abel had been conspicuously quiet and was in and out of his room a great deal. That was the day that he reformulated the introduction to the work he was afraid had been lost in Paris. It was a couple of pages, which he called a theorem, without introduction, and no superfluous remarks, and no applications, only the bare main theme, with the title "Démonstration d'une propriété générale d'une certaine classe de fonctions transcendants." He later admitted that this had been a major exertion. But he accordingly had it sent off to Berlin, and this came to be published in the second issue of the 1829 volume of *Crelle's Journal*. This "Demonstration" would be the last thing he managed to complete.

The plan had been to journey back to his teaching duties in Christiania on January 8th, but he was feverish, felt sharp pains in his chest, and complained about a coldness in his back.

The next morning everyone in the house became apprehensive when they noticed that he was spitting blood when he coughed. They sent for a doctor; and it was in consequence of this turn of events that District Doctor Møller from Arendal found his condition critical and imposed a state of caution and quiet.

These sharp pains in the chest continued for some days. But he felt that Møller's medicine helped, and after two or three days the doctor found him so much better that he considered him convalescent. And since Abel wanted so strongly to take up his work again, Møller acquiesced and allowed him to sit up for a while every day. For the next number of days he got up around midday and wrote feverishly. But after a few days of improved health, the coughing and faintness returned with increased vigour, and he had to stay in bed. After this, he was only on his feet on those occasion when his bed was being made.

District Doctor Møller, Alexander Christian Møller, was sixty-six years of age. As a youth he had been like a son to pharmacist J. H. Maschmann in Christiania, the paternal grandfather of Niels Henrik's friend, C. G. Maschmann. For over thirty years now, Møller had been a very well-liked and respected doctor among both the high and the low of Nedenes District. Møller had also known Abel's father well from the time of The Society for Norway's Well-being, and during the autumn of 1814 they had both been representatives to the Extraordinary Parliament, and had stood together in view of the political realities that now unavoidably must lead Norway into union with Sweden. Otherwise, Møller was well up on the medical science of the day. He had also become a member of the Norwegian Association of the Sciences in Trondheim. It was said that Møller was very good at encouraging and giving heart to his patients, but what he gave Abel to ease the pain is uncertain. Morphine, bromine, chloral hydrate, quinine, salicyl?

In Abel's day the doctors were very much preoccupied with what could be called an inherited "weak chest". Such patients more easily contracted consumption of the lungs than others, and such a weakness, once rooted in a family, could be inherited. But medical science still had no decent treatment in sight. Abel knew the limits of medical science.

To be stamped as having the "consumptive sickness", to be diagnosed as tubercular, was tantamount to being handed a death sentence. Consequently, other words were used, like weak-chestedness, the breast sickness, pneumonia, hectic fever, and the white plague. Science did not know why some fell ill with this disease: inheritance and bad ventilation can scarcely have provided a full explanation.

From his sickbed Abel cursed the medical science of his day which had still not carried out sufficient research to be able to master an illness like tuberculosis. Earlier many had suffered from smallpox, but now smallpox vaccine was common everywhere. Niels Henrik's father had been a pioneer in this field. Father Abel had zealously vaccinated both his own and the parish's children against smallpox. But why had no one solved the puzzle of tuberculosis?

Niels Henrik lay in the stately home of the works owner, Smith, at Froland. He lay in a chamber called "the buck room" after the pattern on the wallpaper on which was printed the picture of a buck come to drink from a wooden tub. This was repeated again and again in between the picture of a flower rendered again and again in the same single image manner, between repeated rows of flowers and borders. And the borders, the bucks and the flowers were rendered in blue on blue, the favourite colour of the period. The room was on the shady side of the house, away from the noise and the life of the works yard. A constantly fired up iron stove was attached to the room's chimney. Candidate Abel could not bear any change in temperature, or so District Doctor Møller had said. From his bed, Abel could look out the window and up at the landscape, at a low, tree-clad hill where the light of the day played over the snow. At night he could see the icy cold moon shadows, and later, the sun that shone through the bare branches of the deciduous trees.

At the end of January and in early February there were encouraging signs from abroad. Abel knew about Crelle's latest effort to get him to Berlin, and a letter now came to Froland from Legendre, dated Paris, the 16th of January, 1829, that improved the prospects. Abel's paper "Remarques sur quelques propriétés générales d'une certaine sorte de fonctions transcendantes" had been published in *Crelle's Journal*, Issue 4, 1828. It was in this eleven-page paper that Abel had taken for his subject the special case of what he had demonstrated in the disappeared Paris Treatise. In a footnote in this "Remarques" piece Abel had indicated the results he had worked out during that autumn of 1826, and he went on to explain that he had delivered this to the Academy in Paris. Now in his letter, Legendre expressed his great enthusiasm upon reading "Remarques", but the venerable old mathematician in Paris seems not to have noticed the footnote. Legendre wrote: "Your paper, 'Remarques sur quelques propriétés générales etc' seems to me to surpass everything that You have hitherto published, not only in the depth of analysis that it has mastered, but also the consequences, generality and beauty of the results. The work does not occupy many pages but it contains many things and is consistently edited with great elegance and insight in almost all parts. If there had been the opportunity to develop this somewhat further, then I would have preferred that You had followed the reverse order and ended with the most general cases. But however this may be, I can only congratulate You for having been able to overcome so many difficulties in such an exemplary manner, for even if it is beyond the powers of an almost eighty year-old, to verify all Your results, I have tested them in any case enough to be impressed with how completely exact they are."

Legendre continued to compare points in Abel's last letter of November 25th, with results that Jacobi had communicated to him, and he concluded the letter with the suggestion that Abel not hesitate to publicize the results that he had mentioned in his letter and that should make it possible to determine when a given equation could be solved by means of square root. Such a theory, Legendre maintained, would bestow great honour, and would be considered as the greatest discovery it was now possible to make in mathematical analysis, and he concluded, "Farewell, my dear sir, You must be pleased of Your success, of the content of Your work. I also hope that one day You will receive a position in society that completely venerates You and Your inspired Genius."

And by way of a short "PS.", Legendre wrote that very recently he had received a letter from Alexandre von Humboldt, who stated that the Minister of Education in Berlin had, on behalf of the King, welcomed and authorized the founding of a seminar for higher mathematics and physics, and that Abel and Jacobi were meant to be professors.

Abel did not manage to answer the letter, and behind all the laudatory words he could not interpret Legendre's words as anything other than that the Paris Treatise was definitely lost forever.

But the prospects in Berlin were better, and it could seem as though Abel had now, from his sickbed, considered that this position in Berlin had, in reality, been decided. He *would* go to Berlin, he had overcome the distaste for having to leave his homeland, and he was pleased. He was pleased at last to be doing something good for his beloved, who had so steadfastly waited for him for so long. "You shall not be called 'madame' or my 'Wife', but it will be 'der Hr. Professor mit seiner Gemahlin,'" he repeated to her again and again. But did he really ever think that he would be well again? Hanna Smith recalls: "From the day that we heard he had been appointed Professor in Berlin, we saw him markedly slip away, and his otherwise very lively jesting almost disappeared."

No *definite* message about a permanent position in Berlin reached Froland as long as Abel lived. Through wishful thinking he must have made himself healthy and in condition to travel to Berlin where, as he had convinced himself, the matter was actually already decided. A great fame and renown, that indeed would have made life easier for him, was, right enough, underway from many directions. But he seems to have been unaware that the French men of science had written to Karl Johan with the plea that the king had to improve his living conditions; he probably did not know either that there was talk in Copenhagen about finding him a professorship there. And the footnote in "Remarques" had led to Bessel, on January 2nd, 1829, saying in a letter to Gauss, "About Abel's theorem as to the possibility of discovering the properties of integrals before they are carried out, I am extremely astounded. I believe a completely new side of integral calculus has been clarified. I would only wish that Abel could follow up with many kinds of applications. It strikes me as totally incomprehensible that such a contribution has been misplaced by the Paris Academy, and consequently remains unknown."

Also, at the beginning of February, an encouraging report came from Christiania. Abel was to get the regular associate professor salary, 600 speciedaler per annum. The request that Abel had sent in about the extra 200 daler before he left for Froland had been granted by royal resolution on February 9th, 1829. Abel wanted very much to get back. He needed the money. The loan to Norges Bank had to be renewed every three months, and he still had 160 speciedaler to pay off.

But District Doctor Møller gave no hope that he would soon be in a condition to go back to work. Møller wrote to the Academic Collegium in Christiania on February 21st: "At the Request of Hr. Docent Abel, and as his Doctor, should the Undersigned write on his Behalf - in as much as he himself is not in a condition to write - and report to the high Collegium Academicum on: the fact that he, shortly after his Arrival at Froland Ironworks, became afflicted with a strong infection of the Lungs and considerable Blood-spitting which indeed after a short Time was stemmed; but which nonetheless, on account of a continuing chronic Cough and great Weakness, he has until the present been hindered from leaving his Bed, and where he must yet remain; moreover he cannot tolerate at all any exposure to even the smallest Alteration of Temperature. The greatest cause for Alarm here is this: that the dry chronic Cough with accompanying Sharp Pains in the Chest, leads one in all Probability to assume that he is suffering from latent Tubercules of the Lung and Bronchitic Passages which can have caused the subsequent consumption in the breast, which seems to be even more deeply based in the Condition of his Constitution. In that dubious State of Health, Herr Docent Abel finds himself in the situation where there is the greatest Possibility that he cannot before Spring make his way back to Christiania, and thus neither can he take care of his incumbent Public Duties, if the Outcome of his Illness is to be that which is most hoped for."

From time to time Abel too had a certain nagging doubt about his recovery, but District Doctor Møller always treated him with comforting hopeful chats, and Abel spoke abstractedly about the happiness of going abroad and moving to Berlin. The diagnosis of tuberculosis had still not been given, but the formulation "latent Tubercules…which can have caused the subsequent consumption in the breast" was perilously close.

In addition to Abel's fiancée, the two eldest daughters of the ironworks owner, Sivert Smith, Hanne and Marie, took turns keeping an eye on him during the day, and providing him with company. But when the coughing had increased and made him sleepless and afraid of being left alone, one of the young women who served in the house sat watch through the night in the "buck room." The sick Abel had mentioned on several occasions that he was very content with this young "night nurse", whom he probably had known from earlier. His fiancée in her state of grief did not possess the strength to be alone with him they said, and moreover, she had her governess duties to perform toward the youngest in the house, four daughters between the ages of seven and fourteen. The youngest, Mette Hedvig, later reported that Abel, from his sickbed, had tried to explain mathematics to them, and he had become impatient when they did not grasp what he was explaining. "Can you not follow this thing? It is so simple and easy!" he was supposed to have said. Otherwise, in relation to Christine, they reported that by this time she was on the edge of collapse, and that at times she even said they could fetch "the Coffin for me at the same time they get Abel's".

From his sickbed, Abel was supposed to have said, "What they told me in Paris is not right - I still don't have consumption."

What had happened in Paris?

There is nothing to indicate that Abel was so ill in the autumn of 1826 that he consulted a doctor while in Paris. He had indeed caught cold, had certainly had a fever and coughed in the course of things, and he was sad, felt sorry for himself, melancholy, as he put it himself, and had been very anxious while waiting for a reaction after he had delivered his manuscript of his great treatise. But how, and from whom, had he been informed that he had consumption?

In the circle of young scientists and scholars around the journal *Ferrusac's Bulletin*, where indeed Abel had been taken in as a mathematical expert and correspondent, and made to feel as heartily welcome as a foreigner could be, in this circle there were also medical doctors who knew the symptoms of this white plague. And there were methods available for diagnosing this illness that was now so prevalent. The so-called percussion method had been known for over fifty years: by placing a finger on the chest and tapping it with another finger, the change that the sickness provoked could be heard in the lungs. Due to the difficulty of reaching through the layers of women's clothing, the French doctor, René Laennec developed the forerunner to the stethoscope, a 33 centimetre-long listening pipe made out of walnut wood. Laennec died the year Abel was in Paris. Among those that Abel met in Ferrusac's library there were quite certainly several familiar with the new developments in medicine who could quite certainly carry out the simple tests. Perhaps Abel's cough had aroused curiosity? Maybe in a social gathering Abel had allowed himself to be talked into an examination? And thus he may have received the diagnosis, that he then, naturally enough, did not want to pay heed to: it was only a party game. He had a touch of influenza with a bit of fever, nothing more. And what if there was something seriously wrong, what could he in any case have done by way of treatment? Even if he had believed the diagnosis on that occasion in Paris, he had neither the will, the desire, nor the means to seek out any doctor at all. In any case, nobody could have helped him. The best thing perhaps was to do what the English writer, Laurence Sterne did before his final book, *A Sentimental Journey through France and Italy*, was written. Sterne wrote this book in a race with tuberculosis, and himself explained his frame of mind as one wherein Death had beat upon his door, and Sterne had bid him come back again, and had tried to do so in such a gentle and carefree voice that Death would most certainly have been in doubt about his errand.

This would also have been what Abel had done. In the twenty months in which he had for working from his homecoming in May, 1827, until he had to set down his pen, he had sent in thirteen papers, the longest 126 large printed pages, and the shortest was the two-page work that he completed on January 6th, 1829.

The sunshine never streamed through the window of the "buck room"; the room was dark and gloomy; he was weak and constantly feverish. Hanna Smith and the others in the house thought he was brave and never gave up hope. During the nights it was another matter. Why did not help come from God The Father, about whom his own father had talked so much, and upon whom his grandfather and a family tree full of priests had all pinned their hopes of consolation and focussed their idle hours? Niels Henrik had himself demonstrated that many deeply ingrained concepts and traditions were based upon weak premises; was God in reality nothing other than a mental construction, totted up by theologians, like infinite power series? He wanted to toss the plaintiff out; nobody could stop him, no God's angels castigated him when he cursed in the full name of the Devil and gave evil its every name and title. The young woman watching over him was making sure that there was liquid available so that he could moisten his lips; terrified, she must have heard his incantations, without being able to help. Everything would soon be taken from him, and he cursed the very time that he had existed.

And then it was daytime again, and his beloved Christine came to his bedside. "You are a noble human being," he was supposed to have told her. And perhaps it was now that he told her that he, through the Smith family, had urged his friend Keilhau to marry her once he, Abel, was gone.

Abel wanted to work in bed but he could not manage to write, lay still and began to speak again in an ominous voice, to blame bitterly all those who had unjustly pushed him aside. More and more he returned to the poverty he had had to suffer from, and the only subject he spoke of positively was Mrs. Hansteen and all that she had done for him. But now he could not even write to her.

The woman who watched over Abel through those nights at Froland was his own age. Afterwards, she refused to talk about what she had experienced at Abel's bedside; whenever asked, she would give way to great emotion. The violent impressions from that sick room haunted her until the end of her days. Consequently she was quite old before she spoke about this to her daughter, who in turn told only her own son, from whose mouth issued a meagre report about Abel: the whole time he had lain in that sickbed he never lost courage, he had never despaired, he had wanted very much to live, and had fought to the end, had not complained about his physical suffering, but had loudly cursed medical science which had not done enough research, nor found a way out of the illness' blind alley. Hence it was only this that the nighwatch-woman's daughter's son could tell about Abel's period of invalidism. In itself this was scarcely enough to disturb another person's peace of mind. The curses that had been so unnerving to the caregiver must have been directed at other authorities.

Those who dealt with Abel during the day said that he was rational and patient to the end, but that his good humour gradually began to ebb away. Hanna Smith's fiancé was away at sea, and Abel said that he hoped to meet this man for then he could explain a mathematical discovery that could revolutionize the seaman's life on the ocean. And when he began to explain what this discovery was about, and the mathematics that was its foundation, and those gathered at his bedside had not understood, he became excited: "You don't understand? It's so simple!" he was said to have repeated over and over. What was it that he saw? A rising sea, and "an ancient mariner" with knowledge, and a fixed compass bearing directed toward a safe harbour? Final salvation for "the flying Dutchman" who had been condemned to slip back and forth across the seas through all eternity without ever being able to find respite? Was it the expectations of taking his science and applying it and making it useful that he saw as his salvation? In any case, it was only in the night that this wild ride seriously seized him; during the day things were simpler.

Those around him noticed that Abel frequently said that Jacob Jacobi was the man who best understood what his own work was all about. But since Abel did not manage to speak very well, grew hoarse, and easily began to cough, no one asked him what this meant, and no one really wanted to probe the names that might disturb his precarious equilibrium.

But when the house was still and empty of sound at night, when the darkness was the undisputed authority, then he spewed out his despair and his rage; it was then that he was ready to refute and curse the whole world. What was it that his father had got him to believe life was? That the justice that did not occur here on earth would occur in the Hereafter? That the soul was a simple, indivisible vessel that lived on after the body died and rotted, and that one could achieve bliss and the happy life so long as one lived in faith and fulfilled one's duties here in this life? But how had his father lived his own life! Drank himself to death, consciously and wilfully, only forty-eight years old! Or was his death predestined? Should he have pulled himself out of the school in Christiania and gone home to Gjerstad to his father's bedside, nine years ago? In any case, the family duties had been loaded onto him, Niels Henrik. And what had his siblings told him about that spring…Could it really be true that their mother had got drunk at the funeral, and in full public view…And now his big brother, Hans Mathias was going around a half-imbecile in his mother's miserable house. How much energy had he, Niels Henrik, not used to keep Peder on the straight and narrow path toward the university entrance exam? And now Peder had managed to produce an illegitimat child in Gjerstad. But at least, if the rumours were true, was it not at least a blessing that Peder had managed to land a position as tutor for the priest in Våler? But things would go well for his sister Elisabeth. She had come to good people and she was indispensable in old Treschow's house. But was it enough to be good if one wanted to be happy? It was no wonder that the young woman watching over him never managed to have any peace of mind about what she had witnessed in that room. To keep a candle burning in such a darkness, to bolster the life-force, and provide the courage necessary to hang on to life in a person who was in the process of dying in such a manner, this must have demanded much more in terms of faith, hope and patience, than any nighttime caregiver was capable of sustaining.

In between all this, in the daytime he could also use strong words about the poverty in which he had always lived; he complained about the sadness of his days, and about the injustice and neglect that he had been exposed to. But if any of the others so much as hinted that he looked tired and ill, then he quickly made light of his sickliness and began to jest and joke.

His speech from the sickbed became increasingly unintelligible to those who were trying to calm and comfort him, and to guide his thoughts into tracks that they could follow. They felt it was not good for him to be reminded that they did not understand the relationships he felt were the most self-evident in the whole wide world.

Only a few days before his death, when he was lying there, pressing his emaciated fingers against one another, he was said to have burst out once again with, "There, you can see that it is not true what they said in Paris - I certainly do not have Consumption!"

He refused to believe that he was so ill. Why should he die when he had barely lived? There was so much that he had not finished doing, it just could not be, that everything would soon be over. How was it that all of those who had misjudged him could know anything for sure about the illness afflicting his body? If only he himself could become well again, if only he himself could bring himself around and explain where the last, most brilliant mathematical idea was to be found.

It would soon be April and he pressed on in the hope that nature, with its sunshine and warmth would work miracles, even in him. But this did not prove to be so. And these phrases that his father had imprinted upon him that death would be an entrance to a new life, to a heavenly bliss, where the justice that did not prevail on earth, would manifest itself in the Hereafter...this must have reverberated scornfully in Niels Henrik's ears, the very worst kind of comfort for those who had lived in poverty.

The night before April 6th was the worst of all. Then with the coming of the dawn he finally began to calm down. Christine and one of the Smith sisters sat with him the whole day, and were with him when he died.

Ironworks-owner Smith organized the funeral and sent in death notices to *Den Norske Rigstidende* and *Morgenbladet*. Christine, who, for all that, did not receive her coffin when her fiancé did, wrote to Mrs. Hansteen's sister in Copenhagen on April 11th and asked her to convey to Mrs. Hansteen "that she had lost a good and gentle Son who had loved her infinitely. My Abel is dead! He died on April 6th at 4 o'clock in the Afternoon. I have lost everything on Earth! Nothing, I have nothing left!" She enclosed a lock of Abel's hair that she had clipped from his head. She placed this inside the letter and signed it, "Broken-hearted, C. Kemp."

And winter now returned to Froland with a terrible snowstorm, a snowstorm none had seen the like of so late into the year.

They had stopped firing up the stove in the "buck room"; pale ice roses formed on the window, and the cold crept along the walls with their blue papering, where buck after buck came to drink, but never found any water in the everlasting wooden tubs between the single flowers with their borders, flowers and borders in an endlessly repeating illusion.

Abel's footnote reference in "Remarques": "J'ai présenté une mémoire sur ces fonctions à l'Académie royale des sciences de Paris vers la fin de l'année 1826" [I presented a memoir on these functions to the Royal Academy of Sciences in Paris at the end of 1826], published in *Crelle's Journal,* Issue 4, 1828, would become the impetus that would drive the Paris Treatise back into the daylight again. On February 9, 1829, Legendre wrote to Jacobi that Abel published "papers in fine order that are pure masterpieces." For his part, Jacobi had also read Abel's "Remarques" and realized what the footnote might imply. On March 14th, somewhat exasperated, he wrote back to Legendre and asked how on earth the Academy in Paris could have overlooked such a treatise, "perhaps the most significant to have been made in mathematics during the current century."

Legendre must have felt somewhat embarassed, for he now tried to look into the matter and acknowledged in a letter to Jacobi on April 8th, that it was correct: Abel had submitted a paper in the autumn of 1826, that Cauchy had kept it, but that he now had urged Cauchy to be allowed to see it.

April 8th was two days after Abel's death, and on this same day - April 8th - Crelle sent his final message, that Abel definitely would be called to Berlin.

Crelle had certainly known all along of Abel lying an invalid at Froland, and in the end he had even used the illness to expedite the matter with regard to Abel's post. Crelle had written to the Ministry in Berlin on April 2nd, and after once again having stressed Abel's rare mathematical mastery that had now found recognition everywhere, Crelle emphasized that this also was another reason to make haste; that is to say, Abel had long been ill, and at the last report, dated March 8th, had been so ill as to no longer be able to write himself. Crelle told the Ministry that Abel had long suffered from a weak chest, and now in addition found himself out in the countryside, far from Christiania, and he was coughing blood. A speedy prospect of a future in a better climate was also imperative: "First of all, it is perhaps only now that this man can be wooed into service here; second, it is perhaps on the whole only by this means that he can be saved for science, and possibly he will soon be lost if he has to stay at that place [where he now is], or in general in his present position." In addition to these words, Crelle had cited Legendre's laudatory letter to von Humboldt, from back in November, and he reminded them of the French Institute's request to Karl Johan on behalf of Abel, but also stressed that Legendre felt it would be better that Abel came to Berlin.

And this time Crelle had received an expeditious and positive response. Already, by April 8th, two days before the funeral at Froland Churchyard, he could write in great enthusiasm to Abel: "Now my dear, precious friend, I can bring You good News. The Ministry of Education has decided to call You to Berlin, and employ You here. I have just this moment heard this, from the gentleman at the Ministry who has been dealing with this matter. There is therefore no doubt. [...] You can now be completely at peace with regard to your future. You shall be among us, and that is a certainty. I am so very pleased that this, which I have so desired, has been fulfilled. It has cost not a few pains, but it has, God be praised, succeeded. [...] You can calmly prepare yourself to travel, so that You are ready to leave the moment You get the official notification, but until that time I must urgently ask You not to say anything until this has occurred. The official communication must certainly come shortly, at the outside, within a few weeks. Above all, see to it that You get well again, and may heaven grant that this letter finds You in recovery. [...] Be of good courage and rest completely assured. You are coming to a good country, a better climate, closer to science and to genuine friends who regard You highly and are very fond of You."

Appendix

Fig. 41. Bust of Abel by Brynjulf Bergslien. Photo: Teigens Fotoatelier A/S.

1.

Chronology

1785

Hans Mathias Abel, Niels Henrik's paternal grandfather, becomes the parish vicar at Gjerstad, after this branch of the Abel family had lived in various parts of the country, moving more or less in response to the economic fortunes of the different regions: the family progenitor, Mathias Abel, came from Abild in Schleswig to participate in the booming economy of Trondheim and the Trøndelag region of central Norway during the 1640s; the next two generations lived on the west coast, the region known as Vestlandet, before timber, iron production and shipping seriously drew prosperity and riches to southern and eastern Norway.

1786–1792

Søren Georg Abel, Niels Henrik's father, is a disciple at the Latin school at Helsingør, Denmark, before becoming a distinguished university student in 1788. He takes the theological public service examination at the University of Copenhagen. The ideas of the Enlightenment, which are asserting themselves everywhere, and the revolutionary ideas from France come to impress themselves on all his later work.

1792–1800

Søren Georg Abel works as chaplain under his father at Gjerstad. In 1800, he marries Anne Marie Simonsen, daughter of the richest man in Risør, the shipping magnate and merchant, Niels Henrik Saxild Simonsen, whose father probably came from Saxild, near Århus, Denmark, to Risør during the boom times of the 1720s.

1800

Søren Georg Abel becomes parish vicar at Finnøy; the cleric and his wife have their first son, Hans Mathias.

1802

On August 5th, *Niels Henrik Abel is born*, at Nedstrand, Finnøy Parish.

1804

Niels Henrik's father succeeds his father as the vicar of Gjerstad Parish.

1804–1815

Childhood in Gjerstad, together, gradually, with five siblings. The war of 1807–14, with its blockade and meagre times, lays its imprint upon the region. Abel Sr. is active with the Reading Society among other things, and becomes the local Member of Parliament for the first, extraordinary Parliament, during the autumn of 1814.

1815–1821

Niels Henrik is a disciple at the Christiania Cathedral School. It turns out that 1818 is a year to be remembered: Niels Henrik gets B. M. Holmboe as his mathematics teacher, and rapidly distinguishes himself as "a mathematical genius". Abel Sr. is castigated as a theologian, is declared dead as a politician, and subsequently as a human being. He dies in May, 1820.

1821

Niels Henrik takes the *examen artium* and is given free accommodations at the student home, Regentsen, and he lives here for four years. He also receives a certain amount of economic support from a section of the teaching staff at the university.

1822

Niels Henrik takes *andeneksamen* (the secondary or preparatory exam), and has probably by this time acquired more mathematics that the teacher of pure mathematics, Søren Rasmussen, possesses. As there is no public service degree examination in the field of general studies, Niels Henrik pursues mathematical studies on his own.

1823

Abel's first article is printed in *Magazinet for Naturvidenskaberne* in Christiania. Professor Rasmussen gives him 100 speciedaler to visit Copenhagen, and here he stays a couple of months with his mother's sister and her husband, Uncle Tuxen; he meets Danish mathematicians, Mrs. Hansteen's family, and the young woman who comes to be his fiancée, Christine Kemp.

1824

After due application, with strong support from the University's teachers, Abel receives a state stipend and a promise of a stipend for travel abroad. The government feels however that he first needs a couple of years more time spent at home "at the University in Languages, and other associated Disciplines, which it is probable that he, in his present youthful Stage of Life has not mastered to the Degree that is considered desirable, so that he may, with full Advantage to his main Science, utilize the contemplated stay at foreign Universities." In April Abel receives a report that he is to be awarded 200 speciedaler for a period of up to two years. At this point in time he decides, on his own account, to have printed his work on the impossibility of solving the general fifth degree equation by means of square root, and he hopes that this paper will be his "entry ticket" to Europe's scientific circles. During the Christmas holidays he becomes engaged to Christine Kemp, who has found a position as governess in Son, on the eastern shore of Oslofjord.

1825

Abel is eager to get abroad and writes on July 1st – with strong recommendations from Professors Søren Rasmussen and Christopher Hansteen – to the King with a request to have his foreign study tour stipend disbursed to him. With 600 silver speciedaler per annum for two years, he leaves Christiania at the beginning of September. The plan is to travel to Paris, the metropole for mathematics, and on the way, to visit Gauss, Europe's leading man of science in mathematics, in Göttingen. But while in Copenhagen, Abel decides to journey to Berlin instead, in the company of friends P. B. Boeck, N. M. Møller, and N. O. Tank, to meet C. G. Maschmann and B. M. Keilhau who are already in Germany. At the end of October, Abel is in Berlin, and stays in the city for four months. Here he meets A. L. Crelle, who becomes the greatest help in his life.

1826

The first issue of *Journal für die reine und angewandte Mathematik* (= *Crelle's Journal*) comes out. It is here that a large part of Abel's work is published. Abel travels on, together with friends, to Leipzig, Freiberg, Dresden, Prague, Vienna, Graz, Trieste, Venice, Verona, Bolzano, Innsbruck, Lucerne, Basel and finally reaches Paris on July 10th. Abel's own excuse for this long detour is: "My Desire to see a little was great, and does one travel only in order to study the strictly Scientific?" Abel finishes an extensive paper, later called the Paris Treatise, and delivers it on October

30th to the French Institute, where it languishes unread. Abel continues on in Paris until the end of the year.

1827

During the first days of January he returns, tired and poverty- stricken to Berlin, and stays here until at the end of April when he sets out on the homeward journey. He has a good stay among friends in Copenhagen, and returns to Christiania by steamship at the end of May. He finds work for his fiancée as a governess at Froland Ironworks. He has great difficulty to renew the stipend that he had had before his study tour abroad. He does private tutoring, but he still has to take out a loan from Norges Bank.

1828

His economic situation improves: Abel receives a temporary associate professorship at the university as a result of Professor Hansteen's absence on a scientific expedition to Siberia. At a rapid tempo and in a steady stream, Abel sends away his mathematical works to Crelle in Berlin. For his part, Crelle works persistently to obtain a position for Abel in Berlin. Abel spends the summer together with his fiancée at Froland; in the autumn he is ill and bedridden for some weeks but despite this, wants to go to Froland for the Christmas holidays.

1829

More or less continually for twelve weeks he lies bedridden in "the buck room" at Froland Works. He is well-cared for by members of the household, and he is under the supervision of the district's best doctor, A. C. Møller, from Arendal. Abel has tuberculosis. He is coughing blood, and he dies on April 6th. Two days later, Crelle writes from Berlin that Abel will receive a permanent position in that city. In Paris, his lost and forgotten treatise is brought out and read with great admiration.

2.

Commentaries and Supplementary Readings

Commentary to Part I

The report of Abel's death was sent from Froland by Sivert Smith on April 7th to Lektor Holmboe in Christiania, and it is highly likely that Holmboe saw to it that the announcement was made in *Morgenbladet* og *Den Norske Rigstidende*. The poet, Conrad Nicolai Schwach subsequently published a memorial poem in *Den Norske Rigstidende*, No. 32, 1829. The last stanza of this poem, not unlike its fellows, exhorts the goddess Urania to hide the memory of Abel in her breast, and when the loss doth stop bleeding, she is, in her maternal way, to meet him upon Eternity's silent Shore. Hans Christian Hammer's poem was published in *Den Norske Huusven*, No.25, 1829. He was probably one of the students who came to Abel for private tutoring. His poem relied considerably upon metaphors of nymphs, and the tree that was felled too soon.

Further reactions to Abel's death

By May 12th, Schumacher (at Altona) had already written to Gauss (in Göttingen): "You must certainly have read of Abel's death in the newspapers. Legendre has published a supplement, in the introduction to which he speaks of Abel in such a manner that it looks like he is ranking him after Jacobi. I know from You that the reverse is the case." On May 19th Gauss replied: "Abel's death, of which I have seen nothing in the newspapers, is an enormous loss to science. Should there possibly be something published or about to be, anywhere, concerning the life details of this most remarkable head, and it falls into Your hands, then I urge You to the utmost to communicate it to me. I would also very much like to have his portrait, if such should be available anywhere. Humboldt, with whom I have discussed him, had the determined desire to do everything to get him to Berlin." Vice-Regent von Platen wrote to Karl Johan on May 6, 1829: "A young mathematician, Abel, très renommé, vient de décéder à 27 ans et sa mort est une perte reèle pour les Sciences."

Crelle's obituary – written in French and dated Berlin, June 20, 1829, and made public in *Journal für die reine und angewandte Mathematik*, 4th issue, 1829 – was translated into Norwegian and published in *Morgenbladet* on November 7, 1829. Large portions were repeated in Holmboe's memorial of 21 pages, in *Magazinet for Naturvidenskaberne*, after a biography of Abel and an account of his works, Crelle ends with the words: "Mais ce ne sont pas les grands talents seuls de Mr Abel qui le rendoient si respectable et qui feront toujours regretter sa perte. Il étoit également distingué par la pureté et la noblesse de son caractère, et par une rare modestie qui le redoit aussi amiable, que son génie étoit extraordinaire. La jalousie du mérite d'autrui lui étoit tout à fait étrangère. It étoit bien éloigné de cette avidité d'argent ou des titres,ou même de renommée, qui porte souvent à abuser de la science en faisant un moyen de parvenir. Il appréciont trop bien la valeur des vérités sublimes qu'il cherchoit, pour les mettre à un prix si bas. Il trouvois la récompense de ses efforts dans leur résults même. Il se réjouissont presque également d'une nouvelle découverte, soit qu'elle eût été faite par lui our par un autre. Les moyens de se faire valoir lui étoient inconnus: il ne faisont rien pour lui-même, mais tout pour sa science chérie. Tout ce qui a été fait pour lui, provient uniquement de ses amis, sans la moindre coopération de sa part. Peut-être une telle insouciance est-elle un peu déplacée dans le monde. Il a sacrifié sa vie pour la science, sans songer à sa propre conservation. Mais personne ne dira qu'un tel sacrifice soit moins généreux que celui qu'on fait pour tout autre grand et noble objet, et auquel on n'hésite pas d'accorder les plus grands honneurs. Gloire donc à la mémoire de cet homme également distingué par les talents les plus extraordinaire et par la pureté de son caractère, d'un de ces êtres rares, que la nature produit à peine une fois dans un siècle!" Berlin, le 20 juin, 1829. Much inaccurate material was gradually written about Abel in newspapers, magazines and encyclopedias in many countries in Europe. Crelle's obituary also contained some inaccuracies – such as having Abel born on August 25th at Frindöe, that he made his debut in 1820, and so on. In the English journal, *The Atheneum* – circa 1830 – they wrote about "the celebrated young Swedish philosopher, Abel." One could read in French and Greek encyclopedias that Abel had been so poor that he had had to trudge home to Norway from Paris on foot. The source of this "footloose" story is probably found in an article written by Guglielmo Libri in *Biographie universelle* (Paris, 1834). (See also commentaries in the chronological bibliography that follows.)

About the Abel Centenary Jubilee in 1902

At the beginning of September, the newspaper *Morgenbladet* printed report-age and materials under the title "The Abel Jubilee". Fridtjof Nansen, chair-

Fig. 42. a Detail of Gustav Vigeland's Abel monument in the Royal Palace Park – The Abel Garden – in Oslo. After additional evaluation, this monument was eventually unveiled on October 17, 1908. The figure is four metres high and it stands upon an eight-metre high plinth of granite. Aschehoug forlag Archive. (Photo: Anders Beer Wilse). **b** Abel, painted by Niels Gude. (Original in the possession of the author of this book, given by Kerstin Voss. Photo: Teigens Fotoatelier A/S.) Down through the years many have painted portraits of Abel, all based upon the original Gørbitz portrait. In Copenhagen in the 1830s, Henchel made a lithograph; Winther in Christiania later made lithographs and Chr. Tønsberg had produced lithographs designed by V. Fassel. J. Jaeger made an Abel portrait for *Acta Mathematica*; Gustav Holter did the same for *Morgenbladet*'s jubilee edition in 1902; Marcelius Førland did a drawing of Abel; Henrik Lund made a painting that is in the Mathematics Library, University of Oslo, Blindern. In Axel Revold's fresco in the Deichmanske Bibliotek, Abel is portrayed in discussion with Sophus Lie, and so on. In Knud Larsen Bergslien's large oil painting, painted around the turn of the century on commission from the academic Collegium, Abel is portrayed with brown eyes and a dark skin tone. Generally it was said that Abel had blue eyes and his passport from 1825 describes him as of medium height, blue eyes and average body build. But in his passport of 1823 one finds that he was rather tall, with brown eyes and a slender body build.

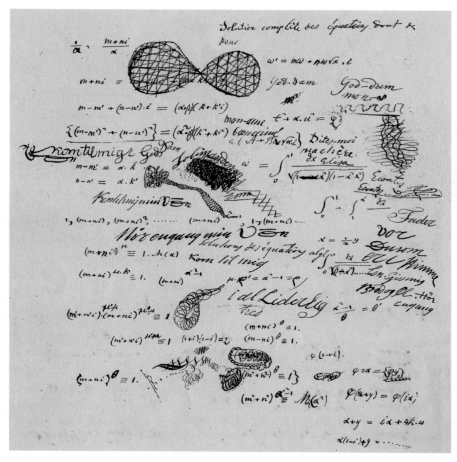

Fig. 43. Page from Abel's Paris notebooks. The reader will note that Abel's mathematical deliberations are interspersed with geometric "doddlings" and snatches of inner dialogue that allow us to glimpse the tensions and excitement of his youthful creative processes. Among the phrases on this page are "Solution complète des équations dont de…God-dam…God-Dam, mon ∞…[complete solutions to the equations in which the…goddamn…goddamn…my (infinity sign)]Fader vor du som er i Himmelen, Giv mig bröd og øl. Hör engang [Our Father who art in Heaven, Give me my bread and beer. Listen for once] …KOM TIL MIG I GUDS NAVN [come to me in God's name]…MON-AMI…BIEN-AIMÉ [come to me my friend, my beloved]…Dites-moi, ma chère, Elisa…ecoutes…ecouter [speak to me, my dear Eliza…listen…listen]…Soliman den Anden [Ottoman Emperor Sulieman II]…KomtilmigminVôn [come to me my Vôn (venn:friend?)]…nå engang min [now, for once, my] solutions des équations algé [solutions to algebraic equations]…KOM TIL MIG…I AL LIDERLIG HED [come to me in all your lewdness]. (Manuscript Collections, National Library of Norway, Oslo Division.)

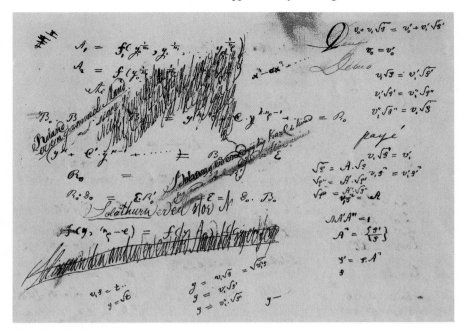

Fig. 44. From another page of Abel's Paris notebooks. Part of the obscure writing reads: Boland is an old Man...Niels Henrik Abel ...Niels Henrik Abel...Bolzano is a clever Fellow from what I have studied...Soliman the Second is a... (Manuscript Collections, National Library of Norway, Oslo Division.)

man of the celebration committee, emphasized that the memorial commemoration would distinguish Norway as a nation of culture. Nansen urged the poet, Bjørnstjerne Bjørnson to write a celebratory cantata, and in a letter he wrote to Bjørnson, Nansen said: "As I see it, it is our duty to make the most out of events in our nation; such as the birth of Abel; in emphasizing this to the whole world, we are emphasizing our right to exist as our own cultural state; but unfortunately, we have only one Abel; the opportunity will not come again for 100 years." Bjørnstjerne Bjørnson's poem on the occasion of the Abel celebration in 1902 was first published in *Den Norske Intelligenz-Seddeler*, on September 5, 1902. It was a lyrical poem with the Norwegian national overtones that Nansen seemed to have sought as the tone appropriate for marking the occasion. This poem, "Niels Henrik Abel" was set to music by the composer, Christian Sinding.

Fig. 45. Dokken at Christianshavn, where Abel lived during the summer of 1823 at the home of his Aunt and Uncle Tuxen. (from *Peder Mandrup Tuxen og hans efterkommere*, Copenhagen, 1883.)

At the National Theatre's celebratory evening, September 7, 1902

Here, there were also foreign mathematicians in the audience (Bäckland, Forsyth, Hilbert, Mittag-Leffler, Newcomb, Picard, Schwartz, Volterra, Weber and Zeuthen) and Fridtjof Nansen gave his speech in English. Ibsen's *Peer Gynt* was performed, and this was followed, above all, by an epilogue recited by Johanne Dybwad, the country's leading actress. The epilogue, on Abel, was written by Elling Holst, and was published in *Morgenbladet* the following day.

At Froland

In August, 1902, there was a wreath-laying ceremony at Abel's grave, and Professor C. A. Bjerknes, who gave a memorial speech, pledged to see that a stone monument was raised in Abel's honour. This monument, a mask of Abel in bronze, was made by the sculptor Gustav Vigeland, and put up in August, 1905.

Fig. 46. **a** This is from one of Abel's work books. It is his characteristic way of ending, of setting the final punctuation to one of his works. This is found at the conclusion of the manuscript versions of many of his articles and at the end of paragraphs in his notebooks. **b** The head of Athena, Abel's sealing wax signet, that he probably acquired in Paris; at least it was used for the first time on a letter from Paris, and later on letters from Froland and Christiania. **c.** Perhaps the single significant trace of Abel's childhood? These initials "NHA" and the year, 1810, were, in any case, found engraved on a windowpane at the Gjerstad vicarage, according to reports, from an outbuilding that was pulled down in 1904. (This window glass in now in the possession of Arne Kveim, Gjerstad. Photo: Dannevig, Arendal.)

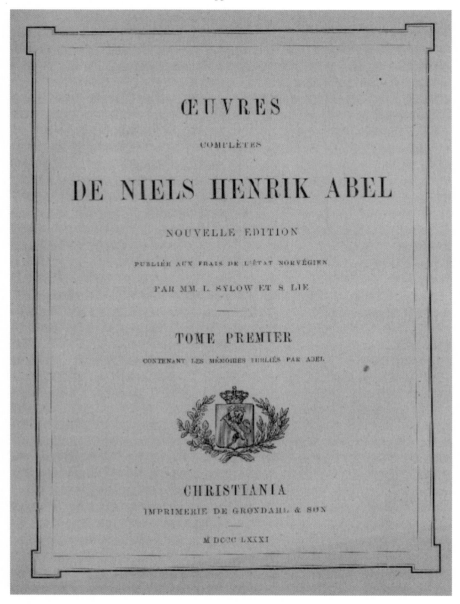

Fig. 47. The first volume of Abel's collected works – in all, 621 quarto pages – containing everything that Abel managed to finish. The other volume (341 quarto pages) contains the works that were left behind at the time of his death.

Fig. 48. a Abel's gravestone at Froland Cemetery was made at the Froland Ironworks and paid for by friends B. M. Keilhau, C. P. Boeck, B. M. Holmboe, N. Treschow, S. Rasmussen, M. Skjelderup, J. H. Hjort and Mrs. Hansteen, and was put in place circa 1830–31. The gravestone of the young German freedom hero, Theodor Körner, was surely the model. Körner had studied mining and metallurgy at Freiberg, been a theatre director in Vienna, and written incendiary poetry on liberation before he died on the battlefield in 1813, aged 22. The location of Abel's grave is 58° 31' 33.956" North, and 8° 39' 07.946" East of Greenwich. **b** A postcard of the monument set up at Froland Works in 1905. (Aust-Agder Archives).

Fig. 49 a–c.
a. Svein MagnusHåvarstein's bas-relief of Abel on the memorial column put up in 1979 at the Finnøy Vicarage. Photo: A. Stubhaug.
b. Ingebrigt Vik's winning model submitted to the great Abel memorial sculpture competition in 1902. It was not until 1966–69 that two bronze castings of Vik's plaster work were made. One is now in the Vik Museum, and the other stands outside Abel House, the mathematics building, at Blindern, University of Oslo. (Vik also made a bust of Abel and a bas-relief which are found in the Vik Museum/Øystese.) Photo: Teigens Fotoatelier A/S.
c. The monument at Gjerstad Vicarage, raised at the initiative of Øystein Ore in 1958. The bust was made by Brynjulf Bergslien; a casting of the same is found in the mathematics library at Blindern, University of Oslo. Photo: Dannevig, Arendal.

Fig. 50. a From 1948 to 1991 Norges Bank honoured Abel by printing his portrait on the front side of the 500 kroner note, in two versions. The first, until 1976, the large green bill with an industrial scene by Reidar Aulie on the back, and the other – smaller and lighter green, with a drawing by Joachim Frich on the back that shows Studentlunden and has the University in the background. **b** On the occasion of the one hundredth anniversary of Abel's death, in 1929, he was honoured with a series of four postage stamps – 10, 15, 20, and 30 øre – green, brown, red and blue – the most commonly used combinations. Before this time only kings, and Henrik Ibsen (in 1928) were pictured on Norwegian postage stamps. (In the same stamp series Bjørnstjerne Bjørnson [1932] and Ludvig Holberg [1934] soon followed, but while all the men of letters had their signatures above their portraits, Abel's name is printed in capital letters.) In 1983 a picture of Vigeland's Abel monument appeared on the Norwegian stamp for letters posted to destinations within Europe: blue-green, 3.50 kr., in a series, with Edvard Grieg, who was portrayed upon the red, 2.50 kr. stamp. **c** *Rue Abel*, Arrondissement 12ᵉ, Paris: a Parisian street name since 1901. Beside this, one finds *Abelsgate* [Abel Street] in Bergen and Trondheim, *Niels Henrik Abels vei* [Niels Henrik Abel Road] in Oslo, *Abelstrappa* [Abel Staircase] and *Niels Abels gate* [Niels Abel Street] in Stavanger. *Abels Hus* [Abel House] is one of the large, twelve-storey buildings on the Blindern Campus, University of Oslo, and at the college, Høgskolen i Agder/Kristiansand there is also an *Abels Hus*. The Research Council has called its largest auditorium Abel Auditorium. (Photo: A. Stubhaug).

At Gjerstad

Here too, there were celebrations on the centenary of Abel's birth. Memories and stories about the Abel family were passed on to a new generation, and Niels Henrik became the village's great native son. The Youth League of Gjerstad decided to put up a headstone on the grave of Abel's father. The grave of the unfortunate Søren Georg Abel had languished in disrepair and had been more or less forgotten. But the monument was put up in 1905, beside the solid iron headstones over the graves of Hans Mathias and Elisabeth Abel, the first Abel generation to live in the village. In general, at the turn-of-the-century putting up monuments was an extremely popular activity among members of Norway's Youth League; despite this alacrity however, the grave of Niels Henrik's alcohol-besotted mother still remains unmarked.

The Abel Garden in the Royal Palace Park

In relation to the one hundredth anniversary, a competition was announced for an Abel monument, as the desire was to have a statue of the young mathematical genius set up in front of the old university's central building. Norway's leading sculptor of the day, Gustav Vigeland, made it known that he would not accept the conventional idea of Abel as a mere "frock coat man" and set to work on his own conception. The jury set up to judge the nineteen submissions, and the winner was Ingebrigt Vik. Nonetheless, it was Vigeland's Abel monument that was constructed.

Gjerstad Vicarage

A bust of Abel, made by Brynjulf Bergslien, was consecrated by Professor Øystein Ore on August 10, 1958. A cast of the same bust has been lodged since 1962 in the Mathematics library at the Blindern campus of the University of Oslo. But the lemniscate, that curved plane figure-eight that Abel used so often in his examples, is engraved only on the plinth of the bust at Gjerstad.

Commentary to Part II:
Family Background

On the Abel name

Abel is a family name as well in Germany and Britain, but without demonstrable links to Abild in Schleswig (which formerly was part of Denmark). But in this case, there was a certain Oluf Madssen who died in 1647 as parish priest at Jydstrup, and Madssen's son, who called himself Jacob Abild, also became a priest and gave rise to a line of people from Abild, Denmark.

A little more on Merchant Simonsen's other affairs

Simonsen was very much involved in the work of getting a permanent school at Risør – approved in 1801, and the new school house was ready by 1805 – and he was present to welcome the school's first teacher, Jens Schanche, who, apart from his teaching duties, preached at the church when the pastor was away preaching at Søndeled. Schanche was cathecized when such a position became available in 1808, and he succeeded Vicar Støren upon the latter's death in 1813. As the parish vicar, Schanche would later have contact with Søren Georg Abel of Gjerstad on several occasions: the two priests substituted for one another, but Schanche soon stymied Abel's plans to have Søndeled included within Gjerstad Parish.

During the war and the blockade of 1807 Simonsen's fortune consisted of 40,000 riksdaler. In 1811 he stated his income as 40,000, and was the richest man in Risør. The next on the list was Henrik Carstensen with 30,000; the magistrate was recorded at mere 3,000, and the vicar, at 1,000 riksdaler.

But along with the end of the war came a rate of inflation that nobody had seen before. In order to cover state expenditures, the financial leadership in Copenhagen allowed a whole series of currencies to become legal tender: mercantile bills, banknotes, bills from loan institutions, tax office notes, assignments, letters of assignment. In 1813 there was a currency shortfall of 142 million riksdaler, and the value of the riksdaler fell dramatically. Prices increased, and public officials who were paid in cash, in the course of a few years received only a fragment of the real value of what they were accustomed to. Those who had borrowed money, did very well. A servant lad who, at the beginning of the war, had borrowed 1,000 riksdaler (from his employer for example) to buy back his father's farm, could a few years later sell a young foal in the market for 1,200 riksdaler.

Those who had placed their money in mortgage loans did tolerably well, while others nonetheless saw their whole fortunes shrink away to nothing. Simonsen was one of those who had large sums of money outstanding, but debt liabilities were unregistered, and the money he managed to collect, was almost worthless. In the course of the year 1813, prices of ordinary commodities increased 800 percent. The daler's value fell to that of a shilling, and even less.

The National Assembly at Eidsvoll in 1814 wanted ardently to put a stop to inflation by guaranteeing a favourable rate of redemption for the notes currently in circulation as riksdaler, by substituting a silver speciedaler. But the guarantee was soon abandonned; the market forces and exchange required an even more thorough restructuring of the monetary system. This was something that everyone felt, but was particularly crucial in the world of trade and commerce and in many of the trading houses along the coast. And the hardest hit were precisely those who, like Simonsen, had based their wealth on exchanges of actual commodities with the help of money, and who in the time of crisis had lost their credit possibilities.

Through these times of economic downturn, Simonsen managed to maintain his social status much longer than his wealth, but he could not do much about his money and he no longer accomplished anything. In 1818, Buvika, his beautiful villa on the seafront, was the last thing to go: he sold the property to Henrik Carstensen, who was one of the few who managed to survive the crisis and build up his fortune and influence again.

Simonsen died in Risør in March, 1820, two months before Søren Georg Abel's death at Gjerstad. Probably there was little or nothing left of Simonsen's money and property. In any case, there was not so much as a bed sheet to pass on to his grandson, Niels Henrik Abel when, at New Year's, 1822, Niels Henrik had indeed asked for a pair of sheets.

Simonsen's wife, Christine Kraft, who at the time of her husband's death was sixty years of age, had to face a degree of social degradation that must have been difficult for her to bear. Only her stepson, the bailiff Daniel Barth Simonsen of Namdalen, sent her money. He died in 1829, and the old records indicate that in 1831 Merchant Simonsen's widow was supported from the poor relief fund. She had become somewhat dimwitted, but every week had been paid out almost two speciedaler, an unusually high sum. The next highest disbursal was under a half daler. Thus it seems that the former glory had not been completely forgone.

A few other family relationships

Niels Henrik Saxild Simonsen had a son by his first wife, Marichen Elisabeth Barth. This son, Daniel Barth Simonsen, became bailiff of Namdalen. This bailiff's daughter, Marie Elisabeth Simonsen, married the chief customs officer in Risør, Gottfried Jørgen Birch, and they had a daughter, Anna Sophie Birch, who married Sophus Lie in Risør in 1874. Sophus Lie is the most internationally famous Norwegian mathematician after Abel.

Jens Kraft, the cousin of Jens Evensen Kraft, and consequently the father of Niels Henrik Abel's maternal grandmother, was born in Halden in 1720. He was professor of mathematics and philosophy at Sorø, Denmark in 1746 and helped the Sorø Academy to achieve prestige and renown. His main work, *Forelæsninger over Mekanik* (1763), was translated into both Latin and German, and used as an eminent work that advanced the collective mathematical and physical knowledge of the age. In addition, Jens Kraft wrote a series of textbooks and papers. In *Betænkninger over Newton og Cartesii Systemata, tilligemed nye Anmærkninger over Lyset,* among other things, he defended Newton's astronomical and physical views against Descartes' – which at the time were predominant in Denmark, and which Holberg too, deemed to be the most corrrect. In the piece *Om Træernes Natur* he observes and describes the rising of sap in plants, and in *Om Sielens Udødelighed* [The Immortality of the Soul] (1754), Kraft links the development of living consciousness with the variations of mathematical functions.

Commentary to Part III:
Childhood at Gjerstad

On Niels Henrik Abel's birth and the Parish of Finnøy

Was perhaps Niels Henrik born prematurely? According to the records, the first-born son of Søren Georg Abel, Hans Mathias, was born barely eight months after the wedding in Risør. This might possibly denote a family tendency toward premature birth, or an adherence to the local customs of amorous, pre-nuptial nocturnal visitation. The question remains, was Niels Henrik also born early?

That Niels Henrik was born at Nedstrand, at the magistrate's estate, and not at Finnøy, is highly likely for three reasons: first, the church records at Finnøy show no church affairs being undertaken at this point in time – something that indicates that the pastor was away. Second, people at Nedstrand said that one of Pastor Abel's children was born there. At least this is what Morten Kjerulf heard while he was priest at Nedstrand from 1829–34. (As a student, Morten Kjerulf had lived at Regentsen while Niels Henrik Abel was there.) The third reason is that Madame A. B. Marstrand, the wife of Magistrate Marstrand at Nedstrand, stood at the top of the list of godparents at Niels Henrik's christening on September 6, 1802.

With his departure from Finnøy on May 26, 1804, Søren Georg Abel wrote in the pastoral record book for the parish: "It is for my successors to judge whether I have accomplished any Good at Findøe, as both God and my Conscience know."

More on the Reading Society

The episode of old Hans Mathias Abel's consternation at finding one of Voltaire's writings among the books of the Reading Society likely has the following background. "Voltairian" was a designation in the Nordic countries that denoted a condition almost synonymous with being a free-thinker. Hans Mathias Abel had been a student in Copenhagen when the terrible earthquake hit Lisbon on November 1, 1755, and shook the whole world. The disaster – at least 30,000 lost their lives – was interpreted by the vast majority as God's revenge on Mammon. For years Lisbon had assembled wealth from its overseas colonies. A long poem came out in Copenhagen, about the "Guilty shall be seized by the neck and pulled down to Hell", and in Norway, the poet C. B. Tullin wrote: "Home with each golden cargo come a thousand vices", and he felt that the name of the man who invented the ship's compass by means of which it was possible to go out across the oceans and bring home things that people did not need, ought to be forgotten forever. But from Paris, Voltaire had written quite differently about the disaster, in "Poème sur le Désastre de Lisbonne", and in his novel *Candide*, the title character, together with his philsopher-teacher, came to Lisbon just as the earthquake was visited upon the city. In the end, the philosopher was hanged for talking against the religious point of view that professes that we are living in the best of all possible worlds, and Candide was whipped for having listened to his teacher with an approving mind.

When H. M. Abel was priest at Skafså in Telemark he wrote a memorial verse to the "blessed Tullin", who with his pen and a transfigured soul, had brought down the atheists with thunder and lightning, and proclaimed "The pure Word of Nature," about God the omnipotent and the good. Besides this, H. M. Abel wrote several poems, both to his wife and to the praise of Skafså village. But he railed strongly against the great Norse poem of the Middle Ages, "The Dream Song". (B. M. Landstad complained in the *Norske Folkeviser* [1853] that H. M. Abel's attitude to the poem "caused it to be consigned to Oblivion. It is certainly going in the same Way in other Places.")

According to Bishop Peder Hansen there were 20 reading societies in Kristiansand Diocese in 1802, but none were as active as the one at Gjerstad. When S. G. Abel returned to Gjerstad in 1804, he also took Vegårshei into the society, and he gave books from his own bookcase to stock up the book collection in this part of the parish. The membership increased, and S. G. Abel now successfully sought out moneyed, non-resident supporting members.

Bishop Christian Sørensen was in Gjerstad on visits in 1812 and 1817, and both times expressed his happiness at the success of the Reading Society. In 1812 the society had a book collection of 163 titles. But in Søren Georg Abel's last year of life, activity fell off, and after his death in 1820, little or nothing is heard of the Reading Society again, until the spring of 1840, when the current vicar of the parish, John Aas – S. G. Abel's successor – took up the task with full vigour once more. Despite the evident gradual substitution of books, there has been a continuity of activity right up to the present, such that the Gjerstad Popular Book Collection/Library can therefore be viewed as one of the oldest in the country.

The custom of nocturnal visits and carryings-on

This can be viewed as an expression of how the youth of the countryside chose mates for themselves in an autonomous and romantic atmosphere. Eilert Sundt, Norway's pioneer sociologist, wrote a great deal about this custom of nocturnal visitations, and historical and demographic researches show that approximately half of the brides of this period were pregnant at the time of their wedding.

Furthermore, at Gjerstad it was later speculated that the priest himself, Søren Georg Abel had impregnated a local village girl. A certain Bjørn Rønningen was thus considered to ressemble so little his brothers: while his brothers were practical and accomplished blacksmiths, but totally uninterested in books, Bjørn was said to be as sharp as a comet in the night sky. From his earliest childhood, Bjørn sat and read and wrote, and he became an unusually brilliant teacher in the village. But he loved to drink, and he came into conflict with Vicar John Aas, who earlier had been his teacher and had prepared him for the teaching post. Bjørn R. ended his days in a ditch beside the village road.

On Niels Henrik's siblings

Hans Mathias, born on November 13, 1800, sent down from the Christiania Cathedral School in 1820, died unmarried at his mother's Gjerstad house in 1842.

Thomas Hammond, born August 14, 1804, died two weeks after birth.

Thomas Hammond, born November 16, 1806, did not show the same abilities and strength of character as his two older brothers, and Abel the father paid for him to be taken into the Houen house in Arendal, and asked Dean Krog of Arendal if he could see that Thomas Hammond became a sailor or took up a trade. For a period of time he was also at the Herlofsen house in Arendal. The Gjerstad church book indicates that he was registered as being at the heart of the Lunde farm, which indicates that he went home to live at his mother's. In 1840 he had a child with Anne Halvorsdatter. It is not known when he died, but highly likely before his mother's death in 1846.

Peder Mandrup Tuxen, born November 12, 1807, was the brother Niels Henrik took such care of in Christiania. In the first period after Niels Henrik's death he got along tolerably well. In 1836 he married Wilhelmine Magdalene Fleischer of Botne, and was chaplain there for his father-in-law. Ten years later he became the vicar of Etne Parish, and was also mayor of the settlement. But around Gjerstad he was better known as "Peder the Priest-Prick". Here he had two children out of wedlock: Ole, born April 4, 1828, and Karen Kristine, born December 11, 1832. The children's mother was Kirsti Olsdatter. (Kirsti Olsdatter was from Salvesbu where her father, Ole Haavorsen Mostad, was a tenant farmer and blacksmith. Kirsti's brother was a regimental blacksmith, and smithing continued in the family – the twelfth smith at Salvesbu died only a short time ago.) The daughter, Karen Kristine, was later married to crofter Knut Ulvmyra, and Knut went on foot over the mountains to Etne Parish to demand a "child support payment" but he quite certainly received little of anything. (Laws and statutes on "child support" existed from 1805, and the provision that "maintenance money" be paid by biological fathers to support the mothers of illegimate children came into effect in 1821 onwards, but these demands quickly became obsolete; only in recent years has it been possible to enforce fathers' support payments by means of the threat of force.) Karen Pedersdatter died young; Knut married again and, according to custom, named the daughter he had by his second wife, after his first wife.) Pastor Abel of Etne Parish and his wife had five children.

Elisabeth Magdalene, born on March 16, 1810 – she whom Niels Henrik had asked Mrs. Hansteen to take in hand in the autumn of 1825 in Christiania – was soon established at the home of Niels Treschow. And when Treschow's daughter married the parish priest at Modum, Elisabeth moved there too, about 1832, and here she met the young, dynamic director of the Blaa-farveværket, the cobalt works, Carl Friederich Böbert, and a year later they were married. Böbert came from Sachsen, and among other things, he had been educated at the mining and metallurgical institute at Freiberg, and seems to have been there when Abel and his friends stayed at the Freiberg Institute in February/March, 1826 (see Chapter 36). Böbert was later appointed director of the silver works at Kongsberg (to the west of Christiania). He became a Member of Parliament, wrote articles, excessive stories, and like Elisabeth, was passionately fond of music. They had four children: a son and a daughter died in their twenties; a second son "went to the dogs" and the remaining daughter, Thekla, married Cabinet Minister Jacob Otto Lange, and it was this Thekla Lange who, in her time, had Gørbitz' Abel portrait in her possession. Elisabeth died in 1873, four years after her husband.

Thor Henrik Carstensen, born at Gjerstad vicarage on December 20, 1814, probably named after the Carstensen brothers, Thor and Henrik, of Risør, was privately tutored and successfully took his *examen artium* in 1840. He became a tutor in his brother's household at Etne, and later at other places in the district. Known to be a particularly talented flutist, he died at Etne in 1870, unmarried.

On the eating of horse meat

The human consumption of horse meat is mentioned in a few places at the beginning of the 1800s. Christine Koren, in her diary on December 26, 1809, mentions eating horse meat, and so does Erik Viborg in writings which were quite certainly published with the endorsements of a doctor. In a letter to his father, Henrik Wergeland wrote that for a period of time he had lived on horse meat. But horse meat never became a common ingredient of the household diet; it was, to be sure, boiled until it became slops, and was used in sausages. The first cookbook to mention horse meat though, was published in 1896. In Henriette Schønberg Erken's *Kogebog for sparsommelige husmødre i by og bygd* [Cookbook for Cost-conscious Housewives of Town and Countryside] (1905) the use of horse meat is described, but even then it was accompanied by clear references to the fact that there was still many prejudices that hindered the normal adoption of this nutritious and inexpensive meat.

A little more on Søren Georg Abel's work at Gjerstad

His devotional book of prayer *De sædvanlige Morgen- og Aftenandagter, samt nogle flere Bønner, omarbeidede* was published at his own cost, in Copenhagen in 1807. It is a 32-page book of prayers and devotions, in which common prayers such as The Lord's Prayer, confessional prayers and prayers for communion were transcribed "according to their Content and Meaning". He dedicated the book to the common folk of Gjerstad and Vegårshei, clearly a follow-up to the previous year's catechismal commentary, *Religions-Spørgsmaal med Svar*. In the foreword he wrote: "I hope now to imbue myself with the same Benevolence wherewith, my respected Friends, I undertook my Questions of Religion, as I give out this little Collection. I very much stress the importance of domestic Devotions, and I so very sincerely want to make You still more venerated in the eyes of God, by advancing this book's Influence over Your Devotions and Your Deeds, in the refinement of which the true Adoration of God is lodged. My Intention, at the very least, aims in that direction. I dedicate this to You publicly, and I state publicly that I owe You this testament: I have undertaken this Betterment with an eternally joyful Heart. Praise be to Providence that I have been allowed to be Your Teacher and Friend!"

Abel's version of "Our Father" (written in accordance with its contents and its meaning) reads: "Good Father, You who are everywhere and are highly exalted over us! We give You the honour You deserve so much. Let the teachings of Your Son be spread further and further, Your Will shall be done here on Earth, just as it must be obeyed by all the Righteous in Heaven. Give us every day what we require to maintain our Lives, and give us a modest Disposition. Forgive us our Sins; we promise once more to forgive our Brothers and our Sisters; defend us from all Wrong-doing, and let us not be enraptured by Evil. You always advise us on how to achieve a state of Bliss, and are all-powerful in advancing this cause; therefore, forever do we honour You. Amen!"

In order to repair the strained relations with the ordinary people of Vegårshei, S. G. Abel was amenable to the idea of putting up a new church building there in 1809–10. The local peoples of Gjerstad and Vegårshei both had purchased their church buildings, for 300 and 200 riksdaler respectively, in 1725, following the Great Nordic War, when the King sold off properties in order to build up the state coffers. Since that time the church at Vegårshei had been patched and repaired innumerable times, but it remained in poor condition, and moreover had become too small. Levies for the building of a new church was based on freely given contributions from the congregation, and now in the time of licenses it was easy for owners of forests to amass large sums by means of their forest operations. And even though the lead builder was unschooled, the new structure was a well-built church, and even though the contract cost was overrun by 800 riksdaler, there was nevertheless a large consecration celebration on August 19, 1810. Moreover, at this consecration, presided over by Dean Krog from Arendal, Vicar Abel gave such a good sermon that he was encouraged to have it printed, and so it was: *Tale, holden ved Vegaardsheiens Kirkes Indvielse 19 August 1810* (Christiania, 1810, 8 pages).

But without the congregation once knowing what the issue was, Vegårshei was made part of Holt Parish in 1812. Bishop Christian Sørensen was disturbed by the rumours surrounding the division, but as the priest in Risør, he himself had been old Hans Mathias Abel's colleague, and would now clearly have understood that Søren Georg Abel was complaining about the long and difficult road from Gjerstad to Vegårshei, and likely would have recalled his old colleague's voice saying, "This Holy Journey will be the life of me!"

Søren Georg Abel's plan to substitute Søndeled for Vegårshei as part of Gjerstad Parish soon proved to have no sound basis, and rapidly came to nothing. Thereafter, S. G. Abel also complained about the decision concerning the fissioning off of Vegårshei. He wrote to Dean Krog in Arendal some years later: "I wonder if it would not be possible to get the Hei back again? If it is possible to play on that String."

Søren Georg Abel remained at Gjerstad Parish for the rest of his life, but he made many attempts to obtain other postings: "My Circumstances and my growing Flock of Children with the Requirements of their Upbringing, make it necessary for me, whenever now a suitable Vacancy arises, to seek a Relocation." At that time he was applying for the Skien posting, "but I yarded in my Sail when I found out how many rivals there were for the post [Hr. Dreyer certainly has applied, and Hr. Schanche in Risør]; indeed there are many worthy applicants." (This was written in a letter to Dean Krog, whom S. G. Abels referred to as "my only ecclesiastical Friend.") Toward the end of 1815, he applied for the Vaaler Parish, southeast of Christiania, along the Swedish border, but withdrew his application some months later, and instead, applied for another one nearby, without having much faith that he would get it. Nicolai Wergeland was among the applicants. In August, 1816, Abel also applied for the vacant Eidsvoll post, that then went to Wergeland, and he had planned to apply for two others. The following year he applied for yet another one in southeast Norway, but withdrew when he came to hear that it was an unusually miserable posting. John Aas writes that in the years after Vegårshei was hived off from Gjerstad, Abel made application for eight different posts, and in the end he got one of them , but by then he was on his deathbed.

In response to an inquiry from Gustav Hjerta, the Swedish adjutant to the viceroy, Søren Georg Abel gave a quite detailed description of the Gjerstad parish in the autumn of 1816: population, climate, grain and field crop production and so on, the number of horses, cattle, sheep, goats, pigs, information on the forest, sawmilling, the fishery, and people's general "Level of Enlightenment." (The information was to be used in a mapping of the united Scandinavia, and a related letter was probably sent to all parish posts around the country, but the mapping never came to anything and it is not known whether Gustav Hjerta's material might yet be in existence).

The remains of Søren Georg Abel's tar oven, which he had built from the remains of the planned tileworks, was turned up in 1930 when the Sørland railway line was constructed. The entrance was 85 cm. long and 1 m. wide, well-built into the mountain, and beside it lay some sun-dried bricks. The local people had the opportunity to view the oven for a few days before the railway tracks were put into place.

Commentary To Part IV:
Disciple In Christiania

The Cathedral School

The School was situated at the corner of Dronningens gate and Tollbodgaten, where Oslo's main post office stands today.

The academic school in Norway, and the Christiania Cathedral School in particular, had in the period between 1799 and 1809 undergone sweeping changes. This happened for many reasons, and many people were involved. Niels Treschow was rector in Christiania when the work of reform began and he sought to ensure that the Christiania Cathedral School would be at the forefront of the reform process. On paper, the major changes appeared thus:

The direction of the school was separated from the Church, the bishop of the diocese would no longer have any definite influence. The old class-learning system of "hearers", vice-rector and rector, had been composed of all-round men, each capable of teaching all subjects to his class, and this was now replaced by a subject-teaching system. The school council was composed of the rector and four senior teachers, and the administrative authority that the rector had held earlier, now lay in the hands of this school council, or school senate as it was also called. It was the school council, with one of the teachers as secretary, that now appointed adjunct teachers, gave them their instructions, worked out the various classes in which the pupils should be placed, filled available places in the school, looked after accounting, and handled complaints. The school council also had responsibility for contact with the parents, as well as responsibility for relations with the higher education authorities, the Board of Trustees, where the chief administrative officer and the bishop had permanent membership. At the top of the school system was the Church Department, and there Niels Treschow sat as minister from 1814 until 1825. (An increase in the number of pupils at the cathedral schools occurred in the years after 1814, in all four of the country's dioceses, and Christiania had the highest number of places for pupils – 120 boys – and during these years several new academic schools were inaugurated: in Drammen in 1816, in Fredrikshald and Skien in 1823. In 1815 Norway's population was 885,431.)

When this pedagogical system of teaching by subject was inaugurated, there had been no provision for an appropriate education of the teachers of the various subjects. In the classical languages and in theology it was not difficult to find qualified candidates, but mathematical teaching, for example, was carried out in many schools by persons with random and deficient education. The background of the teachers of mathematics and the natural sciences often consisted of success in an examination in navigation and practical experience from life at sea, or from land-surveying work, knowledge gained from medical lectures in Copenhagen or from their own private studies. The programme of studies in mathematics, physics and astronomy that led to the secondary examination at the University consisted of whatever knowledge the various teachers could manage to obtain by means of their own endeavours. (Professor Georg Sverdrup founded the philological seminar in 1818, and taught classical philology; as a subject of study. The Norwegian language was first established in 1829 when the University employed a teacher, Rudolf Keyser as "lektor" in history with the additional responsibility of teaching Old Norse – but those who wanted to study modern languages or the natural sciences and mathematics, had to survive as best they could, and were left to their own devices.)

To be sure, a pedagogical seminar was inaugurated in 1800 in Copenhagen. (This was led by Moldenhawer and Professor Sander who taught pedagogy and methods. Sander had earlier been employed at Basedow's in Dessau, and Moldenhawer and Sander were, apart from the Duke of Augustenborg, the spokesmen for the new school reforms.) The seminar had twenty-five pupils, and in addition to pedagogy, there was instruction in languages, religion, anthropology, geography, history, mathematics and natural sciences. But the seminar ceased to function ten years later, and it later came to be considered a misguided experiment. Stoud Platou had been a pupil at this seminar, and Niels Henrik's teacher of religion at the Cathedral School, Christian Døderlein, had also been in the seminar.

Jacob Rosted had become Rector of Christiania Cathedral School in 1803, when Treschow became professor of philosophy in Copenhagen, and Rosted held this position for the next thirty years.

The aim of instruction in the mother tongue – from 1820 called *Norwegian* in the protocols and examination papers was, according to Rector Rosted, to "promote good Taste and the Improvement of the Mind." Positions were determined for the school's disciples, wherein they, by means of the language, would have influence over their fellow beings, and thus the learned had to be able to distinguish between the rough and the refined, the sharp and the dull, the poignant and the flat, the high and low, in terms of expression. Moral feelings, and feelings for the beauties of nature and the workings of art, all had the same source, according to Rector Rosted, and the most important means for developing these feelings was through the reading of well-written and tasteful moral works, moving accounts of refined sensibilities, of virtuous characters and deeds.

During their schooldays, many of Rosted's pupils wrote verse and stories. One of his first pupils, the son of Bishop Frederik Schmidt, had published his verse before he became a university student. And the three who in Norway were known as the poetic three-leaf clover during the 1820s, and who are considered the first of the country's great writers – Henrik Anker Bjerregaard, Maurits Hansen and Conrad Nicolai Schwach – had all been Rosted's pupils and had found in their teacher a helpful friend and benefactor. Bjerregaard had already finished two collections of poetry in manuscript form when he left for Copenhagen in 1809 to become a university student there. Maurits Hansen had found it a great honour and inspiration when the rector praised and read aloud from his (Hansen's) descriptions. And Schwach recalled the rector's appreciative words: "You write like a Man for the Lord!" In order to honour their rector, in 1816 Niels Henrik along with all the other disciples arranged for one of the period's most renowned painters, Captain Jacob Munch, to paint a portrait of Rosted. Since the time this commission was painted, the portrait has been hanging on the wall in the school premises. In 1827 the assembled Parliament commemorated Rosted on the occasion of the fiftieth anniversary of his teaching career, commending him for his educational achievements at the school down through the years. Pupils presented him with a silver cup which was inscribed to him in Latin. Former pupils had collected capital to form "The Rosted Legacy". There was no age of retirement for public servants in those days, and it was said that the grandiose celebration was intended to be a significant hint that the old rector ought now to retire of his own free will. But Rosted himself interpreted the adulation as an opportunity to continue, and it was not until 1832, the year before his death, that he allowed himself to be pensioned off.

The School's achievement files concerning Disciple N. H. Abel

Information on Niels Henrik Abel's school achievements is found in the proficiency records, in which every teacher annually contributed his assessment under the five categories of *Natural Gifts, School Diligence, Handicraft Skills, Progress* and *Morals*. In Niels Henrik's case the categories are filled with standard expressions and standard formulations like "good" and "extremely good", some "remarkably good", and under *Morals*: "extremely good, orderly and conscientious". Those that departed from convention are found in the category *Natural Gifts*; here Holmboe wrote in September, 1818: "an extraordinary mathematical Genius", and around the same time there is the comment from another teacher: "extremely good but in my class he has not exerted himself." And then there is Holmboe's assessment written right across all the categories in 1820: "With the most remarkable Genius he combines an insatiable Ardour for, and Interest in Mathematics, such that he is quite certain, if he lives, to become a great Mathematician."

In addition to the proficiency records, the examination records are extant. Here the marks in each subject have been written down.

Abel's marks in the annual examinations at the Cathedral School

	1816	1817	1818	1819	1820	821	Artium
Religion	2	2	2	3	2	3	3
Arithmetic	1	2	1	1	1	1	1
Geometry	1	1	1	1	1	1	1
History	1	3	4	3	3	3	4
Geography	1	4	4	3	3	3	2
Natural History	2	3		2			
Anthropology		3		2			
Calligraphy	4	3	4	4			
Drawing	3	2	2				
Latin (oral)	2	3	3	2	3	3	3
Latin (composition)		4	4	2		2	3
Greek		1	2	2	2	3	3
Mother tongue (oral)	2	3	2	2	2	3	
Mother tongue (composition)		3	4	3	3	1	3
French (oral)	4	4	4	3	3	3	2
French (composition)					3	3	
German (oral)			3	2	2	3	
German (composition)				1	4		
English				3	2		

There is, in addition, information about a number of other tests, at Christmas and Easter.

Abel's fellow pupils

Those who graduated from Christiania Cathedral School together with Niels Henrik in 1821, and who, at least in the final years, were in the same class, were:

Niels Berg Nielsen: son of the rich mercantile agent, Jacob Nielsen, who had been one of twenty-two persons of "the higher Ranking Class" that Christian Frederik gathered at Eidsvoll in February, 1814, to establish the Norwegian Constitution. He was a cousin of the poet C. N. Schwach, and became a farmer in Vinger.

Jacob Worm Skjelderup: son of Professor M. Skjelderup, took both the philological and juridical public service examinations; became state secretary.

Gustav Adolph Lammers: later a well-known theologian and leader of a separatist movement, the Lammers Movement. (According to some, probably erroneously, he was allegedly the model for the title character in Ibsen's play, *Brand*.)

Hans Steenbuch: vicar of Rollag and Enebakk (probably Professor Steenbuch's brother's son).

Jørgen Olaus Hersleb Walnum: lived at the home of his relative, Professor Hersleb while he was a disciple in Christiania; became the parish vicar of Herø.

Hans Plathe Nilsen: magistrate in Mellom-Jarlsberg.

Frederik William Rode: parish vicar of Rennesø.

Carl Theodor Dahl: lawyer and brigade auditor at Fredrikstad.

Rasmus Rafn Borchsenius: became Government Paymaster (probably a relative of Magistrate Borchsenius, see Chapter 20).

Jacob Andreas Falch: Chief Justice of the Supreme Court, Bergen.

Christian Bull Stoltenberg: the brother of Henrik Stoltenberg (see Chapter 21). Another brother was Mathias Stoltenberg, the deaf, itinerant portrait painter, who was first "discovered" in the decade between 1910–1920.

Carl Gustav Maschmann was also among those who took the examen artium this year. He was privately promoted to university, son of the owner of the Elephant Pharmacy, Professor Hans Henrik Maschmann, who had a large house at Tollbodgaten No.12, which was much frequented by Niels Henrik with the young C. G. Maschmann, who worked in his father's pharmacy from 1822 until 1825, before travelling to Berlin and to Schröder's Pharmaceutical Institute in Neue Kroningsstrasse No.42, where he greeted, and took in hand, Niels Henrik and friends during the autumn of 1825 (see Chapter 34).

Occupation of the Cathedral School's premises by Parliament

That Parliament had to have the city's best meeting hall was obvious to all, but the disruption of the School's activities did not seem to have been thought to be of much consequence. The thinking and the plan from the first had been that Parliament would only assemble for a few months every third year. Parliament should pass laws and grant monies, and it had not been thought very time-consuming. And because before the election the idea was disseminated around this country with its diffuse population density, that it was easier for a parliamentarian to detach himself from winter duties than those of the autumn, the Constitution laid down that Parliamentary sessions would open in February, every third year in the capital so long as the King did not, due to war or plague, deem it be moved to another market town in the Kingdom. The plan was thus that the Members of Parliament would be home again before spring thaw made the roads impassable.

In 1823, the Parliament took the Cathedral School property into its possession, and convened their sessions there until 1854.

A little on the city

Masses of people flocked to Christiania from the villages at the beginning of the nineteenth century. The sewage deposits of the citizenry were removed from Ruseløkke Hill and roads sprung up and were known by nicknames, as Algiers or Tunis; the neighbourhood was a profusion of hastily-constructed small cabins, known from Eiler Sundt's descriptions. This shantytown construction aroused indignation, and in 1813, the Association for the Wellbeing of Christiania City characterized this urban building as "especially irregular and disreputable", with houses of 12m², although most were between 24 and 36m². It was proposed that lots be made available around the Aker Church, and this led to suburban construction along Telthusbakken [Tent House Hill] in 1815. Enerhaugen was parcelled out at the same time, by Jørgen Young. The Briskeby area was parcelled out in 1822, by General Haxthausen. A building lot ordinarily could be had at the cost of twelve work days per annum.

From the Cathedral School

We know about Holmboe's mathematical teaching from his own *Indbydelsesskrift* [Invited Writings] (1822) and from his textbooks (1825 and 1827). In a letter he wrote on September 13, 1849, to the university student, C. A. Bjerknes, who had asked him about how to study mathematics on his own, Holmboe stressed that the best form of instruction that he knew was found in the rules and observations that J.-L. Lagrange had given concerning the study of mathematics. Holmboe wrote that he had taken up these ideas about thirty years earlier – that is, about 1819 – having found them in *Zeitschrift für Astronomie und verwandte Wissenschaften* (a journal published between 1816 and 1818 in Tübingen, by B. von Lindenau and J. G. F. Bohnenberger. In all there were six volumes that contained articles by – in addition to those of the editors – Gauss, von Littrow, Laplace, Bessel and Möbius.) There was as well an introduction of 123 pages in which Lagrange's life was recounted, and later in the journal there was an overview of Lagrange's works.

Lagrange belonged to an old family from Touraine, closely related to Descartes. Lagrange's paternal grandfather was a French artillery captain who had received a royal appointment to the King of Sardinia, and this officer, while in Turin had married a woman from the famous Conti family of Rome. Lagrange's father, the Master of War Taxes for Sardinia, married the daughter of a rich doctor, and while they had eleven children, only the youngest, Joseph-Louis, reached maturity. (There is a statue of J.-L. Lagrange standing in Turin.) (And *à propos* of Lagrange's astronomical works: only four of Jupiter's moons were known at that time, and they were discovered by Galileo Galili in 1610. Then in 1892 a fifth moon was discovered, and today we are aware of all twelve of Jupiter's moons, the last discovered in 1951.) Lagrange wanted very much for one of his masterworks, *Mécanique analytique* (1788), in which he was extremely proud that there was not one single diagram, to be published in Paris because he felt that the formulas given in this work could be rendered with greatest precision and accuracy in Paris. And the person who was given the task of overseeing the printing in Paris in 1788 was the young energetic Legendre, whom Abel as well, almost forty years later, would come to meet and correspond with. Lagrange's *Sur la solution des équations numériques* and *Réflexions sur la résolution algébrique des équations* (Berlin, 1771) gave support to the general question about when it was possible to solve equations by means of algebraic operations; that is, with the rational operations of addition, subtraction, multiplication, division, together with the extraction of roots. This was the question that Abel would subsequently make such an important contribution towards solving.

Lagrange's funeral ceremonies were held at the Panthéon, and Laplace gave the eulogy. Pierre Simon Laplace, who launched the "nebula theory" or the "nebula hypothesis" that gave no role to a personally omniscient God, but rather advanced the idea that this world and the neighbouring planets were thrown out from their source, the sun, and from the state of expanded and super-heated gases, they later precipitated into small, permanent bodies. But was this solar system *stable* or *labile*? This question became Laplace's great project which, by means of mathematical analysis, he worked at all his life. (It was this major work of Laplace's, *Mécanique célest,* in five volumes, that Abel commented upon while he was in Paris: "He who has written such a Book can look back with Pleasure at his scientific life." [See Chapter 42].) Laplace wanted to demonstrate that the solar system was a gigantic *perpetuum mobile,* but whether the solar system really is stable and eternally repetitive in its complex course has still not been definitely proven.

It was Laplace who examined Napoleon in the examination that graduated him from the military academy in 1785, and it was Napoleon who, later on, took Laplace into political life. Laplace later got the reputation for being a politician who always knew which way to turn: among other things, he signed the decree that exiled his former benefactor, Napoleon, to St. Helena.

On the metric system

In Norway the standard weights and measures were the *ale* [an ell: 0.6277m.], the Norwegian and Danish *mil* [mile: 11,295m.], *anker* [barrels], *tønner* [casks], *potter* [jugs] and so on. In 1815–16, Parliament encouraged the government – through the Police Department – to secure a common system of weights and measures throughout the land. The matter was also raised before Parliament in 1818. In Holmboe's *Lærebog i Mathematiken* (1825), the following was written about distances: "The main thing about Measurements of Length is a straight Line, that has approximately the Length of a human Foot, and is therefore called a Foot." Other lengths were 1 alen = 2 feet; 1 favn [fathom] = 3 alen; a Norwegian mile = 18,000 alen = 36,000 feet; and thus, a Danish mile was 2/3 a Norwegian mile.

An assignment in calculation could thus be: How many French metres is five Norwegian miles when a Norwegian mile is 18,000 alen and 1,595 alen is 1000 French metres? (Moreover, 1 metre = 1.595 alen; ie., 1 alen = 0.627 metre = 2 feet; ie., 1 foot = 31.35 cm.)

Commentary to Part V: Student Life

The Mariboe Building

The University's establishment, Regentsen, or the Student Collegium, as it was also called, was set up in the Mariboe Building – first for twelve, and later twenty needy students. The lecturer in technology, Gregers Fougner Lundh, was assigned residence at Mariboe in order to supervise these students. There was criticism in *Nationalbladet* on October 26, 1820, of the fact that he had been given *eight* rooms for himself and his family.

Among Abel's contemporaries, Rudolf Keyser and Jacob Lerche roomed at Regentsen. Abel had Room 6 in 1821, and in 1822, an attic room together with Jens Schmidt (and his brother Peder, their "gris" [little pig], beginning in February, 1822). Later he quite likely shared a room with Halvor Rasch, who became Professor of Zoology, and Arnt Johan Bruun, the future Vicar of Lenvik Parish. From autumn 1823 Abel had Room 3, the only single room. Abel's most zealous fellow-players around the cardtable were certainly, in addition to roommate A. J. Bruun and classmate J. W. Skjelderup, Johan Lyder Brun, later the Vicar of Modum, and Johan Fredrik Holst, later the vicar at Vadsø, as well as at Finnås and Eidsberg. According to Morten Kjerulf, who now also lived at Regentsen, Abel spent most time together with J. F. Holst.

The University had summer holidays from June 15th until August 1st, and Christmas holidays from December 15th to January 15th.

There are two letters of testimony about Abel in the University archives dating from the year that he sat the secondary examination (*Andeneksamen*): The first is from Lektor Søren Bruun Bugge – signed March 30, 1822, stating that Abel had attended his lectures on Horace and Tacitus. On March 31, 1822, Professor Georg Sverdrup attested to the fact that Abel had attended his lectures. Apart from this, there is in the records a petition from the Regentsen students requesting that they be allowed to obtain warm water from the porter at six o'clock in the morning, should they be up and begin their work so early.

The scientific collections, and in particular, those of the natural sciences, required more and more space, and this led to the University's establishment having less and less room at its disposition. The shortage of space at the University property became quite precarious – the collection of Nordic antiquities and the numismatic holdings were squeezed in. The auditoria became too small for the University population, even the cashier had his office away from the property, and there was no toilet for the teaching staff. The student rooms in the University building had no fire safety protection either, and doubts gradually arose about the influence of the arrangement upon the students' diligence and moral conduct.

The Collegium concluded, on August 11, 1832, after having consulted the opinions of the faculty members, gradually to phase out the Regentsen establishment by not filling the rooms with new students as they became vacant.

The University

In the propositions submitted in 1812 concerning the University, the plan was to have twenty-five permanent teachers, professors and lecturers, divided between eight faculties. But in the new "Act Concerning the Foundation of the University", ratified in 1824, the number of faculties was reduced to four: the theological, juridical, medical and philosophical. The philosophical faculty was composed of an assemblage of subjects, from philosophy to classics, modern languages and history, to mining and mineralogy, natural sciences, mathematics and economics. Not until 1860 were the natural sciences separated out with the establishment of the Mathematics-Natural Sciences Faculty; the remaining subjects of the old philosophical faculty now became the Historical-Philosophical Faculty.

Commentary to Part VI:
Europe-Travel

Travel companion Nils Otto Tank
and his incredible life

After Tank's return from Basel in July, 1826, the rebuilding of Fredrikshald and Fredriksten Fortress quickly got underway. Soon young Tank was able to concentrate upon that which he liked best: "My Books and the free Life in the wondrous World of Norwegian Nature." Tank sought out a spiritual presence in God's natural world, and this religious exigency grew so strong that, the same autumn (1826) he journeyed to the Hutterite Brotherhood's headquarters, Herrnhut, a day's journey east of Dresden. The great family inheritance gradually shrank away. The trading house of Tank & Co. was forced to petition for bankruptcy in 1829. But Nils Otto Tank, who had become more and more involved in the activities and way of life of the Hutterites, and their evangelical zeal, collected money for Bibles, involved himself in petty trade, and soon distinguished himself as an economic genius. The prominent leader of the Hutterite Brethren in Christiania, Niels Johannes Holm, wrote in 1833: "Brother Tank seems to have to a special Degree, Judas Iscariot's ability...Money forms itself in his Hands. Can it be that he has been called to be our Economist?" Tank *did* become the economic leader of the Brethren, at their centre in Christiansfeld, Sør-Jylland, Denmark, and under his leadership, the personnel doubled, and in the workshop, where the craftworkers were also given religious instruction, there were floods of orders. After obtaining permission from the congregation, Tank married a certain Mariane Frühof, daughter of a Dutch Hutterite preacher and school director. The young couple moved to Herrnhut, and Tank became ordained as an evangelical preacher. Together with Mariane, he journeyed to the mission station at Paramaribo, in the Dutch colony of Surinam in South America. Mariane became pregnant, gave birth to a daughter, but died of a tropical fever. Tank was in Surinam for five years, and under his leadership, the mission grew to have twenty-nine missionaries. More than a hundred others were at work, building schools and carying out industrial activities; even the station's bakery and grocery shop made great profits. Tank managed to bring a ship full of goods from Europe to Paramaribo, and he obtained 40,000 decares [1 decare = 1000 m^2] of jungle where he planned to set up a forestry business. By making himself useful in practical matters, Tank won the goodwill and friendship of the indigenous Indians. In travelling through the region, he ate and lived as they did, and he also praised the local black slave population's lively and untraditional singing of hymns.

In all his business life and capitalism, Tank worked tirelessly for the education and liberation of the black slaves. His vehement advocacy of the abolition of slavery – expressed in mission tidings and reports – made him unpopular among the plantation owners, and gradually also with the Governor, Minister and King. When, after a tour of inspection around the West Indies and New York, Tank returned to Europe and Amsterdam in 1848, it was clear that there was no desire from the authorities that he return to Surinam.

Tank married again while he was in Amsterdam, this time to Karoline van der Meulen, a girlhood friend of his deceased first wife. Moreover, Tank's new wife soon inherited enormous wealth – from her father's side of the family, generations of art, paintings and furnishings, old books and manuscripts; and from her mother's side the new Mrs. Tank inherited the properties of a certain General von Botzelaar, who in his time had been part of the forces that repulsed Napoleon. Nils Otto Tank was a millionaire. Many also liked to believe that with his geological training, he had been the first to discover gold in Surinam, and that his enormous riches had stemmed from great goldstrikes in Guiana.

In any case, Tank was now in a position to turn the great plan of his life into practice. He wanted to build the New Jerusalem, an actual realm of Jesus where everyone could freely be brothers and sisters. During his honeymoon in Norway in 1849, Tank was urged to support Grundtvigist ideas about building a popular school (a folkhøgskole) for the common people at Hamar, but Tank felt determined that the new Kingdom of God should be established in the pristine land of freedom, America. Furthermore, it was a pity that all the Norwegian women and men who went to America were without knowledge of agriculture and commerce. And when Tank got to know that half of the Hutterite congregation of Stavanger had emigrated to a place called Milwaukee in Wisconsin, it seemed that his concrete plans had taken shape: Tank left with his spouse and daughter for Wisconsin, bought 40,000 decare of land, and invited the whole Norwegian Hutterite community in Milwaukee to accept free land. Forty-two adults and their children eagerly agreed. Schools were started, new enterprises set in operation, every farm was given help with buildings and equipment. During the first year they lived in a kind of enthusiastic communistic fellowship. But not everyone liked Tank's collective thinking. Most of the Norwegians had come to the new land to obtain private land that they could own outright, and more and more of them wondered what the rich-as-Croesus Tank wanted with them out there in the bush: were they involved in gigantic speculations, or something even worse? And when the Norwegian clergyman of the congregation pointed out that Tank, by refusing them written contracts for their parcels of land, wanted to make them into new sharecroppers and *serfs*, all Tank's visions of building a new Jerusalem disappeared in great tumult. Most pulled out, moved north, and founded a town called, to this day, Ephraim. (When the musician, Ole Bull, went to America two years later and bought large areas of land in Pennsylvania to lay out a Norwegian colony, a new Norway, "dedicated to Freedom and Independence" he had much less capital than Tank, and managed to hold his "Oleana" together for only two years.)

After this bitter defeat, Tank became one of the largest investors in the planned Fox-Wisconsin Canal, which was to link the mighty Mississippi River to the Great Lakes of the North American continent. (When the Erie Canal, which linked New York and the Hudson River to the Great Lakes, was built in 1825, the foundation had been laid for trade and shipping, that in turn created many incredible family fortunes in New York – described by, among others, Edith Wharton in *The Age of Innocence* [1921].) But long before the canal from the Mississippi became a reality, the railway network of the country expanded and dominated traffic arteries and trade routes. The canal project was scrapped with enormous losses to Tank and other investors. Shortly thereafter Tank became sick and died, in 1864. His wife lived on another thirty years, and when she died, she left a behest of $100,000 for an orphanage, and in addition, several large sums for various mission activities.

A story or two from Paris

Regarding Abel's stay at the home of the Cottes in rue Ste. Marguerite, No. 41, in Faubourg St. Germain – today, rue Gozlin – we have a humorous story told by Vilhelmine Ullmann, in an undated letter to her mother, Conradine Dunker, Christopher Hansteen's sister. One evening in the observatory, Hansteen had told about Abel living *chez Cotte*, who was evidently a hen-pecked husband. According to Hansteen, Abel often flirted with Madame Cotte, and she willingly accepted his attentions. On one occasion when Monsieur Cotte allowed himself to utter something in a louder and firmer voice than normal, Madame Cotte was said to have retorted: "Monsieur Cotte, si vous prenez le haut ton [high tone], moi je prenais le bâ-ton [low tone/baton]".

It is noteworthy that Abel writes in a letter to Holmboe – after the lively weeks of being together with Keilhau – that he "drank 1 to 1 1/2 Bottles of Wine every Day". It is highly likely that the bottles were a little smaller than normal today – 60 – 70 cl., but the alcoholic content was quite certainly around 11%.

Johan Gørbitz, who helped Abel to settle in during his first weeks in Paris, and who subsequently painted a portrait of him, returned to Norway for the first time in 1836. (The year before, his "teaching master" and atelier-owner,

Antoine Jean Gros, had, in a fit of depression, thrown himself into the Seine – it was said at the time that it had been academic classicism that had made him so dejected and melancholy.) Gørbitz settled down in Christiania and lived a calm and sober life. He was soon the country's most sought-after portraitist. Among others, he painted portraits of Christine Kemp, B. Keilhau, Camilla Collett, A. M. Schweigaard, P. A. Heiberg, Lyder Sagen, Bishop J. Neuman, Professor M. Skjelderup, President W. F. K. Christie, Countess Karen Wedel Jarlsberg, and others.

On the science academies in Paris

The situation after the revolution: already by 1793 the four learned academies in France were disbanded, and in 1795 they were all united within the structure of l'Institut de France. The various academies worked within the Institute, to a certain extent independently, and it was probably *l'Académie des Sciences* that had the greatest influence on the history of science. It had been founded by the humanist, Melchisedec Thévenot and his friends in 1666. This academy worked in all fields of mathematics, the natural sciences, and medicine, and it was this section of the Institute with which Abel was in contact.

The other old academies were *l'Académie Française*, founded in 1635, as a place for forty of the country's greatest poets and stylists. One of its tasks was to compose the official French dictionary, which first came out in 1694. *L'Académie royale des beaux-arts*, founded in the name of the young Louis XIV in 1648, consisted of a certain number of painters, sculptors, architects, copper engravers and musicians. The Academy had a sub-department in Rome where the most talented were sent on generous stipends to cultivate "the sciences of beauty", which were said to compose the one science that served to improve one's heart and sanity. *L'Académie des Inscriptions et Belles-Lettres*, in close cooperation with l'Académie française, was founded in 1663, with the almost overt task of glorifying Louis XIV.

A little more on von Humboldt and Galois, whom Abel did not meet in Paris

Alexander von Humboldt, together with the botanist, Bonpland, had travelled to northern South America in 1797, and after extensive journeys along rivers and in the Andes Mountains, and through Mexico and North America, came home to Europe with great natural historical and ethnographic collections. In the years between 1808 and 1827 he lived in Paris while he was writing up his extensive travel descriptions, in 30 volumes illustrated with 1,425 copper etchings. Thereafter, he settled down in Berlin, and was a central figure – with his all-sided interests and forms of knowledge – in the scientific milieu. (His brother, Karl Wilhelm von Humboldt, was, for a period, Prussia's Minister of Culture; he founded Berlin's university in 1810, and was also a significant researcher in the general field of language theory.) *Morgenbladet* (1822, No.4) carried a report from von Humboldt's expeditions, in a section entitled "The Consumption of Human Beings": "One of the Indians' Paramount Chiefs has a *Harem,* where he confines a great Many Wives, and amuses himself by gradually eating the nicest and fattest of them." *Morgenbladet*'s point was otherwise to show that these savages "whose Virtue and Bliss our writers of novels hailed so highly" were really not so very happy. *Morgenbladet* made the additional point of stressing the story of the missionary, "a great Defender of the Savage", who, once confronted with this story was to have answered, "It is an ugly Custom of these otherwise sweet-tempered and friendly People."

Évariste Galois grew up in Bourg-la-Reine, right outside Paris, where his father – who was a teacher and very talented at writing verses and arranging comedies – was the mayor during the one hundred days during Napoleon's flight from Elba, and where today there is a monument plaque to him on the wall of the city hall. Galois, the father, also continued as mayor after the Battle of Waterloo. He was popular in the local social life with his spiritual verses written in an old-fashioned style. He supported the inhabitants against the priest, and became a thorn in the side of the local clergy. Galois' mother was said to be a religiously devout woman, but down-to-earth and sober-minded. In 1823, young Galois was placed in school in Paris. The first couple of years went quite well; he won the first prize and so forth, but in the third school year he began to fall out of social relations at school, grew disgruntled, and lost his taste for schoolwork. Subsequently, he was refused entrance to the senior class, "la Rhétorique". Galois began to read mathematics, wanted to enter the École Polytechnique, but was refused admittance. His father took his own life in 1829, and then the July Revolution occurred, and so on. (See Chapter 42.)

Commentary to Part VII: The Last Year in Norway

The race with Jacobi

The question of who was first and ought to be awarded the greatest honour and so forth in relation to the theory of the elliptical functions, has been much discussed in mathematical history. The German mathematician, Leo Königsberger, wrote a book on the subject in 1879, and took it up as well in his Jacobi biography (1904). For his part, C. A. Bjerknes, in his biography of Abel (1880), took up as a major issue the presentation of a correct account of the factual course of events. By closely examining the different works' chronological order he firmly maintains that Abel had been first, and that Jacobi, in his book on elliptical function theory, *Fundamenta nova* (1829) does not acknowledge Abel's rightful place, and he implies that this omission has been made in full awareness by the author. Following Bjerknes' work, there seems to have been unanimity on the question of priority and honour. Abel had solutions to a lot more general transformation problems than those that Jacobi had posed in his articles in *Astronomische Nachrichten* in 1827. In the foreword to his "Solution d'un probleme général concernant la transformation des fonctions elliptiques" – what Abel called his "Death-fication of Jacobi", and which he sent off from Christiania to Schumacher on May 27, 1828, Abel states that Jacobi's elegant theorem on the transformation of elliptical functions is only a special case of a much more general theorem that he had laid out in his own "Recherches" (published in *Journalen* in October, 1827). And Abel continues: "But one can consider this theory from an extremely general point of view, in which one poses the problem of the undetermined analysis involved in finding all possible transformations of an elliptical function that can be performed in a given manner [d'une certaine manière]. I have succeeded in completely solving a large number of problems of this sort. Among them, are the following that are of the greatest significance for elliptical function theory..."

The loan from Norges Bank

This loan of 200 daler, granted on October 19, 1827, was paid to Abel on October 29th. It was a three-month term, and the interest was six percent. That is to say, the loan had to be renewed every three months, and indeed, Abel had done so: on January 31, 1827, with an installment of 20 speciedaler; on May 29, 1828, the loan was once more renewed, this time without an installment being made; and again, on September 29, 1828, with an installment of 20 speciedaler, and similarly on January 22, 1829: even while Abel was lying ill at Froland, he paid an installment of 5 speciedaler, probably through B. M. Holmboe. There remained a balance of 155 speciedaler, and this was money that the guarantors, Hansteen and the Holmboe brothers, in the end had to pay.

Commentary to Part VIII:
The Deathbed

Concerning Abel's stay at Froland

When in connection to his work on Abel, C. A. Bjerknes communicated with Hanna Smith (born in 1808, and married to a distant cousin, N. J. C. M. Smith, a ship's captain and lighthouse tender at Utsira) regarding his questions about N. H. Abel and his stay at Froland, Hanna Smith replied in two letters, dated November 9, 1883 and January 28, 1884. These letters are now in the Manuscript Collections at the National Library of Norway, Oslo Division. The first letter is missing at least one page, perhaps several. Hanna Smith writes: "My eldest sister Marie and I, were alternately always with Abel during his Illness, when his poor Dear One was unable to face being alone with him in her Sorrow, and thus we had occasion to take note of the progress of the Illness and the condition of his Moods." H. Smith goes through her memories of Abel in quite some detail. As well, one of the youngest daughters of the Smith family at Froland, Mette Hedvig (born in 1822 and later married to Consul Andreas Beer in Flekkefjord), told, in her old age through Dean Irgens in Arendal, that from his bed, Abel had tried to explain mathematics to those who were looking after him, and that he had become impatient when they did not follow: "Can you really not see this thing – it is really so obvious and so simple!" he was supposed to have said. And here, Abel's complaints about being ignored – and in this connection he mentioned several people by name – is evident in this recollection. One of the sons of the ironworks owner, Hans Smith (born in 1810, and later to become consul in Riga) wrote in a letter (in relation to the one hundredth anniversary in 1902) to Cabinet Minister Jørgen Løvland that the Smith family had always been very fond of Abel. Thus, Consul Hans Smith sent a telegram in 1902 to his nephew, the wholesaler, N. S. Beer, which was published in *Morgenbladet*: "On the occasion of the Niels Henrik Abel Anniversary, I, certainly the only still living youthful friend, bid You on the occasion of the Day's solemn commemoration, give the Abel Committee my sincere Greetings and Thanks for all the work on the memorial celebration. With Sadness I can still recall his period of suffering at our home and the Burial in my Parents' graveyard at Froland."

A few words on Abel's fiancée
Christine Kemp and Mrs. Hansteen

There is much to indicate that while Abel lay in illness at Froland, Christine hoped and actually planned to return to Christiania with him when he became well enough to travel. At least she tried, by means of an announcement in *Den Norske Rigstidende* published on April 9th, three days after Abel's death to work in "a Posting in Norway", but most likely, in Christiania or its nearest environs. Her announcement was as follows: "A young lady from Copenhagen, of good moral Character, schooled in French, German, Drawing and the most common School Subjects, together with knowledge of every kind of handicraft, seeks to obtain a Post in Norway as Governess. The full details can be obtained by addressing a stamped Letter to Miss Kemp at Froland's Ironworks, near Arendal."

These plans now came to nothing; at least she stayed on at Froland for a period of time – the Smith family's youngest child was only seven. She went to Christiania in June, 1829, to hand over to B. M. Holmboe those papers of Abel's that she had in her possession. (A little jewel box fitted out as a traveller's writing desk, that was said to have belonged to Abel, remained at Froland, and even today stands in the "buck room" at the Froland establishment. There are also some other items belonging to Abel that have survived: an ink- and sand-well, the case for a razor knife, a pipe and a watch – for the most part to be found at the home of Stein Abel in Moss. Also in evidence, at the Norwegian Folk Museum at Bygdøy, Oslo, is the silk vest that Abel wore to Crelle's Christmas ball in 1825.)

On January 26, 1830, Christine Kemp wrote to Mrs. Hansteen: "My attempts to write to You, Dearest Mrs. Hansteen, have come to nought for many different Reasons, but today I cannot avoid penning a few lines to describe to You what You perhaps had been afraid was only a dream, namely my Engagement to Keilhau. Do not judge me harshly, but rather, love me if You can, after I have taken a Step that You well know ought to have been most scrupulously contemplated, and hence my Pen cannot write in such a way so as to justify myself. As You know, Keilhau asked me if I would allow him to remain here in the Neighbourhood for a While, something that I, for several Reasons, felt I ought to concede to him, as I was fully aware that this would be the best Means of awakening him from the disappointing Reveries of Phantasy. After the first Visit he paid here, he was invited by Smith to remain a While, something he immediately accepted, as he had learned from Holmboe that I would place no Hindrance in his Way, and consequently, this is the 8th day that he has been here, but since that Time nothing has occurred other than him becoming ever firmer in his self-possessed Resolution, so, by reason of my weak Health, I saw that I was obliged to proceed as quickly as possible to a Decision, as I felt myself too weak to go on any longer with the inner Struggle. Unite Your Prayers to our common Father, Dear Mrs. Hansteen, with my own, that I am able to spread Good Fortune over all his coming Days." And after "Yours affectionately, C. Kemp" there was a PS: "Yesterday, on my 26th birthday, our Engagement happened."

On April 6, 1830, Mrs. Hansteen wrote down in her diary: "Something unpleasant is always happening to me. I spoke with great vehemence about Professor K – my remorse is tremendous. It has chased away Sleep and all my possible Peace of Mind. Much, much too True are the hard words that Treschow once said to me: You have enough remorse, but you do not improve." Professor K. is probably a reference to J. J. Keyser. Mrs. Hansteen came back to Christiania when her husband had returned from his Siberian expedition, but she never felt completely well and at ease in Christiania and Norway. She died in 1840 at the age of fifty-three.

    Tradition has it that Mrs. Hansteen helped her husband adjust weights and measures – something that took place in her kitchen – and she once, in her great zeal, had filed away too much of the weight, giving the control-weight too good a stroke with the file. Thus, in the conversion to kilograms, the Norwegian pound weighed 499 grams, and the Danish, 501 grams. Despite tradition, anecdote actually arose in relation not to Mrs. Hansteen but to the wife of Councillor Saxild. (As for Professor Hansteen's prestige, we see the fact that the first building that was put up for the new University in 1833, was the astronomical observatory. The construction of the buildings on Karl Johans gate did not begin until 1840.)

On Charité Borch, Mrs. Hansteen's sister

Charité Borch married the poet, Frederik Paludan-Müller in 1838, after having cared for him during a serious illness, and she was certainly a good support to him for the rest of his life. They were said to have had a happy but childless marriage. Nevertheless, she has been projected as somewhat sad in later life. In *Danske Digtere* [Danish Poets] (1877), Georg Brandes, who frequently visited Paludan-Müller, writes that "she was by no means ungifted, but exhibited a strange Nature, in which she read Philosophy without grasping much thereof, and constantly grumbled about theological, particularly dogmatic Questions." (One of Paludan-Müller's works has the title "The Death of Abel" (1854) and even though this is an epic about the Biblical Abel, the first death, the first sorrow and the first consolation – faith in eternity – was it also possible that Charité also thought about the other Abel, the friend of her youth? Paludan-Müller, who had became acquainted with Mrs. Andrea Hansteen in Copenhagen during her husband's expedition to Siberia, later remembered this period: "You, Dear Andrea…Every Morning, when on behalf of my Profession, I walked through Vestergade, I would look up at your windows. but You were not there, and although your Spirit was hovering up there, your Picture in my Soul, beckoned to me…your Friendship toward me, so lively, so instructive, and even, so pleasant – the whole House takes on a sweet significance to me, becomes so dear and so holy, that I always have to think of Goethe's words: Die Stelle, die ein guter Mensch betrat, ist eingeweiht.")

A few words on Mrs. Abel, Niels Henrik's mother

It took some work for Mother Abel to lay her hands upon the 1,500 francs of the French Academy's prize that, according to the inscription should go to "the family of Herr Abel, the learned mathematician from Christiania". The French authorities, and in the final analysis, the Finance Department, prolonged the juridical search for satisfactory clarification on who was the legitimate heir to Abel. But Niels Henrik had left no last testament whatsoever, and not all of Niels Henrik's siblings would have written right away to say that their mother should get the money. Mrs. Abel approached both Holmboe and Hansteen, and also sent a message to Count Löwenhielm in Paris stating that she was her son's heir, and Löwenhielm in turn approached F. D. Arago, Secretary of the Academy. In the end, Paris decided that the money should not be divided between Abel's heirs, but rather it should be given personally to his mother, Widow Abel.

Genealogical Table of Niels Henrik Abel's Family

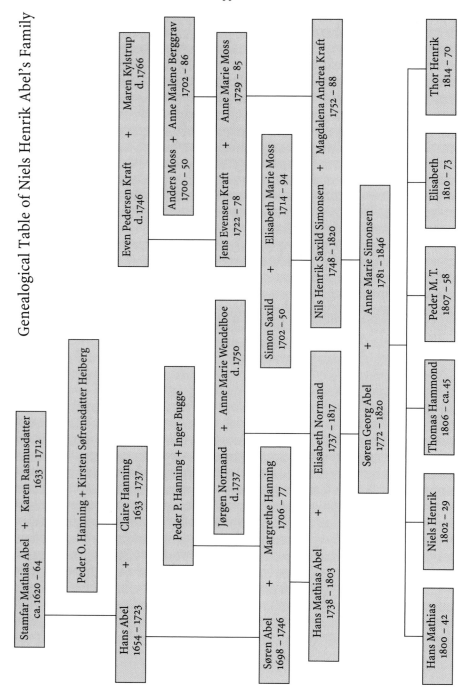

Chronological Bibliography
of Abel's Published Works

Almindelig Methode til at finde Funktioner af een variabel Størrelse, naar en Egenskab af disse Funktioner er udtrykt ved en Ligning mellom to Variable, Magazin for Naturvidenskaberne, 1823, vol. 1, pp.216-29.

Opløsning af et Par Opgaver ved Hjælp af bestemte Integraler, *Magazin for* Naturvidenskaberne, 1823, vol. 2, pp.55–68, 205–15.

Om Maanens Indflydelse paa Pendelens Bevægelse, *Magazin for Naturvidenskaberne*, 1824, vol.1, pp. 219–26. Berigtelse, *Magazin for Naturvidenskaberne*, 1824, vol. 2, pp. 143-44.

Mémoire sur les équations algébriques, où on démontre l'impossibilité de la résolution de l'equation générale du cinquième degré. Self-published, Christiania, 1824 .

(A revision of this work on the fifth degree equation was published in *Crelle's Journal*, 1826, Issue 1. See second entry under *Crelle's Journal*, below.)

Det endelige Integral Σ^n φx udtrykt ved et enkelt bestemt Integral, *Magazin for Naturvidenskaberne*, 1825, vol. 2, pp. 182-89.

Et lidet Bidrag til Læren om adskillige transcendente Functioner, *Det kongelige norske Videnskabers Selskabs Skrifter i det 19 Aarhundre*, vol. II, 1824-27, pp. 177–207 .

Solution d'un problème général concernant la transformation des Fonctions elliptiques, *Astronomische Nachrichtungen* 138, 1828, pp. 365-88. Altona.
(This was "the deathification of Jacobi", and a supplement – "Addition au mémoire précédant" - came out in the same journal, No.147, 1829, pp. 33–44. Altona.)

Recherche de la quantité qui satisfait à la fois à deux équations algébriques données, *Annales Gergonnes de Mathematique*, 1826–27, pp. 204–13.

The following works were published in Berlin, in Crelle's *Journal (Journal für die reine und angewandte Mathematik)*:

Untersuchung der Functionen zweie unabhängig veränderlichen Grössen x und y, wie f(x,y) welche die Eigenschaft haben, dass f (z,f(x,y)) eine symmetrische Function von z, x und y ist. 1826, Issue 1.

Beweis der Unmöglichkeit, algebraische Gleichungen von höheren Graden, als dem vierten, allgemein aufzulsen. 1826, Issue 1.

Beweis eines Ausdruchs, von welchen die Binominal-Formel ein einzelner Fall ist. 1826, Issue 2.

Bemerkungen über die Abhandlung no.4, Seite 37 im ersten Heft dieser Journals. 1826, Issue 2. (A two-page observation about one of the papers published in *Crelle's Journal* - Untersuchung der Wirkung einer Kraft auf drei Puncte - by Ban-Conducteur Kossack in 1826, Issue 1.)

Auflösung einer mechanischen Aufgabe. 1826, Issue 2.

Über die Integration der Differential-Formel $\frac{\delta dx}{\sqrt{R}}$ wenn R und δ ganze Functionen sind. 1826, Issue 3.

Untersuchungen über die Reihe $1 + \frac{m}{1} x + \frac{m\,(m-1)}{1 \cdot 2} x^2 + \frac{m\,(m-1)\,(m-2)}{1 \cdot 2 \cdot 3} x^3 + \dots$ u.s.w. 1826, Issue 4.

Über einige bestimmte Integrale. 1827, Issue 1.

Recherches sur les fonctions elliptiques. 1827, Issue 2.

Aufgaben und Lehrsätze.
(Under this title Abel submitted three small works and a pair of "Lehrsätze" to readers of the journal (1827, Issue 3 and 1828, Issue 2) - among other things, on convergence and divisibility. Abel's treatments and theorems appeared side by side with those of many others.)

Über die Fonctionen, welche der Gleichung $\varphi x + \varphi y = \psi\,(xfy + yfx)$ genug-thun. 1827, Issue 4.

Note sur le mémoire de *M. L. Olivier*, ayant pour 'Remarques sur les séries infinies et leur convergence'. 1828, Issue 1.

Recherches sur les fonctions elliptique. 1828, Issue 2. (Continuation from 1827, Issue 2).

Sur quelques formules elliptiques. 1828, Issue 3.

Remarques sur quelque proprietés générales d'une certaine sorte de fonctions elliptiques. 1828, Issue 4.

Sur le nombre des transformations differentes qu'on peut faire subir à une fonction elliptique par la substitution d'une fonction rationelle dont le degré est un nombre premier donné. 1828, Issue 4.

Théorème général sur la transformation des fonctions elliptiques de la seconde et la troisième espèce. 1828, Issue 4.

Note sur quelques formules elliptiques. 1829, Issue 1.

Mémoire sur une classe particulière d'equations résolubles algébriquement. 1829, Issue 2. (Published just prior to Abel's death on April 6, 1829.)

Théorèmes sur les fonctions elliptiques. 1829, Issue 2.

Démonstration d'une propriété générale d'une certaine classe de fonctions transcendentes. 1829, Issue 2. (Abel wrote this from Froland, on January 6, 1829.)

Precis d'une théorie des fonctions elliptiques. 1829, Issue. 3 - with the continuation in 1829, Issue 4. (Altogether, 100 large pages. This issue, 1829, Issue 3, also contained Crelle's obituary of Abel.)

Papers listed in 1830 and 1831 in Crelle's *Journal:*
Mathematische Bruchstücke aus Herrn N. H. Abels Briefen.
and:
Fernere mathematische Bruchstücke aus Herrn N. H. Abels Briefen. 1830.

Analyse du mémoire précedent. *Bulletin des sciences mathematique, astrono-mie, physique et chimie. Publié par le Baron de Férussac,* vol. 6, p. 347. Paris. 1826. (This was supplementary to the proof for the fifth degree equation, see p. 409–410, above.)

Mémoire sur une propriété générale d'une classe très-étendue des Fonctions transcendentes, *Memoires présentés par divers savants Etranges.* Paris. 1841. (This was the Paris treatise that was eventually found and published.)

There are also the two publications of "The Collected Works":

Oeuvres complètes, avec des notes et développements, rédigées par ordre du roi par B. Holmboe. vol. 1-2. Christiania. 1839.

Oeuvres completes de Niels Henrik Abel, nouvelle edition publiée aux frais de l'etat Norwégien par M. M. L. Sylow et S. Lie. Vol. 1 (621 pp.) and Vol.2 (341 pp.). Christiania. 1881.

Commentary

Thus we find that most of Abel's important works were first published in Crelle's *Journal für die reine und angewandte Mathematik* in Berlin; in other words, he contributed to every single issue from the beginning of 1826, often having several papers in one issue. Also after Abel's death several treatises were published in Berlin: all four issues from 1829 contained works by Abel; in addition, Crelle published Abel material in the volumes of 1830 and 1831. Abel's articles in the volumes for 1826 og 1827 were translated into German by Crelle, and in the course of which small changes were introduced from Abel's original manuscripts in French - the exception is found in Part 2 of the 1827 issue, the paper called "Recherches sur les fonctions elliptiques" which was both written and published in French.

The undertaking to gather together and publish Abel's collected works was undertaken by French men of science and scholarship. The petition to Karl Johan, was signed by Legendre, Poisson, Lacroix and Maurice, and was delivered to the Swedish ambassador in Paris, Count Löwenhielm, on September 15, 1828, quite probably without Abel's knowledge.

When it became known in Paris that Abel was dead, Baron Maurice, on behalf of l'Academie Française, paid an official visit of condolence to Count Löwenhielm, and Maurice pointed out the great scientific significance of having Abel's collected works published, and suggested that His Royal Highness Crown Prince Oscar certainly should be interested in supporting such a project.

A couple of years later, on July 17, 1831, Baron Maurice took up this request on behalf of the four Academy members, and pointed out once more how important it was that Abel's treatises and manuscripts be made accessible. The invitation from Baron Maurice about having all of Abel's work published now sent Löwenhielm to his close friend, Jöns Jacob Berzelius (1779-1848), the world renowned chemist who since 1818 had been secretary to the Royal Swedish Academy of Sciences. Löwenhielm wrote from Paris, asking "would it not be a pity for the sciences, and a loss of advantage to the honour and celebrity of the learned of Scandinavia if these writings, which have awakened the attention of the Institute, and from an unknown master from a distant university has received such a reputation, were to, I would say, remain unknown in manuscript form and then disappear from the corpus of human knowledge?"

Berzelius felt that the university in Christiania was the proper institution to deal with the matter and wrote to Professor Hansteen on September 27, 1831. In this letter it is plain that Berzelius - who had no personal qualifications to understand Abel's mathematical contributions - is somewhat skeptical toward the project. Berzelius was so used to the "talk is cheap" emanations from French mouths that he was not far from considering this only "so much hot air from Paris."

In Christiania Hansteen presented the proposal to the Academic Collegium and a short time later it was decided that Abel's works would be published at state expense, and with B. M. Holmboe as editor.

Those works by Abel that earlier had been published in either German or Norwegian were now translated to French by Holmboe, and part of the unpublished material that Holmboe had taken from Abel's notebooks and the papers he left behind. But still it was impossible to lay hands on the important Paris treatise. Abel's manuscript of this work had been lying on the table at l'Academie Française on June 29, in the summer of 1829, and to be sure, as well, a year later the decision was taken to award Abel a prize for this treatise. But by then the manuscript had disappeared once again, without a trace, and despite repeated searches it was not located. When Holmboe informed the Department in 1838 that he did not have this treatise, the Department itself communicated with the Academy in Paris and requested a copy, but did not receive any reply. Thus it was that Abel's Paris treatise did not find its way into Holmboe's two-volume publication in 1839. The title page declared *Oeuvres completes, rédigée par ordre du roi* This publication, consisting of 800 quarto pages in French, was financed by the Church Department to the sum of 2,360 speciedaler, and in addition to Abel's own works, it included a short biography based on Holmboe's obituary, some letters from Legendre, most of Crelle's obituary, and the letter from l'Institut de France, Academie Royale des sciences, dated at Paris, 24 July, 1830, and signed by Arago, explaining that the Academy's grand prize had been awarded jointly to Abel and Professor Jacobi, and that 1,500 francs would be sent to Abel's next of kin. Moreover, at the end of the second volume Holmboe had included those of Abel's letters that he had managed to assemble, that dealt with mathematical issues. There were four letters to Crelle, one to Legendre and four to Holmboe himself. (In all, Abel wrote Holmboe ten letters from abroad.)

Holmboe's Abel publication engendered general enthusiasm far beyond mathematical circles. The writer Henrik Wergeland wrote in *Statsborgeren* on July 2, 1837: "That the Education Authorities' Fund granted 1,650 speciedaler to defray the Publisher's Costs for the Writings left behind by Abel, and given out as the collected Works in French - Bravo! The Nation lives!" For presentation to King Oscar I, Volume II, was bound in red saffian-dyed leather with a broad border in pressed gold, and a gold-lettered spine from the famous and foremost bookbinder, J. C. Hoppe, ca. 1840, today found in the Universitetsbibliotek, University of Oslo (now within the National Library of Norway, Oslo Division).

Abel's Parisian treatise saw the light of day again, and was published in *Savants étranges* (Paris 1841). However, Abel's manuscript disappeared again a short time later, and was located again only in 1952 in Florence by Professor Viggo Brun (Brun himself has written about the exciting hunt for Abel's Parisian treatise in the 1953 volume of *Norsk Matematisk Tidsskrift,* and in two accounts published in the Oslo newspaper, *Aftenposten:* 11.02.53 and 30.05.53). To make a long story short:

When Abel's manuscript was found again in the summer of 1829, the Academy in Paris decided to publish the treatise itself, even though this would come to 90 printed pages. But then the Academy's secretary, Fourier, died, and the publication was postponed. Then came the July Revolution in 1830, and Cauchy had to go into exile because he refused to swear allegiance to the new king, Louis Philippe. When Cauchy returned eight years later, the Italian, Libri, had entered the picture. Guglielmo Libri had become professor of mathematics in 1832 in Paris, and had the responsibility, among other things, for the publication of Abel's treatise. He became interested in the life story, and himself published a biography of Abel which came out in the great biographical reference work entitled *Biographie Universelle, ancienne et moderne* (Volume 56, Supplement, pp. 22-29. 1834). In his evaluation of Abel's work and contribution, Libri was strongly critical of the treatment Abel had received both in Paris and in Norway, but biographically he purveyed much misinformation. For example, in his article, Libri spread the rumour that Abel had been so poor that he had had to go home from Paris to Christiania on foot! Libri had considerable bookish historical knowledge and was called to index rare hand-written manuscripts spread through the French provincial libraries, in chateaux, churches and learned institutions. But Libri was also "un homme des affaires". He was a large collector of books and manuscripts, and he bought and sold all over Europe - to museums and scholars - but little by little allegations began to multiply that he engaged in swindling and thievery. Following the February Revolution in 1848, the complaints against him were properly investigated. Libri decamped to London and it was said that he took with him as much as he could transport. In Paris long lists were drawn up of documents and manuscripts that had disappeared, and among these, from the archives of the Academy of Sciences. Under the official protocols between the two nations, Libri could not be extradited from England, but in 1850 the French courts found him guilty - *in contumaciam* - and sentenced to ten years in prison, the maximum sentence possible. Libri held book auctions in London and continued his business affairs, often with manuscripts dating from before the year 1200, and among the most recent of the manuscripts were Abel's two last treatises, that Libri must have obtained from Crelle in Berlin. Gradually it become clear to many in London that Libri had not obtained everything by legimate means, and Libri abandonned that city too and spent his last years in the vicinity of his home city of Florence. It was for this reason that Viggo Brun in 1952 decided to search for Abel's manuscript where Libri had ended his days.

The Paris treatise was certainly included in the second and larger edition of Abel's collected works that came out in Christiania in two volumes at the expense of the state in 1881. It was titled *Oeuvres complètes de Niels Henrik Abel,* edited by Ludvig Sylow and Sophus Lie. The initiative for this edition was taken by the Norwegian Academy of Science in 1873, the idea having been launched by Professor Marcus Jacob Monrad after he, by way of Sophus Lie, had come to know that Holmboe's edition from 1839 was sold out, much sought after, and incomplete. The Science Academy applied to the government and Parliament according to the plan worked out by Professors Ole Jakob Broch, Carl Anton Bjerknes and Sophus Lie, for 1,300 speciedaler annually for three years. The money was soon granted and the work set in motion with Sophus Lie and Peter Ludvig Mejdell Sylow as editors. Sylow was a school principal in Fredrikshald (Halden). When the work was completed and had come out on December 9, 1881, Professor Lie pointed out that the major honours for the publication belonged to Sylow, and he added that he hoped Sylow would quickly be able to find a position commensurate with "his broad Knowledge, his sharp Powers of Criticism; and his outstanding mathematical Work." Lie's endeavours to free Principal Sylow from his school labours in Fredrikshald and into the University as a professor probably also motivated an article he wrote for *Aftenposten* on November 25, 1896, under the byline "On Abel, Évariste Galois and Ludvig Sylow." Sylow *did* become a University employee, indeed at the beginning with a lower rate of pay. Sylow's work on Abel's collected works and notes about the treatises resulted in papers that constituted almost a new field of mathematics, and "the Sylowian proposition" is a concept in group theory.

In the course of the work Sylow had an eager discussant in C. A. Bjerknes, who for his part, worked on the Abel biography that was published in 1880. Bjerknes wanted as much as possible of Abel' early works to come out, not only his great treatises with their exemplary stringency, and perhaps there had been more Abel material than what Holmboe used in his edition of 1839? Many of Holmboe's books and papers were destroyed in his house fire on September 8, 1849, also whatever he had in his possession of the papers left behind at the time of Abel's death. In 1883 the last of Abel's six hand-written workbooks - today located in the hand-written manuscript collections of the National Library of Norway, Oslo Division, Manuscript Department - were, for all that, found among Holmboe's surviving books. Volume II in the Sylow-Lie edition contains the writings Abel left behind (Volume II contains 19 items, and Volume I, 29), and of these, the one which in particular is eagerly studied is entitled "Sur la résolution algébrique des équations." This is a reject from a great work on algebraic equations which *can* be be solved by extracting square roots, later called metacyclical - it was here that Abel introduced an arbitrary area of rationality which can be interpreted as germs of the concept of a number field, and so on, and it is here in the introduction that he presented the nearest thing to a programmatic clarification and a formulation of method: "On doit donner au problème une forme telle qu'il soit toujours possible de le résoudre, ce qu'on peut toujours faire d'une problème quelconque. Au lieu de demander une relation dont on ne sait pas si elle existe ou non, il faut demander si une telle relation est en effet possible." (One must give form to a problem such that it is possible to solve it. Given a proper form of approach, any problem can be solved. Instead of looking for a relation that possibly does not exist, one must ask if such a relationship is possible at all.)

Annotated Bibliography

A. Biographical Material on Abel

Bjerknes, C. A. 1880. *Niels Henrik Abel. En skildring af hans Liv og vitenskape-lige Virksomhed.* Stockholm.

(This is the first major biography of Abel. Earlier, biographical information was to be found in the obituaries written by Crelle, Saigey and Holmboe, and in articles written in several foreign lexicons and Norwegian journals. Bjerknes had written about Abel in *Skilling-Magazin* in 1857, and in *Morgenbladet* in 1875. He gave lectures on Abel in the Student Society in 1873, and large portions of the Abel book was published in *Nordisk Tidsskrift* [Letterstedtska Föreningen] in 1878 and 1879. Professor Bjerknes devoted considerable space to the precise examination of the dispute between Abel and Jacobi, and he gave priority to Abel's point of view. The book came out in an expanded form in French [Bordeaux, 1885], in Bordeaux because the translator, the French mathematician G. J. Houël, who was also Bjerkness friend and helper on the Abel biography, lived in Bordeaux. [Moreover, Houël was made a member of the Norwegian Academy of Sciences in 1891-92]. This edition was translated into Japanese in 1990.

A new, edited and shortened edition of Bjerknes book was prepared by his son, Professor Vilhelm Bjerknes in 1929, in relation to the one hundredth anniversary of Abel's death. This edition was entitled, *Niels Henrik Abel. En skildring av hans liv og arbeide.* 1929. Oslo.)

Finne-Grønn, S. H. 1899. *Abel, den store mathematikers slegt.* Christiania.

(At the urging of the brewer, Carl Abel of Fredrikshald [Halden], genealogist and museum director Stian Herlofsen Finne-Grønn wrote this book on the Abel family tree, from the Norwegian progenitor, Mathias Abel, who came from Abild in Schleswig, to Trondheim ca. 1640. The information on Niels Henrik Abel took up 17 of the 200 pages.)

Holmboe, B. M. 1829. *Kort Fremstilling af Niels Henrik Abels liv og viderskaberne Virksomhed.* Christiania.

(Published first in *Magazin for Naturvidenskaberne* 1829, and also a large part appeared in the newspaper *Patrouillen* [21.11.1829].)

Holst, Elling, Carl Størmer and Ludvig Sylow, eds. 1902. *Festskrift ved hundreaarsjubilæet for Niels Henrik Abels fødsel.* Christiania.

(A French edition was published the same year in Paris. Holst wrote a biographical introduction of 108 pages; Sylow wrote on Abel's studies and explorations [53 pages] and Carl Størmer was responsible for preparing the letters to and from Abel, and documents concerning Abel and information pertaining to these [185 pages]).

Mittag-Leffler, Gøsta. 1903. Niels Henrik Abel. In Mittag-Leffler, G. *Ord och Bild.* Stockholm.

(A French edition appeared in 1907. The entry on Abel consisted of 65 pages. At the Abel centenary commemoration in Christiania in 1902 Mittag-Leffler was made honorary doctor at the University, and when he started his journal, *Acta Mathematica* in 1882, he inaugurated it with a portrait of Abel. It was noticed that he also scouted for a young mathematician who could make his journal well-known in the same way that Abel had helped make Crell's Journal a success. Mittag-Leffler then found Henri Poincaré [1852-1912].)

de Peslöuan, Lucas Ch. 1906. *N. H. Abel. Sa vie et son oeuvre.* Paris.

(In addition to Abel's mathematical contributions, one also finds here that Mrs. Hansteen played a particularly crucial role as Abel's best and most important lifelong friend.)

Ore, Øystein. 1954. *Niels Henrik Abel. Et geni og hans samtid.* Oslo.

(This is the longest [317 pages] account of Abels life and has long been the basic biography. Øystein Ore was professor of mathematics at Yale University. An English translation [*Niels Henrik Abel: Mathematician Extraordinaire*]was published by the University of Minnesota Press in 1957.)

B. Some Articles on Abel and his Mathematics

Acta Mathematica. (Vol.27-29). 1902-04. (Niels Henrik Abel *in memorium.*

(Many articles were written on the occasion of the centenary of Abel's birth. Fifty of these pieces one the subject of Abel's mathematics are assembled in these volumes.)

Aubert, Karl Egil. 1979a. Niels Henrik Abel. *Normat,* 4/1979:129-40.
1979b. Abels addisjonsteorem. *Normat,* 4/1979:149-58.

Biermann, Kurt-R. 1959. Crelles Verhältnis zu Gotthold Eisenstein. *Monatsberichte der Deutschen Akademie der Wissenschaften zu Berlin.*
1960. Urteile A. L. Crelles über seine Autoren. Berlin.
1963. Abel und Alexander von Humboldt. *Nordisk Matematisk Tidsskrift* 1963:59-63.
1967. Ein unbekanntes Schreiben von N.H.Abel an A. L. Crelle. *Nordisk matematisk tidsskrift.* 1967:25-32.

Biermann, Kurt-R. and Viggo Brun. 1958. Eine notiz N.H.Abels für A.L.Crelle auf einem manuskript Otto Auberts. *Nordisk matematisk tidsskrift.*1958:84-86.

Brun, Viggo. 1952. Niels Henrik Abel. *Den Kongelige Norske Videnskabers Selskab* bd. xxv, pp. 25-43.
1953. Det gjenfundne manuskript til Abels Pariseravhandling. *Nordisk Matematisk Tidsskrift.* (Also publised as two articles in *Aftenposten:* 11.02.53 and 30.05.53.
1956. Niels Henrik Abel. In *De var fra Norge,* Lorentz Eckhoff, ed., pp. 20-8. Oslo.

Brun, Viggo and Børge Jessen. 1958. Et ungdomsbrev fra Niels Henrik Abel. *Nordisk matematisk tidsskrift.*1958:21-4.

Baas, Nils Andreas. 1964. Sørlendingen og geniet. *Fedrelandsvennen,* 27.08.1964.

Gårding, Lars. 1992. Abel och lösbara ekvationer av primtalsgrad. *Normat,* 1/1992:1-13.

Gårding, Lars and Christian Skau. 1994. Niels Henrik Abel and Solvable Equations. *Archive for History of Exact Sciences* 48(1):81-103.

Heegaard, Poul. 1935. Et brev fra Abel til Degen. *Norsk matematisk tidsskrift,* 1935:33-8.

Killingbergtrø, Hans Georg. 1994. Den 24. juli 1823, klokken 19.05.00. *Normat* 3/1994:129-33.

Kragemo, Helge Bergh. 1929. Tre brever fra Niels Henrik Abel og hans bror til Pastor John Aas. *Norsk matematisk tidsskrift*, 1929:49-52.

Kronen, Torleiv. 1985. Ut over grensene. Norske vitenskapsmenn i Frankrike. Oslo.

Lassen, Kristofer. 1902. Fra Niels Henrik Abels Skoledage. Træk af Kristiania Katedralskoles Historie. *Morgenbladet (1902,* no.515 and 516).

Lange-Nielsen, Fr. 1927. Zur Geschichte des Abelschen Theorems. Der Schicksal der Pariserabhandlung. *Norsk matematisk tidsskrift*, 1927:55-73.

Lange-Nielsen, Fr. 1929a. Abel og Academie des Sciences i Paris. *Norsk matematisk tidsskrift,* 1929:13-17.
  1929b. On Abel in the following journals: *Hjemmet* 12/1929, *Forsikringstidende* 7/1929, *American-Scandinavian Review* 5/1929.
  1929c. Nogen opplysninger om Abels forhold i 1827 og 1828. *Norsk matematisk tidsskrift.* 1929:53-55.
  1953. Niels Henrik Abel. *Nordisk matematisk Tidsskrift,* 1953:65-90.

Lorey, Wilhelm. 1929a. Abels Berufung nach Berlin. *Norsk Matematisk Tidsskrift*,1929:2-13.
  1929b. August Leopold Crelle zum Gedächtnis. *Crelles Journal,* 1929:3-11.

Skau, Christian. 1990. Gjensyn med Abels og Ruffinis bevis for umuligheten av å løse den   generelle ntegradsligningen algebraisk når n = 5. *Normat,* 2/1990:53-84.

Storesletten, Leiv. 1979. Geniet Niels Henrik Abel. *Syn og Segn,* 7/1979:415-25.

Størmer, Carl. 1903. Ein Brief von Niels Henrik Abel an Edmund Külp. In *Oslo Vitenskapsselskaps skrifter,* Mat. nat. klasse, no. 5,1903.

  1929a. Abels opdagelser. Fire forelesninger for de realstuderende i anledning av 100- årsdagen for Abels død. *Norsk matematisk tidsskrift.* (Special supplement of 25 pages).
  1929b. En del nye oplysninger om Abel hentet fra gamle brever. *Norsk Matematisk Tidsskrift,* 1929:1-4.

Sylow, Ludvig. 1915. Om Abels arbeider og planer i hans siste tid, Belyst med dokumenter, som er fremkomne efter den store udgave af hans verker. Christiania.

Tambs-Lyche, R. 1929. Niels Henrik Abel. *Det Kongelige Norske Videnskabers Selskab, Forhandlinger* bd.II:28-31. Trondheim.
   In addition to this list there are lesser biographical articles on Abel in various Norwegian publications from 1830 onwards, and certainly in Norwegian and foreign books of reference.

C. Unpublished Sources

National Library of Norway, Oslo Division, Department of Manuscripts:
Abel's letters from his tour abroad (published in the *festschrift*).
Abel's six work books. His father's teaching book. His father's father's poetry. Mrs. Hansteen's diaries. Morten Kjerulf's letters to C. A. Bjerknes dated Valle, February 25, 1880 and March 22, 1880. Hanna Smith's letters to C. A. Bjerknes dated November 9, 1883 and January 28, 1884. Georg Brochmann's play *Vårfrost* (Spring Frost). The bequeathed papers of Øystein Ore and Viggo Brun. The library's registry of books lent out.

State Archives in Kristiansand:
Bishopric archives and Gjerstad parish archives: letters from Søren Georg Abel to the bishop and the dean. Gjerstad parish records: correspondence 1772-1830, on among other things, university financial files, sales list of the parish farms inventory, Pastor Aas' notes on Vegårshei, S. G. Abel's letters to Dean Krog (much of the information on Niels Henrik's siblings is found here).

Gjerstad Municipal Archives:
Ministry books for Gjerstad Parish. Book list of the Reading Society.

Oslo Cathedral School Archives:
Bound records of disciples' marks, examination results, examiner reports, and reports by the rector.

Parliamentary Archives:
On Søren Georg Abel's activities in the Parliament of 1818 - especially with regard to the Hjorth affair.

State Archives:
Søren Georg Abel's letters to Dean Krog, in all, 33 letters and other documents. A package of papers from the Gjerstad parish files, and the local police station daily journal for the years from 1824 to 1829.

Berg-Kragerø Museum:
S. G. Abel's letter to Merchant J. A. Moss, dated December 10,1809.

Aust-Agder Archives, Arendal:
The file of O. A. Aalholms papers. Jens Vevstad's collections. Gjerstad Reading Society's book list. Minutes and reports of Anders Løvland. Letters to and from Gisle Nilsen/Lars Thorsen. Andreas Vevstad: "Julelesnad 1988."[Christmas Journal].

University Library, Trondheim:
C. N. Schwach's unpublished poetry.

The Royal Library, Manuscript Division, Copenhagen
Abel's letters to F. C. Olsen. Abel's letters to F. Degen. Exchange of letters between P. M. Tuxen and his wife, Elisabeth Simonsen (Niels Henrik's maternal aunt) - in all, about 250 letters dated at Gjerstad, Risør, Copenhagen (some of these are also in the National Library of Norway, Oslo Division, Department of Manuscripts). H. G. von Schmidten's letters, manuscripts.

Danish National Archives for Sjælland:
Church records for Helligånds parish, succession records, etc.

State Archives, Stavanger:
The ministry book for Finnøy parish.

Deichman Library:
Lending records.

Personal Verbal Communication:
Torstein Skaali and Kåre Dalane, Gjerstad.

D. Mathematical History

(Items consulted for this book):

Andersen, Kirsti og Thøger Bang. 1983. *Københavns Universitet 1479-1979*. Volume XII. Copenhagen.

Arvesen, Ole Peder. 1940. *Mennesker og matematikere*. Oslo.

1950. *Gi meg et fast punkt*. Oslo.

1973. *Fra åndens verksteder*. Oslo.

Aubert, Otto. 1836. Indbydelsesskrift til den offentlige Examen ved Christiania Cathedralskole. Christiania.

Bekken, Otto. 1984. Themes from the history of algebra. Kristiansand/Agderdistrikt Høgskole.

1988. Four Lectures. Kristiansand/Agderdistrikt Høgskole.

Belhoste, Bruno. 1991. *Augustin-Louis Cauchy: A Biography*. New York.

Bell, E. T. 1937. *Men of Mathematics*. New York.

1940. *The Development of Mathematics*. New York.

Biermann, Kurt-R. 1960. *Urteile A. L. Crelle über seine Autoren*. Berlin.

Birkeland, Bent. 1993. *Norske matematikere*. Oslo.

Boyer, Carl B. 1968. *A History of Mathematics*. New York.

Brodén, Torsten, Niels Bjerrum and Elis Strömgren. 1925. *Matematiken og de eksakte Naturvidenskaber i deet nittende Aarhundrede*. Copenhagen.

Brun, Viggo. 1964. *Alt er tall*. Oslo.

Christensen, S. A. 1895. *Matematiken Udvikling i Danmark og Norge i det XVIII Aarhundrede*. Odense.

Dieudonné, Jean. 1962. *Algebraic Geometry*. Cambridge, MA.

Eccarius, Wolfgang. 1974. *Der Techniker und Mathematiker August Leopold Crelle (1780-1855) und sein Beitrag zur Förderung und Entwicklung der Mathematik im Deutschland des 19. Jahrhunderts*. Eisennach.

Hag, Per and Ben Johnsen (eds). 1993. *Fra matematikkens spennende verden*. Trondheim.

Holmboe, B. M. 1822. *Forsøg paa en Fremstilling af Mathematikens Principer, samt denne Videnskabs Forhold til Philosophie. Et Inbydelsesskrift til den offentlige Examen ved Christiania Lærde Skole i Juli 1822*. Christiania.

1825/27. *Lærebog i Mathematiken* (Part I, 1825; Part II, 1827). Christiania.

Klein, Felix. 1926-7. *Vorlesungen über die Entwicklung der Mathematik im 19. Jahrhundert*. Berlin.

Kline, Morris. 1972. *Mathematical Thought from Ancient to Modern Times*. Oxford.

Koeningsberger, Leo. 1879. *Zur Geschichte der Theorie der Elliptischen Trancendenten in den Jahren 1826-29*. Leipzig.

Nielsen, Niels. 1910. *Matematiken i Danmark 1801-1908*. Copenhagen.
   1927. *Franske matematikere under revolusjonen*. (Copenhagen University *festskrift*. Copenhagen.
Onstad, Torgeir. 1994. *Fra Babel til Abel. Ligningenes historie*. Oslo.
Ore, Øystein. 1953. *Cardano: The gambling scholar*. New York.
Pedersen, Kirsti Møller. 1979. Caspar Wessel og de komplekse tals repræsen-tation. *Normat*, 1979.
Piene, Kay. 1937. Matematikkens stilling i den høiere skole i Norge efter 1800. *Norsk matematisk tidsskrift*, 2/1937: 52-68.
Rasmussen, Søren. 1812. *Indbydelses-Skrift*. Christiania.
Schmidten, H. G. von. 1827. *Kort Fremstilling af Mathematikens Væsen og Forhold til andre Videnskaber*. Copenhagen.
Sylow, Ludvig. 1920. Évariste Galois. *Norsk matematisk tidsskrift*, 1920.
Tambs-Lyche, R. 1935. Matematikkens stilling i Norge omrkring 1780- årene, belyst ved D. C. Festers virksomhet i Trondheim. *D.K.N.V.S. Forhandlinger Bd VII*. Trondheim.
Thomsen, Klaus og Asger Spangsberg. 1988. *Differentialregningen* i *historisk perspektiv*. Aarhus.
Weil, André. 1971. *Courbes algébriques et variétés abéliennes*. Paris.
Øhrstrøm, Peter. 1985. F. C. H. Arentz matematiske argumenter mod verdens uendelighed i tid og rum. *Normat* 4/1985.

E. General Literature

Reference Works:

Halvorsen, J. B. 1885-1908. *Norsk Forfatter-Lexikon 1814-1860*. 6 vol. Kristiania.
Salmonsen. 1915-28. *Konversations leksikon*. 25 vol. Copenhagen.
Dictionary of Scientific Biography. 1970-76. New York. 14 Vol.
Norsk biografisk leksikon. 1923-69. Oslo. 16 Vol.
Norsk kunstnerleksikon. 1982-6. Oslo. Vol.1-4.

Newspapers and Journals:
Budstikken 1810-18; *Det Norske Nationalblad* 1815-21; *Den norske Rigstidende* 1818, 1824, 1829; *Morgenbladet* 1829, 1830, 1902; *Magazin for Naturvi-denskaberne* 1823-29; *Hermoder* 1821–27; *Den norske Turistforenings Årbok*, 1872, 1874; *Illustreret Nyhedsblad* 1862. Newspapers from the autumn, 1826, Bibliothéque Nationale, Paris: *Moniteur Officiel, Gazette de France, Jour-nal de Paris, Quotidienne, Journal des débats*.

Books and Articles:

Alsvik, Henning. 1940. *Johannes Flintoe.* Oslo.

Andersen, Einar. 1975. *Heinrich Christian Schumacher.* Copenhagen.

Andersen, Per Sveaas. 1960. *Rudolf Keyser.* Oslo/Bergen.

Andersen, Vilh. 1909. *Tider og typer.* Second volume. Copenhagen.

Angell, H. 1914. *Syv-aarskrigen for 17.mai 1807-14.* Ill.by A .Bloch. Christiania.

Ansteinsson, Eli. 1956. *Trønderen Michael Rosing.* Oslo.

Aubert, L. M. B. 1883. *Anton Martin Schweigaards barndom og ungdom.* Christiania.

Aubert, Andreas. 1893. *Professor Dahl.* Christiania.

Bang, A. Chr. 1910. *H. N. Hauge og hans samtid.* Christiania.

Berg, Bjørn Ivar (ed.). 1991. *500 års norsk bergverksdrift.* Kongsberg.

Berggreen, Brit. 1989. *Da Kulturen kom til Norge.* Oslo.

Bergsgård, Arne. 1945. *Året 1814.* Oslo.

Beyer, Edvard and Morten Moi. 1990. *Norsk litteraturkritikkshistorie 1770-1940,* Vol.1. Oslo.

Birkeland, M. 1919-25. *Historiske skrifter,* I-III. Christiania.

Bjarnhof, Karl. 1972. *Støv skal du blive. På spor af Niels Steensen.* Copenhagen.

Bjerknes, Vilhelm. 1925. *C. A. Bjerknes, hans liv og arbeide.* Oslo.

Blanc, T. 1899. *Christiania Theaters historie 1827-1877.* Christiania.

Blom, C. P. 1849. *Fordum og Nu.* Christiania.

Blom, Grethe Authén. 1957. *Fra bergseminar til teknisk høyskole.* Oslo.

Brandes, Georg. 1877. *Danske digtere-portrætter.* (Reissued by Uglebog, Copenhagen. 1966). Copenhagen.

Breen, Else. 1990. *Jens Zetlitz. Et tohundreårsminne.* Oslo.

Brenna, Arne. 1995. Rapport fra unionstiden. *St. Halvard* 4/95:5-30.

Bull, Francis. 1916. Christen Pram og Norge. *Edda,* 418-39.

1932. *Norges Litteratur* vol. II. Oslo.

Christophersen, H. O. 1959. *Marcus Jacob Monrad.* Oslo.

1976. *Over stokk og stein.* Oslo.

1977. *Niels Treschow 1751-1833.* Oslo.

Collett, Alf. 1893. *Gamle Christiania Billeder.* Christiania.

Collins, H. F. 1964. *Talma, A biography of an actor.* London.

Daae, Ludvig. 1857. Albert Peter Lassen. *Illustreret Nyhedsblad* 1857, no.49.

1871. *Det gamle Christiania 1624-1814.* Christiania.

1887. Stemninger i Danmark og Norge i Anledning af og nærmest efter Adskillelsen. *Vidar* 1887.

1894. Aalls brev til Gustav Blom. *Personalhistorisk Tidsskrift* 4th series, Vol.1.

1898. Om den Abelske slektsbok. *Personalhistorisk Tidsskrift* 4th series, 1 vol.

Daae, Ludvig (ed.). 1876. *Breve fra Danske og Norske, især i Tiden nærmest efter Adskillelsen.* Copenhagen.

Dahl, F. C. B. 1905. Breve fra Docent F. P. J. Dahl fra Christiania 1815–17. *Personalhistorisk Tidsskrift,* 5th series, II vol.

Dunker, Conradine. 1909. *Gamle Dage.* Christiania.

Dyrvik, Ståle. 1978. Den lange fredstiden 1720-84. *Norges historie* vol. 8. Knut Mykland, ed. Oslo.

Eriksen, Trond Berg. 1987. *Budbringerens overtak.* Oslo.

Euler, Leonard. 1792-93. *Breve til en Prindsesse i Tyskland over adskillige Gjenstander af Physiken og Philosophien.* Copenhagen.

Faye, Andreas. 1859. *Bidrag til Holts Præsters og Præstegjælds Historie.* Arendal.

1861. *Bidrag til Øiestads Præsters o Præstegjelds Historie.* Arendal.

1867. *Christiansands Stifts Bispe- og Stiftshistorie.* Christiania.

Fet, Jostein. 1987. *Almugens diktar.* Oslo.

Finne-Grønn, S. H. 1897. *Arendals Geistlighed.* Christiania.

Finstad, Håkon. 1989. Da bygdefolket begynte å spise hestekjøtt. *Årsskrift for Kragerø og* Skåtøy *Historielag.*

Flood, J. W. 1889. *Norges Apotekere i 300 År.* Christiania.

Flor, M. R. 1813. *Bidrag til Kundskab om Naturvidenskabens Fremskridt i Norge. Et Inbydelsesskrift.* Christiania.

Gedde, V. 1902. *Barndomserindringer.Ung i 30-årene.* Hamar.

Gran, Gerhard. 1899. *Norges Demring.* Bergen.

Grønningsæter, Tore. 1982. Christopher Hansteen og framveksten av norsk astronomi i begynnelsen av det 19. århundre. University of Oslo, "hovedfag" thesis, unpublished.

Hansen, Maurits. 1819. *Othar af Bretagne.* Christiania. (New edition, Oslo, 1994.)

1907. Breve til Schwach. Meddelt ved Ludvig Daae. *Historiske Samlinger.* Vol. 2. Christiania.

Hansteen, Christopher. 1859. *Bemærkninger og Iagttagelser paa en Reise fra Christiania til* Bergen *og tilbage i Sommeren 1821.* Christiania. (Serialized earlier in *Budstikken* 1821-22, no.51-102.)

Haugstøl, Henrik. 1944. *I diligencens glade tid.* Oslo.

Haugstøl, Henrik (ed.). 1968. *Vår egen by.* Oslo.

Heggelund, Kjell. 1975. Unionstiden med Danmark. In *Norges litteraturhistorie,* Edvard Beyer, ed. Vol.1:343-623. Oslo.

1980. Den lærde litteraturen - og den folkelige. In *Norges kulturhistorie.* Vol.3:253-70. Oslo.

Heggtveit.H. G. 1905. *Den norske kirke i det 18.årh.* Christiania.

Helland, Amund. 1904. *Topografisk-statistisk Beskrivelse over Nedenes Amt*, I and II. Christiania.

Hellerdal, Kirsten. 1977. Det Sønneløvske Læse Sælskab. *Søndeleds Historisk lagets årsskrift*.

1987. "...at dem ej bliver tillagt for lidet, ej heller for meget". Om fattigvesenet i Risør og Søndeled. In *Risør og Søndeled 1837-1987. Glimt fra kommunenes historie gjennom 150 år*. Risør Commune.

Heur, Ludvig. *Udsigt over Helsingør Latinskoles Historie*. Copenhagen.

Hoel, Jacob. 1927. *Fra den gamle bonde-oppposisjon*. Halvdan Koht, ed. Oslo.

Hohlenberg, Johannes. 1995. 1995. *Kulturens forvandling*. Oslo.

Holdt, Jens. 1927. *Niels Johannes Holm*. Copenhagen.

Holmboe, C. A. 1834-5. *Norske Universitets- og Skole-Annaler*. First volume.

Holst, P. C. 1876. *Efterladte Optegnelser*. Christiania.

Huitfeldt, H. J. 1876. *Christianias Theaterhistorie*. Copenhagen.

Hultberg, Helge. 1973. Den unge Henrich Steffens. *Festskrift udgivet af Københavns Universitet*, pp.7-115. Copenhagen.

Hundrup, F. E. 1860. *Lærerstanden ved Helsingørs lærde Skole*. Roeskilde.

Høigård, Einar. 1934. *H. A. Bjerregaard*. Oslo.

1942. *Oslo Katedralskoles historie*. Oslo.

Høverstad,Torstein. 1930. *Norsk Skulesoga 1739-1827*. Oslo.

Irgens, Johannes B. 1978. *Omkring Vestlandske Hovedveg*. Historielaget for Dypvåg, Holt og Tvedestrand. Tvedestrand.

Jæger, Henrik, ed. 1896. *Videnskabernes Literatur i det nittende Aarhundre. Illustreret norsk Literaturhistorie*, Vol. II$_2$. Kristiania.

Karlsen, Jan and Dag Skogheim. 1990. *Tæring. Historia om ein folkesjukdom*. Oslo.

Keihau, B. M. 1857. *Biographie von ihm selbst*. Christiania.

Kiær, F.C. 1888. *Norges Læger*. Christiania.

Krag, Hans. 1968. Prost John Aas arkeologiske notiser. *Agder Historielag Årskrift*.

Krarup, N. B. 1957. *Mellem klassiske filologer*. Copenhagen.

Kveim, Thorleif. 1952. Ei reise for 100 år sia. *Agder Historielag Årskrift*.

Laache, Rolv. 1927. *Henrik Wergeland og hans strid med prokurator Praëm*. Oslo.1941. *Nordmenn og Svensker efter 1814*. Oslo.

Lange, Alexander. 1905. *Optegnelser om sit Liv og sin Samtid*. Christiania.

Langslet, Lars Roar and Hilde Sejersted (eds.). 1994. *Det milde islett. Om dansk-norsk kulturelt samliv*. Oslo.

Lassen, Albert. 1823. Foesøk paa en Fremstilling af Historiens Formaal, samt denne Videnskabs Stof. Et Inbydelsessskrift. Christiania.

Lassen, Hartvig. 1877. *Afhandlinger til Litteraturhistorien*. Christiania.

Lein, Bente Nilsen, Nina Karin Monsen, Janet E. Rasmussen, Anne Wichstrøm and Elisabeth Aasen. 1984. *Furier er også kvinner. Aasta Hansteen 1824-1908.* Oslo.

Lindbæk, Sofie Aubert. 1910. *Landflyktige.* Christiania.

1913. *Fra det Norske Selskabs Kreds.* Christiania.

1914. Om Irgens-Bergh. *Historisk tidsskrift.*

1939. *Hjemmet på Akershus.* Oslo.

Lindstøl, Tallak. 1923. *Risør gjennem 200 Aar.* Risør.

Lunden, Kåre. 1992. *Norsk grålysing.* Oslo.

Lyche, Lise. 1991. *Norges teaterhistorie.* Asker.

Lærum, Ole Didrik. 1995. Helse og Kurbadkultur. In Din Grieg. *Legekunst og apotek fra 1595 og framover. In Extractum: 400 år med apotek.* Bergen.

Løvold, O. A. *Præstehistorier og Sagn fra Ryfylle.* p.38.

Masdalen, Kjell-Olav and Kirsten Hellerdal. 1981. Yrkes- og bosetningsmøn-steret i Risør omrking år 1800. *Søndeled Historielag Årsskrift.*

Molland, Einar. 1979. *Norges kirkehistorie i det 19.århundre.* Vol. I. Oslo.

Munch, Andreas. 1874. *Barndoms- og Ungdoms-Minder.* Christiania.

Munch, Emerentze. 1907. *Optegnelser.* Christiania.

Munthe, Preben. 1992. *Norske økonomer.* Oslo.

Mykland, Knut (ed.). 1978. *Norges historie,* Vol. 8-10. Oslo.

Müller, Carl Arnoldus. 1883. *Katalog over Christiania Katedralskoles Bibliotek.* Christiania.

Møller, Ingeborg. 1948. *Henrik Steffens.* Oslo.

Møller, Nicolai. 1969. *Fra Leibniz til Hegel.* Published at A. H. Winsnes/Thor-leif Dahl's Kulturbibliotek. Oslo.

Nielsen, Yngvar. 1888. *Grev Herman Wedel Jarlsberg og hans samtid.* 1st Part. Christiania.

Nissen, Bernt A. 1945. *Året 1814.* Oslo.

Norborg, Sverre. 1970. *Hans Nielsen Hauge, 1804-24.* Oslo.

Noreng, Harald. 1951. *Christian Braunmann Tullin.* Oslo.

Norman, Victor D. 1993. Risør og skutene. *Risør Magasin.*

Notaker, Henry. 1992. *Hans Allum-husmannsønn, dikter og rabulist.* Oslo.

Nygaard, Knut. 1947. *Jonas Rein.* Oslo.

1960. *Nordmenns syn på Danmark og danskene i 1814 og de første selvstendighetsår.* Oslo.

1966. *Henrik Ankder Bjerregaard.* Oslo.

Næss, Harald. 1993. Nils Otto Tank. *Det norsk-amerikanske historielaget, avd.Norge NAHA*-Norway.

Paasche, Fredrik. 1932. *Norges Littertur* vol. III. Oslo.

Paludan, Julius. 1885. *Det høiere Skolevæsen i Danmark, Norge, Sverige.* Copenhagen.

Pavels, Claus. 1899. *Dagbøger for Aarene 1812-13, og 1817-22.* Published by the Norwegian Historical Association, by Dr. L. Daae. Christiania.

Petersen, Richard. 1881. *Henrik Steffens.* Copenhagen.

Platou, Ludvig Stoud. 1819. *Handbog i Geographien.* First Part (Third edition). Christiania.

1841. Hvorledes fikk Norge et Universitet. *Morgenbladet,* 1841, no.259, supplement.

Pram, Christen. 1964. *Kopibøker fra reiser i Norge 1804-06.* Lillehammer.

Puranen, Bi and Tore Zetterholm. 1987. *Förälskad i Livet.* Wiken.

Pålsson, Erik Kennet. 1988. *Den helige Niels Stensen.* Vejbystrand.

Reich, Ebbe Kløvedal. 1972. *Frederik. En folkebog om N. F. S. Grundtvigs tid og liv.* Copenhagen.

Reiersen, Elsa and Dagfinn Sletten (eds.). 1986. *Mentalitetshistorie. Muligheter og problemer.* Trondheim.

Ringdal, Nils Johan. 1985. *By, bok og borger.* Deichmanske *bibliotek gjennom 200 år.* Oslo.

Rogan, Bjarne. 1968. *Det gamle skysstellet.* Oslo.

Rosted, Jacob. 1810. *Forsøg til en Rhetorik.* Christiania.

Ross, Immanuel: Litterær underholdning i Norge i 1820-aarene (Edda 1918)

Rød, Ole Thordsen. 1948. Lensmann Ole Thordsen Røds opptegnelser. *Aust-Agder Blad,* No.118-22.

Sars, J. E. 1911-12. *Samlede Værker* 1 - 4. Christiania/Copenhagen.

Schmidt, Fredrik. 1866. *Dagbøker* I-II. O. Jacobsen and J. Brandt-Nielsen, eds. Copenhagen.

Schnitler, Carl W. 1914. *Slegten fra 1814.* Oslo.

Schwach, Conrad Nicolai. 1992. *Erindringer af mit Liv.* Arild Stubhaug, ed. Arendal.

Seip, Jens Arup. 1971. *Ole Jakob Broch og hans samtid.* Oslo.

1974. *Utsikt over Norges Historie,*1st pt. Oslo.

Sejersted, Francis. 1978. Den vanskelige frihet. 1814-51. *Norges historie,* vol. 10, Knut Mykland, ed. Oslo.

Skaare, Kolbjørn. 1995. *Norges Mynthistorie* I-II. Oslo.

Skjelderup, Arthur. Barndoms- og Ungdomserindringer af Jens Skjelderup. *Personalhistorisk Tidsskrift,* 7th series, vol.I:157-216.

Sigmund, Einar. 1916. *Filantropismens indflydelse paa den lærde skole i Norge omkring aar* 1800. Christiania.

Sirevåg, Tønnes. 1986. *Niels Treschow. Skolemann med reformprogram.* Oslo.

Stamsøe, Halvor Olsen. 1948. Stortingsmann Halvor Olsen Stamsøes opptegnelser. *Aust-Agder Blad* 1948, No. 123-39.

Steen, Sverre. 1953. *Det frie Norge. På falittens rand.* Oslo.

Steffens, Henrich. 1803. *Indledning til philosophiske Forelæsninger.* Copenhagen.

1967. *Forelæsninger og fragmenter.* Emil Boysen, ed. Thorleif Dahl's Kulturbibliotek. Oslo.

Stiansen, Olav. 1980a. Det Gjerestadske og Sønneløvske Sogneselskab. *Søndeled Historisk lag årsskrift.*

1980b. Et barnedrap, dom og straff ved Egelands Jernværk 1809-10. *Søndeled Historisk lag årsskrift.*

Stortingets forhandlinger. 1818. Stortings-Efterretninger. 1814-32.

Swift, Jonathan. 1795. *Capitain Lemuel Gullivers Reiser.* Stockholm. Danish translation, Copenhagen, 1768.

Søndeled II. 1990. Søndeled og Risør Historielag. Risør.

Sørensen, Knud. 1984. *St. St. Blicher.* Copenhagen.

Thrap, Daniel. 1884. *Bidrag til Den norske Kirkes Historie i det nittende Aarhundrede. Biografiske Skildringer.* Christiania.

Tidemand, Nicolaj. 1881. *Optegnelser om sit Liv og sin Samtid i Norge og Danmark 1766-1828.* Published by C. J. Anker. Christiania.

Torgersen, Johan. 1960. *Naturforskning og Katedral..* Oslo.

Treschow, Niels. 1966. "*Philosophiske Forsøg*" *og andre skrifter.* A. H.Winsnes, ed. Thorleif Dahl's Kulturbibliotek. Oslo.

Try, Hans (ed.). 1986. *Jernverk på Agder.* Kristiansand.

Tuxen, N. E. 1883. *P. M. Tuxen og hans efterkommer.* Copenhagen.

Tysdahl, Bjørn. 1988. *Maurits Hansens fortellerkunst.* Oslo.

Universitetsbibliotekets festskrift. 1911. Kristiania.

Universitetets jubileumsskrift. 1911. Det Kongelige Fredriks Universitet 1811-1911, vol. I-II. Christiania.

Ullmann, Vilhelmine. 1903. *Fra Tyveaarene og lidt mere.* Christiania.

Vedel, Valdemar. 1967. *Guldalderen i dansk Diktning.* Copenhagen.

Vevstad, Andreas. 1984. *Gjerstad og banken.* Gjerstad.

Vevstad, Jens. nd. *Avisartikler gjennom 40 år.* Arendal.

Vibæk, Jens. 1978.. *Danmarks historie,* vol. 10., J. Danstrup and H. Koch (eds.).. Copenhagen.

Vogt, Johanne. 1903. *Statsraad Colletts Hus og hans Samtid.* Oslo.

Voltaire. 1930. *Candide.* (Norwegian edition prepared by Charles Kent). Oslo.

Wahl, Aage. 1928. *Slægten Kemp.* Copenhagen.

Wallem, Fredrik B. 1916. *Det norske Studentersamfund gjennom hundrede aar.* Oslo.

Winsnes, A. H. 1919. *Johan Nordahl Brun.* Christiania/Copenhagen.

Wollstonecraft, Mary. 1796. *Letters written during a short residence in Sweden, Norway, and Denmark.* London. (Norwegian edition: 1976/1995. Per A.

Hartun (trans.). *Min nordiske reise. Beretninger fra et opphold i Sverige, Norge og Danmark 1795.* Oslo.

Wergeland, Nicolai. 1816. *En sandfærdig Beretning.Om Danmarks politiske Forbrydelser imod Kongeriget Norge.* Norway.

Worm-Muller, Jacob. 1922. *Christiania og krisen efter Napoleonskrigene.* Christiania.

Økland, Fridthjof. 1955. *Michael Sars.* Oslo.

Øksnevad, Reidar. 1913. Othar af Bretagne. In *Festskrift til William Nygaard.* Christiania.

Østby, Leif. 1957. *Johan Christian Dahl.* Oslo.

Aall, Jacob. 1859. *Erindringer som Bidrag til Norges Historie fra 1800-15.* Second edition, in one vol. Christiania.

Aarnes, Sigurd Aa. (ed.). 1994. *Laserne.* Oslo.

Aarseth, Asbjørn 1985. *Romantikken som konstruksjon.* Oslo.

Aas, John. 1869. *Gjerestads Præstegjeld og Præster.* Risør.

1955. *Fortegnelse over Ord af Almuesproget i Gjerestad og Wigardsheien.* (Prepared from the Norwegian Dialect Archives by Sigurd Kolsrud. Oslo.

Aasen, Elisabeth. 1993. *Driftige damer.* Oslo.

Fig. 51. *Wallpaper* from Abel's death room at Froland (photo by Dannevig, Arendal): It is a French Grisaille wallpaper, manually produced and painted, most likely from 1820–25. The background color is light gray, and in the paintings of the figures five different colors of gray-blue are used.

Index of Names

Niels Henrik and his father, Søren Georg Abel (1772–1820), have been excluded from this index.

Printing: Mercedes-Druck, Berlin
Binding: Buchbinderei Lüderitz & Bauer, Berlin